SRA

Connecting Math Concepts

Level B Presentation Book 3

COMPREHENSIVE EDITION

A DIRECT INSTRUCTION PROGRAM

McGraw Hill Education

Bothell, WA • Chicago, IL • Columbus, OH • New York, NY

MHEonline.com

Education

Copyright © 2012 The McGraw-Hill Companies, Inc.

All rights reserved. No part of this publication may be
reproduced or distributed in any form or by any means, or
stored in a database or retrieval system, without the prior
written consent of The McGraw-Hill Companies, Inc.,
including, but not limited to, network storage or
transmission, or broadcast for distance learning.

Permission is granted to reproduce the material contained
on page 387 on the condition that such material be
reproduced only for classroom use; be provided to students,
teachers, or families without charge; and be used solely in
conjunction with *Connecting Math Concepts*.

Send all inquiries to:
McGraw-Hill Education
4400 Easton Commons
Columbus, OH 43219

ISBN: 978-0-02-103590-8
MHID: 0-02-103590-3

Printed in the United States of America.

6 7 8 9 10 MER 16 15 14

Table of Contents

Lessons 81–85 Planning Page

	Lesson 81	**Lesson 82**	**Lesson 83**	**Lesson 84**	**Lesson 85**
Student Learning Objectives	**Exercises** 1. Say addition and subtraction facts 2. Learn how to write dollar amounts with a zero in the cents 3. Say two addition and two subtraction facts for number families; find missing numbers in number families 4. **Learn about the greater than and less than (>, <) symbols** 5. Solve addition problems with carrying and with 3 addends 6. Solve subtraction facts 7. Compare two lines for length 8. Solve money problems 9. Complete work independently	**Exercises** 1. Say addition and subtraction facts 2. Learn how to write dollar amounts with a zero in the cents 3. Say two addition and two subtraction facts for number families; find missing numbers in number families 4. Learn about the greater than and less than (>, <) symbols 5. Solve money problems 6. Compare two lines for length 7. **Determine when to carry to the tens column** 8. Write the symbols for word problems in columns and solve 9. **Write the greater than and less than (>, <) symbols** 10. Complete work independently	**Exercises** 1. **Make comparison statements** 2. Say two addition and two subtraction facts for number families; find missing numbers in number families 3. Count forward or backward from a given number 4. Say addition and subtraction facts 5. **Write the greater than and less than (>, <) symbols to point to the greater or lesser number** 6. Determine when to carry to the tens column 7. **Write dollar amounts with 1- or 2-digit cents** 8. **Solve comparison word problems** 9. Solve money problems 10. Complete work independently	**Exercises** 1. Find missing numbers in number families 2. **Learn how to solve dollar problems with a zero in the cents** 3. Make comparison statements 4. Solve subtraction facts 5. Determine when to carry to the tens column 6. Write dollar amounts with 1- or 2-digit cents 7. Solve comparison word problems 8. Write the greater than and less than (>, <) symbols to point to the greater or lesser number 9. Solve addition facts 10. Complete work independently	**Exercises** 1. Say addition and subtraction facts 2. **Learn about cents written with a dollar sign** 3. Make comparison statements 4. Solve money problems 5. Solve subtraction facts 6. Solve addition problems with carrying and with 3 addends 7. Write the greater than and less than (>, <) symbols to point to the greater or lesser number 8. Solve addition facts 9. Solve comparison word problems 10. Complete work independently

Common Core State Standards for Mathematics					
1.OA 3	✔	✔	✔		
1.OA 5	✔	✔	✔	✔	✔
1.OA 6	✔	✔	✔	✔	✔
1.OA 8	✔	✔	✔	✔	✔
1.NBT 1	✔	✔	✔	✔	✔
1.NBT 2	✔	✔	✔	✔	✔
1.NBT 4	✔	✔	✔	✔	✔
1.G 2	✔	✔	✔	✔	✔
Teacher Materials	**Presentation Book 3,** Board Displays CD or chalkboard				
Student Materials	Workbook 2, Pencil				
Additional Practice	• Student Practice Software: Block 4: Activity 1 (1.G.1), Activity 2 (1.NBT.1), Activity 3 (1.NBT.1), Activity 4 (1.NBT.3), Activity 5 (1.NBT.1), Activity 6 (1.NBT.1) • Math Fact Worksheets 41–42 (After Lesson 81), 43 (After Lesson 83), 44–48 (After Lesson 84)				
Mastery Test					

Lesson

EXERCISE 1: FACTS
ADDITION/SUBTRACTION

a. (Display:)   [81:1A]

3 + 2	12 − 6	6 − 4
11 − 2	4 + 3	8 − 3
4 + 6	3 + 6	11 − 5

For each problem, you'll tell me if the answer is the big number or a small number. Then you'll say the fact.

- (Point to **3 + 2**.) Read this problem. Get ready. (Touch.) *3 plus 2.*
- Is the answer the big number or a small number? (Signal.) *The big number.*
- What's 3 plus 2? (Signal.) *5.*
- Say the fact. (Signal.) *3 + 2 = 5.*

b. (Point to **11 − 2**.) Read this problem. Get ready. (Touch.) *11 minus 2.*
- Is the answer the big number or a small number? (Signal.) *A small number.*
- What's 11 minus 2? (Signal.) *9.*
- Say the fact. (Touch.) *11 − 2 = 9.*

c. (Repeat the following tasks for the remaining problems:)

(Point to __.) Read this problem.	Is the answer the big number or a small number?	What's __?		Say the fact.
4 + 6	*The big number.*	4 + 6	10	*4 + 6 = 10*
12 − 6	*A small number.*	12 − 6	6	*12 − 6 = 6*
4 + 3	*The big number.*	4 + 3	7	*4 + 3 = 7*
3 + 6	*The big number.*	3 + 6	9	*3 + 6 = 9*
6 − 4	*A small number.*	6 − 4	2	*6 − 4 = 2*
8 − 3	*A small number.*	8 − 3	5	*8 − 3 = 5*
11 − 5	*A small number.*	11 − 5	6	*11 − 5 = 6*

(Repeat problems that were not firm.)

EXERCISE 2: DOLLARS WITH 1-DIGIT CENTS   `REMEDY`

a. You learned about writing cents numbers that are less than 10. Remember, when you write cents that are less than 10, you write a zero before the number.
- If you want to write 3 cents, you write zero 3 after the dot.
- What do you write after the dot? (Signal.) *Zero 3.*
- If you want to write 4 cents, what do you write after the dot? (Signal.) *Zero 4.*
- If you want to write 2 cents, what do you write after the dot? (Signal.) *Zero 2.*
(Repeat until firm.)

b. I want to write 4 dollars and 7 cents.
- What do I want to write? (Signal.) *4 dollars and 7 cents.*
(Display:) W   [81:2A]

$4.

- Here's the first part of 4 dollars and 7 cents.
- What do I write after the dot? (Signal.) *Zero 7.*
(Add to show:)   [81:2B]

$4.07

- Read this amount. (Signal.) *4 dollars and 7 cents.*
(Change to show:)   [81:2C]

$4.

c. I want to write 4 dollars and 3 cents.
- What do I write after the dot? (Signal.) *Zero 3.*
(Add to show:)   [81:2D]

$4.03

- Read this amount. (Signal.) *4 dollars and 3 cents.*

d. I want to write 4 dollars and 30 cents.
- What do I write after the dot? (Signal.) *30.*
(Change to show:)   [81:2E]

$4.30

- Read this amount. (Signal.) *4 dollars and 30 cents.*

e. I want to write 4 dollars and 7 cents.
• What do I write after the dot? (Signal.) *Zero 7.*
(Change to show:)   [81:2F]

$4.07

• Read this amount. (Signal.) *4 dollars and 7 cents.*

f. I want to write 4 dollars and 70 cents.
• What do I write after the dot? (Signal.) *70.*
(Change to show:)   [81:2G]

$4.70

• Read this amount. (Signal.) *4 dollars and 70 cents.*

g. I want to write 4 dollars and 2 cents.
• What do I write after the dot? (Signal.) *Zero 2.*
(Change to show:)   [81:2H]

$4.02

• Read this amount. (Signal.) *4 dollars and 2 cents.*
Remember, when you write a cents number that is less than 10, you write a zero before the number.

EXERCISE 3: NUMBER FAMILIES
MISSING NUMBER IN FAMILY

a. (Display:)   [81:3A]

$$\underset{\longrightarrow}{4 \quad 2} \underline{\quad}$$

$$\underset{\longrightarrow}{4 \quad 4} \underline{\quad}$$

$$\underset{\longrightarrow}{4 \quad 3} \underline{\quad}$$

We worked with these number families last time.
• (Point to 4→2.) What are the small numbers in this family? (Touch.) *4 and 2.*
What's the big number? (Signal.) *6.*
• (Point to 4→4.) What are the small numbers in this family? (Touch.) *4 and 4.*
What's the big number? (Signal.) *8.*
• (Point to 4→3.) What are the small numbers in this family? (Touch.) *4 and 3.*
What's the big number? (Signal.) *7.*
(Repeat until firm.)

b. (Point to 4→2.) What are the small numbers in this family? (Touch.) *4 and 2.*
• What's the big number? (Signal.) *6.*
• Say the fact that starts with 4. (Signal.) *4 + 2 = 6.*
• Say the other plus fact. (Signal.) *2 + 4 = 6.*
• Say the fact that goes backward down the arrow. (Signal.) *6 – 2 = 4.*
• Say the other minus fact. (Signal.) *6 – 4 = 2.*
(Repeat step b until firm.)

c. (Point to 4→4.) What are the small numbers in this family? (Touch.) *4 and 4.*
• What's the big number? (Signal.) *8.*
There's only one plus fact and one minus fact.
• Say the plus fact. (Signal.) *4 + 4 = 8.*
• Say the minus fact. (Signal.) *8 – 4 = 4.*

d. (Display:)   [81:3B]

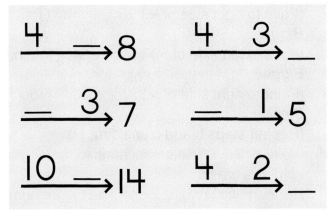

You're going to say the problem for the missing number in each family.

e. (Point to 4→8.) Say the problem for the missing number. Get ready. (Touch.) *8 minus 4.*
• What's 8 minus 4? (Signal.) *4.*
f. (Point to →3→7.) Say the problem for the missing number. Get ready. (Touch.) *7 minus 3.*
• What's 7 minus 3? (Signal.) *4.*
g. (Repeat the following tasks for remaining families:)

(Point to __.)	Say the problem for the missing number.	What's __?	
10→14	14 – 10	14 – 10	4
4 3→	4 + 3	4 + 3	7
→1→5	5 – 1	5 – 1	4
4 2→	4 + 2	4 + 2	6

(Repeat for families that were not firm.)

EXERCISE 4: GREATER THAN/LESS THAN SIGN

REMEDY

a. (Display:) W [81:4A]

> <

These are two new signs.
- (Point to left side of **>**.) This sign has a bigger end (touch) and a (touch right side of >) smaller end.
- (Point to right side of **<**.) This sign has a bigger end (touch) and a (touch left side of <) smaller end.

b. Tell me if the end I touch of each sign is bigger or smaller.
- (Point to left side of **<**.) Which end? (Touch.) *Smaller.*
- (Point to right side of **<**.) Which end? (Touch.) *Bigger.*

c. (Point to left side of **>**.) Which end? (Touch.) *Bigger.*
- (Point to right side of **>**.) Which end? (Touch.) *Smaller.*
(Repeat steps b and c until firm.)

d. We're going to use these signs to tell if numbers are bigger or smaller.
Here's how the signs work: The bigger number is next to the end that's bigger. The smaller number is next to the end that's smaller.
- Which end is next to the bigger number, the end that's bigger or smaller? (Signal.) *Bigger.*
- Which end is next to the smaller number, the end that's bigger or smaller? (Signal.) *Smaller.*

e. (Display:) W [81:4B]

6   3
8   9
7   5

We're going to write one of the signs between the numbers in each row.
- (Point to **6   3**.) Read these numbers. (Touch.) *6 (and) 3.*
- Which number is bigger, 6 or 3? (Signal.) *6.*
So I make the sign so the bigger end is next to the 6.
(Add to show:) [81:4C]

6 > 3

f. Again: Tell me which number is bigger, 6 or 3. (Signal.) *6.*
- Is the bigger end of the sign next to the bigger number? (Signal.) *Yes.*
- Is the smaller end of the sign next to the smaller number? (Signal.) *Yes.*
So this is the right sign.

g. (Point to **8   9**.) Read these numbers. (Touch.) *8 (and) 9.*
- Which number is bigger, 8 or 9? (Signal.) *9.*
- So I make the sign with the bigger end next to which number? (Signal.) *9.*
(Add to show:) [81:4D]

8 < 9

h. Again: Tell me which number is bigger, 8 or 9. (Signal.) *9.*
- Is the bigger end of the sign next to the bigger number? (Signal.) *Yes.*
- Is the smaller end of the sign next to the smaller number? (Signal.) *Yes.*
- So is this sign right? (Signal.) *Yes.*

i. (Point to **7   5**.) Read these numbers. (Touch.) *7 (and) 5.*
- Which number is bigger, 7 or 5? (Signal.) *7.*
- So does the end that's bigger or smaller go next to the 7? (Signal.) *Bigger.*
- Which end goes next to the 5? (Signal.) *(The) smaller (end).*
(Add to show:) [81:4E]

7 < 5

- Is the bigger end of the sign next to the bigger number? (Signal.) *No.*
- So is this sign right? (Signal.) *No.*
(Change to show:) [81:4F]

7 > 5

- Is the bigger end of the sign next to the bigger number? (Signal.) *Yes.*
- Is the smaller end of the sign next to the smaller number? (Signal.) *Yes.*
- So is this sign right? (Signal.) *Yes.*

EXERCISE 5: COLUMN ADDITION
CARRYING

a. (Distribute unopened workbooks to children.)
Open your workbook to Lesson 81 and find part 1.
(Observe children and give feedback.)
(Teacher reference:)

```
a.              b.  16   c.          d.  48
        29          32       46          12
      + 22        +  2     + 16        + 37
```

You're going to work these problems.

- Touch and read problem A. (Signal.)
  29 plus 22.
- Read the problem in the ones column.
  (Signal.) *9 plus 2.*
- What's 9 plus 2? (Signal.) *11.*
- Write the digits for 11 where they should go.
  (Observe children and give feedback.)
- Touch and read the problem for the tens column. (Signal.) *1 plus 2 plus 2.*
- What's 1 plus 2? (Signal.) *3.*
- What's 3 plus 2? (Signal.) *5.*
- Write 5 where it should go.
  (Observe children and give feedback.)
- Start with 29 and read problem A and the answer. (Signal.) *29 + 22 = 51.*
b. Touch and read problem B. (Signal.) *16 plus 32 plus 2.*
- Read the problem in the ones column.
  (Signal.) *6 plus 2 plus 2.*
- What's 6 plus 2? (Signal.) *8.*
- What's 8 plus 2? (Signal.) *10.*
- Write the digits for 10 where they should go.
  (Observe children and give feedback.)
- Touch and read the problem for the tens column. (Signal.) *1 plus 1 plus 3.*
- What's 1 plus 1? (Signal.) *2.*
- What's 2 plus 3? (Signal.) *5.*
- Write 5 where it should go.
  (Observe children and give feedback.)
- Start with 16 and read problem B and the answer. (Signal.) *16 + 32 + 2 = 50.*
c. Touch and read problem C. (Signal.)
  46 plus 16.
- Read the problem in the ones column.
  (Signal.) *6 plus 6.*
- What's the answer? (Signal.) *12.*
- Write the digits for 12 where they should go.
  (Observe children and give feedback.)

- Touch and read the problem for the tens column. (Signal.) *1 plus 4 plus 1.*
- What's 1 plus 4? (Signal.) *5.*
- What's 5 plus 1? (Signal.) *6.*
- Write 6 where it should go.
  (Observe children and give feedback.)
- Read problem C and the answer. (Signal.)
  46 + 16 = 62.
d. Touch and read problem D. (Signal.) *48 plus 12 plus 37.*
- Read the problem in the ones column.
  (Signal.) *8 plus 2 plus 7.*
- What's 8 plus 2? (Signal.) *10.*
- What's 10 plus 7? (Signal.) *17.*
- Write the digits for 17 where they should go.
  (Observe children and give feedback.)
- Touch and read the problem for the tens column. (Signal.) *1 plus 4 plus 1 plus 3.*
- What's 1 plus 4? (Signal.) *5.*
- What's 5 plus 1? (Signal.) *6.*
- What's 6 plus 3? (Signal.) *9.*
- Write 9 where it should go.
  (Observe children and give feedback.)
- Start with 48 and read the problem and the answer. (Signal.) *48 + 12 + 37 = 97.*

EXERCISE 6: FACTS
SUBTRACTION

a. Find part 2 on worksheet 81. ✔
(Teacher reference:)

```
a. 9 – 7      f. 12 – 6
b. 10 – 6     g. 10 – 2
c. 5 – 2      h. 7 – 2
d. 11 – 5     i. 9 – 3
e. 13 – 10    j. 8 – 5
```

You're going to read all these problems and then work them.

- Problem A. (Signal.) *9 minus 7.*
- (Repeat for:) B, *10 – 6;* C, *5 – 2;* D, *11 – 5;* E, *13 – 10;* F, *12 – 6;* G, *10 – 2;* H, *7 – 2;* I, *9 – 3;* J, *8 – 5.*
b. Think of the number families and write the answers to the problems.
  (Observe children and give feedback.)
c. Check your work. You'll read the equation for each problem.
- Problem A. (Signal.) *9 – 7 = 2.*
- Problem B. (Signal.) *10 – 6 = 4.*
- (Repeat for:) C, *5 – 2 = 3;* D, *11 – 5 = 6;* E, *13 – 10 = 3;* F, *12 – 6 = 6;* G, *10 – 2 = 8;* H, *7 – 2 = 5;* I, *9 – 3 = 6;* J, *8 – 5 = 3.*

EXERCISE 7: COMPARISON
LENGTH

REMEDY

a. Find part 3 on worksheet 81. ✔
(Teacher reference:)

R Part F

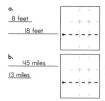

You're going to write problems and figure out how much longer one line is than the other line.

- Touch the lines for problem A. ✔
The top line in the picture says 8 feet long and the bottom line says 18 feet long. The lines aren't really that long, but we'll figure out the answer for lines that are 8 feet and 18 feet long.
- How long is the top line? (Signal.) *8 feet.*
- How long is the bottom line? (Signal.) *18 feet.*
- Touch the line that is longer. ✔
- What's the number for that line? (Signal.) *18.*
- Say the problem for figuring out how many feet longer the bottom line is. (Signal.) *18 minus 8.*
- Write the column problem and work it.
(Observe children and give feedback.)

b. Read the problem and the answer. (Signal.)
18 − 8 = 10.
(Display:)   [81:7A]

Here's what you should have.
- How many feet longer is the bottom line than the top line? (Signal.) *10.*

c. Touch the lines for problem B. ✔
The numbers in the pictures show miles.
- Touch the line that is longer. ✔
- How many miles long is the top line? (Signal.) *45.*

- How many miles long is the bottom line? (Signal.) *13.*
- Say the problem for figuring out how much longer the top line is. (Signal.) *45 minus 13.*
- Write the column problem and work it. (Observe children and give feedback.)

d. Read the problem and the answer for B. (Signal.) *45 − 13 = 32.*
(Display:)   [81:7B]

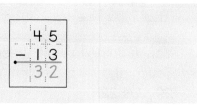

Here's what you should have.
- How many miles longer is the top line than the bottom line? (Signal.) *32.*

EXERCISE 8: MONEY
DOLLAR PROBLEMS

a. Find part 4 on worksheet 81. ✔
(Teacher reference:)

a. $10.46   b. $7.58
+ 3.21   −3.33

You'll touch and read each dollar problem.
- Problem A. Get ready. (Signal.) *10 dollars and 46 cents plus 3 dollars and 21 cents.*
- Problem B. Get ready. (Signal.) *7 dollars and 58 cents minus 3 dollars and 33 cents.*

b. Work the problems in part 4. Remember, write a dollar sign and a dot in each answer.
(Observe children and give feedback.)

c. (Display:)   [81:8A]

a. $10.46   b. $7.58
+ 3.21   −3.33
$13.67 $4.25

Here's what you should have.
- Look at the answer for problem A. Everybody, what's 10 dollars and 46 cents plus 3 dollars and 21 cents? (Signal.) *13 dollars and 67 cents.*
- Look at the answer for problem B. Everybody, what's 7 dollars and 58 cents minus 3 dollars and 33 cents? (Signal.) *4 dollars and 25 cents.*

EXERCISE 9: INDEPENDENT WORK

a. Find part 5 on worksheet 81. ✔
   (Teacher reference:)

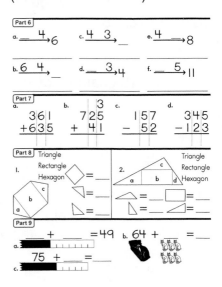

$$\underset{6\quad 4}{}\rightarrow 10$$

$$\underset{6\quad 6}{}\rightarrow 12$$

You'll write the facts for both number families on the space below them. You'll write two facts below the first family. You'll write four facts for the other family.

b. Turn to the other side of worksheet 81 and find part 6. ✔
   (Teacher reference:)

Part 6
a. __ 4 → 6   c. 4 3 → __   e. 4 → 8
b. 6 4 → __   d. __ 3 → 4   f. __ 5 → 11

Part 7
a. 361   b.   3   c. 157   d. 345
   +635   725   − 52   −123
                + 41

Part 8   Triangle
1. Rectangle
   Hexagon   ◇ = __   2. △ᶜ Triangle
                           a b d Rectangle
                                    Hexagon
      c
   b ▽ = __   △ = __ ▢ = __
 a   △ = __   △ = __ ◁ = __

Part 9
a. __ + __ = 49   b. 64 + __ = __
   75 + __ = __
c.

You'll write the problem and the answer for the missing number in each family and complete the family.
In part 7, you'll work the column problems.
In part 8, you'll circle the name for the whole figure and complete the equation for each part.
In part 9, you'll complete the equations for the rulers and the objects.

c. Complete worksheet 81. Remember to write the facts for part 5 on the other side of the worksheet.
   (Observe children and mark incorrect responses on children's worksheets as you give feedback.)

Connecting Math Concepts

Lesson

EXERCISE 1: FACTS

ADDITION/SUBTRACTION

REMEDY

a. (Display:) [82:1A]

7 – 4	6 – 4	4 + 4
14 – 4	9 – 6	10 – 6
6 + 4	8 – 3	8 – 4

For each problem, you'll tell me if the answer is the big number or a small number. Then you'll say the fact.

- (Point to **7 – 4.**) Read this problem. Get ready. (Touch.) *7 minus 4.*
- Is the answer the big number or a small number? (Signal.) *A small number.*
- What's 7 minus 4? (Signal.) *3.*
- Say the fact. (Signal.) *7 – 4 = 3.*

b. (Point to **14 – 4.**) Read this problem. Get ready. (Touch.) *14 minus 4.*
- Is the answer the big number or a small number? (Signal.) *A small number.*
- What's 14 minus 4? (Signal.) *10.*
- Say the fact. (Signal.) *14 – 4 = 10.*

c. (Repeat the following tasks for the remaining problems:)

(Point to __.) Read this problem.	Is the answer the big number or a small number?	What's __?	Say the fact.
6 + 4	*The big number.*	6 + 4 10	*6 + 4 = 10*
6 – 4	*A small number.*	6 – 4 2	*6 – 4 = 2*
9 – 6	*A small number.*	9 – 6 3	*9 – 6 = 3*
8 – 3	*A small number.*	8 – 3 5	*8 – 3 = 5*
4 + 4	*The big number.*	4 + 4 8	*4 + 4 = 8*
10 – 6	*A small number.*	10 – 6 4	*10 – 6 = 4*
8 – 4	*A small number.*	8 – 4 4	*8 – 4 = 4*

(Repeat problems that were not firm.)

EXERCISE 2: DOLLARS WITH 1-DIGIT CENTS

a. You learned about writing cents numbers that are less than 10. Remember, when you write cents that are less than 10, you write a zero before the number.
- If you want to write 3 cents, you write zero 3 after the dot.
- What do you write after the dot? (Signal.) *Zero 3.*
- If you want to write 9 cents, what do you write after the dot? (Signal.) *Zero 9.*
- If you want to write 5 cents, what do you write after the dot? (Signal.) *Zero 5.*
  (Repeat until firm.)

b. I want to write 4 dollars and 2 cents.
- What do I want to write? (Signal.) *4 dollars and 2 cents.*
  (Display:) W [82:2A]

  $4.

  Here's the first part of 4 dollars and 2 cents.
- What do I write after the dot? (Signal.) *Zero 2.*
  (Add to show:) [82:2B]

  $4.02

- Read this amount. (Signal.) *4 dollars and 2 cents.*
  (Change to show:) [82:2C]

  $4.

c. I want to write 4 dollars and 5 cents.
- What do I write after the dot? (Signal.) *Zero 5.*
  (Add to show:) [82:2D]

  $4.05

- Read this amount. (Signal.) *4 dollars and 5 cents.*

d. I want to write 4 dollars and 50 cents.
- What do I write after the dot? (Signal.) *50.*
  (Change to show:) [82:2E]

  $4.50

- Read this amount. (Signal.) *4 dollars and 50 cents.*

e. I want to write 4 dollars and 9 cents.
• What do I write after the dot? (Signal.) *Zero 9.*
(Change to show:)   [82:2F]

$4.09

• Read this amount. (Signal.) *4 dollars and 9 cents.*
f. I want to write 4 dollars and 90 cents.
• What do I write after the dot? (Signal.) *90.*
(Change to show:)   [82:2G]

$4.90

• Read this amount. (Signal.) *4 dollars and 90 cents.*
g. I want to write 4 dollars and 4 cents.
• What do I write after the dot? (Signal.) *Zero 4.*
(Change to show:)   [82:2H]

$4.04

• Read this amount. (Signal.) *4 dollars and 4 cents.*
Remember, when you write a cents number that is less than 10, you write a zero before the number.

EXERCISE 3: NUMBER FAMILIES
MISSING NUMBER IN FAMILY

a. (Display:)   [82:3A]

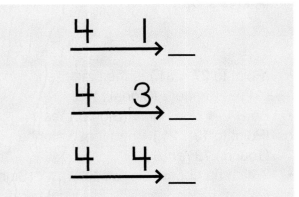

We worked with these number families before.
• (Point to 4 →1.) What are the small numbers in this family? (Touch.) *4 and 1.*
What's the big number? (Signal.) *5.*
• (Point to 4 →3.) What are the small numbers in this family? (Touch.) *4 and 3.*
What's the big number? (Signal.) *7.*
• (Point to 4 →4.) What are the small numbers in this family? (Touch.) *4 and 4.*
What's the big number? (Signal.) *8.*
(Repeat until firm.)

b. (Point to 4 →1.) What are the small numbers in this family? (Touch.) *4 and 1.*
• What's the big number? (Signal.) *5.*
• Say the fact that starts with 4. (Signal.) *4 + 1 = 5.*
• Say the other plus fact. (Signal.) *1 + 4 = 5.*
• Say the fact that goes backward down the arrow. (Signal.) *5 – 1 = 4.*
• Say the other minus fact. (Signal.) *5 – 4 = 1.*
(Repeat step b until firm.)
c. (Point to 4 →3.) What are the small numbers in this family? (Touch.) *4 and 3.*
• What's the big number? (Signal.) *7.*
• Say the fact that starts with 4. (Signal.) *4 + 3 = 7.*
• Say the other plus fact. (Signal.) *3 + 4 = 7.*
• Say the fact that goes backward down the arrow. (Signal.) *7 – 3 = 4.*
• Say the other minus fact. (Signal.) *7 – 4 = 3.*
d. (Display:)   [82:3B]

6 →4_	4 →4_	2 →4
4 →7	2 →6	4 →3_
4 →14	4 →10	4 →5

You're going to say the problem for the missing number in each family.
e. (Point to 6 →4_.) Say the problem for the missing number. Get ready. (Touch.) *6 plus 4.*
• What's 6 plus 4? (Signal.) *10.*
f. (Point to 4 →7.) Say the problem for the missing number. Get ready. (Touch.) *7 minus 4.*
• What's 7 minus 4? (Signal.) *3.*
g. (Repeat the following tasks for remaining families:)

(Point to __.) Say the problem for the missing number.		What's __?	
4 →14	14 – 4	14 – 4	10
4 →4_	4 + 4	4 + 4	8
2 →6	6 – 2	6 – 2	4
4 →10	10 – 4	10 – 4	6
2 →4	4 – 2	4 – 2	2
4 →3_	4 + 3	4 + 3	7
4 →5	5 – 4	5 – 4	1

(Repeat families that were not firm.)

EXERCISE 4: GREATER THAN/LESS THAN SIGN

a. (Display:) W   [82:4A]

<
>

You learned about these signs last time.

b. (Point to **<**.) Remember: The bigger end of the sign goes next to the bigger number. The smaller end of the sign goes next to the smaller number.
- (Point to left.) Is this end of the sign bigger or smaller? (Touch.) *Smaller.*
- So does this side go next to the number that's bigger or smaller? (Touch.) *Smaller.*
- (Point to right.) Does the bigger or smaller number go next to this end of the sign? (Touch.) *Bigger (number).*

c. (Point to left of **>**.) Does the bigger or smaller number go next to this end of the sign? (Touch.) *Bigger (number).*
- (Point to right.) Does the bigger or smaller number go next to this end of the sign? (Touch.) *Smaller (number).*

d. (Display:) W   [82:4B]

6	9
12	1
27	31
13	21

We're going to write one of the signs between the numbers in each row.
- (Point to **6   9**.) Read these numbers. (Touch.) *6 (and) 9.*
- Which number is bigger, 6 or 9? (Signal.) *9.* So I make the sign with the bigger end next to the 9.
(Add to show:)   [82:4C]

6 > 9

- (Point to **6 > 9**.) Is the bigger end next to the bigger number? (Signal.) *No.*
- So is this sign right? (Signal.) *No.*
(Change to show:)   [82:4D]

6 < 9

- (Point to **6 < 9**.) Is the bigger end next to the bigger number? (Signal.) *Yes.*
- (Point to **6 < 9**.) Is the smaller end next to the smaller number? (Signal.) *Yes.*
- So is this sign right? (Signal.) *Yes.*

e. (Point to **12   11**.) Read these numbers. (Touch.) *12 (and) 11.*
- Which number is bigger, 12 or 11? (Signal.) *12.* So I make the sign with the bigger end next to the 12.
(Add to show:)   [82:4E]

12 < 11

- (Point to **12 < 11**.) Is the bigger end next to the bigger number? (Signal.) *No.*
- So is this sign right? (Signal.) *No.*
(Change to show:)   [82:4F]

12 > 11

- (Point to **12 > 11**.) Is the bigger end next to the bigger number? (Signal.) *Yes.*
- (Point to **12 > 11**.) Is the smaller end next to the smaller number? (Signal.) *Yes.*
- So is this sign right? (Signal.) *Yes.*

f. (Point to **27   31**.) Read these numbers. (Touch.) *27 (and) 31.*
- Which number is bigger, 27 or 31? (Signal.) *31.* So I make the sign with the bigger end next to the 31.
(Add to show:)   [82:4G]

27 < 31

- (Point to **27 < 31**.) Is the bigger end next to the bigger number? (Signal.) *Yes.*
- So is this sign right? (Signal.) *Yes.*

g. (Point to **13   21**.) Read these numbers. (Touch.) *13 (and) 21.*
- Which number is bigger, 13 or 21? (Signal.) *21.* So I make the sign with the bigger end next to the 21.
(Add to show:)   [82:4H]

13 < 21

- (Point to **13 < 21**.) Is the bigger end next to the bigger number? (Signal.) *Yes.*
- So is this sign right? (Signal.) *Yes.*

EXERCISE 5: MONEY
DOLLAR PROBLEMS

a. (Distribute unopened workbooks to children.)
    Open your workbook to Lesson 82 and find
    part 1.
    (Observe children and give feedback.)
    (Teacher reference:)

a. $14.99   b. $1.60
   – 4.62   +7.25

- Touch and read problem A. Get ready.
  (Signal.) *14 dollars and 99 cents minus
  4 dollars and 62 cents.*
- Touch and read problem B. Get ready.
  (Signal.) *1 dollar and 60 cents plus 7 dollars
  and 25 cents.*

b. Work both problems. Remember, write a dollar
    sign and a dot in the answer before working
    each problem.
    (Observe children and give feedback.)

c. (Display:) W   [82:5A]

    a. $ 1 4.9 9
       –   4.6 2
       $ 1 0.3 7

    Here's what you should have for problem A.
- (Point to **$10.37**.) Read the answer. (Signal.)
  10 dollars and 37 cents.

d. (Display:) W   [82:5B]

    b. $ 1.6 0
       +7.2 5
       $ 8.8 5

    Here's what you should have for problem B.
- (Point to **$8.85**.) Read the answer. (Signal.)
  8 dollars and 85 cents.

EXERCISE 6: COMPARISON
LENGTH   REMEDY

a. Find part 2 on worksheet 82. ✔
    (Teacher reference:)   R Part G

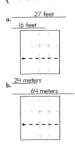

    You're going to write problems and figure out
    how much longer one line is than the other line.
- Touch the lines for problem A. ✔
  The top line in the picture is 27 feet long and
  the bottom line is 16 feet long.
- How long is the top line? (Signal.) *27 feet.*
- How long is the bottom line? (Signal.) *16 feet.*
- Touch the line that is longer. ✔
- What's the number for that line? (Signal.) *27.*
- Say the problem for figuring out how many feet
  longer the top line is. (Signal.) *27 minus 16.*
- Write the column problem and work it.
  Remember to make the equals bar.
  (Observe children and give feedback.)
  (Display:) W   [82:6A]

    a.   2 7
       – 1 6
         1 1

    Here's what you should have.
- How many feet longer is the top line than the
  bottom line? (Signal.) *11.*

b. Touch the lines for problem B. ✔
    The numbers in the picture show meters.
- Touch the line that is longer. ✔
- How many meters long is the bottom line?
  (Signal.) *64.*
- Say the problem. (Signal.) *64 minus 24.*
- Write the column problem and work it.
  (Observe children and give feedback.)
  (Display:) W   b.   [82:6B]

       6 4
     – 2 4
       4 0

    Here's what you should have.
- How many meters longer is the bottom line
  than the top line? (Signal.) *40.*

EXERCISE 7: COLUMN ADDITION
CARRYING DISCRIMINATION

   REMEDY

a. Find part 3 on worksheet 82. ✔
(Teacher reference:)   R Part J

$$\begin{array}{ccc} a. & b. & c. \\ 11 & 37 & 38 \\ +59 & +42 & +42 \end{array}$$

You have to carry to work some of these
problems.

• Problem A is 11 plus 59. Say the problem for
the ones. Get ready. (Signal.) *1 plus 9.*
• What's the answer? (Signal.) *10.*
• Do you write anything in the tens column?
(Signal.) *Yes.*
b. Problem B is 37 plus 42. Say the problem for
the ones. Get ready. (Signal.) *7 plus 2.*
• What's the answer? (Signal.) *9.*
• Do you write anything in the tens column?
(Signal.) *No.*
c. Problem C is 38 plus 42. Say the problem for
the ones. Get ready. (Signal.) *8 plus 2.*
• What's the answer? (Signal.) *10.*
• Do you write anything in the tens column?
(Signal.) *Yes.*
d. Work all the problems in part 3.
(Observe children and give feedback.)
e. Check your work.
• Problem A. 11 plus 59. What's the answer?
(Signal.) *70.*
• Problem B. 37 plus 42. What's the answer?
(Signal.) *79.*
• Problem C. 38 plus 42. What's the answer?
(Signal.) *80.*

EXERCISE 8: WORD PROBLEMS (COLUMNS)

a. Find part 4. ✔
(Teacher reference:)

You're going to write the symbols for word
problems in columns and work them.
• Touch where you'll write symbols for
problem A. ✔
b. Listen to problem A: A dentist saw 12 patients
on Monday. She saw 37 more patients on
Tuesday. How many patients did she see
altogether?

• Listen again and write the symbols for the
whole problem: A dentist saw 12 patients
on Monday. She saw 37 more patients on
Tuesday. How many patients did she see
altogether?
(Observe children and give feedback.)
• Everybody, touch and read the problem for A.
Get ready. (Signal.) *12 plus 37.*
c. Work problem A. Put your pencil down when
you know how many patients the dentist saw
altogether.
(Observe children and give feedback.)
(Teacher reference:)   [82:8A]

a. $$\begin{array}{r} 1\ 2 \\ +\ 3\ 7 \\ \hline 4\ 9 \end{array}$$

• Read the problem and the answer you wrote
for A. Get ready. (Signal.) *12 + 37 = 49.*
• How many patients did she see altogether?
(Signal.) *49.*
d. Touch where you'll write symbols for
problem B. ✔
• Listen to problem B: Big Jim weighed 248
pounds. Then he lost 32 pounds. How many
pounds does Jim weigh now?
(Observe children and give feedback.)
• Listen again and write the symbols for the
whole problem: Big Jim weighed 248 pounds.
Then he lost 32 pounds. How many pounds
does Jim weigh now?
(Observe children and give feedback.)
• Everybody, touch and read the problem for B.
Get ready. (Signal.) *248 – 32.*
e. Work problem B. Put your pencil down when
you know how many pounds Jim weighs now.
(Observe children and give feedback.)
(Teacher reference:)   [82:8B]

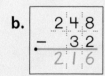

b. $$\begin{array}{r} 2\ 4\ 8 \\ -\ \ 3\ 2 \\ \hline 2\ 1\ 6 \end{array}$$

• Read the problem and the answer you wrote
for B. Get ready. (Signal.) *248 – 32 = 216.*
• How many pounds does Jim weigh now?
(Signal.) *216.*

EXERCISE 9: GREATER THAN/LESS THAN SIGN
WRITING

a. Find part 5 on worksheet 82. ✔
   (Teacher reference:)

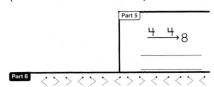

   There are only two facts for the number family in part 5. You'll write the facts for the family below it as part of your independent work.

b. Find part 6. ✔
   You're going to practice making the signs that have a bigger end and a smaller end.

c. Touch the first dotted sign. ✔

• Start with the big dot and make the sign.
   (Observe children and give feedback.)

d. Make the rest of the signs.
   (Observe children and give feedback.)

EXERCISE 10: INDEPENDENT WORK

a. Find part 7. ✔
   (Teacher reference:)

   a. $600 + 10 + 5 =$ c. $\_\_ + \_\_ + \_\_ = 574$
   b. $80 + 0 =$ d. $\_\_ + \_\_ + \_\_ = 209$

   You'll complete the place-value addition equations in part 7.

b. Turn to the other side of worksheet 82 and find part 8. ✔
   (Teacher reference:)

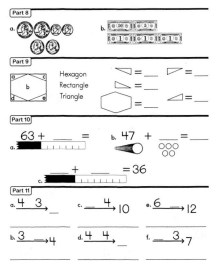

   You'll count for the coins and count for the bills. Then you'll write an equals and a number to show how many cents or dollars each group is worth.
   In part 9, you'll circle the name for the whole figure and complete the equation for each part.
   In part 10, you'll complete the equations for the rulers and the objects.
   In part 11, you'll write the problem and the answer for the missing number in each family and complete the family.

c. Complete worksheet 82. Remember to write the facts below the family in part 5 and complete the equations for part 7 on the other side of your worksheet.
   (Observe children and mark incorrect responses on children's worksheets as you give feedback.)

Lesson

EXERCISE 1: COMPARISON STATEMENTS
LENGTH

a. (Display:)   [83:1A]

$$7$$

$$8$$

These lines are supposed to show meters.
- Is the top line or the bottom line longer?
  (Signal.) *The bottom line.*
- Which line is shorter? (Signal.) *The top line.*
- Say the problem for figuring out how many
  meters longer the bottom line is. (Signal.)
  8 minus 7.
- How much longer is the bottom line? (Signal.)
  One meter.

b. The bottom line is 1 meter longer than the top
  line. Say the sentence about the bottom line.
  (Signal.) *The bottom line is 1 meter longer
  than the top line.*

  My turn to say the sentence about the top
  line: The top line is 1 meter shorter than the
  bottom line.
- Say the sentence about the top line.
  (Signal.) *The top line is 1 meter shorter than
  the bottom line.*
- Say the sentence about the bottom line.
  (Signal.) *The bottom line is 1 meter longer
  than the top line.*
  (Repeat until firm.)

c. (Display:)   [83:1B]

$$10$$

$$2$$

These lines are supposed to show yards.
- Is the top line or the bottom line longer?
  (Signal.) *The top line.*
- Which line is shorter? (Signal.) *The bottom line.*
- Say the problem for figuring out how many
  yards longer the top line is. (Signal.) *10 minus 2.*
- How much longer is the top line? (Signal.)
  8 yards.

d. The top line is 8 yards longer than the
  bottom line.
- Say the sentence about the top line. (Signal.) *The
  top line is 8 yards longer than the bottom line.*
- Say the sentence about the bottom line.
  (Signal.) *The bottom line is 8 yards shorter
  than the top line.*
  (Repeat until firm.)

e. (Display:)   [83:1C]

$$17$$

$$7$$

These lines are supposed to show inches.
- Is the top line or the bottom line longer?
  (Signal.) *The top line.*
- Which line is shorter? (Signal.) *The bottom line.*
- Say the problem for figuring out how many
  inches longer the top line is. (Signal.)
  17 minus 7.
- How much longer is the top line? (Signal.)
  10 inches.

f. The top line is 10 inches longer than the
  bottom line.
- Say the sentence about the top line. (Signal.)
  *The top line is 10 inches longer than the
  bottom line.*
- Say the sentence about the bottom line.
  (Signal.) *The bottom line is 10 inches shorter
  than the top line.*
  (Repeat until firm.)

EXERCISE 2: NUMBER FAMILIES
MISSING NUMBER IN FAMILY

a. (Display:) [83:2A]

$$4 \quad 3 \longrightarrow \_$$
$$10 \quad 4 \longrightarrow \_$$
$$4 \quad 4 \longrightarrow \_$$

We worked with these number families before.
- (Point to $4 \quad 3 \rightarrow$.) What are the small numbers in this family? (Touch.) *4 and 3.*
  What's the big number? (Signal.) *7.*
- (Point to $10 \quad 4 \rightarrow$.) What are the small numbers in this family? (Touch.) *10 and 4.*
  What's the big number? (Signal.) *14.*
- (Point to $4 \quad 4 \rightarrow$.) What are the small numbers in this family? (Touch.) *4 and 4.*
  What's the big number? (Signal.) *8.*
- (Repeat until firm.)

b. (Point to $4 \quad 3 \rightarrow$.) What are the small numbers in this family? (Touch.) *4 and 3.*
- What's the big number? (Signal.) *7.*
- Say the fact that starts with 4. (Signal.) *4 + 3 = 7.*
  Say the other plus fact. (Signal.) *3 + 4 = 7.*
- Say the fact that goes backward down the arrow. (Signal.) *7 – 3 = 4.*
  Say the other minus fact. (Signal.) *7 – 4 = 3.*
- (Repeat step b until firm.)

c. (Point to $10 \quad 4 \rightarrow$.) What are the small numbers in this family? (Touch.) *10 and 4.*
- What's the big number? (Signal.) *14.*
- Say the fact that starts with 10. (Signal.) *10 + 4 = 14.*
- Say the other plus fact. (Signal.) *4 + 10 = 14.*
- Say the fact that goes backward down the arrow. (Signal.) *14 – 4 = 10.*
- Say the other minus fact. (Signal.) *14 – 10 = 4.*
- (Repeat step c until firm.)

d. (Display:) [83:2B]

$\_\xrightarrow{\ 3\ } 7$	$4 \quad 4 \longrightarrow \_$	$4 \quad 3 \longrightarrow \_$
$6 \longrightarrow 10$	$2 \longrightarrow 4$	$6 \quad 3 \longrightarrow \_$
$\_\xrightarrow{\ 2\ } 10$	$\_\xrightarrow{\ 6\ } 12$	$4 \longrightarrow 8$

You're going to say the problem for the missing number in each family.

e. (Point to $\_\xrightarrow{3} 7$.) Say the problem for the missing number. Get ready. (Touch.) *7 minus 3.*
- What's 7 minus 3? (Signal.) *4.*
f. (Point to $6 \rightarrow 10$.) Say the problem for the missing number. Get ready. (Touch.) *10 minus 6.*
- What's 10 minus 6? (Signal.) *4.*
g. (Repeat the following tasks for remaining families:)

(Point to __.) Say the problem for the missing number	What's __?	
$\_\xrightarrow{2} 10$	$10 - 2$	$10 - 2$ 8
$4 \quad 4 \rightarrow \_$	$4 + 4$	$4 + 4$ 8
$2 \rightarrow 4$	$4 - 2$	$4 - 2$ 2
$\_\xrightarrow{6} 12$	$12 - 6$	$12 - 6$ 6
$4 \quad 3 \rightarrow \_$	$4 + 3$	$4 + 3$ 7
$6 \quad 3 \rightarrow \_$	$6 + 3$	$6 + 3$ 9
$4 \rightarrow 8$	$8 - 4$	$8 - 4$ 4

(Repeat families that were not firm.)

EXERCISE 3: MIXED COUNTING

a. You're going to do some counting.
- Start with 40 and count by fives to 80. Get 40 going. *Fortyyy.* Count. (Tap.) *45, 50, 55, 60, 65, 70, 75, 80.*
- Start with 40 and plus tens to 100. Get 40 going. *Fortyyy.* Plus tens. (Tap.) *50, 60, 70, 80, 90, 100.*
b. Count backward by tens from 100. Get 100 going. *One huuundred.* Count backward. (Tap.) *90, 80, 70, 60, 50, 40, 30, 20, 10.*
c. Start with 24 and plus tens to 94. Get 24 going. *Twenty-fouuur.* Plus tens. (Tap.) *34, 44, 54, 64, 74, 84, 94.*
d. Start with 73 and count backward to 63. Get 73 going. *Seventy-threee.* Count backward. (Tap.) *72, 71, 70, 69, 68, 67, 66, 65, 64, 63.*

e. Count by hundreds to 1000. Get ready.
   (Tap.) *100, 200, 300, 400, 500, 600, 700, 800, 900, 1000.*
f. You have 75. When you plus fives, what's the next number? (Signal.) *80.*
• You have 75. When you minus ones, what's the next number? (Signal.) *74.*
• You have 75. When you plus tens, what's the next number? (Signal.) *85.*
• You have 75. When you plus 25s, what's the next number? (Signal.) *100.*

EXERCISE 4: FACTS
ADDITION/SUBTRACTION

a. (Display:)   [83:4A]

6 + 6	4 + 6	4 + 4
9 − 3	8 − 3	11 − 9
3 + 4	6 + 5	7 − 3

For each problem, you'll tell me if the answer is the big number or a small number. Then you'll say the fact.
• (Point to **6 + 6.**) Read this problem. Get ready. (Touch.) *6 plus 6.*
• Is the answer the big number or a small number? (Signal.) *The big number.*
• What's 6 plus 6? (Signal.) *12.*
• Say the fact. (Signal.) *6 + 6 = 12.*
b. (Point to **9 − 3.**) Read this problem. Get ready. (Touch.) *9 minus 3.*
• Is the answer the big number or a small number? (Signal.) *A small number.*
• What's 9 minus 3? (Signal.) *6.*
• Say the fact. (Signal.) *9 − 3 = 6.*
c. (Repeat the following tasks for the remaining problems:)

(Point to __.) Read this problem.	Is the answer the big number or a small number?	What's __?		Say the fact.
3 + 4	The big number.	3 + 4	7	3 + 4 = 7
4 + 6	The big number.	4 + 6	10	4 + 6 = 10
8 − 3	A small number.	8 − 3	5	8 − 3 = 5
6 + 5	The big number.	6 + 5	11	6 + 5 = 11
4 + 4	The big number.	4 + 4	8	4 + 4 = 8
11 − 9	A small number.	11 − 9	2	11 − 9 = 2
7 − 3	A small number.	7 − 3	4	7 − 3 = 4

(Repeat problems that were not firm.)

EXERCISE 5: GREATER THAN/LESS THAN SIGN   `REMEDY`

a. (Display:) W   [83:5A]

28	19
5	7
35	42
81	69

We're going to write one of the signs with a bigger and a smaller end between the numbers in each row.
b. (Point to **28   19.**) Read these numbers. (Touch.) *28 (and) 19.*
• Which number is bigger, 28 or 19? (Signal.) *28.* So I make the sign with the bigger end next to the 28.
   (Add to show:)   [83:5B]

28 > 19

• (Point to **28 > 19.**) Is the bigger end next to the bigger number? (Signal.) *Yes.*
• So is this sign right? (Signal.) *Yes.*
c. (Point to **5   7.**) Read these numbers. (Touch.) *5 (and) 7.*
• Which number is bigger, 5 or 7? (Signal.) *7.* So I make the sign with the bigger end next to the 7.
   (Add to show:)   [83:5C]

5 > 7

• (Point to **5 > 7.**) Is the bigger end next to the bigger number? (Signal.) *No.*
• So is this sign right? (Signal.) *No.*
   (Change to show:)   [83:5D]

5 < 7

• (Point to **5 < 7.**) Is the bigger end next to the bigger number? (Signal.) *Yes.*
• Is the smaller end next to the smaller number? (Signal.) *Yes.*
• So is this sign right? (Signal.) *Yes.*

d. (Point to **35   42.**) Read these numbers. (Touch.) *35 (and) 42.*
- Which number is bigger, 35 or 42? (Signal.) *42.*
- So I make the sign with the bigger end next to which number? (Signal.) *42.*
(Add to show:) [83:5E]

$$35 < 42$$

- (Point to **35 < 42.**) Is the bigger end next to the bigger number? (Signal.) *Yes.*
- So is this sign right? (Signal.) *Yes.*
e. (Point to **81   69.**) Read these numbers. (Touch.) *81 (and) 69.*
- Which number is bigger, 81 or 69? (Signal.) *81.*
- So I make the sign with the bigger end next to which number? (Signal.) *81.*
(Add to show:) [83:5F]

$$28 > 19$$
$$5 < 7$$
$$35 < 42$$
$$81 > 69$$

- (Point to **81 > 69.**) Is the bigger end next to the bigger number? (Signal.) *Yes.*
- So is this sign right? (Signal.) *Yes.*
f. (Distribute unopened workbooks to children.)
- Open your workbook to Lesson 83 and find part 1.
(Observe children and give feedback.)
(Teacher reference:) R Part H

a. 19⁢⁢20
b. 33⁢⁢22
c. 53⁢⁢48

These problems show two numbers. Both signs with a bigger end and a smaller end are partly shown. You're going to make the right one for each problem.
- Touch problem A. ✔
- Read the numbers. (Signal.) *19 (and) 20.*
- Which is bigger? (Signal.) *20.*

- Put your pencil on the dot for the sign with the bigger end next to 20.
(Observe children and give feedback.)
- Make the sign. ✔
(Display:) [83:5G]

a. 19 < 20

Here's what you should have.
g. Read the numbers for problem B. (Signal.) *33 (and) 22.*
- Which is bigger? (Signal.) *33.*
- Put your pencil on the dot for the sign with the bigger end next to 33. ✔
- Make the sign. ✔
(Add to show:) [83:5H]

b. 33 > 22

Here's what you should have.
h. Read the numbers for problem C. (Signal.) *53 (and) 48.*
- Which is bigger? (Signal.) *53.*
- Put your pencil on the dot for the sign with the bigger end next to 53. ✔
- Make the sign. ✔
(Add to show:) [83:5I]

c. 53 > 48

Here's what you should have.

EXERCISE 6: COLUMN ADDITION
CARRYING DISCRIMINATION

a. Find part 2 on worksheet 83. ✔
(Teacher reference:)

a.
```
  26
+ 53
```
b.
```
  58
+ 12
```
c.
```
  14
  63
+ 12
```
d.
```
  43
   6
+ 22
```

You have to carry to work some of these problems.
- Work the problems. Put your pencil down when you've completed part 2.
(Observe children and give feedback.)

b. Check your work.
- Problem A is 26 plus 53. Did you carry? (Signal.) *No.*
- What's the answer? (Signal.) *79.*
c. Problem B is 58 plus 12. Did you carry? (Signal.) *Yes.*
- What's the answer? (Signal.) *70.*
d. Problem C is 14 plus 63 plus 12. Did you carry? (Signal.) *No.*
- What's the answer? (Signal.) *89.*
e. Problem D is 43 plus 6 plus 22. Did you carry? (Signal.) *Yes.*
- What's the answer? (Signal.) *71.*

EXERCISE 7: WRITING DOLLARS WITH 1-DIGIT OR 2-DIGIT CENTS

a. (Display:) W   [83:7A]

$$\$7.$$

I want to write 7 dollars and 6 cents.
- What do I write after the dot? (Signal.) *Zero 6.* (Add to show:)   [83:7B]

$$\$7.06$$

- Read this amount. (Signal.) *7 dollars and 6 cents.*
b. (Change to show:)   [83:7C]

$$\$7.$$

I want to write 7 dollars and 11 cents.
- What do I write after the dot? (Signal.) *11.*
c. (Add to show:)   [83:7D]

$$\$7.11$$

- Read this amount. (Signal.) *7 dollars and 11 cents.*
d. I want to write 7 dollars and 8 cents.
- What do I write after the dot? (Signal.) *Zero 8.* (Change to show:)   [83:7E]

$$\$7.08$$

- Read this amount. (Signal.) *7 dollars and 8 cents.*

e. Find part 3 on your worksheet. ✔ (Teacher reference:)

a. $5.___ b. $5.___ c. $5.___ d. $5.___

The first part of each amount is 5 dollars. You're going to write cents after the dot.
- Touch A. ✔
- The amount for A is 5 dollars and 4 cents. What amount? (Signal.) *5 dollars and 4 cents.*
- Complete A so it shows 5 dollars and 4 cents. (Display:) W   [83:7F]

a. $5.04

Here's what you should have.
f. Touch B. ✔
- The amount for B is 5 dollars and 40 cents. What amount? (Signal.) *5 dollars and 40 cents.*
- Complete B so it shows 5 dollars and 40 cents. (Display:) W   [83:7G]

b. $5.40

Here's what you should have.
g. Touch C. ✔
- The amount for C is 5 dollars and 60 cents. What amount? (Signal.) *5 dollars and 60 cents.*
- Complete C so it shows 5 dollars and 60 cents. (Display:) W   [83:7H]

c. $5.60

Here's what you should have.
h. Touch D. ✔
- The amount for D is 5 dollars and 6 cents. What amount? (Signal.) *5 dollars and 6 cents.*
- Complete D so it shows 5 dollars and 6 cents. (Display:) W   [83:7I]

c. $5.06

Here's what you should have.

EXERCISE 8: WORD PROBLEMS
COMPARISONS

REMEDY

a. Find part 4. ✔
   (Teacher reference:)

R Part A

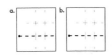

You've said problems for figuring out how much longer or how much shorter lines are. Problem A tells about two lines. You're going to write the problem and figure out how much longer one line is than the other line.

- Touch where you'll write the column problem for A. ✔
- Listen to problem A: The red line is 89 inches long. The blue line is 38 inches long.
- Listen to problem A again and answer some questions. The red line is 89 inches long. How long is the red line? (Signal.) *89 inches.*
- Listen: The blue line is 38 inches long. How long is the blue line? (Signal.) *38 inches.*
- Which line is longer, the red line or the blue line? (Signal.) *The red line.*
- Say the problem for figuring out how much longer the red line is than the blue line. (Signal.) *89 minus 38.*
- Write the problem for A and work it. Put your pencil down when you know how many inches longer the red line is than the blue line. **(Observe children and give feedback.)**

b. Check your work.
- Read the problem and the answer for A. Get ready. (Signal.) *89 − 38 = 51.*
- The red line is how much longer than the blue line? (Signal.) *51 inches.*

c. Touch where you'll write the column problem for B. ✔
   You're going to write the problem and figure out how much more one dog weighs than another dog weighs.
- Listen to problem B: The brown dog weighs 24 pounds. The black dog weighs 66 pounds.

- Listen to problem B again and answer some questions. The brown dog weighs 24 pounds. How much does the brown dog weigh? (Signal.) *24 pounds.*
- Listen: The black dog weighs 66 pounds. How much does the black dog weigh? (Signal.) *66 pounds.*
- Which dog is heavier, the brown dog or the black dog? (Signal.) *The black dog.*
- Say the problem for figuring out how much heavier the black dog is than the brown dog. (Signal.) *66 minus 24.*
- Write the problem for B and work it. Put your pencil down when you know how many pounds heavier the black dog is than the brown dog. **(Observe children and give feedback.)**

d. Check your work.
- Read the problem and the answer for B. Get ready. (Signal.) *66 − 24 = 42.*
- The black dog is how much heavier than the brown dog? (Signal.) *42 pounds.*

EXERCISE 9: MONEY
DOLLAR PROBLEMS

a. Find part 5. ✔
   (Teacher reference:)

a. $6.45 b. $25.49
   +3.12   −14.17

You'll work these problems when you do your independent work.
Remember, write a dollar sign and a dot for each answer.
You will write the number for dollars before the dot.
- What will you write after the dot? **(Call on a child.)** *The number for cents.*
Yes, you'll write the number for cents after the dot.
- Everybody, what will you write after the dot? (Signal.) *The number for cents.*

EXERCISE 10: INDEPENDENT WORK

a. Find part 6. ✔
    (Teacher reference:)

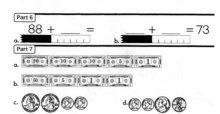

    You'll complete the equations for these rulers.

•   Find part 7. ✔
    You'll write an equals and the number of
    dollars or cents for each group.

b. Turn to the other side of worksheet 83 and find
    part 8. ✔
    (Teacher reference:)

Part 8

a. $\underline{4 \quad 3}_{\rightarrow 7}$ b. $\underline{6 \quad 5}_{\rightarrow 11}$ c. $\underline{6 \quad 3}_{\rightarrow 9}$
____ ____ ____
____ ____ ____
____ ____ ____
____ ____ ____

Part 9

Hexagon $\triangleleft = \underline{\ }$ $\triangledown = \underline{\ }$
Rectangle
Triangle
          $\triangle = \underline{\ }$ $\triangle = \underline{\ }$

Part 10

a. $\underline{4}_{\rightarrow 8}$ c. $\underline{7 \quad 2}_{\rightarrow \underline{\ }}$ e. $\underline{10 \quad 4}_{\rightarrow \underline{\ }}$
____ ____ ____

b. $\underline{\quad 2}_{\rightarrow 7}$ d. $\underline{\quad 4}_{\rightarrow 10}$ f. $\underline{4}_{\rightarrow 7}$
____ ____ ____

In part 8 you'll write four facts below each family.
In part 9, you'll circle the name for the whole
figure. Then you'll complete the equations for
the parts.
In part 10, below each family, you'll write
the fact for the missing number. Then you'll
complete the family.

c. Complete worksheet 83. Remember to
    work parts 5, 6, and 7 on the other side of
    worksheet 83.
    **(Observe children and mark incorrect responses
    on children's worksheets as you give feedback.)**

Lesson 84

EXERCISE 1: NUMBER FAMILIES
MISSING NUMBER IN FAMILY

a. (Display:)   [84:1A]

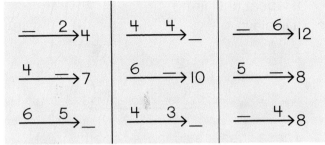

You're going to say the problem for the missing number in each family.

b. (Point to ⎯2→4.) Say the problem for the missing number. Get ready. (Touch.)
4 minus 2.
- What's 4 minus 2? (Signal.) *2.*

c. (Point to 4⎯→7.) Say the problem for the missing number. (Touch.) *7 minus 4.*
- What's 7 minus 4? (Signal.) *3.*

d. (Repeat the following tasks for the remaining families:)

(Point to __.) Say the problem for the missing number.	What's __?	
6 5→_	6 + 5	6 + 5	11
4 4→_	4 + 4	4 + 4	8
6⎯→10	10 − 6	10 − 6	4
4 3→_	4 + 3	4 + 3	7
⎯→12	12 − 6	12 − 6	6
5⎯→8	8 − 5	8 − 5	3
⎯4→8	8 − 4	8 − 4	4

(Repeat for families that were not firm.)

EXERCISE 2: MONEY
DOLLAR PROBLEMS WITH 1-DIGIT CENTS

a. (Display:) W   [84:2A]

$$\begin{array}{r} \$\ 2.07 \\ +\ 3.01 \\ \hline \end{array}$$

- Read the problem. Get ready. (Signal.)
2 dollars and 7 cents plus 3 dollars and 1 cent.
(Repeat until firm.)

b. Read the problem for the cent ones column. Get ready. (Signal.) *7 plus 1.*
- What's the answer? (Signal.) *8.*
- What are the two things I write in the answer before I write 8? (Call on different children: *a dollar sign; a dot.*)
(Add to show:)   [84:2B]

$$\begin{array}{r} \$\ 2.07 \\ +\ 3.01 \\ \hline \$\ \ \ .\ \ 8 \end{array}$$

- Read the problem for the cent tens column. Get ready. (Signal.) *Zero plus zero.*
- What's the answer? (Signal.) *Zero.*
(Add to show:)   [84:2C]

$$\begin{array}{r} \$\ 2.07 \\ +\ 3.01 \\ \hline \$\ \ \ .08 \end{array}$$

- Read the problem for the dollars. Get ready. (Signal.) *2 plus 3.*
- What's the answer? (Signal.) *5.*
(Add to show:)   [84:2D]

$$\begin{array}{r} \$\ 2.07 \\ +\ 3.01 \\ \hline \$\ 5.08 \end{array}$$

- Read the problem and the answer. Get ready. (Signal.) *2 dollars and 7 cents plus 3 dollars and 1 cent equals 5 dollars and 8 cents.*

c. (Display:) W   [84:2E]

$$\begin{array}{r} \$\ 9.03 \\ -\ 7.00 \\ \hline \end{array}$$

- Read the problem. Get ready. (Signal.)
9 dollars and 3 cents minus 7 dollars.
(Repeat until firm.)

d. Read the problem for the cent ones column.
    Get ready. (Signal.) *3 minus zero.*
- What's the answer? (Signal.) *3.*
- What are the two things I write in the answer before I write 3? (Call on different children: *a dollar sign; a dot.*)
  (Add to show:)   [84:2F]

$$\begin{array}{r} \$\ 9{.}0\ 3 \\ -\ 7{.}0\ 0 \\ \hline \$\quad .\quad 3 \end{array}$$

- Read the problem for the cent tens column.
  Get ready. (Signal.) *Zero minus zero.*
- What's the answer? (Signal.) *Zero.*
  (Add to show:)   [84:2G]

$$\begin{array}{r} \$\ 9{.}0\ 3 \\ -\ 7{.}0\ 0 \\ \hline \$\quad .0\ 3 \end{array}$$

- Read the problem for the dollars. Get ready.
  (Signal.) *9 minus 7.*
- What's the answer? (Signal.) *2.*
  (Add to show:)   [84:2H]

$$\begin{array}{r} \$\ 9{.}0\ 3 \\ -\ 7{.}0\ 0 \\ \hline \$\ 2{.}0\ 3 \end{array}$$

- Read the problem and the answer. Get ready.
  (Signal.) *9 dollars and 3 cents minus 7 dollars equals 2 dollars and 3 cents.*

EXERCISE 3: COMPARISON STATEMENTS
LENGTH

a. (Display:)   [84:3A]

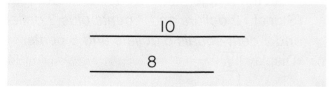

These lines are supposed to show miles.
- Is the top line or the bottom line longer?
  (Signal.) *The top line.*
- Which line is shorter? (Signal.) *The bottom line.*
- Raise your hand when you know how many miles longer the top line is. ✔
- How many miles longer is the top line?
  (Signal.) *2.*

b. The top line is 2 miles longer than the bottom line.
- Say the sentence about the top line.
  (Signal.) *The top line is 2 miles longer than the bottom line.*
- Say the sentence about the bottom line.
  (Signal.) *The bottom line is 2 miles shorter than the top line.*
  (Repeat until firm.)

c. (Display:)   [84:3B]

$$\underline{\qquad 2 \qquad}$$
$$\underline{\qquad\quad 7 \qquad\quad}$$

These lines are supposed to show feet.
- Raise your hand when you know which line is longer and how many feet longer it is. ✔
- Everybody, which line is longer? (Signal.)
  The bottom line.
- How many feet longer is the bottom line than the top line? (Signal.) *5.*
- Say the sentence about the bottom line.
  (Signal.) *The bottom line is 5 feet longer than the top line.*
- Say the sentence about the top line.
  (Signal.) *The top line is 5 feet shorter than the bottom line.*
  (Repeat until firm.)

d. (Display:)   [84:3C]

$$\underline{\qquad\qquad 18 \qquad\qquad}$$
$$\underline{\qquad 10 \qquad}$$

These lines are supposed to show feet.
- Raise your hand when you know which line is longer and how many feet longer it is. ✔
- Everybody, which line is longer? (Signal.) *The top line.*
- How many feet longer is the top line than the bottom line? (Signal.) *8.*
- Say the sentence about the top line.
  (Signal.) *The top line is 8 feet longer than the bottom line.*
- Say the sentence about the bottom line.
  (Signal.) *The bottom line is 8 feet shorter than the top line.*
  (Repeat until firm.)

EXERCISE 4: FACTS
SUBTRACTION

a. (Distribute unopened workbooks to children.)
• Open your workbook to Lesson 83 and find part 1.
   (Observe children and give feedback.)
   (Teacher reference:)

a. $7 - 3$	d. $14 - 4$	g. $11 - 5$
b. $6 - 4$	e. $10 - 4$	h. $8 - 4$
c. $9 - 3$	f. $11 - 9$	i. $12 - 6$

• These are minus problems you'll work. Think of the number families for each problem and write the answers. Put your pencil down when you've completed the equations in part 1.
   (Observe children and give feedback.)
b. Check your work. You'll read the equation for each problem.
• Problem A. (Signal.) $7 - 3 = 4$.
• (Repeat for:) B, $6 - 4 = 2$; C, $9 - 3 = 6$; D, $14 - 4 = 10$; E, $10 - 4 = 6$; F, $11 - 9 = 2$; G, $11 - 5 = 6$; H, $8 - 4 = 4$; I, $12 - 6 = 6$.

EXERCISE 5: COLUMN ADDITION
CARRYING DISCRIMINATION

a. (Display:) W   [84:5A]

$$\begin{array}{r} 2 \\ + 4\ 8 \\ \hline \end{array}$$

This problem adds a one-digit number and a two-digit number.
• What's the one-digit number? (Signal.) 2.
• What's the two-digit number? (Signal.) 48.

• Read the problem for the ones. (Signal.) 2 plus 8.
• What's the answer? (Signal.) 10.
• What do I write in the tens column? (Signal.) 1.
• What do I write in the ones column? (Signal.) Zero.
   (Add to show:)   [84:5B]

$$\begin{array}{r} ^{1} \\ 2 \\ + 4\ 8 \\ \hline 0 \end{array}$$

• Say the problem for the tens. (Signal.) 1 plus 4.
• What's the answer? (Signal.) 5.
   (Add to show:)   [84:5C]

$$\begin{array}{r} ^{1} \\ 2 \\ + 4\ 8 \\ \hline 5\ 0 \end{array}$$

• What's 2 plus 48? (Signal.) 50.
b. Find part 2 on worksheet 84. ✔
   (Teacher reference:)

a. $\begin{array}{r} 5\ 4 \\ +\ \ 6 \\ \hline \end{array}$	b. $\begin{array}{r} 3 \\ +2\ 4 \\ \hline \end{array}$	c. $\begin{array}{r} 6 \\ +3\ 5 \\ \hline \end{array}$

You have to carry to work some of these problems. Work the problems. Put your pencil down when you've finished part 2.
   (Observe children and give feedback.)
c. Check your work.
• Problem A is 54 plus 6. What's the answer? (Signal.) 60.
• Problem B is 3 plus 24. What's the answer? (Signal.) 27.
• Problem C is 6 plus 35. What's the answer? (Signal.) 41.

EXERCISE 6: WRITING DOLLARS WITH 1-DIGIT OR 2-DIGIT CENTS

REMEDY

a. Find part 3 on worksheet 84. ✔
(Teacher reference:)

R Part C

a. $18.___
b. $18.___
c. $18.___

You're going to complete these dollar amounts. Part of each amount is shown.

• Touch A. ✔
• Complete A so it shows 18 dollars and 9 cents.
(Observe children and give feedback.)
(Display:) W   [84:6A]

> **a. $18.09**

Here's what you should have.

b. Touch B. ✔
• Complete B so it shows 18 dollars and 8 cents.
(Observe children and give feedback.)
(Display:)   [84:6B]

> **b. $18.08**

Here's what you should have.

c. Touch C. ✔
• Complete C so it shows 18 dollars and 80 cents.
(Observe children and give feedback.)
(Display:)   [84:6C]

> **c. $18.80**

Here's what you should have.

EXERCISE 7: WORD PROBLEMS
COMPARISONS

a. Find part 4. ✔
(Teacher reference:)

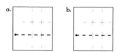

For problem A, I'm going to tell you how tall two doors are. You're going to write the problem and figure out how much taller one door is than the other door.

• Touch where you'll write the column problem for A. ✔
• Listen to problem A: The red door is 77 inches tall. The white door is 89 inches tall.
• Listen to problem A again and answer some questions. The red door is 77 inches tall. How tall is the red door? (Signal.) *77 inches.*
• Listen: The white door is 89 inches tall. How tall is the white door? (Signal.) *89 inches.*
• Which door is taller, the red door or the white door? (Signal.) *The white door.*
• Say the problem for figuring out how much taller the white door is than the red door. (Signal.) *89 minus 77.*
• Write the problem for A and work it. Remember to make the equals bar. Put your pencil down when you know how many inches taller the white door is than the red door. (Observe children and give feedback.)

b. Check your work.
• Read the problem and the answer for A. Get ready. (Signal.) *89 – 77 = 12.*
• The white door is how much taller than the red door? (Signal.) *12 inches.*

c. Touch where you'll write the column problem for B. ✔
You're going to write the problem and figure out how much more one person weighs than another person.
• Listen to problem B: Billy weighs 62 pounds. Tim weighs 87 pounds.
• Listen to problem B again and answer some questions. Billy weighs 62 pounds. How much does Billy weigh? (Signal.) *62 pounds.*
• Listen: Tim weighs 87 pounds. How much does Tim weigh? (Signal.) *87 pounds.*
• Who is heavier, Billy or Tim? (Signal.) *Tim.*
• Say the problem for figuring out how much heavier Tim is than Billy. (Signal.) *87 minus 62.*
• Write the problem for B and work it. Put your pencil down when you know how many pounds heavier Tim is than Billy. (Observe children and give feedback.)

d. Check your work.
• Read the problem and the answer for B. Get ready. (Signal.) *87 – 62 = 25.*
• Tim is how much heavier than Billy? (Signal.) *25 pounds.*

EXERCISE 8: GREATER THAN/LESS THAN SIGN

a. Find part 5. ✔
(Teacher reference:)

a. 74 ✖ 68
b. 28 ✖ 31
c. 16 ✖ 52
d. 45 ✖ 36

You're going to tell me the bigger number in each problem. Then you're going to make the sign between the numbers for each problem. Both signs are partly shown.
• Touch problem A. ✔
• Read the numbers for problem A. (Signal.) *74 (and) 68.*
• Which number is bigger, 74 or 68? (Signal.) *74.*
• Touch the dot for the sign with the bigger end next to 74. ✔
b. Read the numbers for problem B. (Signal.) *28 (and) 31.*
• Which number is bigger, 28 or 31? (Signal.) *31.*
• Touch the dot for the sign with the bigger end next to 31. ✔
c. Read the numbers for problem C. (Signal.) *16 (and) 52.*
• Which number is bigger, 16 or 52? (Signal.) *52.*
• Touch the dot for the sign with the bigger end next to 52. ✔
d. Read the numbers for problem D. (Signal.) *45 (and) 36.*
• Which number is bigger, 45 or 36? (Signal.) *45.*
• Touch the dot for the sign with the bigger end next to 45. ✔
(Repeat steps a through d that were not firm.)
e. Go back to problem A. ✔
• Read the numbers to yourself. Then make the sign with the bigger end next to the bigger number for problem A.
(Observe children and give feedback.)
• Check your work.
(Display:)   [84:8A]

a.   74 > 68

Here's what you should have. The bigger end of the sign is next to 74.

f. Read the numbers for problem B to yourself. Then make the sign with the bigger end next to the bigger number.
(Observe children and give feedback.)
• Check your work. Did you make the bigger end of the sign for problem B next to 28 or 31? (Signal.) *31.*
(Display:)   [84:8B]

b.   28 < 31

Here's what you should have for B.
g. Read the numbers for problem C to yourself. Then make the sign with the bigger end next to the bigger number.
(Observe children and give feedback.)
• Check your work. Did you make the bigger end of the sign for problem C next to 16 or 52? (Signal.) *52.*
(Display:)   [84:8C]

c.   16 < 52

Here's what you should have for C.
h. Read the numbers for problem D to yourself. Then make the sign with the bigger end next to the bigger number.
(Observe children and give feedback.)
• Check your work. Did you make the bigger end of the sign for problem D next to 45 or 36? (Signal.) *45.*
(Display:)   [84:8D]

d.   45 > 36

Here's what you should have for D.

EXERCISE 9: FACTS

ADDITION

a. Find part 6. ✔
(Teacher reference:)

a. 4 + 3   d. 9 + 2   g. 4 + 10
b. 2 + 8   e. 3 + 6   h. 5 + 6
c. 4 + 4   f. 6 + 6   i. 3 + 5

Think of the number families for each problem and write the answers. Put your pencil down when you've completed the equations in part 6.
(Observe children and give feedback.)

b. Check your work. You'll touch and read the equation for each problem.

- Problem A. (Signal.) *4 + 3 = 7.*
- (Repeat for:) B, *2 + 8 = 10;* C, *4 + 4 = 8;* D, *9 + 2 = 11;* E, *3 + 6 = 9;* F, *6 + 6 = 12;* G, *4 + 10 = 14;* H, *5 + 6 = 11;* I, *3 + 5 = 8.*

EXERCISE 10: INDEPENDENT WORK

a. Turn to the other side of worksheet 84 and find part 7. ✔
(Teacher reference:)

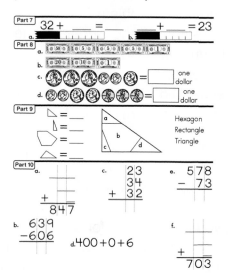

You'll complete the equations for the rulers. In part 8, you'll write an equals and the number of dollars to show what each group of bills is worth. You'll circle the words one dollar or write the cents for each group of coins.
In part 9, you'll circle the name for the whole figure. Then you'll complete the equations for the parts.
In part 10, you'll complete the equations.

b. Complete worksheet 84.
(Observe children and mark incorrect responses on children's worksheets as you give feedback.)

Lesson

EXERCISE 1: FACTS
ADDITION/SUBTRACTION

a. (Display:) [85:1A]

6 + 4	9 − 6	11 − 6
12 − 6	8 − 4	10 − 8
7 − 4	3 + 4	3 + 6

For each problem, you'll tell me if the answer is the big number or a small number. Then you'll say the fact.

- (Point to **6 + 4**.) Is the answer the big number or a small number? (Signal.) *The big number.*
- Say the fact. (Signal.) *6 + 4 = 10.*

b. (Point to **12 − 6**.) Is the answer the big number or a small number? (Signal.) *A small number.*

- Say the fact. (Signal.) *12 − 6 = 6.*

c. (Repeat the following tasks for the remaining problems:)

(Point to __.) Is the answer the big number or a small number?		Say the fact.
7 − 4	A small number.	7 − 4 = 3
9 − 6	A small number.	9 − 6 = 3
8 − 4	A small number.	8 − 4 = 4
3 + 4	The big number.	3 + 4 = 7
11 − 6	A small number.	11 − 6 = 5
10 − 8	A small number.	10 − 8 = 2
3 + 6	The big number.	3 + 6 = 9

(Repeat problems that were not firm.)

EXERCISE 2: MONEY
ONLY CENTS

a. (Display:) [W] [85:2A]

$ .56	$ .90
$ .18	$ .09
$ .05	

These are cents without dollars.

- (Point to **.56**.) This is 56 cents. What is it? (Touch.) *56 cents.*

b. (Point to **.18**.) What is this? (Touch.) *18 cents.*
- (Point to **.05**.) What is this? (Touch.) *5 cents.*
- (Point to **.90**.) What is this? (Touch.) *90 cents.*
- (Point to **.09**.) What is this? (Touch.) *9 cents.*
- (Point to **.56**.) What is this? (Touch.) *56 cents.*
(Repeat step b until firm.)

EXERCISE 3: COMPARISON STATEMENTS
LENGTH

a. (Display:) [85:3A]

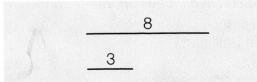

These lines are supposed to show inches.
- Which line is longer? (Signal.) *The top line.*
- Which line is shorter? (Signal.) *The bottom line.*
- Raise your hand when you know how much longer the top line is. ✔
- How much longer is the top line? (Signal.) *5 inches.*
- Say the sentence about the top line. (Signal.) *The top line is 5 inches longer than the bottom line.*
- Say the sentence about the bottom line. (Signal.) *The bottom line is 5 inches shorter than the top line.*
(Repeat until firm.)

b. (Display:)   [85:3B]

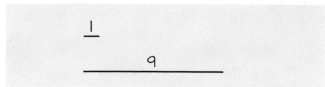

These lines are supposed to show inches.

- Which line is longer? (Signal.) *The bottom line.*
- Which line is shorter? (Signal.) *The top line.*
- Raise your hand when you know how much longer the bottom line is. ✔
- How much longer is the bottom line? (Signal.) *8 inches.*
- Say the sentence about the bottom line. (Signal.) *The bottom line is 8 inches longer than the top line.*
- Say the sentence about the top line. (Signal.) *The top line is 8 inches shorter than the bottom line.*
   (Repeat until firm.)

c. (Display:)   [85:3C]

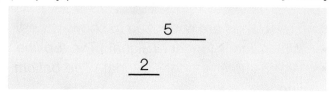

These lines are supposed to show inches.

- Which line is longer? (Signal.) *The top line.*
- Which line is shorter? (Signal.) *The bottom line.*
- Raise your hand when you know how much longer the top line is. ✔
- How much longer is the top line? (Signal.) *3 inches.*
- Say the sentence about the top line. (Signal.) *The top line is 3 inches longer than the bottom line.*
- Say the sentence about the bottom line. (Signal.) *The bottom line is 3 inches shorter than the top line.*
   (Repeat until firm.)

EXERCISE 4: MONEY
DOLLAR PROBLEMS

a. (Distribute unopened workbooks to children.)
- Open your workbook to Lesson 85 and find part 1.
   (Observe children and give feedback.)
   (Teacher reference:)

a. $27.89   b. $ 2.45
   − 3.83   + 12.32

- Touch and read problem A. Get ready. (Signal.) *27 dollars and 89 cents minus 3 dollars and 83 cents.*
- Touch and read problem B. Get ready. (Signal.) *2 dollars and 45 cents plus 12 dollars and 32 cents.*

b. Work both problems. Remember, write a dollar sign and a dot in the answer before working each problem.
   (Observe children and give feedback.)

c. Check your work. You'll touch and read the problem and the answer.
- Problem A. (Signal.) *27 dollars and 89 cents minus 3 dollars and 83 cents equals 24 dollars and 6 cents.*
- Problem B. (Signal.) *2 dollars and 45 cents plus 12 dollars and 32 cents equals 14 dollars and 77 cents.*

d. (Display:) W   [85:4A]

a. $27.89   b. $ 2.45
    − 3.83   + 12.32
    $24.06 $14.77

Here's what you should have for problems A and B.

EXERCISE 5: FACTS
SUBTRACTION

a. Find part 2 on worksheet 85. ✔
(Teacher reference:)

a. $6-4$ d. $8-4$ g. $9-6$
b. $19-9$ e. $11-5$ h. $10-4$
c. $7-2$ f. $7-3$ i. $11-9$

I'll read each minus problem. You'll tell me the answer. Then you'll complete the facts.
- Problem A. 6 minus 4. What's the answer? (Signal.) *2.*
- Problem B. 19 minus 9. What's the answer? (Signal.) *10.*
- Problem C. 7 minus 2. What's the answer? (Signal.) *5.*
- Problem D. 8 minus 4. What's the answer? (Signal.) *4.*
- Problem E. 11 minus 5. What's the answer? (Signal.) *6.*
- Problem F. 7 minus 3. What's the answer? (Signal.) *4.*
- Problem G. 9 minus 6. What's the answer? (Signal.) *3.*
- Problem H. 10 minus 4. What's the answer? (Signal.) *6.*
- Problem I. 11 minus 9. What's the answer? (Signal.) *2.*
(Repeat problems that were not firm.)

b. Complete the fact for each problem. (Observe children and give feedback.)
c. Check your work. You'll read the fact for each problem.
- Problem A. (Signal.) *6 – 4 = 2.*
- (Repeat for:) B, *19 – 9 = 10;* C, *7 – 2 = 5;* D, *8 – 4 = 4;* E, *11 – 5 = 6;* F, *7 – 3 = 4;* G, *9 – 6 = 3;* H, *10 – 4 = 6;* I, *11 – 9 = 2.*

EXERCISE 6: COLUMN ADDITION
CARRYING

a. Find part 3 on worksheet 85. ✔
(Teacher reference:)

a.
```
    8   b.
  1 2      7 6
+ 4 7    + 1 6
```

For both problems the answer for the ones column has 2 digits.
- Work these problems.
(Observe children and give feedback.)

b. Check your work.
- Problem A. 8 plus 12 plus 47. What's the answer? (Signal.) *67.*
- Problem B. 76 plus 16. What's the answer? (Signal.) *92.*

EXERCISE 7: GREATER THAN/LESS THAN SIGN

a. Find part 4. ✔
(Teacher reference:)

a. $11 \gtrless 10$ d. $82 \gtrless 79$
b. $89 \gtrless 90$ e. $15 \gtrless 14$
c. $52 \gtrless 60$ f. $100 \gtrless 200$

You're going to tell me the bigger number for each problem. Then you're going to make the sign between the numbers for each problem. Both signs are partly shown.
- Touch problem A. ✔
- Read the numbers for problem A. (Signal.) *11 (and) 10.*
- Which number is bigger, 11 or 10? (Signal.) *11.*
- Touch the dot for the sign with the bigger end next to 11. ✔

b. Read the numbers for problem B. (Signal.) *89 (and) 90.*
- Is 89 or 90 bigger? (Signal.) *90.*

c. (Repeat the following tasks for remaining problems:)

Read the numbers for problem __.		Is __ or __ bigger?	
C	52 (and) 60	52, 60	60
D	82 (and) 79	82, 79	82
E	15 (and) 14	15, 14	15
F	100 (and) 200	100, 200	200

(Repeat problems that were not firm.)

d. Go back to problem A. ✔
You're going to make the sign for each problem.
- Read the numbers to yourself. Then make the sign with the bigger end next to the bigger number. Put your pencil down when you've made the sign for all of the problems in part 3. (Observe children and give feedback.)

e. Check your work. You'll read the number in each problem that has the bigger end of the sign next to it.
• Problem A. Read the number with the bigger end next to it. (Signal.) *11.*
• (Repeat for:) B, *90;* C, *60;* D, *82;* E, *15;* F, *200.*
(Display:)   [85:7A]

a. 11 > 10	d. 82 > 79
b. 89 < 90	e. 15 > 14
c. 52 < 60	f. 100 < 200

Here are the signs you should have made for all the problems.

EXERCISE 8: FACTS
ADDITION

a. Find part 5. ✔
(Teacher reference:)

a. 4 + 4	e. 4 + 6
b. 3 + 5	f. 3 + 4
c. 2 + 9	g. 10 + 4
d. 6 + 6	h. 6 + 5
	i. 3 + 6

These are plus problems. Think of the number families and write answers to these problems.
(Observe children and give feedback.)
b. Check your work. You'll touch and read the fact for each problem.
• Problem A. (Signal.) *4 + 4 = 8.*
• (Repeat for:) B, *3 + 5 = 8;* C, *2 + 9 =11;* D, *6 + 6 = 12;* E, *4 + 6 = 10;* F, *3 + 4 = 7;* G, *10 + 4 = 14;* H, *6 + 5 = 11;* I, *3 + 6 = 9.*

EXERCISE 9: WORD PROBLEMS
COMPARISONS   REMEDY

a. Find part 6. ✔
(Teacher reference:)   R Part B

For problem A, I'm going to tell you how old two people are. You're going to write the problem and figure out how much older one person is than the other person.
• Touch where you'll write the column problem for A. ✔

• Listen to problem A: Rita and Donna are friends. Rita is 86 years old. Donna is 54 years old.
• Listen to problem A again and answer some questions. Rita is 86 years old. How old is Rita? (Signal.) *86 years (old).*
• Listen: Donna is 54 years old. How old is Donna? (Signal.) *54 years (old).*
• Which person is older, Rita or Donna? (Signal.) *Rita.*
• Say the problem for figuring out how much older Rita is than Donna. (Signal.) *86 minus 54.*
• Write the problem for A and work it. Remember to make the equals bar. Put your pencil down when you know how many years older Rita is than Donna.
(Observe children and give feedback.)
b. Check your work.
• Read the problem and the answer for A. Get ready. (Signal.) *86 – 54 = 32.*
• Rita is how much older than Donna? (Signal.) *32 years.*
c. Touch where you'll write the column problem for B. ✔
You're going to write the problem and figure out how much more one person weighs than another person.
• Listen to problem B: Mr. Jones weighs 249 pounds. His wife weighs 142 pounds.
• Listen to problem B again and answer some questions. Mr. Jones weighs 249 pounds. How much does Mr. Jones weigh? (Signal.) *249 pounds.*
• Listen: His wife weighs 142 pounds. How much does his wife weigh? (Signal.) *142 pounds.*
• Who is heavier, Mr. Jones or his wife? (Signal.) *Mr. Jones.*
• Say the problem for figuring out how much heavier Mr. Jones is than his wife. (Signal.) *249 minus 142.*
• Write the problem for B and work it. Put your pencil down when you know how many pounds heavier Mr. Jones is than his wife.
(Observe children and give feedback.)
d. Check your work.
• Read the problem and the answer for B. Get ready. (Signal.) *249 – 142 = 107.*
• Mr. Jones is how much heavier than his wife? (Signal.) *107 pounds.*

EXERCISE 10: INDEPENDENT WORK

a. Turn to the other side of worksheet 85 and find part 7. ✔

(Teacher reference:)

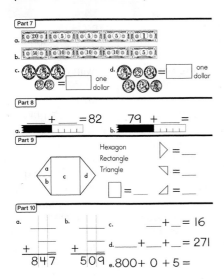

For each group in part 7, write an equals and the dollars or cents it's worth.

In part 8, you'll complete the equations for the rulers.

In part 9, you'll circle the name for the whole figure. Then you'll complete the equations for the parts.

In part 10, you'll complete the equations.

b. Complete worksheet 85.

(Observe children and mark incorrect responses on children's worksheets as you give feedback.)

Lessons 86–90 Planning Page

	Lesson 86	**Lesson 87**	**Lesson 88**	**Lesson 89**	**Lesson 90**
Student Learning Objectives	**Exercises** 1. Say addition and subtraction facts 2. Make comparison statements **3. Learn how to solve 3-digit addition problems with carrying** **4. Compose 2-dimensional figures** 5. Solve subtraction facts 6. Write dollar amounts with 1- or 2-digit cents 7. Solve addition facts 8. Write the greater than and less than (>, <) symbols to point to the greater or lesser number 9. Solve money problems 10. Complete work independently	**Exercises** 1. Say addition and subtraction facts 2. Make comparison statements **3. Learn how to carry to solve money problems** 4. Solve subtraction facts 5. Compose 2-dimensional figures **6. Solve 3-digit addition problems with carrying** 7. Write the greater than and less than (>, <) symbols to point to the greater or lesser number 8. Solve addition facts 9. Write dollar amounts with 1- or 2-digit cents 10. Complete work independently	**Exercises** 1. Say addition and subtraction facts 2. Make comparison statements **3. Count the value of coins added to dollar amounts** 4. Solve 3-digit addition problems with carrying 5. Compose 2-dimensional figures 6. Solve subtraction facts **7. Write the greater than, less than, and equal signs (>, <, =) to compare numbers** 8. Solve addition facts 9. Complete work independently	**Exercises** **1. Say addition and subtraction facts with the same small number;** find missing numbers in number families 2. Make comparison statements 3. Say addition and subtraction facts 4. Count the value of coins added to dollar amounts 5. Count forward or backward from a given number **6. Solve 3-digit addition problems with carrying and with 3 addends** 7. Write the greater than, less than, and equal signs (>, <, =) to compare numbers 8. Solve subtraction facts 9. Write the symbols for word problems in columns and solve 10. Complete work independently	**Exercises** 1. Say addition and subtraction facts with the same small number; find missing numbers in number families **2. Learn how to solve 3-digit addition problems with carrying multiple digits** 3. Make comparison statements 4. Solve addition and subtraction facts 5. Write the greater than, less than, and equal signs (>, <, =) to compare numbers 6. Write the symbols for word problems in columns and solve **7. Count and write the value of coins added to dollar amounts** 8. Solve subtraction facts 9. Complete work independently

Common Core State Standards for Mathematics					
1.OA 3	✔				
1.OA 5	✔		✔	✔	✔
1.OA 6	✔	✔	✔	✔	✔
1.OA 8	✔	✔	✔	✔	✔
1.NBT 1	✔	✔	✔	✔	✔
1.NBT 2			✔	✔	
1.NBT 4	✔	✔	✔	✔	✔
1.G 2	✔	✔	✔	✔	✔
Teacher Materials	Presentation Book 3, Board Displays CD or chalkboard				
Student Materials	Workbook 2, Pencil				
Additional Practice	• Student Practice Software: Block 4: Activity 1 (1.G.1), Activity 2 (1.NBT.1), Activity 3 (1.NBT.1), Activity 4 (1.NBT.3), Activity 5 (1.NBT.1), Activity 6 (1.NBT.1)				
Mastery Test					Student Assessment Book (Present Mastery Test 9 following Lesson 90.)

Lesson

EXERCISE 1: FACTS
ADDITION/SUBTRACTION

a. (Display:)   [86:1A]

2 + 7	11 − 6	8 − 4
8 − 2	5 + 2	6 + 6
4 + 3	3 + 6	10 − 6

For each problem, you'll tell me if the answer is the big number or a small number. Then you'll say the fact.

• (Point to **2 + 7**.) Is the answer the big number or a small number? (Signal.) *The big number.*
• Say the fact. (Signal.) *2 + 7 = 9.*
b. (Point to **8 − 2**.) Is the answer the big number or a small number? (Signal.) *A small number.*
• Say the fact. (Signal.) *8 − 2 = 6.*
c. (Repeat the following tasks for the remaining problems:)

(Point to __.) Is the answer the big number or a small number?	Say the fact.
4 + 3	*The big number.*	*4 + 3 = 7*
11 − 6	*A small number.*	*11 − 6 = 5*
5 + 2	*The big number.*	*5 + 2 = 7*
3 + 6	*The big number.*	*3 + 6 = 9*
8 − 4	*A small number.*	*8 − 4 = 4*
6 + 6	*The big number.*	*6 + 6 = 12*
10 − 6	*A small number.*	*10 − 6 = 4*

(Repeat problems that were not firm.)

EXERCISE 2: COMPARISON STATEMENTS
LENGTH

a. You know how to figure out how much longer a line is. You start with the number for the line that is longer and minus the number for the line that is shorter. You do the same thing to figure out how much shorter a line is.
(Display:)   [86:2A]

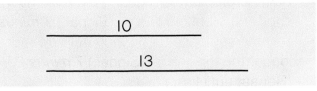

This picture shows two lines. The numbers show how many yards long each line is supposed to be.
• What's the number for the line that is longer? (Signal.) *13.*
• What's the number for the line that is shorter? (Signal.) *10.*
• Say the problem for figuring out how much shorter the top line is. (Signal.) *13 minus 10.*
• Say the problem for figuring out how much longer the bottom line is. (Signal.) *13 minus 10.*
(Repeat until firm.)
b. What's 13 minus 10? (Signal.) *3.*
• So how many yards shorter is the top line than the bottom line? (Signal.) *3 (yards).*
• How much longer is the bottom line than the top line? (Signal.) *3 (yards).*
• Say the sentence about the bottom line. (Signal.) *The bottom line is 3 yards longer than the top line.*
• Say the sentence about the top line. (Signal.) *The top line is 3 yards shorter than the bottom line.*
(Repeat until firm.)

c. (Display:)   [86:2B]

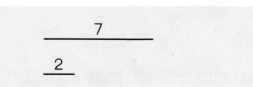

$$\frac{7}{2}$$

This picture shows two lines. The numbers show how many meters long each line is supposed to be.

- What's the number for the line that is longer? (Signal.) *7.*
- What's the number for the line that is shorter? (Signal.) *2.*
- Say the problem for figuring out how much shorter the bottom line is. (Signal.) *7 minus 2.*
- Say the sentence for figuring out how much longer the top line is. (Signal.) *7 minus 2.*
  (Repeat until firm.)
d. What's 7 minus 2? (Signal.) *5.*
- So how many meters shorter is the bottom line than the top line? (Signal.) *5 (meters).*
- How much longer is the top line than the bottom line? (Signal.) *5 (meters).*
- Say the sentence about the top line. (Signal.) *The top line is 5 meters longer than the bottom line.*
- Say the sentence about the bottom line. (Signal.) *The bottom line is 5 meters shorter than the top line.*
  (Repeat until firm.)
e. (Display:)   [86:2C]

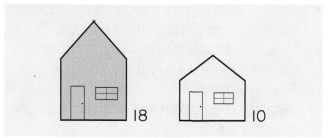

The number next to each house tells how many feet tall that house is.

- (Point to 🏠.) How tall is the grey house? (Signal.) *18 feet.*
- (Point to 🏠.) How tall is the white house? (Signal.) *10 feet.*
- What's the number for the house that is taller? (Signal.) *18.*
- What's the number for the house that is shorter? (Signal.) *10.*
- Say the problem for figuring out how much taller the grey house is. (Signal.) *18 minus 10.*

- What's the answer? (Signal.) *8.*
  Yes, the grey house is 8 feet taller than the white house.
- Say the sentence about the grey house. (Signal.) *The grey house is 8 feet taller than the white house.*
- Say the sentence about the white house. (Signal.) *The white house is 8 feet shorter than the grey house.*
  (Repeat until firm.)

EXERCISE 3: COLUMN ADDITION
CARRYING WITH 3-DIGIT NUMBERS   REMEDY

a. (Display:) W   [86:3A]

$$\begin{array}{r} 267 \\ +361 \\ \hline \end{array}$$

- Everybody, read the problem. Get ready. (Signal.) *267 plus 361.*
- Read the problem for the ones. Get ready. (Signal.) *7 plus 1.*
- What's the answer? (Signal.) *8.*
  (Add to show:)   [86:3B]

$$\begin{array}{r} 267 \\ +361 \\ \hline 8 \end{array}$$

- Read the problem for the tens. Get ready. (Signal.) *6 plus 6.*
- What's the answer? (Signal.) *12.*
- What are the digits of 12? (Signal.) *1 and 2.*
  So I write 1 in the hundreds column and 2 in the tens column.
  Watch.
  (Add to show:)   [86:3C]

$$\begin{array}{r} {}^{1} \\ 267 \\ +361 \\ \hline 28 \end{array}$$

- Read the problem for the hundreds. Get ready. (Signal.) *1 plus 2 plus 3.*
- Tell me the answer. Get ready. (Signal.) *6.*
  (Add to show:)   [86:3D]

$$\begin{array}{r} {}^{1} \\ 267 \\ +361 \\ \hline 628 \end{array}$$

- What's 267 plus 361? (Signal.) *628.*

b. (Display:)   [86:3E]

$$166$$
$$+325$$

- Everybody, read the problem. Get ready. (Signal.) *166 plus 325.*
- Read the problem for the ones. Get ready. (Signal.) *6 plus 5.*
- What's the answer? (Signal.) *11.*
- What do I write in the tens column? (Signal.) *1.*
- What do I write in the ones column? (Signal.) *1.*
  (Add to show:)   [86:3F]

$$\overset{1}{1}66$$
$$+325$$
$$1$$

- Read the problem for the tens. Get ready. (Signal.) *1 plus 6 plus 2.*
- Tell me the answer. Get ready. (Signal.) *9.*
  (Add to show:)   [86:3G]

$$\overset{1}{1}66$$
$$+325$$
$$91$$

- Read the problem for the hundreds. Get ready. (Signal.) *1 plus 3.*
- What's the answer? (Signal.) *4.*
  (Add to show:)   [86:3H]

$$\overset{1}{1}66$$
$$+325$$
$$491$$

- What's 166 plus 325? (Signal.) *491.*

EXERCISE 4: 2-DIMENSIONAL FIGURES
COMPOSITION

a. (Display:)   [86:4A]

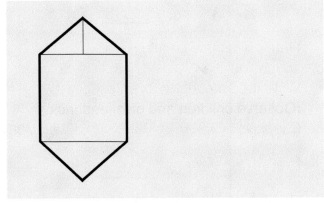

- (Point to .) What's the name of this whole shape? (Touch.) *(A) hexagon.*
  The whole shape has parts. You'll tell me the shape of each part.
- (Run finger around ◺.) What shape? (Signal.) *(A) triangle.*
- (Run finger around ◹.) What shape? (Signal.) *(A) triangle.*
- (Run finger around ☐.) What shape? (Signal.) *(A) square.*
- (Run finger around ▽.) What shape? (Signal.) *(A) triangle.*

b. (Display:)   [86:4B]

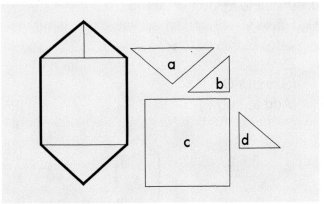

- (Point to **a** through **d**.) Here are the parts of a hexagon. You have to figure out where each part goes.
- (Point to ▽ᵃ.) What shape is part A? (Touch.) *(A) triangle.*
- (Point to ⬠. Is this part A? (Touch.) No.
- (Point to .) Is this part A? (Touch.) *No.*
- (Point to ▮.) Is this part A? (Touch.) *No.*
- (Point to ⬠.) Is this part A? (Touch.) *Yes.*
  (Add to show:)   [86:4C]

c. (Point to ⬠.) Look at parts B, C, and D.
- Raise your hand when you can tell me the letter of this part. ✔
- Everybody, what's the letter of this part? (Signal.) *D.*
(Add to show:)   [86:4D]

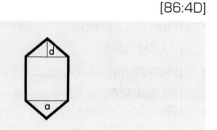

d. (Point to ⬛.) Look at the parts.
- Raise your hand when you can tell me the letter of this part. ✔
- Everybody, what's the letter of this part? (Signal.) *C.*
(Add to show:)   [86:4E]

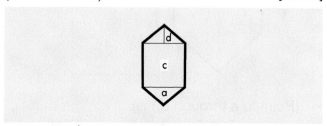

e. (Point to ⬠.) Look at the parts.
- Raise your hand when you can tell me the letter of this part. ✔
- Everybody, what's the letter of this part? (Signal.) *B.*
(Add to show:)   [86:4F]

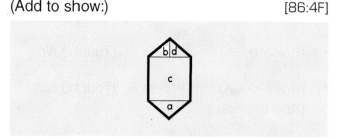

EXERCISE 5: FACTS
SUBTRACTION

a. (Distribute unopened workbooks to children.)
- Open your workbook to Lesson 86 and find part 1.
(Observe children and give feedback.)
(Teacher reference:)

a. 7 − 4	e. 8 − 5
b. 9 − 3	f. 4 − 3
c. 8 − 4	g. 11 − 5
d. 7 − 2	h. 10 − 4

b. These are minus problems you'll work. Think of the number families for each problem and write the answers. Put your pencil down when you've completed the equations in part 1.
(Observe children and give feedback.)

c. Check your work. You'll read the equation for each problem.
- Problem A. (Signal.) *7 − 4 = 3.*
- (Repeat for:) B, *9 − 3 = 6;* C, *8 − 4 = 4;* D, *7 − 2 = 5;* E, *8 − 5 = 3;* F, *4 − 3 = 1;* G, *11 − 5 = 6;* H, *10 − 4 = 6.*

EXERCISE 6: WRITING DOLLARS WITH 1-DIGIT OR 2-DIGIT CENTS

a. (Display:) W   [86:6A]

$ .09	$ .89
$ .19	$ .01
$ .90	

You're going to read these cents amounts. These are cents without dollars.
- (Point to **.09.**) What is this? (Touch.) *9 cents.*
- (Point to **.19.**) What is this? (Touch.) *19 cents.*
- (Point to **.90.**) What is this? (Touch.) *90 cents.*
- (Point to **.89.**) What is this? (Touch.) *89 cents.*
- (Point to **.01.**) What is this? (Touch.) *1 cent.*

b. Find part 2 on worksheet 86. ✔
(Teacher reference:)

a. $. _____
b. $. _____
c. $. _____
d. $. _____

You're going to write dollars and cents amounts. Some amounts will show only cents with no dollars. The dollar sign and the dot are shown for each amount you'll write.
- Touch line A. ✔
Write the amount for 88 cents.
(Observe children and give feedback.)
(Display:)   [86:6B]

> a. $ .88

Here's what you should have.

c. Touch line B. ✔
- Write the amount for 5 cents.
(Observe children and give feedback.)
(Display:)   [86:6C]

> b. $ .05

Here's what you should have.

d. Touch line C. ✔
• Write the amount for 17 cents.
 (Observe children and give feedback.)
 (Display:) [86:6D]

c. $ .17

Here's what you should have.
e. Touch line D. ✔
• Write the amount for 3 dollars and 42 cents.
 (Observe children and give feedback.)
 (Display:) [86:6E]

d. $ 3.42

Here's what you should have.

EXERCISE 7: FACTS
ADDITION

a. Find part 3. ✔
 (Teacher reference:)

a. 3+4   g. 2+4
b. 5+6   h. 6+6
c. 10+7   i. 8+2
d. 4+4   j. 2+9
e. 6+4   k. 3+6
f. 5+3

These are plus problems so the answer is the big number in the family. I'll read each problem and you'll tell me the answer. Then you'll complete each fact.
• Problem A. 3 plus 4. What's the answer? (Signal.) 7.
• (Repeat the following tasks for remaining problems:)

Problem __.		What's the answer?
B	5 + 6	11
C	10 + 7	17
D	4 + 4	8
E	6 + 4	10
F	5 + 3	8
G	2 + 4	6
H	6 + 6	12
I	8 + 2	10
J	2 + 9	11
K	3 + 6	9

(Repeat problems that were not firm.)
b. Complete the equations in part 3. Put your pencil down when you're finished.
 (Observe children and give feedback.)

c. Check your work. You're going to touch and read the fact for each problem.
• Problem A. (Signal.) 3 + 4 = 7.
• (Repeat for:) B, 5 + 6 = 11; C, 10 + 7 = 17; D, 4 + 4 = 8; E, 6 + 4 = 10; F, 5 + 3 = 8; G, 2 + 4 = 6; H, 6 + 6 = 12; I, 8 + 2 = 10; J, 2 + 9 = 11; K, 3 + 6 = 9.

EXERCISE 8: GREATER THAN/LESS THAN SIGN

a. Find part 4. ✔
 (Teacher reference:)

a. 47 ⊠ 39
b. 82 ⊠ 91
c. 57 ⊠ 48
d. 22 ⊠ 17
e. 107 ·· 111
f. 65 ·· 56

You're going to tell me the bigger number for each problem. Then you're going to make the sign between the numbers for each problem. Some of the signs are partly shown. For some of the signs, only the big dot is shown.
• Touch problem A. ✔
• Read the numbers for problem A. (Signal.) 47 (and) 39.
• Which number is bigger, 47 or 39? (Signal.) 47.
b. Read the numbers for problem B. (Signal.) 82 (and) 91.
• Is 82 or 91 bigger? (Signal.) 91.
c. (Repeat the following tasks for remaining problems:)

Read the numbers for problem __.		Is __ or __ bigger?	
C	57 (and) 48	57, 48	57
D	22 (and) 17	22, 17	22
E	107 (and) 111	107, 111	111
F	65 (and) 56	65, 56	65

(Repeat for problems that were not firm.)
d. Go back to problem A. ✔
 You're going to make the sign for each problem.
• Read the numbers to yourself. Then make the sign with the bigger end next to the bigger number. Put your pencil down when you've made the sign for all of the problems in part 3.
 (Observe children and give feedback.)

e. Check your work. You'll read the number in each problem that has the bigger end of the sign next to it.

- Problem A. Read the number with the bigger end next to it. (Signal.) *47.*
- (Repeat for:) B, *91;* C, *57;* D, *22;* E, *111;* F, *65.*
(Display:)   [86:8A]

> a. 47 ⋊ 39
>
> b. 82 ⋉ 91
>
> c. 57 ⋉ 48
>
> d. 22 ⋊ 17
>
> e. 107 ⋖ 111
>
> f. 65 ⋗ 56

Here are the signs you should have made for all of the problems.

EXERCISE 9: MONEY
DOLLAR PROBLEMS

a. Find part 5. ✔
(Teacher reference:)

a. $ 4.66   b. $ .74   c. $ 9.81
  − 1.02   + 2.02   − 9.01

Work the problems. Remember to write a dollar sign and a dot in the answer.
(Observe children and give feedback.)

b. Check your work.
- Read the problem and the answer for A. Get ready. (Signal.) *4 dollars and 66 cents minus 1 dollar and 2 cents equals 3 dollars and 64 cents.*
- Read the problem and the answer for B. Get ready. (Signal.) *74 cents plus 2 dollars and 2 cents equals 2 dollars and 76 cents.*
- Read the problem and the answer for C. Get ready. (Signal.) *9 dollars and 81 cents minus 9 dollars and 1 cent equals 80 cents.*

EXERCISE 10: INDEPENDENT WORK

a. Find part 6 on your worksheet. ✔
(Teacher reference:)

You'll complete the equation for the boxes and the equation for the ruler.

b. Turn to the other side of worksheet 86 and find part 7. ✔
(Teacher reference:)

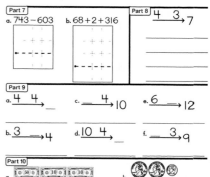

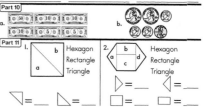

You'll write each problem in a column and work it.
In part 8, you'll write the four facts for the family.
In part 9, below each family, you'll write the fact for the missing number and you'll complete the family.
In part 10, you'll write an equals and the number of dollars or cents each group is worth.
In part 11, you'll circle the shape of the whole figure and complete the equations to show the parts of it.

c. Complete worksheet 86. Remember to work part 6 on the other side of your worksheet.
(Observe children and mark incorrect responses on children's worksheets as you give feedback.)

Lesson 87

EXERCISE 1: FACTS
ADDITION/SUBTRACTION

a. (Display:) [87:1A]

5 + 10	6 − 2	3 + 3
6 − 5	5 + 3	9 − 3
3 + 4	12 − 6	6 + 5

For each problem, you'll tell me if the answer is the big number or a small number. Then you'll say the fact.

- (Point to **5 + 10**.) Is the answer the big number or a small number? (Signal.) *The big number.*
- Say the fact. (Signal.) *5 + 10 = 15.*
b. (Point to **6 − 5**.) Is the answer the big number or a small number? (Signal.) *A small number.*
- Say the fact. (Signal.) *6 − 5 = 1.*
c. (Repeat the following tasks for the remaining problems:)

(Point to __.) Is the answer the big number or a small number?		Say the fact.
3 + 4	*The big number.*	*3 + 4 = 7*
6 − 2	*A small number.*	*6 − 2 = 4*
5 + 3	*The big number.*	*5 + 3 = 8*
12 − 6	*A small number.*	*12 − 6 = 6*
3 + 3	*The big number.*	*3 + 3 = 6*
9 − 3	*A small number.*	*9 − 3 = 6*
6 + 5	*The big number.*	*6 + 5 = 11*

(Repeat problems that were not firm.)

EXERCISE 2: COMPARISON STATEMENTS

a. (Display:) [87:2A]

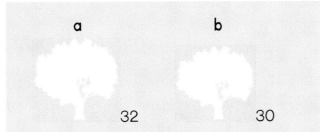

This picture shows two trees. The numbers show how many feet tall they are.

- (Point to **a.**) How many feet tall is tree A? (Signal.) *32.*
- (Point to **b.**) How many feet tall is tree B? (Signal.) *30.*
- What's the number for the tree that is taller? (Signal.) *32.*
- What's the number for the tree that is shorter? (Signal.) *30.*
- Say the problem for figuring out how many feet taller tree A is. (Signal.) *32 minus 30.*
- How many feet taller is tree A? (Signal.) *2 feet.* Yes, tree A is 2 feet taller than tree B.
- Say the sentence about tree A. (Signal.) *Tree A is 2 feet taller than tree B.*
- Say the sentence about tree B. (Signal.) *Tree B is 2 feet shorter than tree A.* (Repeat until firm.)

b. (Display:) [87:2B]

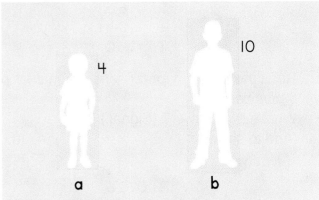

The number for each person tells me how many years old that person is.

- (Point to **a.**) How many years old is person A? (Signal.) *4.*
- (Point to **b.**) How many years old is person B? (Signal.) *10.*
- Say the problem for figuring out how many years older person B is. (Signal.) *10 minus 4.*
- How much older is person B? (Signal.) *6 years.*
c. Here's the sentence about person A: Person A is 6 years younger than person B.
- Say the sentence about person A. (Signal.) *Person A is 6 years younger than person B.*
- Say the sentence about person B. (Signal.) *Person B is 6 years older than person A.*
- One more time: Say the sentence about person A. (Signal.) *Person A is 6 years younger than person B.* (Repeat until firm.)

EXERCISE 3: MONEY
COLUMN PROBLEMS

a. (Display:) W   [87:3A]

$$\begin{array}{r} \$10.66 \\ +\ \ .04 \\ \hline \end{array}$$

- Read the problem. Get ready. (Signal.)
  10 dollars and 66 cents plus 4 cents.
  (Repeat until firm.)
b. Read the problem for the ones. Get ready.
  (Signal.) *6 plus 4.*
- What's the answer? (Signal.) *10.*
- What do I write in the tens column? (Signal.) *1.*
- What do I write in the ones column?
  (Signal.) *Zero.*
  I also write the dollar sign and the dot.
  (Add to show:)   [87:3B]

$$\begin{array}{r} ^{1} \\ \$10.66 \\ +\ \ .04 \\ \hline \$\ \ .\ 0 \end{array}$$

- Read the problem for the tens. (Signal.) *1 plus
  6 plus zero.*
- What's the answer? (Signal.) *7.*
  (Add to show:)   [87:3C]

$$\begin{array}{r} ^{1} \\ \$10.66 \\ +\ \ .04 \\ \hline \$\ \ .70 \end{array}$$

I'll write the answer for the dollars.
(Add to show:)   [87:3D]

$$\begin{array}{r} ^{1} \\ \$10.66 \\ +\ \ .04 \\ \hline \$10.70 \end{array}$$

- Read the problem and the answer. Get ready.
  (Signal.) *10 dollars and 66 cents plus 4 cents
  equals 10 dollars and 70 cents.*

EXERCISE 4: FACTS
SUBTRACTION

a. (Distribute unopened workbooks to children.)
- Open your workbook to Lesson 87 and find
  part 1.
  (Observe children and give feedback.)
  (Teacher reference:)

a. 7 − 3	d. 11 − 6	g. 14 − 10
b. 10 − 4	e. 5 − 3	h. 7 − 5
c. 9 − 3	f. 8 − 3	i. 8 − 4

These are minus problems so the answer is
a small number in the family. I'll read each
problem and you'll tell me the answer. Then
you'll complete each fact.
- Problem A. 7 minus 3. What's the answer?
  (Signal.) *4.*
- (Repeat the following tasks for remaining
  problems:)

Problem __.		What's the answer?
B	10 − 4	6
C	9 − 3	6
D	11 − 6	5
E	5 − 3	2
F	8 − 3	5
G	14 − 10	4
H	7 − 5	2
I	8 − 4	4

(Repeat problems that were not firm.)
b. Complete the equations in part 1. Put your
  pencil down when you're finished.
  (Observe children and give feedback.)
c. Check your work. You're going touch and read
  the fact for each problem.
- Problem A. (Signal.) *7 − 3 = 4.*
- (Repeat for:) B, *10 − 4 = 6;* C, *9 − 3 = 6;*
  D, *11 − 6 = 5;* E, *5 − 3 = 2;* F, *8 − 3 = 5;*
  G, *14 − 10 = 4;* H, *7 − 5 = 2;* I, *8 − 4 = 4.*

EXERCISE 5: 2-DIMENSIONAL FIGURES
COMPOSITION
REMEDY

a. (Display:) [87:5A]

- (Point to ◻.) What's the name of this whole shape? (Touch.) *(A) rectangle.*
- Raise your hand when you know the shape of all the parts inside the rectangle. ✔
- What's the name of all the parts? (Signal.) *Triangles.*
(Add to show:) [87:5B]

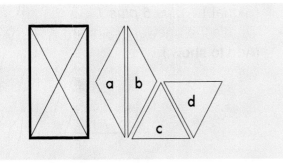

b. (Point to **a** through **d.**) Here are the parts that are in the rectangle. You have to figure out where each shape goes.
- (Point to ◻.) Look at this part. Raise your hand when you can tell me the letter of this part. ✔
- Everybody, what's the letter of this part? (Signal.) *D.*

c. (Point to ◹.) Look at this part. Raise your hand when you can tell me the letter of this part. ✔
- Everybody, what's the letter of this part? (Signal.) *B.*

d. (Point to ◩.) Look at this part. Raise your hand when you can tell me the letter of this part. ✔
- Everybody, what's the letter of this part? (Signal.) *A.*

e. (Point to ◪.) Look at this part. Raise your hand when you can tell me the letter of this part. ✔
- Everybody, what's the letter of this part? (Signal.) *C.*

f. Find part 2 on your worksheet. ✔
(Teacher reference:) R Part 0

- Touch the whole shape in part 2. ✔
- What is the shape? (Signal.) *(A) triangle.*
g. The parts of the triangle are shown next to it.
- Touch part A. ✔
- What shape is part A? (Signal.) *(A) rectangle.*
- What shape is part B? (Signal.) *(A) triangle.*
- What shape is part C? (Signal.) *(A) triangle.*
h. Write the letters for each part of the large shape. Don't get fooled.
(Observe children and give feedback.)
i. Check your work.
(Display:) [87:5C]

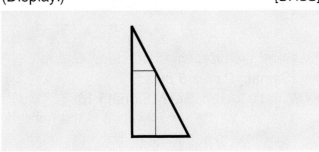

- (Point to ◹.) What letter did you write for this part? (Signal.) *B.*

- (Point to ◤.) What letter did you write for this part? (Signal.) *A.*

- (Point to ◿.) What letter did you write for this part? (Signal.) *C.*
(Add to show:) [87:5D]

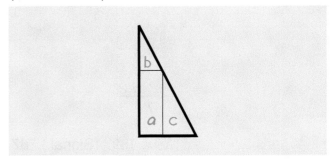

Here's what you should have.

EXERCISE 6: COLUMN ADDITION
CARRYING WITH 3-DIGIT NUMBERS

REMEDY

a. (Display:) W   [87:6A]

```
  5 2 4
    8 1
+ 1 6 2
```

- Everybody, read the problem. Get ready.
  (Signal.) *524 plus 81 plus 162.*
- Read the problem for the ones. Get ready.
  (Signal.) *4 plus 1 plus 2.*
- What's the answer? (Signal.) *7.*
  (Add to show:)   [87:6B]

```
  5 2 4
    8 1
+ 1 6 2
      7
```

- Read the problem for the tens. Get ready.
  (Signal.) *2 plus 8 plus 6.*
- What's the answer? (Signal.) *16.*
  16 has two digits so I write 1 in the hundreds
  column and 6 in the tens column.
  Watch.
  (Add to show:)   [87:6C]

```
  ¹
  5 2 4
    8 1
+ 1 6 2
    6 7
```

- Read the problem for the hundreds. Get ready.
  (Signal.) *1 plus 5 plus 1.*
- What's the answer? (Signal.) *7.*
  (Add to show:)   [87:6D]

```
  ¹
  5 2 4
    8 1
+ 1 6 2
  7 6 7
```

- What's 524 plus 81 plus 162? (Signal.) *767.*

b. (Display:)   [87:6E]

```
  2 5 9
+ 7 1 2
```

- Everybody, read the problem. Get ready.
  (Signal.) *259 plus 712.*
- Read the problem for the ones. Get ready.
  (Signal.) *9 plus 2.*
- What's the answer? (Signal.) *11.*
- What do I write in the tens column? (Signal.) *1*
- What do I write in the ones column? (Signal.) *1.*
  (Add to show:)   [87:6F]

```
    ¹
  2 5 9
+ 7 1 2
      1
```

- Read the problem for the tens. Get ready.
  (Signal.) *1 plus 5 plus 1.*
- What's the answer? (Signal.) *7.*
  (Add to show:)   [87:6G]

```
    ¹
  2 5 9
+ 7 1 2
    7 1
```

- Read the problem for the hundreds. Get ready.
  (Signal.) *2 plus 7.*
- What's the answer? *9.*
  (Add to show:)   [87:6H]

```
    ¹
  2 5 9
+ 7 1 2
  9 7 1
```

- What's 259 plus 712? (Signal.) *971.*
c. Remember, if the answer in the tens column
  has two digits, you write the first digit in the
  hundreds column.

d. Find part 3. ✔
(Teacher reference:) R Part K

a.
```
 | 2 2
+4 8 6
```

- Touch and read problem A. Get ready.
(Signal.) *122 plus 486.*
- Work the problem in the ones column.
Then stop.
(Observe children and give feedback.)
(Display:) W   [87:6I]

a.
```
   | 2 2
 + 4 8 6
       8
```

Here's what you should have.
- Work the problem in the tens column. Then stop.
(Observe children and give feedback.)
(Add to show:)   [87:7J]

a.
```
   |
   | 2 2
 + 4 8 6
     0 8
```

Here's what you should have.
- You wrote 1 in the hundreds column and zero in the tens column. Now work the problem in the hundreds column.
(Observe children and give feedback.)
(Add to show:)   [87:6K]

a.
```
   |
   | 2 2
 + 4 8 6
   6 0 8
```

Here's what you should have.
- Everybody, read the problem we started with and the answer. (Signal.) *122 plus 486 equals 608.*

EXERCISE 7: GREATER THAN/LESS THAN SIGN

a. Find part 4. ✔
(Teacher reference:)

a. 64 ⋇ 71
b. 37 ·· 28
c. 36 ·· 19
d. 78 ·· 90
e. 27 ·· 25
f. 58 ·· 61

For each problem, make the sign with the bigger end next to the bigger number. Both signs for the first problem are partly shown. Only the big dot is shown for the rest of the signs. Put your pencil down when you've made the sign for each problem.
(Observe children and give feedback.)
b. Check your work. You'll read the numbers for each problem. Then tell me the number with the bigger end of the sign next to it.
- Read the numbers for problem A. (Signal.)
64 (and) 71.
- Which number has the bigger end next to it?
(Signal.) *71.*
c. (Repeat the following tasks for remaining problems:)

Read the numbers for problem __.		Which number has the bigger end next to it?
B	37 (and) 28	37
C	36 (and) 19	36
D	78 (and) 90	90
E	27 (and) 25	27
F	58 (and) 61	61

(Display:)   [87:7A]

a. 64 ⋚ 71

b. 37 ⟩ 28

c. 36 ⟩ 19

d. 78 ⟨ 90

e. 27 ⟩ 25

f. 58 ⟨ 61

Here are the signs you should have made for all of the problems.

EXERCISE 8: FACTS
ADDITION

a. Find part 5. ✔
(Teacher reference:)

```
        e.6+3
a. 4+3    f.4+4
b.10+5    g.3+2
c. 2+5    h.3+5
```

Think of the number families for each problem and write the answers. Put your pencil down when you've completed the equations in part 5.
(Observe children and give feedback.)

b. Check your work. You'll touch and read the equation for each problem.
- Problem A. (Signal.) *4 + 3 = 7.*
- (Repeat for:) B, *10 + 5 = 15;* C, *2 + 5 = 7;* D, *5 + 6 = 11;* E, *6 + 3 = 9;* F, *4 + 4 = 8;* G, *3 + 2 = 5;* H, *3 + 5 = 8;* I, *6 + 6 = 12.*

EXERCISE 9: WRITING DOLLARS WITH 1-DIGIT OR 2-DIGIT CENTS
REMEDY

a. Find part 6 on worksheet 87. ✔
(Teacher reference:)   R Part D

```
a. $    .
b. $    .
c. $    .
```

You're going to write dollars and cents amounts. Some amounts will show only cents with no dollars. The dollar sign and the dot are shown for each amount you'll write.
- Touch A. ✔
- The amount for A is 9 cents. What amount? (Signal.) *9 cents.*
- Write it. ✔
(Observe children and give feedback.)
(Display:) W   [87:9A]

a. $ .09

Here's what you should have.

b. Touch B. ✔
- The amount for B is 15 dollars and 60 cents. What amount? (Signal.) *15 dollars and 60 cents.*

- Write it. ✔
(Observe children and give feedback.)
(Display:)   [87:9B]

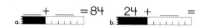
b. $15.60

Here's what you should have.

c. Touch C. ✔
The amount for C is 6 cents. What amount? (Signal.) *6 cents.*
- Write it. ✔
(Observe children and give feedback.)
(Display:)   [87:9C]

c. $ .06

Here's what you should have.

EXERCISE 10: INDEPENDENT WORK

a. Find part 7. ✔
(Teacher reference:)

```
a. ___ + ___ = 84    b. 24 + ___ =
```

You'll complete the equations for the rulers.

b. Turn to the other side of worksheet 87 and find part 8. ✔
(Teacher reference:)

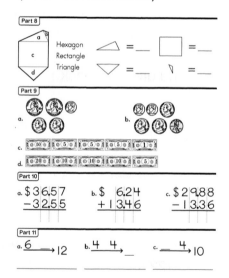

You'll circle the shape of the whole figure and complete the equations to show the parts of it.
In part 9, you'll write an equals and the number of dollars or cents each group is worth.
In part 10, you'll work the dollar problems.
In part 11, you'll write the fact for the missing number and you'll complete the family.

c. Complete worksheet 87. Remember to work part 7 on the other side of your worksheet.
(Observe children and mark incorrect responses on children's worksheets as you give feedback.)

Lesson

EXERCISE 1: FACTS
ADDITION/SUBTRACTION REMEDY

a. (Display:)   [88:1A]

8 − 4	7 + 2	4 + 6
8 + 2	7 − 3	4 + 4
8 − 5	7 − 5	11 − 5

For each problem, you'll tell me if the answer is the big number or a small number. Then you'll say the fact.

- (Point to **8 − 4.**) Is the answer the big number or a small number? (Signal.) *A small number.*
- Say the fact. (Signal.) *8 − 4 = 4.*
b. (Point to **8 + 2.**) Is the answer the big number or a small number? (Signal.) *The big number.*
- Say the fact. (Signal.) *8 + 2 = 10.*
c. (Repeat the following tasks for the remaining problems:)

(Point to __.) Is the answer the big number or a small number?	Say the fact.
8 − 5	A small number.	8 − 5 = 3
7 + 2	The big number.	7 + 2 = 9
7 − 3	A small number.	7 − 3 = 4
7 − 5	A small number.	7 − 5 = 2
4 + 6	The big number.	4 + 6 = 10
4 + 4	The big number.	4 + 4 = 8
11 − 5	A small number.	11 − 5 = 6

(Repeat problems that were not firm.)

EXERCISE 2: COMPARISON STATEMENTS REMEDY

a. (Display:)   [88:2A]

The number for each dog shows how many years old that dog is.

- (Point to black dog.) My turn: How old is the black dog? 8 years.

- (Point to white dog.) How old is the white dog? 5 years.
- Your turn: Which dog is older, the black dog or the white dog? (Signal.) *The black dog.*
- Say the problem for figuring out how much older the black dog is. (Signal.) *8 minus 5.*
- Say the problem for figuring how much younger the white dog is. (Signal.) *8 minus 5.*
- Tell me how much older the black dog is. Get ready. (Signal.) *3 years.*
- Say the sentence about the black dog. (Signal.) *The black dog is 3 years older than the white dog.*
- Say the sentence about the white dog. (Signal.) *The white dog is 3 years younger than the black dog.*
 (Repeat until firm.)
b. (Display:)   [88:2B]

The number for each person tells how many years old that person is.

- (Point to boy.) How old is the boy? (Signal.) *10 years.*
- (Point to girl.) How old is the girl? (Signal.) *19 years.*
- Say the problem for figuring out how much younger the boy is. (Signal.) *19 minus 10.*
- Say the problem for figuring how much older the girl is. (Signal.) *19 minus 10.*
- Tell me how much younger the boy is. Get ready. (Signal.) *9 years.*
- Say the sentence about the boy. (Signal.) *The boy is 9 years younger than the girl.*
- Say the sentence about the girl. (Signal.) *The girl is 9 years older than the boy.*
 (Repeat until firm.)

c. (Display:) [88:2C]

The number for each flower tells how many feet tall the flower is.

- (Point to **a.**) How tall is flower A? (Signal.) *5 feet.*
- (Point to **b.**) How tall is flower B? (Signal.) *2 feet.*
- Say the problem for figuring out how much taller flower A is. (Signal.) *5 minus 2.*
- Say the problem for figuring out how much shorter flower B is. (Signal.) *5 minus 2.*
- Tell me how much taller flower A is. Get ready. (Signal.) *3 feet.*
- Say the sentence about flower A. (Signal.) *Flower A is 3 feet taller than flower B.*
- Say the sentence about flower B. (Signal.) *Flower B is 3 feet shorter than flower A.* (Repeat until firm.)

EXERCISE 3: MONEY
DOLLARS PLUS COINS

REMEDY

a. (Display:) [88:3A]

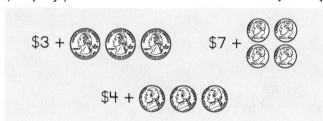

These problems show dollar amounts plus some coins. We're going to figure out how much money there is for each problem.

- (Point to **$3.**) How many dollars for this problem? (Signal.) *3.*
- My turn to start with 3 dollars and add the quarters. (Touch $3.) 3 dollars. (Touch each quarter.) 3 dollars and 25 cents, 3 dollars and 50 cents, 3 dollars and 75 cents.
- Your turn to start with 3 dollars and add the quarters. (Touch $3.) *3 dollars.* (Touch each quarter.) *3 dollars and 25 cents, 3 dollars and 50 cents, 3 dollars and 75 cents.* (Repeat until firm.)

- (Point to **$3.**) How much is this group worth? (Signal.) *3 dollars and 75 cents.*
b. (Point to **$7.**) How many dollars are in this group? (Signal.) *7.*
- What do you add for each dime? (Signal.) *10 cents.*
- Your turn to start with 7 dollars and add the dimes. (Touch $7.) *7 dollars.* (Touch each dime.) *7 dollars and 10 cents, 7 dollars and 20 cents, 7 dollars and 30 cents, 7 dollars and 40 cents.* (Repeat until firm.)
- (Point to **$7.**) How much is this group worth? (Signal.) *7 dollars and 40 cents.*
c. (Point to **$4.**) How many dollars are in this row? (Signal.) *4.*
- What do you add for each nickel? (Signal.) *5 cents.*
- Your turn to start with 4 dollars and add the nickels. (Touch $4.) *4 dollars.* (Touch each nickel.) *4 dollars and 5 cents, 4 dollars and 10 cents, 4 dollars and 15 cents.* (Repeat until firm.)
- (Point to **$4.**) How much is this group worth? (Signal.) *4 dollars and 15 cents.*

EXERCISE 4: COLUMN ADDITION
CARRYING WITH 3-DIGIT NUMBERS

a. (Distribute unopened workbooks to children.)
- Open your workbook to Lesson 88 and find part 1. (Observe children and give feedback.) (Teacher reference:)

a. $$\begin{array}{r} 394 \\ +112 \\ \hline \end{array}$$ b. $$\begin{array}{r} 106 \\ +856 \\ \hline \end{array}$$

You have to carry to work these problems.
- Touch and read problem A. Get ready. (Signal.) *394 plus 112.* (Repeat until firm.)
b. Read the problem for the ones. (Signal.) *4 plus 2.*
- What's the answer? (Signal.) *6.*
- Read the problem for the tens. (Signal.) *9 plus 1.*
- What's the answer? (Signal.) *10.*
That answer has two digits, 1 and zero. You'll write the 1 in the hundreds column.
- Touch where you'll write 1. ✔
- What will you write in the tens column? (Signal.) *Zero.*
- Touch where you'll write zero. ✔ (Repeat until firm.)

c. Touch and read problem B. Get ready.
(Signal.) *106 plus 856.*
- Read the problem for the ones. (Signal.)
6 plus 6.
- What's the answer? (Signal.) *12.*
12 has two digits, 1 and 2.
- What will you write in the tens column?
(Signal.) *1.*
- What will you write in the ones column?
(Signal.) *2.*
(Repeat until firm.)

d. Go back and work problem A. Remember, you carry a digit to the next column.
(Observe children and give feedback.)
(Display:) [88:4A]

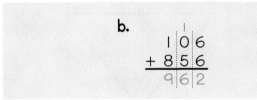

Here's what you should have.
The answer to the problem is 506.

e. Work problem B. Remember, you carry a digit to the next column.
(Observe children and give feedback.)
(Display:) [88:4B]

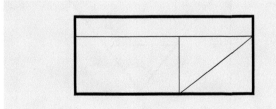

Here's what you should have.
The answer to the problem is 962.

EXERCISE 5: 2-DIMENSIONAL FIGURES
COMPOSITION

a. (Display:) [88:5A]

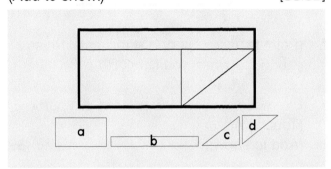

- (Point to ▱.) What's the name of this whole shape? (Touch.) *(A) rectangle.*
This rectangle has four parts.
- Raise your hand when you know how many parts are triangles. ✔
- How many parts are triangles? (Signal.) *Two.*
b. Raise your hand when you know how many parts are rectangles. ✔
- How many parts are rectangles? (Signal.) *Two.*
c. Here are the parts that are in the rectangle.
(Add to show:) [88:5B]

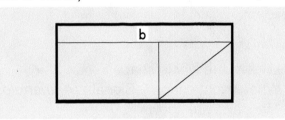

You have to figure out where each part goes.
- (Point to ▱.) Look at this part. Raise your hand when you can tell me the letter of this part. ✔
- Everybody, what's the letter of this part?
(Touch.) *B.*
(Add to show:) [88:5C]

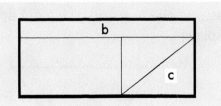

d. (Point to ▱.) Look at this part. Raise your hand when you can tell me the letter of this part. ✔
- Everybody, what's the letter of this part?
(Touch.) *C.*
(Add to show:) [88:5D]

e. (Point to 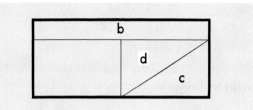.) Look at this part. Raise your hand when you can tell me the letter of this part. ✔

• Everybody, what's the letter of this part? (Touch.) *D.*
(Add to show:) [88:5E]

f. (Point to 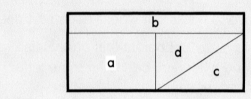.) Look at this part. Raise your hand when you can tell me the letter of this part. ✔

• Everybody, what's the letter of this part? (Touch.) *A.*
(Add to show:) [88:5F]

g. Find part 2 on your worksheet. ✔
(Teacher reference:)

• Look at the whole shape in part 2. ✔
• What is that shape? (Signal.) *(A) triangle.*
h. The parts are shown.
• Touch part A. ✔
• What shape is part A? (Signal.) *(A) triangle.*
• What shape is part B? (Signal.) *(A) rectangle.*
• What shape is part C? (Signal.) *(A) triangle.*
• What shape is part D? (Signal.) *(A) triangle.*
i. Write the letters for each part inside the whole shape. Don't get fooled.
(Observe children and give feedback.)
j. (Display:) [88:5G]

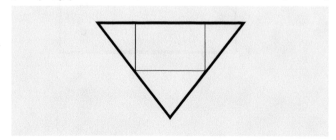

• (Point to 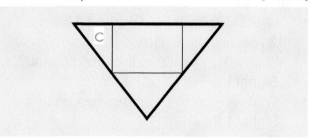.) What letter did you write for this part? (Signal.) *C.*
(Add to show:) [88:5H]

Here it is.

• (Point to 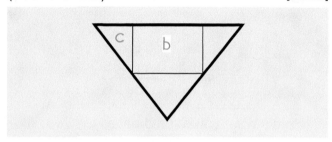.) What letter did you write for this part? (Signal.) *B.*
(Add to show:) [88:5I]

Here it is.

• (Point to 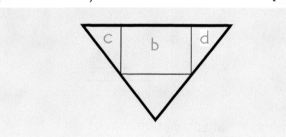.) What letter did you write for this part? (Signal.) *D.*
(Add to show:) [88:5J]

Here it is.

• (Point to 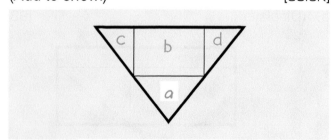.) What letter did you write for this part? (Signal.) *A.*
(Add to show:) [88:5K]

Here it is.

EXERCISE 6: FACTS
SUBTRACTION

[REMEDY]

a. Find part 3 on worksheet 88. ✔
(Teacher reference:)

[R] *Part P*

a. 8 – 4	e. 8 – 6	i. 11 – 9
b. 3 – 2	f. 11 – 6	j. 7 – 3
c. 9 – 6	g. 7 – 4	k. 8 – 3
d. 6 – 2	h. 12 – 6	l. 7 – 5

I'll read each minus problem. You'll tell me the answer. Then you'll complete the facts.

• Problem A. 8 minus 4. What's the answer? (Signal.) *4.*
• Problem B. 3 minus 2. What's the answer? (Signal.) *1.*
• Problem C. 9 minus 6. What's the answer? (Signal.) *3.*
• Problem D. 6 minus 2. What's the answer? (Signal.) *4.*
• Problem E. 8 minus 6. What's the answer? (Signal.) *2.*
• Problem F. 11 minus 6. What's the answer? (Signal.) *5.*
• Problem G. 7 minus 4. What's the answer? (Signal.) *3.*
• Problem H. 12 minus 6. What's the answer? (Signal.) *6.*
• Problem I. 11 minus 9. What's the answer? (Signal.) *2.*
• Problem J. 7 minus 3. What's the answer? (Signal.) *4.*
• Problem K. 8 minus 3. What's the answer? (Signal.) *5.*
• Problem L. 7 minus 5. What's the answer? (Signal.) *2.*
(Repeat problems that were not firm.)

b. Complete the fact for each problem.
(Observe children and give feedback.)

c. Check your work. You'll read the fact for each problem.
• Problem A. (Signal.) *8 – 4 = 4.*
• (Repeat for:) B, *3 – 2 = 1;* C, *9 – 6 = 3;* D, *6 – 2 = 4;* E, *8 – 6 = 2;* F, *11 – 6 = 5;* G, *7 – 4 = 3;* H, *12 – 6 = 6;* I, *11 – 9 = 2;* J, *7 – 3 = 4;* K, *8 – 3 = 5;* L, *7 – 5 = 2.*

EXERCISE 7: GREATER THAN/LESS THAN/EQUALS SIGN

[REMEDY]

a. Find part 4. ✔
(Teacher reference:)

[R] *Part I*

a. 46	53
b. 16	16
c. 57	48
d. 24	24
e. 117	130
f. 43	35

For each problem, you're going to tell me if the first number is more than, less than, or equal to the other number. Then you'll make the sign with the bigger end next to the bigger number, or you'll make an equals sign.

• Read the numbers for problem A. (Signal.) *46 (and) 53.*
• Is 46 more than, less than, or equal to 53? (Signal.) *Less than.*
• 46 is less than 53, so do you make the sign with the bigger or smaller end next to 46? (Signal.) *Smaller.*
• Which number will the bigger end be next to? (Signal.) *53.*
• Make the sign. ✔

b. Read the numbers for problem B. (Signal.) *16 (and) 16.*
• Is 16 more than, less than, or equal to 16? (Signal.) *Equal to.*
• Write an equals sign for problem B to show the sides are equal. ✔

c. Read the numbers for problem C. (Signal.) *57 (and) 48.*
• Is 57 more than, less than, or equal to 48? (Signal.) *More than.*
• 57 is more than 48, so do you make the sign with the bigger or smaller end next to 57? (Signal.) *Bigger.*
• Which number will the smaller end be next to? (Signal.) *48.*
• Make the sign for problem C. ✔

d. Read the numbers for problem D. (Signal.) *24 (and) 24.*
• Is 24 more than, less than, or equal to 24? (Signal.) *Equal to.*
• Write an equals sign for problem D to show the sides are equal. ✔

e. Read the numbers for problem E. (Signal.)
117 (and) 130.

- Is 117 more than, less than, or equal to 130?
(Signal.) *Less than.*
- 117 is less than 130, so do you make the sign
with the bigger or smaller end next to 117?
(Signal.) *Smaller.*
- Which number will the bigger end be next to?
(Signal.) *130.*
- Make the sign for problem E. ✔

f. Read the numbers for problem F. (Signal.)
43 (and) 35.

- Is 43 more than, less than, or equal to 35?
(Signal.) *More than.*
- 43 is more than 35, so do you make the sign
with the bigger or smaller end next to 43?
(Signal.) *Bigger.*
- Which number will the smaller end be next to?
(Signal.) *35.*
- Make the sign for problem F. ✔
(Display:)   [88:7A]

> **a.** $46 < 53$
>
> **b.** $16 = 16$
>
> **c.** $57 > 48$
>
> **d.** $24 = 24$
>
> **e.** $117 < 130$
>
> **f.** $43 > 35$

Here are the signs you should have made for
all of the problems.

EXERCISE 8: FACTS

ADDITION   **REMEDY**

a. Find part 5. ✔
(Teacher reference:)   **R Part Q**

a. $5 + 2$	g. $8 + 10$
b. $3 + 6$	h. $4 + 6$
c. $4 + 4$	i. $4 + 2$
d. $5 + 3$	j. $6 + 6$
e. $2 + 7$	k. $9 + 1$
f. $3 + 4$	l. $5 + 6$

b. Think of the number families and write
answers to these plus problems.
(Observe children and give feedback.)

c. Check your work. You'll read the fact for each
problem.

- Problem A. (Signal.) $5 + 2 = 7.$
- (Repeat for:) B, $3 + 6 = 9$; C, $4 + 4 = 8$;
D, $5 + 3 = 8$; E, $2 + 7 = 9$; F, $3 + 4 = 7$;
G, $8 + 10 = 18$; H, $4 + 6 = 10$; I, $4 + 2 = 6$;
J, $6 + 6 = 12$; K, $9 + 1 = 10$; L, $5 + 6 = 11.$

EXERCISE 9: INDEPENDENT WORK

a. Find part 6. ✔
(Teacher reference:)

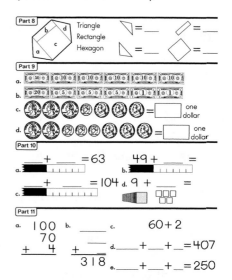

To work the problems in part 6, you'll have to
carry.
In part 7, you'll work money problems.

b. Turn to the other side of worksheet 88 and find
part 8. ✔
(Teacher reference:)

You'll circle the shape of the whole figure and
complete equations to show the parts of it.
In part 9, you'll write an equals and the
number of dollars to show what the group of
bills is worth. You'll circle the words one dollar
or write the cents for each group of coins.
In part 10, you'll complete the equations for
the rulers and the equation for the boxes.
In part 11, you'll complete the place value
equations.

c. Complete worksheet 88. Remember to
work the carrying problems and the money
problems in parts 5 and 6 on the other side of
your worksheet.
(Observe children and mark incorrect responses
on children's worksheets as you give feedback.)

Lesson  89

EXERCISE 1: NUMBER FAMILIES
SMALL NUMBER OF 3, 5 · REMEDY

a. (Display:) [89:1A]

$$\xrightarrow[\quad 3 \quad\quad 3 \quad]{} \_$$

$$\xrightarrow[\quad 5 \quad\quad 5 \quad]{} \_$$

The big number is missing in these families. These are new families that have small numbers of 3 and small numbers of 5.

- (Point to $\xrightarrow[3 \quad 3]{}$.) What are the small numbers in this family? (Signal.) *3 and 3.*
- The big number is 6. What's the big number? (Signal.) *6.*
- (Point to $\xrightarrow[5 \quad 5]{}$.) What are the small numbers in this family? (Signal.) *5 and 5.*
- The big number is 10. What's the big number? (Signal.) *10.*

b. (Point to $\xrightarrow[3 \quad 3]{}$.) What are the small numbers in this family? (Signal.) *3 and 3.*
- What's the big number? (Signal.) *6.*
c. (Point to $\xrightarrow[5 \quad 5]{}$.) What are the small numbers in this family? (Signal.) *5 and 5.*
- What's the big number? (Signal.) *10.*
(Repeat steps b and c until firm.)

d. There's only one plus fact and one minus fact for each family.
- (Point to $\xrightarrow[3 \quad 3]{}$.) Say the plus fact. (Signal.) *3 + 3 = 6.*
- Say the minus fact. (Signal.) *6 – 3 = 3.*
e. (Point to $\xrightarrow[5 \quad 5]{}$.) Say the plus fact. (Signal.) *5 + 5 = 10.*
- Say the minus fact. (Signal.) *10 – 5 = 5.*
(Repeat until firm.)

f. (Display:) [89:1B]

$\xrightarrow[\quad 5 \quad]{} 10$	$\xrightarrow[6 \quad 3]{} \_$	$\xrightarrow[\quad 5 \quad]{} 8$
$\xrightarrow[10 \quad 5]{} \_$	$\xrightarrow[4 \quad 4]{} \_$	$\xrightarrow[\quad 6 \quad]{} 12$
$\xrightarrow[\quad 3 \quad]{} 6$	$\xrightarrow[\quad 4 \quad]{} 10$	$\xrightarrow[4 \quad 3]{} \_$

You're going to say the problem for the missing number in each family.

g. (Point to $\xrightarrow[\quad 5 \quad]{} 10$.) Say the problem for the missing number. Get ready. (Touch.) *10 minus 5.*
- What's 10 minus 5? (Signal.) *5.*

h. (Point to $\xrightarrow[10 \quad 5]{} \_$.) Say the problem for the missing number. (Touch.) *10 plus 5.*
- What's 10 plus 5? (Signal.) *15.*

i. (Repeat the following tasks for remaining families:)

(Point to __.) Say the problem for the missing number.		What's __?	
$\xrightarrow[\quad 3 \quad]{} 6$	6 – 3	6 – 3	3
$\xrightarrow[6 \quad 3]{} \_$	6 + 3	6 + 3	9
$\xrightarrow[4 \quad 4]{} \_$	4 + 4	4 + 4	8
$\xrightarrow[\quad 4 \quad]{} 10$	10 – 4	10 – 4	6
$\xrightarrow[\quad 5 \quad]{} 8$	8 – 5	8 – 5	3
$\xrightarrow[\quad 6 \quad]{} 12$	12 – 6	12 – 6	6
$\xrightarrow[4 \quad 3]{} \_$	4 + 3	4 + 3	7

(Repeat for families that were not firm.)

EXERCISE 2: COMPARISON STATEMENTS

a. (Display:) [89:2A]

8   10

The number for each cat tells how many years old that cat is.

- (Point to white cat.) How old is the white cat? (Signal.) *8 years (old).*
(Point to black cat.) How old is the black cat? (Signal.) *10 years (old).*
- Say the problem for figuring out how much younger the white cat is. (Signal.) *10 minus 8.*
- Say the problem for figuring how much older the black cat is. (Signal.) *10 minus 8.*
- Tell me how much younger the white cat is. Get ready. (Signal.) *2 years.*
- Say the sentence about the white cat. (Signal.) *The white cat is 2 years younger than the black cat.*
- Say the sentence about the black cat. (Signal.) *The black cat is 2 years older than the white cat.*
(Repeat until firm.)

b. (Display:) [89:2B]

|| 5

The number for each insect tells how many days old that insect is.

- (Point to fly.) How old is the fly? (Signal.) *11 days.*
- (Point to ant.) How old is the ant? (Signal.) *5 days.*
- Say the problem for figuring out how much older the fly is. (Signal.) *11 minus 5.*
- Say the sentence for figuring how much younger the ant is. (Signal.) *11 minus 5.*
- Tell me how much older the fly is. Get ready. (Signal.) *6 days.*
- Say the sentence about the fly. (Signal.) *The fly is 6 days older than the ant.*
- Say the sentence about the ant. (Signal.) *The ant is 6 days younger than the fly.*
 (Repeat until firm.)

c. (Display:) [89:2C]

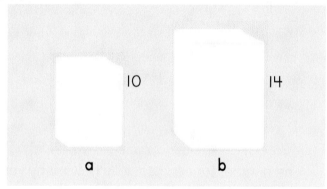

10 14

a b

The number for each book tells how many inches tall it is.

- (Point to **a.**) How tall is book A? (Signal.) *10 inches.*
- (Point to **b.**) How tall is book B? (Signal.) *14 inches.*
- Say the problem for figuring out how much shorter book A is. (Signal.) *14 minus 10.*

- Say the sentence for figuring how much taller book B is. (Signal.) *14 minus 10.*
- Tell me how much shorter book A is. Get ready. (Signal.) *4 inches.*
- Say the sentence about book A. (Signal.) *Book A is 4 inches shorter than book B.*
- Say the sentence about book B. (Signal.) *Book B is 4 inches taller than book A.*
 (Repeat until firm.)

EXERCISE 3: FACTS
ADDITION/SUBTRACTION

a. (Display:) [89:3A]

5 + 5	7 − 2	4 + 4
10 − 2	3 + 6	9 − 6
4 + 3	14 − 10	7 − 4

For each problem, you'll tell me if the answer is the big number or a small number. Then you'll say the fact.

- (Point to **5 + 5.**) Is the answer the big number or a small number? (Signal.) *The big number.*
- Say the fact. (Signal.) *5 + 5 = 10.*

b. (Point to **10 − 2.**) Is the answer the big number or a small number? (Signal.) *A small number.*
- Say the fact. (Signal.) *10 − 2 = 8.*

c. (Repeat the following tasks for the remaining problems:)

(Point to __.) Is the answer the big number or a small number?		Say the fact.
4 + 3	The big number.	4 + 3 = 7
7 − 2	A small number.	7 − 2 = 5
3 + 6	The big number.	3 + 6 = 9
14 − 10	A small number.	14 − 10 = 4
4 + 4	The big number.	4 + 4 = 8
9 − 6	A small number.	9 − 6 = 3
7 − 4	A small number.	7 − 4 = 3

(Repeat problems that were not firm.)

EXERCISE 4: MONEY

DOLLARS PLUS COINS

REMEDY

a. (Display:)   [89:4A]

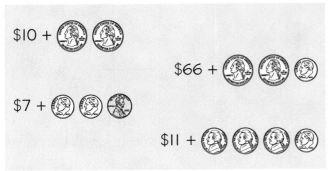

These problems show dollar amounts plus some coins. We're going to figure out how much money there is for each problem.

- (Point to **$10.**) How many dollars for this problem? (Touch.) *10.*
- Start with 10 dollars and add for the quarters. (Touch $10.) *10 dollars.* Add for the quarters. (Touch quarters.) *10 dollars and 25 cents, 10 dollars and 50 cents.* (Repeat until firm.)
- How much is this problem worth? (Signal.) *10 dollars and 50 cents.*

b. (Point to **$66.**) How many dollars for this problem? (Touch.) *66.*
- Start with 66 dollars and add for the quarters, then add for the dime. (Touch $66.) *66 dollars.* Add for the quarters. (Touch quarters.) *66 dollars and 25 cents, 66 dollars and 50 cents.* Add for the dime. (Touch.) *66 dollars and 60 cents.* (Repeat until firm.)
- How much is this problem worth? (Signal.) *66 dollars and 60 cents.*

c. (Point to **$7.**) How many dollars for this problem? (Touch.) *7.*
- Start with 7 dollars and add for the dimes. Then add for a penny. (Touch $7.) *7 dollars.* Add for the dimes. (Touch dimes.) *7 dollars and 10 cents, 7 dollars and 20 cents.* Add for the penny. (Touch.) *7 dollars and 21 cents.* (Repeat until firm.)
- How much is this problem worth? (Signal.) *7 dollars and 21 cents.*

d. (Point to **$11.**) How many dollars for this problem? (Touch.) *11.*
- Start with 11 dollars and add for the nickels, then add for the dime. (Touch $11.) *11 dollars.* Add for the nickels. (Touch nickels.) *11 dollars and 5 cents, 11 dollars and 10 cents, 11 dollars and 15 cents.* Add for the dime. (Touch.) *11 dollars and 25 cents.* (Repeat until firm.)
- How much is this problem worth? (Signal.) *11 dollar and 25 cents.*

EXERCISE 5: MIXED COUNTING

a. Let's do some counting.
- Start with 80 and plus tens to 180. Get 80 going. *Eightyyy.* Plus tens. (Tap.) *90, 100, 110, 120, 130, 140, 150, 160, 170, 180.*
- Start with 14 and plus tens to 74. Get 14 going. *Fourteeen.* Plus tens. (Tap.) *24, 34, 44, 54, 64, 74.*
- Start with 17 and plus tens to 77. Get 17 going. *Seventeeen.* Plus tens. (Tap.) *27, 37, 47, 57, 67, 77.*
- Start with 11 and plus tens to 71. Get 11 going. *Elevennn.* Plus tens. (Tap.) *21, 31, 41, 51, 61, 71.*

b. Start with 100 and count backward by tens. Get 100 going. *One huuundred.* Count backward. (Tap.) *90, 80, 70, 60, 50, 40, 30, 20, 10.*

c. Start with 50 and count by fives to 100. Get 50 going. *Fiftyyy.* Count. (Tap.) *55, 60, 65, 70, 75, 80, 85, 90, 95, 100.*

d. Start with 100 and count by hundreds to 1000. Get ready. (Tap.) *100, 200, 300, 400, 500, 600, 700, 800, 900, 1000.*

e. Start with 300 and count by 25s to 400. Get 300 going. *Three huuundred.* Count. (Tap.) *325, 350, 375, 400.*

EXERCISE 6: COLUMN ADDITION
CARRYING WITH 3-DIGIT NUMBERS
(3 ADDENDS)

REMEDY

a. (Distribute unopened workbooks to children.)
- Open your workbook to Lesson 89 and find part 1.
  (Observe children and give feedback.)
  (Teacher reference:)

R Part L

```
a.  2 9 7
      1 2
   +3 7 0
```

```
b.     8
    4 1 2
   +1 5 3
```

You have to carry to work these problems.
- Touch and read problem A. Get ready.
  (Signal.) *297 plus 12 plus 370.*
  (Repeat until firm.)
b. Read the problem for the ones. (Signal.)
  7 plus 2 plus zero.
- What's the answer? (Signal.) *9.*
- Read the problem for the tens. (Signal.) *9 plus 1 plus 7.*
- What's 9 plus 1? (Signal.) *10.*
- What's 10 plus 7? (Signal.) *17.*
c. 17 has two digits, 1 and 7. You'll write the 1 in the hundreds column.
- Touch where you'll write 1. ✔
- What number will you write in the tens column? (Signal.) *7.*
- Touch where you'll write 7. ✔
  (Repeat step c until firm.)
d. Touch and read problem B. Get ready.
  (Signal.) *8 plus 412 plus 153.*
- Read the problem for the ones. (Signal.)
  8 plus 2 plus 3.
- What's 8 plus 2? (Signal.) *10.*
- What's 10 plus 3? (Signal.) *13.*
e. 13 has two digits, 1 and 3. What will you write in the tens column? (Signal.) *1.*
- What will you write in the ones column? (Signal.) *3.*
  (Repeat step e until firm.)
f. Go back and work problem A. Remember, you carry a digit to the next column.
  (Observe children and give feedback.)

g. Check your work. Problem A is 297 plus 12 plus 370. What's the answer? (Signal.) *679.*
  (Display:)   [89:6A]

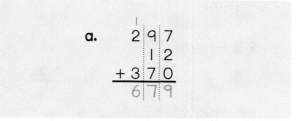

Here's what you should have.
h. Work problem B. Remember, you carry a digit to the next column.
  (Observe children and give feedback.)
i. Check your work. Problem B is 8 plus 412 plus 153. What's the answer? (Signal.) *573.*
  (Display:)   [89:6B]

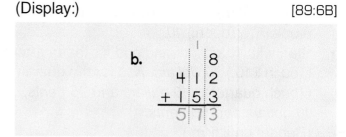

Here's what you should have.

EXERCISE 7: GREATER THAN/LESS THAN/EQUALS SIGN

a. Find part 2. ✔
  (Teacher reference:)

```
a. 67   67
b. 76   67
c. 35   42
d. 86   86
e. 162  154
f. 14   20
```

For each problem, you're going to tell me if the first number is more than, less than, or equal to the other number. Then you'll make the sign with the bigger end next to the bigger number, or you'll make an equals sign.
- Read the numbers for problem A. (Signal.)
  67 (and) 67.
- Is 67 more than, less than, or equal to 67? (Signal.) *Equal to.*
- Write an equals sign for problem A to show the equation 67 equals 67. ✔

b. Read the numbers for problem B. (Signal.)
76 (and) 67.

• Is 76 more than, less than, or equal to 67?
(Signal.) *More than.*

• 76 is more than 67, so do you make the sign with the bigger or smaller end next to 76?
(Signal.) *Bigger.*

• Which number will the smaller end be next to?
(Signal.) *67.*

• Make the sign for problem B. ✔

c. Read the numbers for problem C. (Signal.)
35 (and) 42.

• Is 35 more than, less than, or equal to 42?
(Signal.) *Less than.*

• 35 is less than 42, so do you make the sign with the bigger or smaller end next to 35?
(Signal.) *Smaller.*

• Which number will the bigger end be next to?
(Signal.) *42.*

• Make the sign for problem C. ✔

d. Read the numbers for problem D. (Signal.)
86 (and) 86.

• Is 86 more than, less than, or equal to 86?
(Signal.) *Equal to.*

• Write an equals sign for problem D to show the equation 86 equals 86. ✔

e. Read the numbers for problem E. (Signal.)
162 (and) 154.

• Is 162 more than, less than, or equal to 154?
(Signal.) *More than.*

• 162 is more than 154, so do you make the sign with the bigger or smaller end next to 162? (Signal.) *Bigger.*

• Which number will the smaller end be next to?
(Signal.) *154.*

• Make the sign for problem E. ✔

f. Read the numbers for problem F. (Signal.)
14 (and) 20.

• Is 14 more than, less than, or equal to 20?
(Signal.) *Less than.*

• 14 is less than 20, so do you make the sign with the bigger or smaller end next to 14?
(Signal.) *Smaller.*

• Which number will the bigger end be next to?
(Signal.) *20.*

• Make the sign for problem F. ✔
(Display:) [89:7A]

a. $67 = 67$

b. $76 > 67$

c. $35 < 42$

d. $86 = 86$

e. $162 > 154$

f. $14 < 20$

Here are the signs you should have made for all of the problems.

EXERCISE 8: FACTS
SUBTRACTION

a. Find part 3. ✔
(Teacher reference:)

a. $6-2$	g. $12-6$
b. $6-3$	h. $8-3$
c. $7-4$	i. $7-3$
d. $8-4$	j. $6-3$
e. $10-4$	k. $11-5$
f. $10-5$	l. $10-5$

These are minus problems you'll work. Think of the number families for each problem and write the answers. Put your pencil down when you've completed the equations in part 3.
(Observe children and give feedback.)

b. Check your work. You'll read the equation for each problem.

• Problem A. (Signal.) *6 – 2 = 4.*
• (Repeat for:) B, *6 – 3 = 3;* C, *7 – 4 = 3;* D, *8 – 4 = 4;* E, *10 – 4 = 6;* F, *10 – 5 = 5;* G, *12 – 6 = 6;* H, *8 – 3 = 5;* I, *7 – 3 = 4;* J, *6 – 3 = 3;* K, *11 – 5 = 6;* L, *10 – 5 = 5.*

EXERCISE 9: WORD PROBLEMS (COLUMNS)

a. Find part 4. ✔
    (Teacher reference:)

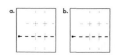

You're going to write the symbols for word
problems in columns and work them.

b. Touch where you'll write symbols for problem A.
    Listen to problem A: Sam had 12 shoes. His
    brother had 28 shoes. How many shoes did
    the men have altogether?

• Listen again and write the symbols for the
    whole problem: Sam had 12 shoes. His
    brother had 28 shoes.
    (Observe children and give feedback.)

• Everybody, touch and read the problem for A.
    Get ready. (Signal.) *12 plus 28.*

c. Work problem A. Put your pencil down when
    you know how many shoes the men had
    altogether.
    (Observe children and give feedback.)
    (Teacher reference:)

• Read the problem and the answer you wrote
    for A. Get ready. (Signal.) *12 + 28 = 40.*

• How many shoes did the men have
    altogether? (Signal.) *40.*

d. Touch where you'll write the symbols for
    problem B. ✔
    Listen to problem B: There were 174 worms in
    the garden. Birds ate 52 of those worms. How
    many worms were still in the garden?

• Listen again and write the symbols for the
    whole problem: There were 174 worms in the
    garden. Birds ate 52 of those worms.
    (Observe children and give feedback.)

• Everybody, touch and read the problem for B.
    Get Ready. (Signal.) *174 minus 52.*

e. Work problem B. Put your pencil down when
    you know how many worms were still in the
    garden.
    (Observe children and give feedback.)
    (Teacher reference:)

• Read the problem and the answer you wrote
    for B. Get ready. (Signal.) *174 − 52 = 122.*

• How many worms were still in the garden?
    (Signal.) *122.*

EXERCISE 10: INDEPENDENT WORK

a. Find part 5. ✔
    (Teacher reference:)

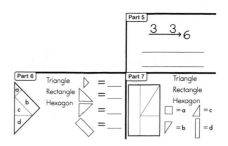

This is one of the families you learned today.
You'll write both facts for the family.

In part 6, you'll circle the name for the whole
shape. Then you'll complete each equation by
writing the letter to show which part it is.

In part 7, you'll circle the name for the whole
shape. Then you'll write the letter for each part.

b. Turn to the other side of worksheet 89 and find part 8. ✔
(Teacher reference:)

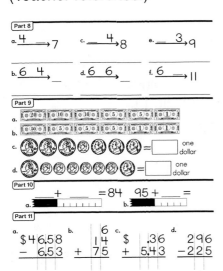

Below each family, you'll write the fact for the missing number and you'll complete the family.
In part 9, you'll write an equals and the number of dollars to show what each group of bills is worth. You'll circle the words one dollar or write the cents for each group of coins.
In part 10, you'll complete the equations for the rulers.
In part 11, you'll work the money problems and the other column problems.

c. Complete worksheet 89. Remember to work parts 5, 6, and 7 on the other side of your worksheet.
(Observe children and mark incorrect responses on children's worksheets as you give feedback.)

Lesson 90

EXERCISE 1: NUMBER FAMILIES
MISSING NUMBER IN FAMILY

a. (Display:) [90:1A]

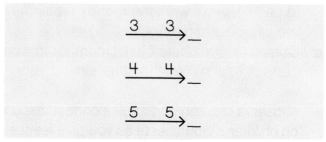

The big number is missing in these families. These families have small numbers of 3, 4, and 5.

- (Point to $\xrightarrow{3\quad 3}$.) What are the small numbers in this family? (Signal.) *3 and 3.*
- The big number is 6. What's the big number? (Signal.) *6.*
- (Point to $\xrightarrow{4\quad 4}$.) What are the small numbers in this family? (Signal.) *4 and 4.*
- The big number is 8. What's the big number? (Signal.) *8.*
- (Point to $\xrightarrow{5\quad 5}$.) What are the small numbers in this family? (Signal.) *5 and 5.*
- The big number is 10. What's the big number? (Signal.) *10.*

b. (Point to $\xrightarrow{3\quad 3}$.) What are the small numbers in this family? (Signal.) *3 and 3.*
- What's the big number? (Signal.) *6.*

c. (Point to $\xrightarrow{4\quad 4}$.) What are the small numbers in this family? (Signal.) *4 and 4.*
- What's the big number? (Signal.) *8.*

d. (Point to $\xrightarrow{5\quad 5}$.) What are the small numbers in this family? (Signal.) *5 and 5.*
- What's the big number? (Signal.) *10.*
(Repeat steps b through d until firm.)

e. There's only one plus fact and one minus fact for each family.
- (Point to $\xrightarrow{3\quad 3}$.) Say the plus fact. (Signal.) *3 + 3 = 6.*
- Say the minus fact. (Signal.) *6 − 3 = 3.*
(Repeat until firm.)

f. (Point to $\xrightarrow{4\quad 4}$.) Say the plus fact. (Signal.) *4 + 4 = 8.*
- Say the minus fact. (Signal.) *8 − 4 = 4.*
(Repeat step f until firm.)

g. (Point to $\xrightarrow{5\quad 5}$.) Say the plus fact. (Signal.) *5 + 5 = 10.*
- Say the minus fact. (Signal.) *10 − 5 = 5.*
(Repeat until firm.)

h. (Display:) [90:1B]

$$\xrightarrow{5\qquad} 10 \qquad \xrightarrow{4\qquad} 8 \qquad \xrightarrow{\quad 4} 10$$
$$\xrightarrow{\quad 3} 6 \qquad \xrightarrow{6\quad 3} \underline{\ } \qquad \xrightarrow{3\quad 3} \underline{\ }$$
$$\xrightarrow{5\quad 3} \underline{\ } \qquad \xrightarrow{5\quad 5} \underline{\ } \qquad \xrightarrow{4\qquad} 7$$

You're going to say the problem for the missing number in each family.

i. (Point to $\xrightarrow{5\quad} 10$.) Say the problem for the missing number. Get ready. (Touch.) *10 minus 5.*
- What's 10 minus 5? (Signal.) *5.*

j. (Point to $\xrightarrow{\quad 3} 6$.) Say the problem for the missing number. (Touch.) *6 minus 3.*
- What's 6 minus 3? (Signal.) *3.*

k. (Repeat the following tasks for remaining families:)

(Point to __.) Say the problem for the missing number.		What's __?	
$\xrightarrow{5\quad 3} \underline{\ }$	5 + 3	5 + 3	8
$\xrightarrow{4\quad} 8$	8 − 4	8 − 4	4
$\xrightarrow{6\quad 3} \underline{\ }$	6 + 3	6 + 3	9
$\xrightarrow{5\quad 5} \underline{\ }$	5 + 5	5 + 5	10
$\xrightarrow{\quad 4} 10$	10 − 4	10 − 4	6
$\xrightarrow{3\quad 3} \underline{\ }$	3 + 3	3 + 3	6
$\xrightarrow{4\quad} 7$	7 − 4	7 − 4	3

(Repeat for families that were not firm.)

EXERCISE 2: COLUMN ADDITION
CARRYING (MULTIPLE DIGITS)   REMEDY

a. Here's a new kind of problem. You have to carry a digit from the ones column and carry a digit from the tens column.
   (Display:) W   [90:2A]

$$\begin{array}{r} 2\,9\,6 \\ +\,5\,4\,6 \\ \hline \end{array}$$

- Read the problem. Get ready. (Signal.) *296 plus 546.*
- Read the problem for the ones. (Signal.) *6 plus 6.*
- What's the answer? (Signal.) *12.*
  That answer has two digits, 1 and 2. We write 2 in the ones column.
  (Add to show:)   [90:2B]

$$\begin{array}{r} 2\,9\,6 \\ +\,5\,4\,6 \\ \hline 2 \end{array}$$

- What do we write in the tens column?
  (Signal.) *1.*
  (Add to show:)   [90:2C]

$$\begin{array}{r} {}^{1}\;\; \\ 2\,9\,6 \\ +\,5\,4\,6 \\ \hline 2 \end{array}$$

- Read the problem for the tens column. Get ready. (Signal.) *1 plus 9 plus 4.*
- What's 1 plus 9? (Signal.) *10.*
- What's 10 plus 4? (Signal.) *14.*
  That answer has two digits, 1 and 4. We write 4 in the tens column.
  (Add to show:)   [90:2D]

$$\begin{array}{r} {}^{1}\;\; \\ 2\,9\,6 \\ +\,5\,4\,6 \\ \hline 4\,2 \end{array}$$

- What do we write in the hundreds column?
  (Signal.) *1.*
  (Add to show:)   [90:2E]

$$\begin{array}{r} {}^{1}\;{}^{1}\; \\ 2\,9\,6 \\ +\,5\,4\,6 \\ \hline 4\,2 \end{array}$$

- Read the problem for the hundreds. Get ready.
  (Signal.) *1 plus 2 plus 5.*
- What's the answer? (Signal.) *8.*
  (Add to show:)   [90:2F]

$$\begin{array}{r} {}^{1}\;{}^{1}\; \\ 2\,9\,6 \\ +\,5\,4\,6 \\ \hline 8\,4\,2 \end{array}$$

- Read the problem and the answer. Get ready.
  (Signal.) *296 + 546 = 842.*

b. (Display:) W   [90:2G]

$$\begin{array}{r} 1\,4\,2 \\ +\,7\,6\,8 \\ \hline \end{array}$$

- Read the problem. Get ready. (Signal.)
  142 plus 768.
- Read the problem for the ones. (Signal.)
  2 plus 8.
- What's the answer? (Signal.) *10.*
  That answer has two digits, 1 and zero. We write zero in the ones column.
  (Add to show:)   [90:2H]

$$\begin{array}{r} 1\,4\,2 \\ +\,7\,6\,8 \\ \hline 0 \end{array}$$

- What do we write in the tens column? (Signal.) *1.*
  (Add to show:)   [90:2I]

$$\begin{array}{r} {}^{1}\;\; \\ 1\,4\,2 \\ +\,7\,6\,8 \\ \hline 0 \end{array}$$

- Read the problem for the tens column. Get ready. (Signal.) *1 plus 4 plus 6.*

- What's 1 plus 4 plus 6? (Signal.) *11.*
  That answer has two digits, 1 and 1. We write
  a 1 in the tens column.
  (Add to show:)   [90:2J]

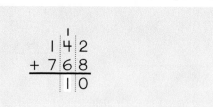

- What do we write in the hundreds column?
  (Signal.) *1.*
  (Add to show:)   [90:2K]

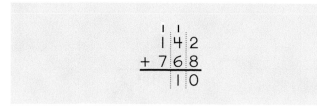

- Read the problem for the hundreds. Get ready.
  (Signal.) *1 plus 1 plus 7.*
- What's the answer? (Signal.) *9.*
  (Add to show:)   [90:2L]

- Read the problem and the answer. Get ready.
  (Signal.) *142 + 768 = 910.*

EXERCISE 3: COMPARISON STATEMENTS   REMEDY

a. (Display:)   [90:3A]

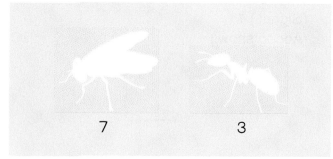

These numbers tell how many pounds each
thing weighs. The thing that is heavier has
a bigger number of pounds.
- (Point to bread.) How much does the bread
  weigh? (Signal.) *2 pounds.*
- Yes, the bread weighs 2 pounds. Say the
  sentence. (Signal.) *The bread weighs
  2 pounds.*

- (Point to cat.) How much does the cat weigh?
  (Signal.) *11 pounds.*
- Yes, the cat weighs 11 pounds. Say the
  sentence. (Signal.) *The cat weighs 11 pounds.*
- Once more: How much does the bread weigh?
  (Signal.) *2 pounds.*
- How much does the cat weigh? (Signal.)
  11 pounds.
- Which weighs more, the bread or the cat?
  (Signal.) *The cat.*
- Say the problem for figuring out how much
  more the cat weighs. Get ready. (Signal.)
  11 minus 2.
- Say the problem for figuring out how much
  less the bread weighs. Get ready. (Signal.)
  11 minus 2.
- What's the answer? (Signal.) *9.*
  Here's the sentence about the bread. The
  bread weighs 9 pounds less than the cat.
- Say the sentence. (Signal.) *The bread weighs
  9 pounds less than the cat.*
- Say the sentence about the cat. (Signal.) *The
  cat weighs 9 pounds more than the bread.*
  (Repeat until firm.)
b. (Display:)   [90:3B]

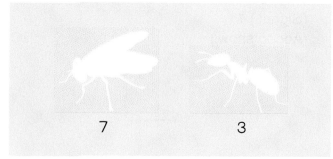

The number for each insect tells how many
days old that insect is.
- (Point to fly.) How old is the fly? (Signal.)
  7 days.
- (Point to ant.) How old is the ant? (Signal.)
  3 days.
- Say the problem for figuring out how much
  older the fly is. (Signal.) *7 minus 3.*
- Say the problem for figuring out how much
  younger the ant is. (Signal.) *7 minus 3.*
- How much older is the fly? (Signal.) *4 days.*
- Say the sentence about the fly. (Signal.) *The
  fly is 4 days older than the ant.*
- Say the sentence about the ant. (Signal.) *The
  ant is 4 days younger than the fly.*
  (Repeat until firm.)

EXERCISE 4: FACTS
ADDITION/SUBTRACTION

a. (Distribute unopened workbooks to children.)
- Open your workbook to Lesson 90 and find part 1.
  (Observe children and give feedback.)
  (Teacher reference:)

a. 7 − 4	d. 3 + 3	g. 8 − 5
b. 6 + 5	e. 9 − 3	h. 3 + 4
c. 10 − 5	f. 4 + 2	i. 8 − 4

- Touch and read problem A. Get ready. (Signal.) *7 minus 4.*
- Is the answer the big number or a small number? (Signal.) *A small number.*
- What's 7 minus 4? (Signal.) *3.*

b. (Repeat the following tasks for problems B through I:)

Touch and read problem __.		Is the answer the big number or a small number?	What's __?	
B	6 + 5	*The big number.*	6 + 5	*11*
C	10 − 5	*A small number.*	10 − 5	*5*
D	3 + 3	*The big number.*	3 + 3	*6*
E	9 − 3	*A small number.*	9 − 3	*6*
F	4 + 2	*The big number.*	4 + 2	*6*
G	8 − 5	*A small number.*	8 − 5	*3*
H	3 + 4	*The big number.*	3 + 4	*7*
I	8 − 4	*A small number.*	8 − 4	*4*

(Repeat problems that were not firm.)

c. Write the answer to each fact. Put your pencil down when you've completed the equations for part 1.
  (Observe children and give feedback.)

d. Check your work.
- Touch and read fact A. (Signal.) *7 − 4 = 3.*
- (Repeat for:) B, *6 + 5 = 11;* C, *10 − 5 = 5;* D, *3 + 3 = 6;* E, *9 − 3 = 6;* F, *4 + 2 = 6;* G, *8 − 5 = 3;* H, *3 + 4 = 7;* I, *8 − 4 = 4.*

EXERCISE 5: GREATER THAN/LESS THAN/EQUALS SIGN

a. Find part 2. ✔
  (Teacher reference:)

a.	315	296
b.	84	92
c.	602	602
d.	55	48
e.	93	101
f.	41	41

For each problem, you'll make the sign with the bigger end next to the bigger number, or you'll make an equals sign.
- Read the numbers for problem A. (Signal.) *315 (and) 296.*
- Is 315 more than, less than, or equal to 296? (Signal.) *More than.*
- 315 is more than 296, so do you make the sign with the bigger or smaller end next to 315? (Signal.) *Bigger.*
- Make the sign. Then make the sign for the rest of the problems. Put your pencil down when you've finished part 2.
  (Observe children and give feedback.)

b. Check your work. You'll read the numbers and tell about the sign for each problem.
- Read the numbers for problem B. (Signal.) *84 (and) 92.*
- Did you write an equation? (Signal.) *No.*
- Is the bigger or smaller end of the sign next to 84? (Signal.) *Smaller.*

c. Read the numbers for problem C. (Signal.) *602 (and) 602.*
- Did you write an equation? (Signal.) *Yes.*
- Read the equation. (Signal.) *602 equals 602.*

d. Read the numbers for problem D. (Signal.) *55 (and) 48.*
- Did you write an equation? (Signal.) *No.*
- Is the bigger or smaller end of the sign next to 55? (Signal.) *Bigger.*

e. Read the numbers for problem E. (Signal.) *93 (and) 101.*
- Did you write an equation? (Signal.) *No.*
- Is the bigger or smaller end of the sign next to 93? (Signal.) *Smaller.*

f. Read the numbers for problem F. (Signal.)
41 (and) 41.
- Did you write an equation? (Signal.) *Yes.*
- Read the equation. (Signal.) *41 equals 41.*
(Display:)   [90:5A]

a.	315 > 296
b.	84 < 92
c.	602 = 602
d.	55 > 48
e.	93 < 101
f.	41 = 41

Here are the signs you should have made for all of the problems.

EXERCISE 6: WORD PROBLEMS (COLUMNS)

a. Find part 3. ✔
(Teacher reference:)

You're going to write the symbols for word problems in columns and work them.
b. Touch where you'll write symbols for problem A. ✔
Listen to problem A: There were 57 bottles on the table. Jim took 23 of those bottles. How many bottles were still on the table?
- Listen again and write the symbols for the whole problem: There were 57 bottles on the table. Jim took 23 of those bottles.
(Observe children and give feedback.)
- Everybody, touch and read the problem for A. Get ready. (Signal.) *57 minus 23.*

c. Work problem A. Put your pencil down when you know how many bottles were still on the table.
(Observe children and give feedback.)
(Teacher reference:)

- Read the problem and the answer you wrote for A. Get ready. (Signal.) *57 – 23 = 34.*
- How many bottles were still on the table? (Signal.) *34.*
d. Touch where you'll write symbols for problem B. ✔
Listen to problem B: Jan had 16 cards. Then her dad gave her 26 more cards. How many cards did she end up with?
- Listen again and write the symbols for the whole problem: Jan had 16 cards. Then her dad gave her 26 more cards.
(Observe children and give feedback.)
- Everybody, touch and read the problem for B. Get ready. (Signal.) *16 plus 26.*
e. Work problem B. Put your pencil down when you know how many cards Jan ended up with.
(Observe children and give feedback.)
(Teacher reference:)

- Read the problem and the answer you wrote for B. Get ready. (Signal.) *16 + 26 = 42.*
- How many cards did Jan end up with? (Signal.) *42.*

f. Touch where you'll write the symbols for
    problem C. ✔
    Listen to problem C: There were 257 birds
    on a wire. Then 204 of those birds flew away.
    How many birds ended up on the wire?
•   Listen again and think of the number you write
    first: There were 257 birds on a wire.
•   Listen to the next part and think of the sign
    and the number for that part: Then 204 of
    those birds flew away.
•   Listen again and tell me the symbols for the
    whole problem: There were 257 birds on a
    wire. Then 204 of those birds flew away.
•   Tell me the symbols. Get ready. (Signal.)
    257 minus 204.
    (Repeat until firm.)
    Yes, you'll write 257 minus 204 equals.
g. Write 257 minus 204 equals and work problem
    C. Put your pencil down when you know how
    many birds ended up on the wire.
    (Observe children and give feedback.)
    (Teacher reference:)

•   Read the problem and the answer you wrote
    for B. Get ready. (Signal.) 257 – 204 = 53.
•   How many birds ended up on the wire?
    (Signal.) 53.

EXERCISE 7: MONEY
DOLLARS PLUS COINS

a. Find part 4 on worksheet 62. ✔
    (Teacher reference:)

    a. $12 + 🪙🪙🪙🪙
    b. $3 + 🪙🪙
    c. $5 + 🪙🪙🪙🪙

    These problems show dollar amounts and
    coins. You're going to figure out how much
    money there is for each problem.
•   Touch problem A. ✔
•   How many dollars are in problem A? (Signal.) 12.

•   What do you add for each nickel? (Signal.)
    5 cents.
•   You'll touch 12 dollars and say the amount.
    Then you'll touch and count for the nickels.
•   Touch and say the dollar amount. Twelve
    dollars. Count for the nickels. (Tap 4.) 12 dollars
    and 5 cents, 12 dollars and 10 cents, 12 dollars
    and 15 cents, 12 dollars and 20 cents.
    (Repeat until firm.)
•   How much is problem A worth? (Signal.)
    12 dollars and 20 cents.
b. Touch problem B. ✔
•   How many dollars are in problem B? (Signal.) 3.
•   You'll touch 3 dollars and say the amount.
    Then you'll touch and count for the quarters.
•   Touch and say the dollar amount. 3 dollars.
    Count for the quarters. (Tap 2.) 3 dollars and
    25 cents, 3 dollars and 50 cents.
    (Repeat until firm.)
•   How much is problem B worth? (Signal.)
    3 dollars and 50 cents.
c. Touch problem C. ✔
•   How many dollars are in problem C? (Signal.) 5.
•   You'll touch 5 dollars and say the amount.
    Then you'll touch and count for the dimes.
    Then touch and count for the pennies.
•   Touch and say the dollar amount. 5 dollars.
    Count for the dimes. (Tap 2.) 5 dollars and
    10 cents, 5 dollars and 20 cents. Count for
    the pennies. (Tap 2.) 5 dollars and 21 cents,
    5 dollars and 22 cents.
    (Repeat until firm.)
•   How much is problem C worth? (Signal.)
    5 dollars and 22 cents.
d. Go back and work problem A. Write an
    equals, a dollar sign, and the symbols for the
    dollars and cents.
    (Observe children and give feedback.)
•   Everybody, what did you write for problem A?
    (Signal.) (Equals) 12 dollars and 20 cents.
e. Write an equals and the dollar and cents
    amount for problem B.
    (Observe children and give feedback.)
•   Everybody, what did you write for problem B?
    (Signal.) (Equals) 3 dollars and 50 cents.
f. Write an equals and the dollar and cents
    amount for problem C.
    (Observe children and give feedback.)
•   Everybody, what did you write for problem C?
    (Signal.) (Equals) 5 dollars and 22 cents.

EXERCISE 8: FACTS
SUBTRACTION

a. Find part 5. ✔
    (Teacher reference:)

a. $6 - 2$	e. $12 - 6$	i. $8 - 4$
b. $6 - 3$	f. $9 - 6$	j. $11 - 5$
c. $14 - 10$	g. $10 - 5$	k. $11 - 9$
d. $7 - 3$	h. $10 - 6$	l. $8 - 5$

b. These are minus problems you'll work. Think
    of the number families for each problem and
    write the answers. Put your pencil down when
    you've completed the equations in part 5.
    (Observe children and give feedback.)

c. Check your work. You'll read the equation for
    each problem.

• Problem A. (Signal.) $6 - 2 = 4$.
• (Repeat for:) B, $6 - 3 = 3$; C, $14 - 10 = 4$;
  D, $7 - 3 = 4$; E, $12 - 6 = 6$; F, $9 - 6 = 3$;
  G, $10 - 5 = 5$; H, $10 - 6 = 4$; I, $8 - 4 = 4$;
  J, $11 - 5 = 6$; K, $11 - 9 = 2$; L, $8 - 5 = 3$.

EXERCISE 9: INDEPENDENT WORK

a. Turn to the other side of worksheet 90 and find
    part 6. ✔
    (Teacher reference:)

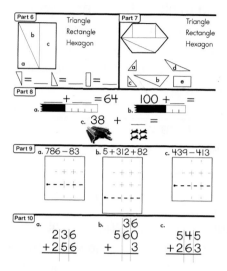

You'll circle the name for the whole shape.
Then you'll complete each equation by writing
the letter to show which part is it.
In part 7, you'll circle the name for the whole
shape. Then you'll write the letter for each part.
In part 8, you'll complete the equations for the
rulers and the equation for the dogs.
In part 9, you'll write problems in columns and
work them.
In part 10, to work the problems, you'll have
to carry.

b. Complete worksheet 90.
    (Observe children and mark incorrect responses
    on children's worksheets as you give feedback.)

Mastery Test 9

Teacher Presentation

a. Find Test 9 in your test booklet. ✔
- Touch part 1. ✔
  (Teacher reference:)

  You're going to write symbols for problems that compare two things. Problem A tells about two lines. You're going to write the problem and figure out how much longer one line is than the other line.
- Touch where you'll write the column problem for A. ✔
- Listen to problem A: The red line is 89 inches long. The blue line is 38 inches long. How much longer is the red line than the blue line?
- Listen to problem A again and write the symbols for the problem as I say it: The red line is 89 inches long. The blue line is 38 inches long. How much longer is the red line than the blue line?
- Work problem A. Remember to make the equals bar. Put your pencil down when you've written the number for how many inches longer the red line is than the blue line.
  (Observe children.)
b. Touch where you'll write the column problem for B. ✔
- Listen to problem B: Mr. Jones weighs 249 pounds. Mrs. Jones weighs 142 pounds. How much lighter is Mrs. Jones than Mr. Jones?
- Listen to problem B again and write the symbols for the problem as I say it: Mr. Jones weighs 249 pounds. Mrs. Jones weighs 142 pounds. How much lighter is Mrs. Jones than Mr. Jones?
- Work problem B. Remember to make the equals bar. Put your pencil down when you've written the number for how many pounds lighter Mrs. Jones is than Mr. Jones.
  (Observe children.)
c. Touch where you'll write the column problem for C. ✔
- Listen to problem C: Rita and Donna are friends. Rita is 86 years old. Donna is 54 years old. How much older is Rita than Donna?
- Listen to problem C again and write the symbols for the problem as I say it: Rita and Donna are friends. Rita is 86 years old. Donna is 54 years old. How much older is Rita than Donna?
- Work problem C. Remember to write the equals bar. Put your pencil down when you've written the number for how much older Rita is than Donna.
  (Observe children.)
d. Touch part 2 on test sheet 9. ✔
  (Teacher reference:)

a. ___.___ b. ___.___ c. ___.___ d. ___.___ e. ___.___

  You're going to write dollars and cents amounts. Some amounts will show only cents with no dollars. For each amount, the dot is already shown but you'll write the dollar sign.
- The amount for A is 4 dollars and 2 cents. What amount? (Signal.) *4 dollars and 2 cents.*
- Write 4 dollars and 2 cents for A.
  (Observe children.)
e. The amount for B is 15 dollars and 60 cents. What amount? (Signal.) *15 dollars and 60 cents.*
- Write 15 dollars and 60 cents for B.
  (Observe children.)
f. The amount for C is 18 dollars and 8 cents. What amount? (Signal.) *18 dollars and 8 cents.*
- Write 18 dollars and 8 cents for C.
  (Observe children.)
g. The amount for D is 6 cents. What amount? (Signal.) *6 cents.*
- Write 6 cents for D.

h. The amount for E is 9 cents. What amount? (Signal.) *9 cents.*

•   Write 9 cents for E. (Observe children.)

i. Touch part 3 on test sheet 9. ✔ (Teacher reference:)

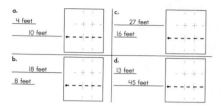

For each problem, you'll write the column problem to figure out how much longer one line is than the other line.

j. Touch part 4 on test sheet 9. ✔ (Teacher reference:)

a. 7   5   e. 3   2
b. 8   9   f. 19   20
c. 6   3   g. 57   48
d. 4   4   h. 16   16

For each problem, you'll make the sign to show which number is more or if the sides are equal.

•   Read the numbers for A to yourself. If the numbers are equal, make an equals sign between them. If the sides are not equal, make the sign with the bigger end next to the bigger number. (Observe children.)

•   Read the numbers for B to yourself. If the numbers are equal, make an equals sign between them. If the sides are not equal, make the sign with the bigger end next to the bigger number. (Observe children.)

•   Read the numbers for C to yourself. If the numbers are equal, make an equals sign between them. If the sides are not equal, make the sign with the bigger end next to the bigger number. (Observe children.)

•   Read the numbers for D to yourself. If the numbers are equal, make an equals sign between them. If the sides are not equal, make the sign with the bigger end next to the bigger number. (Observe children.)

•   For the rest of the problems in part 4, if the numbers are equal make an equals sign between them. If the sides are not equal, make the sign with the bigger end next to the bigger number. Put your pencil down when you've completed the signs for all of the problems in part 4. (Observe children.)

k. Turn to the other side of Test 9 and touch part 5. ✔ (Teacher reference:)

a.   1 1
    +   5 9

b.   3 7
    +   4 2

c.   2 9 7
        3 1 2
    +   7 0

d.   4 1 2
              8
    + 1 5 3

e.   8 1
        1 2 4
    + 5 6 2

You'll work the column problems in part 5.

l. Touch part 6. ✔ (Teacher reference:)

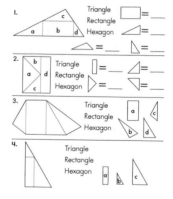

Next to each figure are the names: triangle, rectangle, and hexagon. You'll circle the word for the shape for each whole figure. Then, for figures 1 and 2, you'll write the letter to complete the equation for each part. For figures 3 and 4, you'll write the letter in each part for the whole figure.

m. Touch part 7. ✔

(Teacher reference:)

a. $4 + 3$
b. $4 - 3$
c. $4 + 6$
d. $8 - 4$
e. $9 - 3$
f. $4 + 4$
g. $7 - 4$
h. $10 - 4$
i. $7 - 3$

You'll complete the equation for each problem in part 7.

• Your turn: Work the rest of the problems on Test 9.

(Direct students where to put their assessment books when they are finished.)

Scoring Notes

a. Collect test booklets. Use the Answer Key and Passing Criteria Table to score the tests.

Passing Criteria Table — Mastery Test 9			
Part	Score	Possible Score	Passing Score
1	4 for each problem	12	8
2	2 for each dollar amount	10	8
3	3 for each problem	12	8
4	2 for each problem	16	14
5	4 for each problem	20	16
6	3 for each problem	12	9
7	2 for each equation	18	16
	Total	100	

b. Complete the Mastery Test 9 Remedy Summary Sheet to determine whether group remedies are needed. Reproducible Remedy Summary Sheets are at the back of the Answer Key and at the back of the *Teacher's Guide.*

• If ¼ or more of the students did not pass a test part, present the remedy for that part before beginning Lesson 91. The Remedy Table follows and is also at the end of the Mastery Test 9 Answer Key. Remedies worksheets follow Mastery Test 9 in the *Student Assessment Book.*

Remedy Table — Mastery Test 9				
Part	Test Items	Remedy		Student Material Remedies Worksheet
		Lesson	Exercise	
1	Word Problems (Comparison)	83	8	Part A
		85	9	Part B
		88	2	—
		90	3	—
2	Writing Dollar Amounts	80	2	—
		81	2	—
		84	6	Part C
		87	9	Part D
3	Comparison (Length)	79	2	—
		80	6	Part E
		81	7	Part F
		82	6	Part G
4	Greater Than, Less Than, Equal To	81	4	—
		83	5	Part H
		88	7	Part I
5	Column Addition (Carrying)	82	7	Part J
		86	3	—
		87	6	Part K
		89	6	Part L
6	Shapes (Composition/ Decomposition)	76	5	Part M
		78	5	Part N
		87	5	Part O
7	Facts (Mix)	80	1	—
		82	1	—
		88	1	—
		88	6	Part P
		88	8	Part Q

Retest

Retest individual students on any part failed.

Lessons 91–95 Planning Page

	Lesson 91	Lesson 92	Lesson 93	Lesson 94	Lesson 95
Student Learning Objectives	**Exercises** 1. Find missing numbers in number families **2. Tell time from a digital display** 3. Count forward or backward from a given number **4. Say two addition and two subtraction facts for number families with 2- and 3-digit numbers** **5. Solve 3-digit addition problems with carrying multiple digits** 6. Solve addition and subtraction facts 7. Write the greater than, less than, and equal signs (>, <, =) to compare numbers 8. Count and write the value of coins added to dollar amounts 9. Solve subtraction facts 10. Complete work independently	**Exercises** 1. Find missing numbers in number families 2. Count forward or backward from a given number 3. Say two addition and two subtraction facts for number families with 2- and 3-digit numbers 4. Tell time from a digital display 5. Solve addition and subtraction facts **6. Write the value of coins and bills** 7. Solve comparison word problems 8. Solve 3-digit addition problems with carrying multiple digits 9. Write the symbols for word problems in columns and solve 10. Complete work independently	**Exercises** 1. Tell time from a digital display **2. Count by twos** 3. Find missing numbers in number families **4. Order objects by length or height** 5. Solve addition and subtraction facts **6. Find missing numbers in number families with 2-digit numbers** **7. Count and write the value of coins and bills** 8. Solve 3-digit addition problems with carrying 9. Complete work independently	**Exercises** 1. Count by twos 2. Say addition and subtraction facts 3. Count forward or backward from a given number 4. Order objects by length or height 5. Solve addition facts 6. Solve 3-digit addition problems with carrying **7. Find and write missing numbers in number families with 2-digit or 3-digit numbers** **8. Tell time from a digital display and write digital time** 9. Solve subtraction facts 10. Complete work independently	**Exercises** 1. Say addition and subtraction facts **2. Identify 2-dimensional faces on 3-dimensional objects** 3. Count by twos 4. Order objects by length or height 5. Solve subtraction facts 6. Write the symbols for word problems in columns and solve 7. Find and write missing numbers in number families with 2-digit or 3-digit numbers **8. Write digital time** 9. Solve addition facts 10. Complete work independently
Common Core State Standards for Mathematics					
1.OA 3	✔	✔	✔	✔	✔
1.OA 5	✔	✔	✔		✔
1.OA 6	✔	✔	✔	✔	✔
1.OA 8	✔	✔	✔	✔	✔
1.NBT 1	✔	✔	✔	✔	✔
1.NBT 4	✔	✔	✔	✔	✔
1.MD 1			✔	✔	✔
1.MD 3	✔	✔	✔	✔	✔
1.G 1					✔
1.G 2	✔	✔	✔	✔	✔
Teacher Materials	Presentation Book 3, Board Displays CD or chalkboard				
Student Materials	Workbook 2, Pencil				
Additional Practice	• Student Practice Software: Block 4: Activity 1 (1.G.1), Activity 2 (1.NBT.1), Activity 3 (1.NBT.1), Activity 4 (1.NBT.3), Activity 5 (1.NBT.1), Activity 6 (1.NBT.1) • Math Fact Worksheets 49–50 (After Lesson 91), 51 (After Lesson 93), 52–56 (After Lesson 94)				
Mastery Test					

Lesson  91

EXERCISE 1: NUMBER FAMILIES
MISSING NUMBER IN FAMILY

a. (Display:) [91:1A]

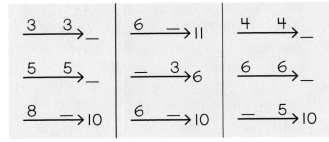

You're going to say the problem for the missing number in each family.

b. (Point to $\xrightarrow[3]{3}$_.) Say the problem for the missing number. Get ready. (Touch.) *3 plus 3.*
• What's 3 plus 3? (Signal.) *6.*
c. (Point to $\xrightarrow[5]{5}$_.) Say the problem for the missing number. (Touch.) *5 plus 5.*
• What's 5 plus 5? (Signal.) *10.*
d. (Repeat the following tasks for the remaining families:)

(Point to __.) Say the problem for the missing number.		What's __?	
$\xrightarrow{8}$10	10 – 8	10 – 8	2
$\xrightarrow{6}$11	11 – 6	11 – 6	5
$\xrightarrow{3}$6	6 – 3	6 – 3	3
$\xrightarrow{6}$10	10 – 6	10 – 6	4
$\xrightarrow[4]{4}$_	4 + 4	4 + 4	8
$\xrightarrow[6]{6}$_	6 + 6	6 + 6	12
$\xrightarrow{5}$10	10 – 5	10 – 5	5

(Repeat for families that were not firm.)

EXERCISE 2: TELLING TIME
READING DIGITAL DISPLAY   REMEDY

a. (Display:) [91:2A]

These are clocks that show hours and minutes.

• (Point to **12:30.**) The hour number for this clock is 12. The minute number is 30. What's the hour number for this clock? (Signal.) *12.*
• What's the minute number for this clock? (Signal.) *30.*
b. (Point to **4:00.**) The hour number for this clock is 4. The minute number is zero. What's the hour number for this clock? (Signal.) *4.*
• What's the minute number for this clock? (Signal.) *Zero.*
c. (Point to **10:00.**) What's the hour number for this clock? (Signal.) *10.*
• What's the minute number? (Signal.) *Zero.*
d. (Repeat the following tasks with the remaining clocks:)

(Point to __.) What's the hour number for this clock?		What's the minute number?
5:30	5	30
11:30	11	30
7:00	7	Zero

(Repeat for clocks that were not firm.)
e. These clocks are easy to read. First you read the hour number. What do you read first? (Signal.) *The hour number.*
• Then, if the minute number is zero, you say **o'clock.** What do you say if the minute number is zero? (Signal.) *O'clock.*
• If the minute number is 30, you say **30.** What do you say if the minute number is 30? (Signal.) *30.*
f. (Point to **12:30.**) Look at the minutes for this clock. Is the minute number zero? (Signal.) *No.*
• So do you say o'clock after the hour number? (Signal.) *No.*
• What do you say for the minutes? (Signal.) *30.*
g. (Point to **4:00.**) Is the minute number for this clock zero? (Signal.) *Yes.*
• So do you say o'clock after the hour number? (Signal.) *Yes.*
• What do you say for the minutes? (Signal.) *O'clock.*

h. (Repeat the following tasks for the remaining clocks:)

(Point to __.) Is the minute number for this clock zero?	So do you say o'clock after the hour number?	What do you say for the minutes?
10:00	Yes	Yes	O'clock
5:30	No	No	30
11:30	No	No	30
7:00	Yes	Yes	O'clock

(Repeat for clocks that were not firm.)

i. (Point to **12:30**.) I'm going to read the time for each clock. (Touch clocks as you say:) 12 thirty, 4 o'clock, 10 o'clock, 5 thirty, 11 thirty, 7 o'clock.
Your turn to read the time for each clock.

- (Point to **12:30**.) What time? (Touch.) *12 thirty.*
- (Repeat for remaining clocks: 4 o'clock, 10 o'clock, 5 thirty, 11 thirty, 7 o'clock.) (Repeat clocks that were not firm.)

=== **INDIVIDUAL TURNS** ===

I'll call on individual students to read the clocks.

- (Point to **11:30**.) What's the time for this clock? (Call on a student.) *11 thirty.*
- (Point to **7:00**.) What's the time for this clock? (Call on a student.) *7 o'clock.*
- (Point to **10:00**.) What's the time for this clock? (Call on a student.) *10 o'clock.*
- (Point to **12:30**.) What's the time for this clock? (Call on a student.) *12 thirty.*

EXERCISE 3: MIXED COUNTING

a. Let's do some counting.
- Start with 50 and count by fives to 100. Get 50 going. *Fiftyyy.* Count. (Tap.) *55, 60, 65, 70, 75, 80, 85, 90, 95, 100.*
- Start with 200 and plus tens to 300. Get 200 going. *Two huuundred.* Plus tens. (Tap.) *210, 220, 230, 240, 250, 260, 270, 280, 290, 300.*
- Start with 100 and count by hundreds to 1000. Get 100 going. *One huuundred.* Count. (Tap.) *200, 300, 400, 500, 600, 700, 800, 900, 1000.*
- Start with 400 and count by 25s to 500. Get 400 going. *Four huuundred.* Count. (Tap.) *425, 450, 475, 500.*

b. Start with 80 and count backward by ones to 70. Get 80 going. *Eightyyy.* Count backward. (Tap.) *79, 78, 77, 76, 75, 74, 73, 72, 71, 70.*

c. Start with 11 and plus tens to 71. Get 11 going. *Elevennn.* Plus tens. (Tap.) *21, 31, 41, 51, 61, 71.*
- Start with 19 and plus tens to 79. Get 19 going. *Nineteeen.* Plus tens. (Tap.) *29, 39, 49, 59, 69, 79.*
- Start with 12 and plus tens to 72. Get 12 going. *Tweeelve.* Plus tens. (Tap.) *22, 32, 42, 52, 62, 72.*

d. Start with 100 and count backward by tens. Get 100 going. *One huuundred.* Count backward. (Tap.) *90, 80, 70, 60, 50, 40, 30, 20, 10.*

e. You have 85. When you minus fives, what's the next number? (Signal.) *80.*
- You have 85. When you plus tens, what's the next number? (Signal.) *95.*
- You have 85. When you minus ones, what's the next number? (Signal.) *84.*

EXERCISE 4: NUMBER FAMILIES
2-DIGIT AND 3-DIGIT NUMBERS

a. (Display:) [91:4A]

Here's a number family with big numbers. The small numbers in this family are 14 and 32.
- What are the small numbers in this family? (Signal.) *14 and 32.*
- What's the big number? (Signal.) *46.*
(Repeat until firm.)
Listen: I'm going to say a plus equation for this family.
So I start with a small number.
- Listen: 14 plus 32 equals 46. Say that equation. (Signal.) *14 + 32 = 46.*
I'm going to say the other plus fact for this family.
- Tell me what number I start with. (Signal.) *32.*
Here's the fact: 32 plus 14 equals 46.
- Now we'll say the minus equations. Do we start with the big number or a small number? (Signal.) *The big number.*
- Say the minus equation that goes backward down the arrow. Get ready. (Signal.) *46 − 32 = 14.*
- Say the other minus equation. (Signal.) *46 − 14 = 32.*
(Repeat until firm.)

b. Let's say those equations again.
- Say the equation that goes forward along the arrow. (Signal.) *14 + 32 = 46.*
- Say the other plus equation. (Signal.) *32 + 14 = 46.*
- Say the minus equation that goes backward down the arrow. (Signal.) *46 – 32 = 14.*
- Say the other minus equation. Get ready. (Signal.) *46 – 14 = 32.*

c. (Display:)   [91:4B]

$$\underline{100 \quad\quad 62}_{\,\searrow} 162$$

- What's the first small number in this family? (Signal.) *100.*
- What's the other small number? (Signal.) *62.*
- What's the big number? (Signal.) *162.*
- Say the equation that starts with the first small number. Get ready. (Signal.) *100 + 62 = 162.*
- Say the other plus equation. Get ready. (Signal.) *62 + 100 = 162.*
- Say the equation that goes backward along the arrow. Get ready. (Signal.) *162 – 62 = 100.*
- Say the other minus equation. Get ready. (Signal.) *162 – 100 = 62.*
(Repeat until firm.)

EXERCISE 5: COLUMN ADDITION
CARRYING (MULTIPLE DIGITS)   REMEDY

a. (Display:)   [91:5A]

$$\begin{array}{r} 471 \\ +329 \\ \hline \end{array}$$

You have to carry to work this problem.
- Read the problem. Get ready. (Signal.) *471 plus 329.*
- Read the problem for the ones. (Signal.) *1 plus 9.*
- What's the answer? (Signal.) *10.*
- The answer has two digits. What do we write in the tens column? (Signal.) *1.*
- What do we write in the ones column? (Signal.) *Zero.*
(Add to show:)   [91:5B]

$$\begin{array}{r} \overset{1}{4}71 \\ +329 \\ \hline 0 \end{array}$$

- Read the problem for the tens column. Get ready. (Signal.) *1 plus 7 plus 2.*
- What's the answer? (Signal.) *10.*
- That answer has two digits. What do we write in the hundreds column? (Signal.) *1.*
- What do we write in the tens column? (Signal.) *Zero.*
(Add to show:)   [91:5C]

$$\begin{array}{r} \overset{1}{4}\overset{1}{7}1 \\ +329 \\ \hline 00 \end{array}$$

- Read the problem for the hundreds. Get ready. (Signal.) *1 plus 4 plus 3.*
- What's the answer? (Signal.) *8.*
(Add to show:)   [91:5D]

$$\begin{array}{r} \overset{1}{4}\overset{1}{7}1 \\ +329 \\ \hline 800 \end{array}$$

- Read the whole problem and the answer. Get ready. (Signal.) *471 + 329 = 800.*

b. (Distribute unopened workbooks to children.)
- Open your workbook to Lesson 91 and find part 1.
(Observe children and give feedback.)
(Teacher reference:)   R | Part M

a. $\begin{array}{r} 1\,86 \\ +7\,16 \\ \hline \end{array}$

b. $\begin{array}{r} 2\,96 \\ +4\,75 \\ \hline \end{array}$

- Touch and read problem A. Get ready. (Signal.) *186 plus 716.*
You have to carry more than one time to work this problem. Remember, if the answer has two digits you carry to the next column.
- Work problem A. Put your pencil down when you're finished.
(Observe children and give feedback.)

c. Check your work.
- Read problem A and the answer. (Signal.)
  186 + 716 = 902.
  (Display:) W   [91:5E]

$$\begin{array}{r} \overset{1}{}\overset{1}{} \\ 1\,8\,6 \\ +\ 7\,1\,6 \\ \hline 9\,0\,2 \end{array}$$

  Here's what you should have.
  The answer is 902.
d. Touch and read problem B. Get ready.
  (Signal.) *296 plus 475.*
- Work problem B. Put your pencil down when
  you're finished.
  (Observe children and give feedback.)
e. Check your work.
- Read problem B and the answer. (Signal.)
  296 + 475 = 771.
  (Display:) W   [91:5F]

$$\begin{array}{r} \overset{1}{}\overset{1}{} \\ 2\,9\,6 \\ +\ 4\,7\,5 \\ \hline 7\,7\,1 \end{array}$$

  Here's what you should have.
  The answer is 771.

EXERCISE 6: FACTS
PLUS/MINUS MIX

a. Find part 2 on worksheet 91. ✔
  (Teacher reference:)

a. 12−6	f. 6−4
b. 5+5	g. 3+3
c. 3+5	h. 5+6
d. 9−2	i. 4+4
e. 4+6	j. 10−5

- Touch and read problem A. Get ready. (Signal.)
  12 minus 6.
- Is the answer the big number or a small
  number? (Signal.) *A small number.*
- What's 12 minus 6? (Signal.) *6.*

b. (Repeat the following tasks for problems B
  through J:)

Touch and read problem __.	Is the answer the big number or a small number?	What's __?	
B	*The big number.*	5 + 5	10
C	*The big number.*	3 + 5	8
D	*A small number.*	9 − 2	7
E	*The big number.*	4 + 6	10
F	*A small number.*	6 − 4	2
G	*The big number.*	3 + 3	6
H	*The big number.*	5 + 6	11
I	*The big number.*	4 + 4	8
J	*A small number.*	10 − 5	5

(Repeat problems that were not firm.)
c. Write the answer to each problem. Put your
  pencil down when you've completed the
  equations for part 2.
  (Observe children and give feedback.)
d. Check your work.
- Touch and read fact A. (Signal.) *12 − 6 = 6.*
- (Repeat for:) B, *5 + 5 = 10*; C, *3 + 5 = 8*;
  D, *9 − 2 = 7*; E, *4 + 6 = 10*; F, *6 − 4 = 2*;
  G, *3 + 3 = 6*; H, *5 + 6 = 11*; I, *4 + 4 = 8*;
  J, *10 − 5 = 5.*

EXERCISE 7: GREATER THAN/LESS THAN/EQUALS SIGN

a. Find part 3. ✔
  (Teacher reference:)

  a. 62 27
  b. 13 13
  c. 354 435
  d. 624 641
  e. 207 207
  f. 81 90

  For each problem, you'll make the sign with
  the bigger end next to the bigger number, or
  you'll make an equals sign. Put your pencil
  down when you've finished part 3.
  (Observe children and give feedback.)
b. Check your work. You'll read the numbers and
  tell about the sign for each problem.
- Read the numbers for problem A. (Signal.)
  62 (and) 27.
- Did you write an equation? (Signal.) *No.*
- Is the bigger or smaller end of the sign next to
  62? (Signal.) *Bigger.*
c. Read the numbers for problem B. (Signal.)
  13 (and) 13.
- Did you write an equation? (Signal.) *Yes.*
- Read the equation. (Signal.) *13 equals 13.*

d. Read the numbers for problem C. (Signal.)
354 (and) 435.
- Did you write an equation? (Signal.) *No.*
- Is the bigger or smaller end next to 354?
(Signal.) *Smaller.*
e. Read the numbers for problem D. (Signal.)
624 (and) 641.
- Did you write an equation? (Signal.) *No.*
- Is the bigger or smaller end next to 624?
(Signal.) *Smaller.*
f. Read the numbers for problem E. (Signal.)
207 (and) 207.
- Did you write an equation? (Signal.) *Yes.*
- Read the equation. (Signal.) *207 equals 207.*
g. Read the numbers for problem F. (Signal.)
81 (and) 90.
- Did you write an equation? (Signal.) *No.*
- Is the bigger or smaller end next to 81?
(Signal.) *Smaller.*
(Display:)   [91:7A]

a.   62 > 27

b.   13 = 13

c.   354 < 435

d.   624 < 641

e.   207 = 207

f.   81 < 90

Here are the signs you should have made for
all of the problems.

EXERCISE 8: MONEY
DOLLARS PLUS COINS   REMEDY

a. Find part 4 on worksheet 91. ✔
(Teacher reference:)   R Part I

a. $39 + ⊚⊚⊚
b. $72 + ⊚⊚⊚⊚
c. $15 + ⊚⊚ ⊚⊚
d. $28 + ⊚⊚⊚⊚⊚

These problems show a dollar amount and
coins. You're going to figure out how much
money there is for each problem.
- Touch problem A. ✔
- How many dollars are in problem A?
(Signal.) *39.*
- What do you add for each quarter? (Signal.)
25 cents.

- You'll touch 39 dollars and say the amount.
Then you'll touch and count for the quarters.
- Touch and say the dollar amount. *39 dollars.*
Count for the quarters. (Tap 3.) *39 dollars and
25 cents, 39 dollars and 50 cents, 39 dollars
and 75 cents.*
(Repeat until firm.)
- How much is problem A worth? (Signal.)
39 dollars and 75 cents.
b. Touch problem B. ✔
- How many dollars are in problem B?
(Signal.) *72.*
- You'll touch 72 dollars and say the amount.
Then you'll touch and count for the dimes.
Then you'll touch and count for the pennies.
- Touch and say the dollar amount. *72 dollars.*
Count for the dimes. (Tap 2.) *72 dollars and
10 cents, 72 dollars and 20 cents.* Count for
the pennies. (Tap 2.) *72 dollars and 21 cents,
72 dollars and 22 cents.*
(Repeat until firm.)
- How much is problem B worth? (Signal.)
72 dollars and 22 cents.
c. Touch problem C. ✔
- How many dollars are in problem C?
(Signal.) *15.*
- You'll touch 15 dollars and say the amount.
Then you'll touch and count for the quarters.
Then touch and count for the dimes.
- Touch and say the dollar amount. *15 dollars.*
Count for the quarters. (Tap 2.) *15 dollars and
25 cents, 15 dollars and 50 cents.* Count for
the dimes. (Tap 2.) *15 dollars and 60 cents,
15 dollars and 70 cents.*
(Repeat until firm.)
- How much is problem C worth? (Signal.)
15 dollars and 70 cents.
d. Touch problem D. ✔
- How many dollars are in problem D?
(Signal.) *28.*
- You'll touch 28 dollars and say the amount.
Then you'll touch and count for the nickel.
Then touch and count for the pennies.
- Touch and say the dollar amount. *28 dollars.*
Count for the nickel. (Tap.) *28 dollars and
5 cents.* Touch and count for the pennies.
(Tap 4.) *28 dollars and 6 cents, 28 dollars
and 7 cents, 28 dollars and 8 cents,
28 dollars and 9 cents.*
(Repeat until firm.)
- How much is problem D worth? (Signal.)
28 dollars and 9 cents.

e. Go back and work problem A. Write an equals, a dollar sign, and the symbols for the dollars and cents.
(Observe children and give feedback.)
• Everybody, what did you write for problem A? (Signal.) *(Equals) 39 dollars and 75 cents.*
f. Write an equals, a dollar sign, and the dollar and cent amount for problem B.
(Observe children and give feedback.)
• Everybody, what did you write for problem B? (Signal.) *(Equals) 72 dollars and 22 cents.*
g. Write an equals, a dollar sign, and the dollar and cent amount for problem C.
(Observe children and give feedback.)
• Everybody, what did you write for problem C? (Signal.) *(Equals) 15 dollars and 70 cents.*
h. Write an equals, a dollar sign, and the dollar and cent amount for problem D.
(Observe children and give feedback.)
• Everybody, what did you write for problem D? (Signal.) *(Equals) 28 dollars and 9 cents.*

EXERCISE 9: FACTS

SUBTRACTION

a. Find part 5. ✔
(Teacher reference:)

e. $9 - 6$
a. $8 - 4$ f. $10 - 5$
b. $10 - 6$ g. $8 - 2$
c. $6 - 3$ h. $13 - 10$
d. $11 - 5$ i. $10 - 8$

These are minus problems you'll work. Think of the number families for each problem and write the answers. Put your pencil down when you've completed the equations in part 5.
(Observe children and give feedback.)
b. Check your work. You'll read the equation for each problem.
• Problem A. (Signal.) $8 - 4 = 4.$
• (Repeat for:) B, $10 - 6 = 4$; C, $6 - 3 = 3$; D, $11 - 5 = 6$; E, $9 - 6 = 3$; F, $10 - 5 = 5$; G, $8 - 2 = 6$; H, $13 - 10 = 3$; I, $10 - 8 = 2.$

EXERCISE 10: INDEPENDENT WORK

a. Find part 6. ✔
(Teacher reference:)

a. ___ + ___ = 103

b. $116 +$ ___ =

You'll complete the equations for the rulers.
b. Turn to the other side of your worksheet and find part 7. ✔
(Teacher reference:)

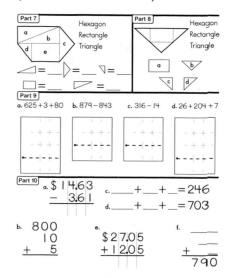

You'll circle the name for the whole shape. Then you'll complete each equation by writing the letter to show which part it is.
In part 8, you'll circle the name for the whole shape. Then you'll write the letter for each part.
In part 9, you'll write each problem in a column and work it.
In part 10, you'll complete the place-value equations and work the money problems.
c. Complete worksheet 91. Remember to work part 6 on the other side of your worksheet.
(Observe children and mark incorrect responses on children's worksheets as you give feedback.)

Lesson

EXERCISE 1: NUMBER FAMILIES
MISSING NUMBER IN FAMILY

a. (Display:) [92:1A]

5 5 →_	= 4 →8	6 →10
= 3 →6	6 3 →_	8 →10
4 →7	3 →5	3 3 →_

You're going to say the problem for the missing number in each family.

b. (Point to 5 5 →_.) Say the problem for the missing number. Get ready. (Touch.) *5 plus 5.*
• What's 5 plus 5? (Signal.) *10.*

c. (Point to = 3 →6.) Say the problem for the missing number. (Touch.) *6 minus 3.*
• What's 6 minus 3? (Signal.) *3.*

d. (Repeat the following tasks for the remaining families:)

(Point to __.) Say the problem for the missing number.		What's __?	
4 →7	7 – 4	7 – 4	3
= 4 →8	8 – 4	8 – 4	4
6 3 →_	6 + 3	6 + 3	9
3 →5	5 – 3	5 – 3	2
6 →10	10 – 6	10 – 6	4
8 →10	10 – 8	10 – 8	2
3 3 →_	3 + 3	3 + 3	6

(Repeat for families that were not firm.)

EXERCISE 2: MIXED COUNTING

a. Let's do some counting.
• Count by 25s to 100. Get ready. (Tap.) *25, 50, 75, 100.*
• Count by hundreds to 1000. Get 100 going. *One huuundred.* Count. (Tap.) *200, 300, 400, 500, 600, 700, 800, 900, 1000.*
• Start with 35 and plus tens to 75. Get 35 going. *Thirty-fiiive.* Plus tens. (Tap.) *45, 55, 65, 75.*
• Start with 36 and plus tens to 76. Get 36 going. *Thirty-siiix.* Plus tens. (Tap.) *46, 56, 66, 76.*

b. Start with 80 and count backward by ones to 70. Get 80 going. *Eightyyy.* Count backward. (Tap.) *79, 78, 77, 76, 75, 74, 73, 72, 71, 70.*

c. Start with 500 and count by 25s to 600. Get 500 going. *Five huuundred.* Count. (Tap.) *525, 550, 575, 600.*

d. Start with 100 and count backward by tens. Get 100 going. *One huuundred.* (Tap.) *90, 80, 70, 60, 50, 40, 30, 20, 10.*

e. You have 75. When you plus 25s, what's the next number? (Signal.) *100.*
• You have 75. When you minus ones, what's the next number? (Signal.) *74.*
• You have 75. When you plus ones, what's the next number? (Signal.) *76.*
• You have 75. When you plus fives, what's the next number? (Signal.) *80.*
• You have 75. When you plus tens, what's the next number? (Signal.) *85.*

EXERCISE 3: NUMBER FAMILIES
2-DIGIT AND 3-DIGIT NUMBERS   REMEDY

a. (Display:) [92:3A]

$$\underrightarrow{15 \quad 62}\,77$$

Here's a number family with 2-digit numbers.
• What's the first small number in this family? (Signal.) *15.*
• What's the other small number? (Signal.) *62.*
• What's the big number? (Signal.) *77.*

b. You're going to say four equations for this family.
• Start with the first small number and say the equation. (Signal.) *15 + 62 = 77.*
• Say the other plus equation. (Signal.) *62 + 15 = 77.*
• Say the equation that goes backward along the arrow. (Signal.) *77 – 62 = 15.*
• Say the other minus equation. (Signal.) *77 – 15 = 62.*

c. (Display:) [92:3B]

$$\underline{26 \quad 170}_{\,\longrightarrow\,} 196$$

- What's the first small number in this family? (Signal.) *26.*
- What's the other small number? (Signal.) *170.*
- What's the big number? (Signal.) *196.*
d. You're going to say four equations for this family.
- Start with the first small number and say the equation. (Signal.) *26 + 170 = 196.*
- Say the other plus equation. (Signal.) *170 + 26 = 196.*
- Say the equation that goes backward along the arrow. (Signal.) *196 – 170 = 26.*
- Say the other minus equation. (Signal.) *196 – 26 = 170.*

EXERCISE 4: TELLING TIME
READING DIGITAL DISPLAY

a. (Display:) [92:4A]

Last time you learned how to read the time for clocks that showed zero minutes or 30 minutes.
- First you read the hour number. What do you read first? (Signal.) *The hour number.*
- Then, if the minute number is zero, you say o'clock. What do you say if the minute number is zero? (Signal.) *O'clock.*
- If the minute number is 30, you say 30. What do you say if the minute number is 30? (Signal.) *30.*
b. (Point to **9:00.**) Look at the minutes for this clock. Is the minute number zero? (Signal.) *Yes.*
- So do you say o'clock after the hour number? (Signal.) *Yes.*
- What do you say for the minutes? (Signal.) *O'clock.*
- Read the time for this clock. (Signal.) *9 o'clock.*

c. (Point to **2:30.**) Is the minute number for this clock zero? (Signal.) *No.*
- So do you say o'clock after the hour number? (Signal.) *No.*
- What do you say for the minutes? (Signal.) *30.*
- Read the time for this clock. (Signal.) *2 thirty.*
d. (Repeat the following tasks with remaining clocks:)

(Point to __.) Is the minute number for this clock zero?	So do you say o'clock after the hour number?	What do you say for the minutes?	Read the time for this clock.
11:00	Yes	Yes	O'clock	11 o'clock
10:30	No	No	Thirty	10 thirty
1:30	No	No	Thirty	1 thirty
5:00	Yes	Yes	O'clock	5 o'clock

(Repeat for clocks that were not firm.)
e. Read the time for these clocks again.
- (Point to **9:00.**) What time? (Touch.) *9 o'clock.*
- (Repeat for remaining clocks: 2 thirty, 11 o'clock, 10 thirty, 1 thirty, 5 o'clock.)

INDIVIDUAL TURNS

I'll call on individual students to read one or two clocks.
- (Point to **11:00.**) What's the time for this clock? (Call on a student.) *11 o'clock.*
- (Point to **1:30.**) What's the time for this clock? (Call on a student.) *1 thirty.*
- (Point to **9:00.**) What's the time for this clock? (Call on a student.) *9 o'clock.*
- (Point to **10:30.**) What's the time for this clock? (Call on a student.) *10 thirty.*

EXERCISE 5: FACTS
PLUS/MINUS MIX [REMEDY]

a. (Distribute unopened workbooks to children.)
- Open your workbook to Lesson 92 and find part 1.
(Observe children and give feedback.)
(Teacher reference:) [R Part K]

a. 6 – 3	d. 3 + 4	h. 10 – 5
b. 5 + 5	e. 10 – 6	i. 5 + 10
c. 8 – 5	f. 5 – 3	j. 8 – 4
	g. 3 + 3	

- Touch and read problem A. Get ready. (Signal.) *6 minus 3.*
- Is the answer the big number or a small number? (Signal.) *A small number.*
- What's 6 minus 3? (Signal.) *3.*

b. (Repeat the following tasks for problems B through J:)

Touch and read problem __.	Is the answer the big number or a small number?	What's __?	
B	The big number.	5 + 5	10
C	A small number.	8 – 5	3
D	The big number.	3 + 4	7
E	A small number.	10 – 6	4
F	A small number.	5 – 3	2
G	The big number.	3 + 3	6
H	A small number.	10 – 5	5
I	The big number.	5 + 10	15
J	A small number.	8 – 4	4

(Repeat problems that were not firm.)

c. Write the answer to each problem. Put your pencil down when you've completed the equations for part 1.
(Observe children and give feedback.)

d. Check your work.
  • Touch and read fact A. (Signal.) *6 – 3 = 3.*
  • (Repeat for:) B, *5 + 5 = 10;* C, *8 – 5 = 3;* D, *3 + 4 = 7;* E, *10 – 6 = 4;* F, *5 – 3 = 2;* G, *3 + 3 = 6;* H, *10 – 5 = 5;* I, *5 + 10 = 15;* J, *8 – 4 = 4.*

EXERCISE 6: MONEY
BILLS AND COINS

a. (Display:) [92:6A]

Here is a group of bills and coins.
  • Raise your hand when you know how many dollars there are. ✔
  • How many dollars? (Signal.) *40.*
    Yes, 40. So I write 40 dollars.
    (Add to show:) W [92:6B]

$$= \$40$$

b. Now we figure out how many cents there are.
  • Raise your hand when you know how many cents there are. ✔
  • How many cents? (Signal.) *53.*
    Yes, 53 cents. So I write the dot and 53.
    (Add to show:) [92:6C]

$$= \$40.53$$

We figured out how much these bills and coins are worth.
  • How much are they worth? (Signal.) *40 dollars and 53 cents.*
    Yes, 40 dollars and 53 cents.
    Remember, write the dollars. Then write the cents.

c. Find part 2 on worksheet 92. ✔
(Teacher reference:)

  • Touch problem A. ✔
  • Write an equals. Then write the dollar sign and the number of dollars. Stop when you've done that much.
    (Observe children and give feedback.)
  • Everybody, how much are the bills worth in problem A? (Signal.) *35 dollars.*
    (Display:) W [92:6D]

$$a. \quad = \$35$$

Here's what you should have written so far. There are 35 dollars.

d. Now figure out the cents, write the dot and the number.
    (Observe children and give feedback.)
  • Everybody, how much are the coins worth in problem A? (Signal.) *60 cents.*
    (Add to show:) [92:6E]

$$a. \quad = \$35.60$$

Here's everything you should have written for problem A.
  • Everybody, read the amount for A. Get ready. (Signal.) *35 dollars and 60 cents.*

e. Write an equals and the dollars and cents for problem B. Put your pencil down when you're finished. ✔

- Problem B. Did you write an equals, a dollar sign, and a dot? (Signal.) *Yes.*
- Read the amount for problem B. (Signal.) *15 dollars and 8 cents.*
(Display:) W̲   [92:6F]

> a. = $15.08

Here's what you should have written for problem B.

EXERCISE 7: WORD PROBLEMS
COMPARISONS

a. Find part 3. ✔
(Teacher reference.)

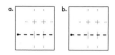

For problem A, I'm going to tell you how heavy two dogs are. You're going to write the problem and figure out how much heavier one dog is than the other dog.

- Listen to problem A: A poodle weighs 21 pounds. A Labrador weighs 134 pounds.
- Listen to problem A again and answer some questions. A poodle weighs 21 pounds. How heavy is the poodle? (Signal.) *21 pounds.*
- Listen: A Labrador weighs 134 pounds. How heavy is the Labrador? (Signal.) *134 pounds.*
- Which dog is heavier, the poodle or the Labrador? (Signal.) *The Labrador.*
- Say the problem for figuring out how much heavier the Labrador is than the poodle. (Signal.) *134 minus 21.*
- Write the problem for A and work it. Put your pencil down when you know how many pounds heavier the Labrador is than the poodle. (Observe children and give feedback.)

b. Check your work.
- Read the problem and the answer for A. Get ready. (Signal.) *134 – 21 = 113.*
- The Labrador is how much heavier than the poodle? (Signal.) *113 pounds.*

c. Touch where you'll write the column problem for B. ✔
You're going to write the problem and figure out how much younger one tree is than another tree.

- Listen to problem B: An oak tree was 458 years old. A pine tree was 153 years old. How much younger was the pine tree than the oak tree?
- Listen to problem B again and answer some questions. An oak tree was 458 years old. How old was the oak tree? (Signal.) *458 years (old).*
- Listen: The pine tree was 153 years old. How old was the pine tree? (Signal.) *153 years (old).*
- Say the problem for figuring out how much younger the pine tree was than the oak tree. (Signal.) *458 minus 153.*
- Write the problem for B and work it. Put your pencil down when you know how much younger the pine tree was than the oak tree. (Observe children and give feedback.)

d. Check your work.
- Read the problem and the answer for B. Get ready. (Signal.) *458 – 153 = 305.*
- How much younger was the pine tree than the oak tree? (Signal.) *305 years.*

EXERCISE 8: COLUMN ADDITION
CARRYING (MULTIPLE DIGITS)

a. Find part 4. ✔
(Teacher reference.)

```
a.  488   b.  196
   +222      +706
```

- Touch and read problem A. Get ready. (Signal.) *488 plus 222.*
You have to carry more than one time to work this problem. Remember, if the answer for one column has two digits you carry to the next column.
- Work the problem. Put your pencil down when you're finished.
(Observe children and give feedback.)

- Problem A: What's 488 plus 222?
(Signal.) *710.*
(Display:) W̄   [92:8A]

       a.   488
          + 222
          710

   Here's what you should have.

b. Touch and read problem B. Get ready. (Signal.)
196 plus 706.
- Work the problem. Put your pencil down when you're finished.
(Observe children and give feedback.)
- Problem B: What's 196 plus 706? (Signal.) *902.*
(Display:) W̄   [92:8B]

      b.   196
         + 706
         902

   Here's what you should have.

EXERCISE 9: WORD PROBLEMS (COLUMNS)

a. Find part 5. ✔
(Teacher reference.)

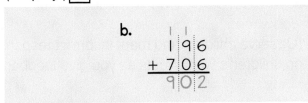

You're going to write symbols for word problems in columns and work them.

b. Touch where you'll write the symbols for problem A. ✔
Listen to problem A: There were 38 cars on the street. Then 42 more cars parked on the street. How many cars ended up on the street?
- Listen again and write the symbols for the whole problem: There were 38 cars on the street. Then 42 more cars parked on the street.
(Observe children and give feedback.)
- Everybody, touch and read the problem for A. Get ready. (Signal.) *38 plus 42.*

c. Work problem A. Put your pencil down when you know how many cars ended up on the street.
(Observe children and give feedback.)
(Teacher reference:)

- Read the problem and the answer you wrote for A. Get ready. (Signal.) *38 + 42 = 80.*
- How many cars ended up on the street? (Signal.) *80.*
d. Touch where you'll write the symbols for problem B. ✔
Listen to problem B. Jan had 284 dollars. Then she spent 164 of those dollars. How many dollars did she end up with?
- Listen again and write the symbols for the whole problem: Jan had 284 dollars. Then she spent 164 of those dollars.
(Observe children and give feedback.)
- Everybody, touch and read the problem for B. Get ready. (Signal.) *284 minus 164.*
e. Work problem B. Put your pencil down when you know how many dollars Jan ended up with.
(Observe the children and give feedback.)
(Teacher reference:)

- Read the problem and the answer you wrote for B. Get ready. (Signal.) *284 − 164 = 120.*
- How many dollars did Jan end up with? (Signal.) *120.*

EXERCISE 10: INDEPENDENT WORK

a. Find part 6. ✔
   (Teacher reference:)

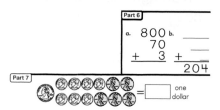

   You'll complete the place-value equations.
   In part 7, you circle the words one dollar or
   write the cents for the group of coins.

b. Turn to the other side of your worksheet and
   find part 8. ✔
   (Teacher reference:)

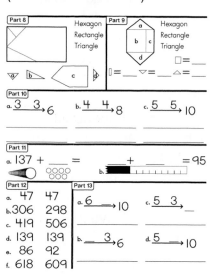

You'll circle the name for the whole shape.
Then you'll write the letter for each part.
In part 9, you'll circle the name for the whole
shape. Then you'll complete each equation by
writing the letter to show which part is it.
In part 10, these are families you learned
today. You'll write both facts for each family.
In part 11, you'll complete the equation for the
circles and the equation for the ruler.
In part 12, you'll write an equals sign, or a
sign that has a bigger end next to one of the
numbers.
In part 13, below each family, you'll write
the fact for the missing number and you'll
complete the family.

c. Complete worksheet 92. Remember to
   work parts 6 and 7 on the other side of your
   worksheet.
   (Observe children and mark incorrect responses
   on children's worksheets as you give feedback.)

Lesson  93

EXERCISE 1: TELLING TIME
READING DIGITAL DISPLAY

a. (Display:) [93:1A]

d. Read the time for these clocks again.
- (Point to **9:20.**) What time? (Touch.) *9 twenty.*
- (Repeat for remaining clocks: 2 o'clock, 11 forty-six, 7 twenty-eight, 4 fifty-one, 12 o'clock.)

━━━━━━━━ INDIVIDUAL TURNS ━━━━━━━━

I'll call on individual students to read one or two clocks.
- (Point to **12:00.**) What's the time for this clock? (Call on a student.) *12 o'clock.*
- (Point to **11:46.**) What's the time for this clock? (Call on a student.) *11 forty-six.*
- (Point to **4:51.**) What's the time for this clock? (Call on a student.) *4 fifty-one.*
- (Point to **2:00.**) What's the time for this clock? (Call on a student.) *2 o'clock.*

s.

s. For

Then

ro

al.) *20.*

hat's

ty.

b. ro

- (Signal.) *2.*
- What do you say for the minutes? (Signal.) *O'clock.*
- What's the time for this clock? (Signal.) *2 o'clock.*
c. (Repeat the following tasks for the remaining clocks:)

(Point to __.) Does this clock show zero minutes?	What's the hour number for this clock?	What do you say for the minutes?	What's the time for this clock?
11:46	No	11	46	11 forty-six
7:28	No	7	28	7 twenty-eight
4:51	No	4	51	4 fifty-one
12:00	Yes	12	O'clock	12 o'clock

(Repeat for clocks that were not firm.)

EXERCISE 2: COUNT BY TWOS

a. (Display:) [93:2A]

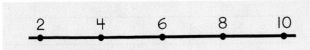

These are numbers for counting by twos.
When you count by twos, the first number is 2.
My turn to start with 2 and say some numbers.
(Point and touch:) 2, 4 ,6.
- Say those numbers with me. Get ready.
(Touch.) **2, 4, 6.**
(Repeat until firm.)
b. Your turn: Count by twos to 6. Get ready.
(Touch.) *2, 4, 6.*
(Repeat step b until firm.)
c. My turn to count by twos to 10.
(Point and touch:) 2, 4, 6, 8, 10.
Once more: 2, 4, 6, 8, 10.
- Your turn: Count by twos to 10. Get ready.
(Touch.) *2, 4, 6, 8, 10.*
(Repeat until firm.)
d. (Do not show display.) I can count by twos to 10 without looking at the numbers. Here I go: 2, 4, 6, 8, 10.
- Your turn: Count by twos to 10. (Tap.) *2, 4, 6, 8, 10.*
(Repeat until firm.)

EXERCISE 3: NUMBER FAMILIES
MISSING NUMBER IN FAMILY

a. (Display:)   [93:3A]

You're going to say the problem for the missing number in each family.

b. (Point to 3 →3→_.) Say the problem for the missing number. Get ready. (Signal.) *3 plus 3.*
- What's 3 plus 3? (Signal.) *6.*

c. (Point to 5 →5→_.) Say the problem for the missing number. (Signal.) *5 plus 5.*
- What's 5 plus 5? (Signal.) *10.*

d. (Repeat the following tasks for the remaining families:)

(Point to __.) Say the problem for the missing number.		What's __?	
4 →4→_	4 + 4	4 + 4	8
=→3→6	6 – 3	6 – 3	3
6 →5→_	6 + 5	6 + 5	11
=→3→7	7 – 3	7 – 3	4
6 →10	10 – 6	10 – 6	4
=→4→8	8 – 4	8 – 4	4
6 →12	12 – 6	12 – 6	6

(Repeat for families that were not firm.)

EXERCISE 4: RELATIVE SIZE
ORDERING BY HEIGHT

a. (Display:)   [93:4A]

- (Point to Billy.) This is Billy.
  (Point to Jenny.) This is Jenny.
- Who is taller—Billy or Jenny? (Signal.) *Jenny.*

b. (Display:)   [93:4B]

- (Point to Jenny.) Who is this? (Touch.) *Jenny.*
- (Point to mother.) This is her mother.
- Who is taller—Jenny or her mother? (Signal.) *Her mother.*

c. Billy is not in this picture.
- Who is taller—Billy or Jenny? (Signal.) *Jenny.*
- So if Billy were in the picture, who would be taller—Billy or Jenny's mother? (Signal.) *Jenny's mother.*
  Yes, Jenny's mother is taller than Jenny. Jenny is taller than Billy. So Jenny's mother is taller than Billy.

d. Here's a hard question.
- Who's the tallest—Billy, Jenny, or Jenny's mother? (Signal.) *Jenny's mother.*
- Who is the next tallest—Billy or Jenny? (Signal.) *Jenny.*
- Who is the shortest? (Signal.) *Billy.*
  Yes, Jenny's mother is the tallest, Jenny is the next tallest, and Billy is the shortest.

e. (Display:)   [93:4C]

- Here's the picture. Who is the tallest? (Signal.) *Jenny's mother.*
- Who is the next tallest? (Signal.) *Jenny.*
- Who is the shortest? (Signal.) *Billy.*
  (Repeat until firm.)
  Good ordering Billy, Jenny, and her mother.

f. (Display:) [93:4D]

(Point to Juan's father.) This is Juan's father.
- (Point to Juan.) This is Juan.
- Who is taller—Juan's father or Juan? (Signal.) *Juan's father.*

g. (Display:) [93:4E]

- (Point to Juan.) Who is this? (Touch.) *Juan.* (Point to Maria.) This is Maria.
- Who is taller—Juan or Maria? (Signal.) *Juan.*

h. Juan's father is not in this picture.
- Who is taller—Juan's father or Juan? (Signal.) *Juan's father.*
- So if Juan's father were in the picture, who would be taller—Juan's father or Maria? (Signal.) *Juan's father.*
Yes, Juan's father is taller than Juan. Juan is taller than Maria. So Juan's father is taller than Maria.

i. Here's a hard question.
- Who is the tallest—Juan's father, Juan, or Maria? (Signal.) *Juan's father.*
- Who is the next tallest—Juan or Maria? (Signal.) *Juan.*
- Who is the shortest? (Signal.) *Maria.*
Yes, Juan's father is the tallest, Juan is the next tallest, and Maria is the shortest.

j. (Display:) [93:4F]

Here's the picture.
- Who is the tallest? (Signal.) *Juan's father.*
- Who is the next tallest? (Signal.) *Juan.*
- Who is the shortest? (Signal.) *Maria.*
- (Repeat until firm.)
Good ordering Juan's father, Juan, and Maria from tallest to shortest.

EXERCISE 5: FACTS
PLUS/MINUS MIX REMEDY

a. (Distribute unopened workbooks to children.)
- Open your workbook to Lesson 93 and find part 1.
(Observe children and give feedback.)
(Teacher reference:) R Part L

a. 11−5	f. 4+3
b. 6+3	g. 6−3
c. 5+5	h. 6+2
d. 12−6	i. 8−4
e. 10−5	j. 4+6

- Touch and read problem A. Get ready. (Signal.) *11 minus 5.*
- Is the answer the big number or a small number? (Signal.) *A small number.*
- What's 11 minus 5? (Signal.) *6.*

b. (Repeat the following tasks for problems B through I:)

Touch and read problem __.	Is the answer the big number or a small number?	What's __?	
B	The big number.	6 + 3	9
C	The big number.	5 + 5	10
D	A small number.	12 − 6	6
E	A small number.	10 − 5	5
F	The big number.	4 + 3	7
G	A small number.	6 − 3	3
H	The big number.	6 + 2	8
I	A small number.	8 − 4	4
J	The big number.	4 + 6	10

(Repeat problems that were not firm.)

c. Write the answer to each fact. Put your pencil down when you've completed the equations for part 1.
(Observe children and give feedback.)

d. Check your work.
- Touch and read fact A. (Signal.) *11 − 5 = 6.*
- (Repeat for:) B, *6 + 3 = 9;* C, *5 + 5 = 10;* D, *12 − 6 = 6;* E, *10 − 5 = 5;* F, *4 + 3 = 7;* G, *6 − 3 = 3;* H, *6 + 2 = 8;* I, *8 − 4 = 4;* J, *4 + 6 = 10.*

EXERCISE 6: NUMBER FAMILIES
2-DIGIT NUMBERS WITH MISSING NUMBER
REMEDY

a. (Display:) [93:6A]

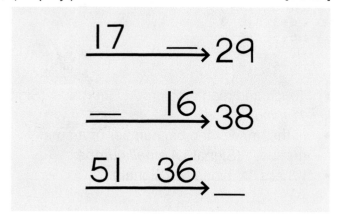

You're going to say the problem for each family.
- (Point to **17**.) A small number is 17. The big number is 29.
- Is a small number missing in this family? (Signal.) *Yes.*
- So do you minus to find the missing number? (Signal.) *Yes.*
- Say the problem. (Signal.) *29 minus 17.*

b. (Point to **16**.) A small number is 16. The big number is 38.
- Is a small number missing in this family? (Signal.) *Yes.*
- So do you minus to find the missing number? (Signal) *Yes.*
- Say the problem. (Signal.) *38 minus 16.*

c. (Point to **51**.) The small numbers are 51 and 36.
- Is a small number missing in this family? (Signal.) *No.*
- So do you minus to find the missing number? (Signal.) *No.*
- Say the problem. (Signal.) *51 plus 36.*
(Repeat families that were not firm.)

d. Find part 2 on your worksheet. ✔
(Teacher reference:)   **R** **Part D**

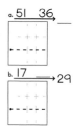

These are two of the families we just worked with. You're going to write the column problem for each family and figure out the missing number.
- Touch family A. ✔
- Is the big number or a small number missing in this family? (Signal.) *The big number.*
- Say the problem you'll write. (Signal.) *51 plus 36.*
(Repeat until firm.)
- Write the problem and figure out the missing number.
(Observe children and give feedback.)
- Everybody, read the problem and the answer. (Signal.) *51 + 36 = 87.*
- Write 87 where it goes in family A. ✔

e. Touch family B. ✔
- Is the big number or a small number missing in this family? (Signal.) *A small number.*
- Say the problem you'll write. (Signal.) *29 minus 17.*
(Repeat until firm.)
- Write the problem and figure out the missing number.
(Observe children and give feedback.)
- Everybody, read the problem and the answer for B. (Signal.) *29 − 17 = 12.*
- Write 12 where it goes in family B. ✔

EXERCISE 7: MONEY
BILLS AND COINS
REMEDY

a. Find part 3 on worksheet 93. ✔
   (Teacher reference:)
   R Part J

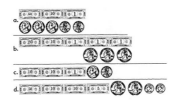

- Touch problem A. ✔
- Write equals and the dollars and cents.
  (Observe children and give feedback.)
- Read the dollar amount for problem A. (Signal.)
  61 dollars and 17 cents.
  (Display:)   [93:7A]

> **a.** = $61.17

  Here's what you should have written for
  problem A.
b. Work the rest of the problems in part 3.
  (Observe children and give feedback.)
c. Check your work.
- Read the dollar amount for problem B. (Signal.)
  33 dollars and 75 cents.
  (Display:)   [93:7B]

> **b.** = $33.75

  Here's what you should have written for
  problem B.
- Read the dollar amount for problem C. (Signal.)
  21 dollars and 6 cents.
  (Display:)   [93:7C]

> **c.** = $21.06

  Here's what you should have written for
  problem C.
- Read the dollar amount for problem D. (Signal.)
  75 dollars and 70 cents.
  (Display:)   [93:7D]

> **d.** = $75.70

  Here's what you should have written for
  problem D.
  Who got everything right?

EXERCISE 8: COLUMN ADDITION
CARRYING (MULTIPLE DIGITS)
REMEDY

a. Find part 4. ✔
   (Teacher reference:)
   R Part N

a. $\begin{array}{r} 457 \\ +162 \end{array}$ b. $\begin{array}{r} 341 \\ 9 \\ + 34 \end{array}$ c. $\begin{array}{r} 742 \\ +168 \end{array}$

You have to carry more than one time to work
one of these problems.
- Touch and read problem A. Get ready.
  (Signal.) *457 plus 162.*
- Work the problem. Raise your hand when
  you're finished.
  (Observe children and give feedback.)
- Everybody, what does 457 plus 162 equal?
  (Signal.) *619.*
  (Display:)   [93:8A]

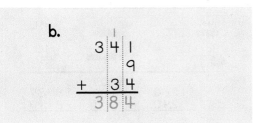

> **a.**
> $\begin{array}{r} \overset{1}{4}5\overset{}{7} \\ +162 \\ \hline 619 \end{array}$

  Here's what you should have.
b. Touch and read problem B. Get ready.
  (Signal.) *341 plus 9 plus 34.*
  Remember, if the answer has two digits you
  carry to the next column.
- Work the problem. Raise your hand when
  you're finished.
  (Observe children and give feedback.)
- Everybody, what's 341 plus 9 plus 34?
  (Signal.) *384.*
  (Display:)   [93:8B]

> **b.**
> $\begin{array}{r} 3\overset{1}{4}1 \\ 9 \\ + 34 \\ \hline 384 \end{array}$

  Here's what you should have.

c. Touch and read problem C. Get ready.
(Signal.) *742 plus 168.*
- Work the problem. Raise your hand when you're finished.
(Observe children and give feedback.)
- Everybody, what does 742 plus 168 equal?
(Signal.) *910.*
(Display:)   [93:8C]

$$
\begin{array}{r}
{\scriptstyle 1\ 1} \\
7\ 4\ 2 \\
+\ 1\ 6\ 8 \\
\hline
9\ 1\ 0
\end{array}
$$

c.

Here's what you should have.

EXERCISE 9: INDEPENDENT WORK

a. Find part 5. ✔
(Teacher reference:)

$$\underline{4\quad3}_{,7}$$

You'll write the four facts for the family.

b. Turn to the other side of your worksheet and find part 6. ✔
(Teacher reference:)

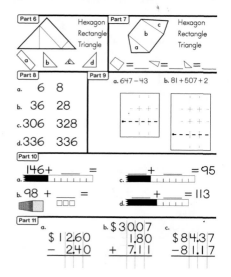

You'll circle the name for the whole shape. Then you'll write the letter for each part.
In part 7, you'll circle the name for the whole shape. Then you'll complete each equation by writing the letter to show which part it is.
In part 8, you'll write an equals sign, or a sign that has a bigger end next to one of the numbers.
In part 9, you'll write each problem in a column and work it.
In part 10, you'll complete the equations for the rulers and the equation for the boxes.
In part 11, you'll work problems and figure out the dollar amounts.

c. Complete worksheet 93. Remember to work part 5 on the other side of your worksheet.
(Observe children and mark incorrect responses on children's worksheets as you give feedback.)

Lesson 94

EXERCISE 1: COUNT BY TWOS

a. (Display:) [94:1A]

These are numbers for counting by twos.

- What's the first number you say when you count by twos? (Signal.) *2.*
  My turn to count by twos to 10.
  (Point and touch:) *2, 4 ,6, 8, 10.*
- Say those numbers with me. Get ready.
  (Touch.) *2, 4, 6, 8, 10.*
  (Repeat until firm.)
b. Your turn: Count by twos to 10. Get ready.
  (Touch as students count.) *2, 4, 6, 8, 10.*
  (Repeat until firm.)
c. (Do not show display.)
  I can count by twos to 10 without looking at the numbers.
  Here I go. 2, 4, 6, 8, 10.
  Once more: 2, 4, 6, 8, 10.
- Your turn: Count by twos to 10. (Tap 5.) *2, 4, 6, 8, 10.*
  (Repeat until firm.)

=== INDIVIDUAL TURNS ===

(Call on individual students to count by twos to 10.)

EXERCISE 2: FACTS
ADDITION/SUBTRACTION

a. (Display:) [94:2A]

3 + 3	9 − 1	3 + 10
8 − 3	6 − 4	12 − 6
5 − 2	4 + 2	5 + 3
5 + 6	7 − 3	10 − 5

For each problem, you'll tell me if the answer is the big number or a small number. Then you'll say the fact.

- (Point to **3 + 3.**) Is the answer the big number or a small number? (Signal.) *The big number.*
- Say the fact. (Signal.) *3 + 3 = 6.*
b. (Point to **8 − 3.**) Is the answer the big number or a small number? (Signal.) *A small number.*
- Say the fact. (Signal.) *8 − 3 = 5.*
c. (Repeat the following tasks for the remaining problems:)

(Point to __.) Is the answer the big number or a small number?		Say the fact.
5 − 2	A small number.	5 − 2 = 3
5 + 6	The big number.	5 + 6 = 11
9 − 1	A small number.	9 − 1 = 8
6 − 4	A small number.	6 − 4 = 2
4 + 2	The big number.	4 + 2 = 6
7 − 3	A small number.	7 − 3 = 4
3 + 10	The big number.	3 + 10 = 13
12 − 6	A small number.	12 − 6 = 6
5 + 3	The big number.	5 + 3 = 8
10 − 5	A small number.	10 − 5 = 5

(Repeat problems that were not firm.)

EXERCISE 3: MIXED COUNTING

a. Let's do some counting.
- Count by 25s to 200. Get ready. (Tap.) *25, 50, 75, 100, 125, 150, 175, 200.*
- Start with 30 and count by fives to 70. Get 30 going. *Thirtyyy.* Count. (Tap.) *35, 40, 45, 50, 55, 60, 65, 70.*
- Start with 30 and plus tens to 70. Get 30 going. *Thirtyyy.* Plus tens. (Tap.) *40, 50, 60, 70.*
- Count by hundreds to 1000. Get 100 going. *One huuundred.* Count. (Tap.) *200, 300, 400, 500, 600, 700, 800, 900, 1000.*
b. Start with 100 and count backward by tens. Get 100 going. *One huuundred.* Count backward. (Tap.) *90, 80, 70, 60, 50, 40, 30, 20, 10.*
c. You have 60. When you plus fives, what's the next number? (Signal.) *65.*
- You have 60. When you plus twos, what's the next number? (Signal.) *62.*
- You have 60. When you plus tens, what's the next number? (Signal.) *70.*
- You have 60. When you minus ones, what's the next number? (Signal.) *59.*

EXERCISE 4: RELATIVE SIZE
ORDERING BY HEIGHT AND LENGTH

a. (Display:) [94:4A]

- (Point to Lana.) This is Lana.
  (Point to Jenny.) This is her sister Jenny.
- Who is taller—Lana or Jenny? (Signal.) *Lana.*

b. (Display:) [94:4B]

- (Point to Jenny.) Who is this? (Touch.) *Jenny.*
  (Point to Juan.) This is Juan.
- Who is taller—Jenny or Juan? (Signal.) *Jenny.*
c. Lana is not in this picture.
- Who is taller—Lana or Jenny? (Signal.) *Lana.*
- So if Lana were in this picture, who would be taller—Lana or Juan? (Signal.) *Lana.*
  Yes, Lana is taller than Jenny. Jenny is taller than Juan. So Lana is taller than Juan.
d. Here's a hard question.
- Who's the tallest—Lana, Jenny, or Juan? (Signal.) *Lana.*
- Who is the next tallest—Jenny or Juan? (Signal.) *Jenny.*
- Who is the shortest? (Signal.) *Juan.*
  Yes, Lana is the tallest, Jenny is next tallest, and Juan is the shortest.
  Would you like to see a picture of all three of them?

e. (Display:) [94:4C]

- Who is the tallest? (Signal.) *Lana.*
- Who is the next tallest? (Signal.) *Jenny.*
- Who is the shortest? (Signal.) *Juan.*
  (Repeat until firm.)
  Good ordering Lana, Jenny, and Juan from tallest to shortest.
f. (Display:) [94:4D]

This picture shows a marker and a pen.
- Which is longer—the marker or the pen? (Signal.) *The pen.*

g. (Display:) [94:4E]

Here's the pen and a pencil.
- Which is longer—the pen or the pencil? (Signal.) *The pencil.*
- So if the marker were in this picture, which would be longer—the marker or the pencil? (Signal.) *The pencil.*
  Yes, the pencil is longer than the pen. The pen is longer than the marker. So the pencil is longer than the marker.
h. Which is the longest—the marker, the pen, or the pencil? (Signal.) *The pencil.*
- Which is the next longest—the marker or the pen? (Signal.) *The pen.*
- Which is the shortest? (Signal.) *The marker.*
  (Repeat until firm.)

i. (Display:) [94:4F]

Here's the marker, the pen, and the pencil.
- Which is the longest? (Signal.) *The pencil.*
- Which is the next longest? (Signal.) *The pen.*
- Which is the shortest? (Signal.) *The marker.*
Good ordering objects by their lengths.

EXERCISE 5: FACTS
ADDITION

a. (Distribute unopened workbooks to children.)
- Open your workbook to Lesson 94 and find part 1.
(Observe children and give feedback.)
(Teacher reference:)

$$a. 4+3 \quad d. 6+6 \quad h. 6+5$$
$$b. 4+6 \quad e. 5+5 \quad i. 3+3$$
$$c. 3+5 \quad f. 3+6 \quad j. 2+2$$
$$g. 4+4$$

- These are plus problems you'll work. Think of the number families for each problem and write the answers. Put your pencil down when you've completed the equations in part 1.
(Observe children and give feedback.)
b. Check your work. You'll read the equation for each problem.
- Problem A. (Signal.) *4 + 3 = 7.*
- (Repeat for:) B, *4 + 6 = 10;* C, *3 + 5 = 8;* D, *6 + 6 = 12;* E, *5 + 5 = 10;* F, *3 + 6 = 9;* G, *4 + 4 = 8;* H, *6 + 5 = 11;* I, *3 + 3 = 6;* J, *2 + 2 = 4.*

EXERCISE 6: COLUMN ADDITION
CARRYING (MULTIPLE DIGITS)

a. Find part 2 on your worksheet. ✔
(Teacher reference:)

$$a. \begin{array}{r} 358 \\ +442 \\ \hline \end{array} \quad b. \begin{array}{r} 261 \\ 4 \\ +62 \\ \hline \end{array} \quad c. \begin{array}{r} 532 \\ +48 \\ \hline \end{array}$$

You have to carry more than one time to work one of these problems.

- Touch and read problem A. Get ready. (Signal.) *358 plus 442.*
Remember, if the answer has two digits you carry to the next column.
- Work the problem. Raise your hand when you're finished.
(Observe children and give feedback.)
(Display:) [94:6A]

$$a. \begin{array}{r} \overset{1}{3}58 \\ +442 \\ \hline 800 \end{array}$$

Here's what you should have.
- Everybody, what's 358 plus 442? (Signal.) *800.*
b. Touch and read problem B. Get ready. (Signal.) *261 plus 4 plus 62.*
- Work the problem. Raise your hand when you're finished.
(Observe children and give feedback.)
(Display:) [94:6B]

$$b. \begin{array}{r} 2\overset{1}{6}1 \\ 4 \\ +62 \\ \hline 327 \end{array}$$

Here's what you should have.
- Everybody, what's 261 plus 4 plus 62? (Signal.) *327.*
c. Touch and read problem C. Get ready. (Signal.) *532 plus 48.*
- Work the problem. Raise your hand when you're finished.
(Observe children and give feedback.)
(Display:) [94:6C]

$$c. \begin{array}{r} 5\overset{1}{3}2 \\ +48 \\ \hline 580 \end{array}$$

Here's what you should have.
- Everybody, what's 538 plus 48? (Signal.) *580.*

EXERCISE 7: NUMBER FAMILIES
2-DIGIT AND 3-DIGIT NUMBERS WITH MISSING NUMBER

a. (Display:) [94:7A]

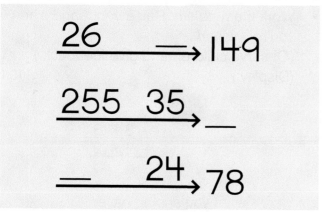

You're going to say the problem for each family.

- (Point to **26**.) The small number is 26. The big number is 149.
- Is a small number missing in this family? (Signal.) *Yes.*
- So do you minus to find the missing number? (Signal.) *Yes.*
- Say the problem. (Signal.) *149 minus 26.*

b. (Point to **255**.) The small numbers are 255 and 35.

- Is a small number missing in this family? (Signal.) *No.*
- So do you minus to find the missing number? (Signal.) *No.*
- Say the problem. (Signal.) *255 plus 35.*

c. (Point to **24**.) The small number is 24. The big number is 78.

- Is a small number missing in this family? (Signal.) *Yes.*
- So do you minus to find the missing number? (Signal.) *Yes.*
- Say the problem. (Signal.) *78 minus 24.* (Repeat families that were not firm.)

d. Find part 3 on your worksheet. ✔ (Teacher reference:)

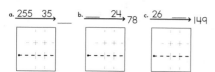

These are the families we just worked with. You're going to write the column problem for each family and figure out the missing number.

- Touch family A. ✔
- Is the big number or a small number missing? (Signal.) *The big number.*
- Say the problem you'll write. (Signal.) *255 plus 35.* (Repeat until firm.)

- Write the problem and figure out the missing number. (Observe children and give feedback.)
- Everybody, read the problem and the answer. (Signal.) *255 + 35 = 290.*
- Write 290 where it goes in family A. ✔

e. Touch family B. ✔
- Is the big number or a small number missing? (Signal.) *A small number.*
- Say the problem you'll write. (Signal.) *78 minus 24.* (Repeat until firm.)
- Write the problem and figure out the missing number. (Observe children and give feedback.)
- Everybody, read the problem for B and the answer. (Signal.) *78 – 24 = 54.*
- Write 54 where it goes in family B. ✔

f. Touch family C. ✔
- Is the big number or a small number missing? (Signal.) *A small number.*
- Say the problem you'll write. (Signal.) *149 minus 26.* (Repeat until firm.)
- Write the problem and figure out the missing number. (Observe children and give feedback.)
- Everybody, read the problem for C and the answer. (Signal.) *149 – 26 = 123.*
- Write 123 where it goes in family C. ✔

EXERCISE 8: TELLING TIME
READING AND WRITING   REMEDY

a. (Display:) [94:8A]

You're going to read the time from these clocks. Then I'm going to tell you times, and you're going to write them.

- (Point to **4:16**.) Does this clock show zero minutes? (Signal.) *No.*
- What's the hour number for this clock? (Signal.) *4.*
- What do you say for the minutes? (Signal.) *16.*
- What's the time for this clock? (Signal.) *4 sixteen.*

b. (Point to **10:35**.) Does this clock show zero minutes? (Signal.) *No.*
• What's the hour number for this clock? (Signal.) *10.*
• What do you say for the minutes? (Signal.) *35.*
• What's the time for this clock? (Signal.) *10 thirty-five.*
c. (Repeat the following tasks with remaining clocks:)

(Point to __.) Does this clock show zero minutes?	What's the hour number for this clock?	What do you say for the minutes?	What's the time for this clock?
1:00	Yes	1	O'clock	1 o'clock
6:30	No	6	30	6 thirty
10:00	Yes	10	O'clock	10 o'clock
12:48	No	12	48	12 forty-eight

(Repeat for clocks that were not firm.)
d. Read the time for these clocks again.
• (Point to **4:16**.) What time? (Touch.) *4 sixteen.*
• (Repeat for remaining clocks: 10 thirty-five, 1 o'clock, 6 thirty, 10 o'clock, 12 forty-eight.)
e. Listen: If a clock shows zero minutes, what do you say for the minutes? (Signal.) *O'clock.*
• So if you say o'clock, how many minutes does the clock show? (Signal.) *Zero.*
  When you write a time that says o'clock, you write zero, zero after the dots.
• When you write a time that says o'clock, what do you write after the dots? (Signal.) *Zero, zero.*
  (Repeat step e until firm.)
f. Find part 4 on your worksheet. ✔
   (Teacher reference:)   R | Part A

   a. __:__
   b. __:__
   c. __:__
   d. __:__

   Now I'll say times and you'll write them. You'll write the hour number before the dots. You'll write the minute number after the dots.
• If I say o'clock, what will you write after the dots? (Signal.) *Zero, zero.*
g. Touch where you'll write time A. ✔
• Listen: 11 o'clock. What's time A? (Signal.) *11 o'clock.*
• What's the hour number for 11 o'clock? (Signal.) *11.*
• What's the number of minutes for 11 o'clock? (Signal.) *Zero.*

• So what will you write after the dots? (Signal.) *Zero, zero.*
• Write 11 o'clock in space A.
  (Observe children and give feedback.)
  (Display:)   [94:8B]

a.	11:00

Here's what you should have written for 11 o'clock.
h. Time B is 2 fifteen. What's time B? (Signal.) *2 fifteen.*
• What's the hour number for 2 fifteen? (Signal.) *2.*
• What's the number of minutes for 2 fifteen? (Signal.) *15.*
• Write 2 fifteen in space B.
  (Observe children and give feedback.)
  (Display:)   [94:8C]

b.	2:15

Here's what you should have written for 2 fifteen.
i. Time C is 1 o'clock. What's time C? (Signal.) *1 o'clock.*
• What's the hour number for 1 o'clock? (Signal.) *1.*
• What's the number of minutes for 1 o'clock? (Signal.) *Zero.*
• Write 1 o'clock in space C. Remember what to write for the minutes.
  (Observe children and give feedback.)
  (Display:)   [94:8D]

c.	1:00

Here's what you should have written for 1 o'clock.
j. Time D is 12 thirty. What's time D? (Signal.) *12 thirty.*
• What's the hour number for 12 thirty? (Signal.) *12.*
• What's the number of minutes for 12 thirty? (Signal.) *30.*
• Write 12 thirty in space D.
  (Observe children and give feedback.)
  (Display:)   [94:8E]

d.	12:30

Here's what you should have written for 12 thirty.

k. Now you'll read the times you wrote for part 4.
- • Read time A. (Signal.) *11 o'clock.*
- • (Repeat for:) B, *2 fifteen;* C, *1 o'clock;*
  D, *12 thirty.*

EXERCISE 9: FACTS
SUBTRACTION

a. Find part 5 on worksheet 94. ✔
(Teacher reference:)

a. 7−5   f. 6−3
b. 8−2   g. 10−6
c. 11−6 h. 12−6
d. 15−5 i. 8−4
e. 9−3   j. 7−2

I'll read each minus problem. You'll tell me the answer. Then you'll complete the facts.
- Problem A. 7 minus 5. What's the answer? (Signal.) *2.*
- Problem B. 8 minus 2. What's the answer? (Signal.) *6.*
- Problem C. 11 minus 6. What's the answer? (Signal.) *5.*
- Problem D. 15 minus 5. What's the answer? (Signal.) *10.*
- Problem E. 9 minus 3. What's the answer? (Signal.) *6.*
- Problem F. 6 minus 3. What's the answer? (Signal.) *3.*
- Problem G. 10 minus 6. What's the answer? (Signal.) *4.*
- Problem H. 12 minus 6. What's the answer? (Signal.) *6.*
- Problem I. 8 minus 4. What's the answer? (Signal.) *4.*
- Problem J. 7 minus 2. What's the answer? (Signal.) *5.*
(Repeat problems that were not firm.)
b. Complete the fact for each problem.
(Observe children and give feedback.)
c. Check your work. You'll read the fact for each problem.
- • Problem A. (Signal.) *7 − 5 = 2.*
- • (Repeat for:) B, *8 − 2 = 6;* C, *11 − 6 = 5;*
  D, *15 − 5 = 10;* E, *9 − 3 = 6;* F, *6 − 3 = 3;*
  G, *10 − 6 = 4;* H, *12 − 6 = 6;* I, *8 − 4 = 4;*
  J, *7 − 2 = 5.*

EXERCISE 10: INDEPENDENT WORK

a. Find part 6. ✔
(Teacher reference:)

This family has big numbers. You'll write the four facts for the family.
b. Turn to the other side of your worksheet and find part 7. ✔
(Teacher reference:)

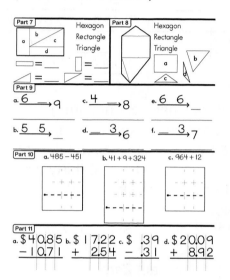

You'll circle the name for the whole shape. Then you'll complete each equation by writing the letter to show which part it is.
In part 8, you'll circle the name for the whole shape. Then you'll write the letter for each part.
In part 9, below each family, you'll write the fact for the missing number and you'll complete the family.
In part 10, you'll write each problem in a column and work it.
In part 11, you'll work each problem and figure out the dollar amount.
c. Complete worksheet 94. Remember to work part 6 on the other side of your worksheet.
(Observe children and mark incorrect responses on children's worksheets as you give feedback.)

Lesson

EXERCISE 1: FACTS
ADDITION/SUBTRACTION

a. (Display:) [95:1A]

10 – 6	6 + 6	6 – 4
5 + 5	8 – 2	2 + 8
7 – 4	4 + 4	11 – 5

For each problem, you'll tell me if the answer is the big number or a small number. Then you'll say the fact.

- (Point to **10 – 6.**) Is the answer the big number or a small number? (Signal.) *A small number.*
- Say the fact. (Signal.) *10 – 6 = 4.*

b. (Point to **5 + 5.**) Is the answer the big number or a small number? (Signal.) *The big number.*
- Say the fact. (Signal.) *5 + 5 = 10.*

c. (Repeat the following tasks for the remaining problems:)

(Point to __.) Is the answer the big number or a small number?		Say the fact.
7 – 4	A small number.	7 – 4 = 3
6 + 6	The big number.	6 + 6 = 12
8 – 2	A small number.	8 – 2 = 6
4 + 4	The big number.	4 + 4 = 8
6 – 4	A small number.	6 – 4 = 2
2 + 8	The big number.	2 + 8 = 10
11 – 5	A small number.	11 – 5 = 6

(Repeat problems that were not firm.)

EXERCISE 2: 3-D OBJECTS WITH 2-D FACES [REMEDY]

a. (Display:) [95:2A]

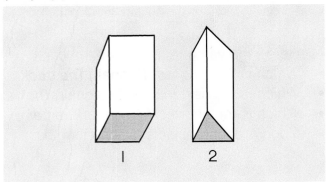

(Point to **1.**) This is an object. The bottom face of the object is shaded.

- (Touch ◢.) This is the bottom face of the object. What face is this? (Signal.) *The bottom face.*
- (Touch →▯.) This is a side face of the object. What face is this? (Signal.) *A side face.*
- (Touch ▯←.) This is another side face. How many side faces can you see? (Signal.) 2.

b. Look at the bottom face.
- Raise your hand when you know if the bottom face is a triangle, a rectangle, or a hexagon. ✔
- What's the bottom face? (Signal.) *(A) rectangle.*

c. (Point to **1.**) Look at the side faces.
- Raise your hand when you know if each side face is a triangle, a rectangle, or a hexagon. ✔
- What shape is each side face? (Signal.) *(A) rectangle.*

d. (Point to **2.**) This is another object. The bottom face of the object is shaded.
- (Touch ▲.) What face is this? (Signal.) *The bottom face.*
- (Touch →▲.) This is a side face of the object. What face is this? (Signal.) *A side face.*
- How many side faces of this object can you see? (Signal.) 2.

e. Raise your hand when you know if the bottom face is a triangle, a rectangle, or a hexagon. ✔
- What is the bottom face? (Signal.) *(A) triangle.*

f. Raise your hand when you know if each side face is a triangle, a rectangle, or a hexagon. ✔
- What shape is each side face? *(A) rectangle.*

g. Once more: What shape is the bottom face of this object? (Signal.) *(A) triangle.*
- What shape are the side faces of this object? (Signal.) *Rectangles.*

EXERCISE 3: COUNT BY TWOS

a. (Display:) [95:3A]

These are numbers for counting by 2.
- What's the first number you say when you count by twos? (Signal.) *2.*
My turn to count by twos to 10.
(Point and touch:) 2, 4, 6, 8, 10.
- Say those numbers with me. Get ready.
(Touch.) *2, 4, 6, 8, 10.*
(Repeat until firm.)

b. Your turn: Count by twos to 10. Get ready.
(Touch as students count.) *2, 4, 6, 8, 10.*
(Repeat until firm.)

c. (Do not show display.)
- Your turn: Count by twos to 10. (Tap 5.) *2, 4, 6, 8, 10.*
(Repeat until firm.)

d. You have 46. When you plus ones, what's the next number? (Signal.) *47.*
- You have 46. When you minus ones, what's the next number? (Signal.) *45.*
- You have 46. When you plus tens, what's the next number? (Signal.) *56.*
- You have 46. When you minus tens, what's the next number? (Signal.) *36.*

═══════ INDIVIDUAL TURNS ═══════

(Call on individual students to count by twos to 10.)

EXERCISE 4: RELATIVE SIZE
ORDERING BY LENGTH

a. (Display:) [95:4A]

This picture shows a van and a truck.
- Which is longer—the van or the truck? (Signal.) *The truck.*

b. (Display:) [95:4B]

Here's the van and a car.
- Which is longer—the van or the car? (Signal.) *The van.*
- So if the truck were in this picture, which would be longer—the car or the truck? (Signal.) *The truck.*
Yes, the truck is longer than the van. The van is longer than the car. So the truck is longer than the car.

c. Which is the longest—the van, the truck, or the car? (Signal.) *The truck.*
- Which is the next longest—the van or the car? (Signal.) *The van.*
- Which is the shortest? (Signal.) *The car.*
(Repeat step c until firm.)

d. (Display:) [95:4C]

Here's the van, the truck, and the car.
- Which is the longest? (Signal.) *The truck.*
- Which is the next longest? (Signal.) *The van.*
- Which is the shortest? (Signal.) *The car.*
Good ordering objects by their lengths.

e. (Display:) [95:4D]

This picture shows a shoe and a sandal.
- Which is longer—the shoe or the sandal?
  (Signal.) *The sandal.*

f. (Display:) [95:4E]

Here's the sandal and a boot.
- Which is longer—the sandal or the boot?
  (Signal.) *The boot.*
- So if the shoe were in this picture, which would be longer—the shoe or the boot?
  (Signal.) *The boot.*
  Yes, the shoe is shorter than the sandal. The sandal is shorter than the boot. So the shoe is shorter than the boot, which means the boot is longer than the shoe.

g. Which is the longest—the shoe, the sandal, or the boot? (Signal.) *The boot.*
- Which is the next longest—the shoe or the sandal? (Signal.) *The sandal.*
- Which is the shortest? (Signal.) *The shoe.*

h. (Display:) [95:4F]

Here's the shoe, the sandal, and the boot.

- Which is the longest? (Signal.) *The boot.*
- Which is the next longest? (Signal.) *The sandal.*
- Which is the shortest? (Signal.) *The shoe.*
  Good ordering objects by their lengths.

EXERCISE 5: FACTS
SUBTRACTION

a. (Distribute unopened workbooks to children.)
- Open your workbook to Lesson 95 and find part 1.
  (Observe children and give feedback.)
  (Teacher reference:)

a. $10 - 6$
b. $8 - 4$
c. $9 - 3$
d. $11 - 5$
e. $6 - 3$
f. $10 - 5$
g. $7 - 4$
h. $12 - 6$
i. $8 - 5$
j. $10 - 2$

These are minus problems so the answer is a small number in the family. I'll read each problem and you'll tell me the answer. Then you'll complete each fact.
- Problem A. 10 minus 6. What's the answer?
  (Signal.) *4.*
- (Repeat the following tasks for remaining problems:)

Problem __.		What's the answer?
B	$8 - 4$	4
C	$9 - 3$	6
D	$11 - 5$	6
E	$6 - 3$	3
F	$10 - 5$	5
G	$7 - 4$	3
H	$12 - 6$	6
I	$8 - 5$	3
J	$10 - 2$	8

(Repeat problems that were not firm.)

b. Complete the equations in part 1. Put your pencil down when you're finished.
  (Observe children and give feedback.)

c. Check your work. You're going to touch and read the fact for each problem.
- Problem A. (Signal.) *10 – 6 = 4.*
- (Repeat for:) B, *8 – 4 = 4;* C, *9 – 3 = 6;* D, *11 – 5 = 6;* E, *6 – 3 = 3;* F, *10 – 5 = 5;* G, *7 – 4 = 3;* H, *12 – 6 = 6;* I, *8 – 5 = 3;* J, *10 – 2 = 8.*

EXERCISE 6: WORD PROBLEMS (COLUMNS)

a. Find part 2 on worksheet 95. ✔
(Teacher reference.)

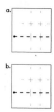

You're going to write the symbols for word problems in columns and work them.

b. Touch where you'll write the symbols for problem A. ✔
Listen to problem A: A room had 56 people in it. Then 26 of those people left the room. How many people ended up in the room?

• Listen again and write the symbols for the whole problem: A room had 56 people in it. Then 26 of those people left the room.
(Observe children and give feedback.)

• Everybody, touch and read the problem for A. Get ready. (Signal.) *56 minus 26.*

c. Work problem A. Put your pencil down when you know how many people ended up in the room.
(Observe children and give feedback.)
(Teacher reference:)

• Read the problem and the answer you wrote for A. Get ready. (Signal.) *56 – 26 = 30.*

• How many people ended up in the room? (Signal.) *30.*

d. Touch where you'll write the symbols for problem B. ✔
Listen to problem B. Little Billy had 117 dollars. His sister had 290 dollars. How many dollars did they have altogether?

• Listen again and write the symbols for the whole problem: Little Billy had 117 dollars. His sister had 290 dollars.
(Observe children and give feedback.)

• Everybody, touch and read the problem for B. Get ready. (Signal.) *117 plus 290.*

e. Work problem B. Put your pencil down when you know how many dollars the children had altogether.
(Observe children and give feedback.)
(Teacher reference:)

• Read the problem and the answer you wrote for A. Get ready. (Signal.) *117 + 290 = 407.*

• How many dollars did the children have altogether? (Signal.) *407.*

EXERCISE 7: NUMBER FAMILIES
2-DIGIT AND 3-DIGIT NUMBERS WITH MISSING NUMBER

a. Find part 3 on worksheet 95. ✔
(Teacher reference.)

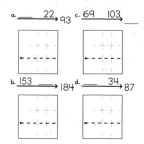

These are number families with 2-digit or 3-digit numbers. You're going to say the problem for each number family. Then you'll write a column problem for each family and work it.

• Family A. Is the big number or a small number missing in this family? (Signal.) *A small number.*

• Say the problem for the missing number. (Signal.) *93 minus 22.*
(Repeat until firm.)

b. Family B. What's missing, the big number or a small number? (Signal.) *A small number.*

• Say the problem for the missing number. (Signal.) *184 minus 153.*

c. (Repeat the following tasks for the remaining families:)

Family __. What's missing, the big number or a small number?		Say the problem for the missing number.
C	The big number.	69 plus 103
D	A small number.	87 minus 34

(Repeat families that were not firm.)

d. Go back to family A and write the column problem and figure out the missing number. Put your pencil down when you've completed family A.
(Observe children and give feedback.)
• Check your work. Read the column problem and the answer for family A. (Signal.) 93 – 22 = 71.
(Display:)   [95:7A]

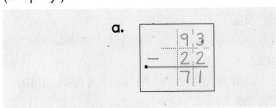

Here's what you should have for family A.

e. Write the column problem and figure out the missing number for family B. Put your pencil down when you've completed family B.
(Observe children and give feedback.)
• Check your work. Read the column problem and the answer for family B. (Signal.)
184 – 153 = 31.
(Display:)   [95:7B]

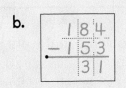

Here's what you should have for family B.

f. Write the column problem and figure out the missing number for family C. Put your pencil down when you've completed family C.
(Observe children and give feedback.)
• Check your work. Read the column problem and the answer for family C. (Signal.)
69 + 103 = 172.
(Display:)   [95:7C]

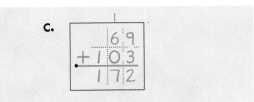

Here's what you should have for family C.

g. Write the column problem and figure out the missing number for family D. Put your pencil down when you've completed family D.
(Observe children and give feedback.)
• Check your work. Read the column problem and the answer for family D. (Signal.) 87 – 34 = 53.
(Display:)   [95:7D]

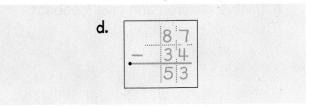

Here's what you should have for family D.

EXERCISE 8: TELLING TIME
READING AND WRITING

a. Find part 4 on your worksheet. ✔
(Teacher reference:)

a. ___:___
b. ___:___
c. ___:___
d. ___:___
e. ___:___
f. ___:___

I'll say a time for each space. You'll write the time. The dots for each time are already shown. Remember, the number for hours goes before the dots. The number for minutes goes after the dots.
• If I say o'clock, what will you write after the dots? (Signal.) Zero, zero.
b. Time A is 11 nineteen. What's time A? (Signal.) 11 nineteen.
• What's the hour number for 11 nineteen? (Signal.) 11.
• What's the number of minutes for 11 nineteen? (Signal.) 19.
• Write 11 nineteen in space A. ✔
(Observe children and give feedback.)
(Display:)   [95:8A]

a.   11:19

Here's what you should have written for 11 nineteen.

c. Time B is 8 o'clock. What's time B? (Signal.)
8 o'clock.
- What's the hour number for 8 o'clock?
(Signal.) *8.*
- What do you write after the dots for 8 o'clock?
(Signal.) *Zero, zero.*
- Write 8 o'clock in space B. ✔
(Observe children and give feedback.)
(Display:)   [95:8B]

b. ___8:00___

Here's what you should have written for
8 o'clock.
d. Time C is 9 twenty-four. What's time C?
(Signal.) *9 twenty-four.*
- Write 9 twenty-four in space C. ✔
(Observe children and give feedback.)
(Display:)   [95:8C]

c. ___9:24___

Here's what you should have written for
9 twenty-four.
e. Time D is 1 thirty. What's time D? (Signal.)
1 thirty.
- Write 1 thirty in space D. ✔
(Observe children and give feedback.)
(Display:)   [95:8D]

d. ___1:30___

Here's what you should have written for 1 thirty.
f. Time E is 12 o'clock. What's time E? (Signal.)
12 o'clock.
- Write 12 o'clock in space E. ✔
(Observe children and give feedback.)
(Display:)   [95:8E]

e. ___12:00___

Here's what you should have written for
12 o'clock.
g. Time F is 3 fifty-nine. What's time F? (Signal.)
3 fifty-nine.
- Write 3 fifty-nine in space F. ✔
(Observe children and give feedback.)
(Display:)   [95:8F]

f. ___3:59___

Here's what you should have written for 3
fifty-nine.

h. Now you'll read the times you wrote for part 4.
- Read time A. (Signal.) *11 nineteen.*
- (Repeat for:) B, *8 o'clock;* C, *9 twenty-four;*
D, *1 thirty;* E, *12 o'clock;* F, *3 fifty-nine.*

EXERCISE 9: FACTS
ADDITION

a. Find part 5. ✔
(Teacher reference:)

a. $4 + 3$	f. $2 + 9$
b. $5 + 5$	g. $6 + 6$
c. $6 + 3$	h. $4 + 4$
d. $3 + 5$	i. $5 + 6$
e. $3 + 3$	j. $6 + 4$

These are plus problems so the answer is the
big number in the family. I'll read each problem
and you'll tell me the answer. Then you'll
complete each fact.
- Problem A. 4 plus 3. What's the answer?
(Signal.) *7.*
b. (Repeat the following tasks for remaining
problems:)

Problem __.		What's the answer?
B	$5 + 5$	10
C	$6 + 3$	9
D	$3 + 5$	8
E	$3 + 3$	6
F	$2 + 9$	11
G	$6 + 6$	12
H	$4 + 4$	8
I	$5 + 6$	11
J	$6 + 4$	10

(Repeat problems that were not firm.)
c. Complete the equations in part 5. Put your
pencil down when you're finished.
(Observe children and give feedback.)
d. Check your work. You're going to touch and
read the fact for each problem.
- Problem A. (Signal.) *4 + 3 = 7.*
- (Repeat for:) B, *5 + 5 = 10;* C, *6 + 3 = 9;*
D, *3 + 5 = 8;* E, *3 + 3 = 6;* F, *2 + 9 = 11;*
G, *6 + 6 = 12;* H, *4 + 4 = 8;* I, *5 + 6 = 11;*
J, *6 + 4 = 10.*

EXERCISE 10: INDEPENDENT WORK

a. Turn to the other side of your worksheet and find part 6. ✔
(Teacher reference:)

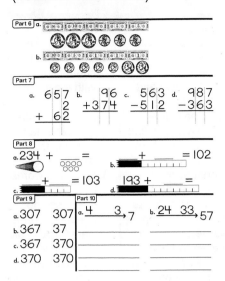

You'll write the dollars and cents amount for each group of bills and coins.

In part 7, you'll work the column problems. You'll have to carry more than one time for the addition problems.

In part 8, you'll complete the equation for the circles and the equations for the rulers.

In part 9, you'll write an equals sign, or a sign that has a bigger end next to one of the numbers.

In part 10, you'll write four facts for each family. Family B has 2-digit numbers.

b. Complete worksheet 95.
(Observe children and mark incorrect responses on children's worksheets as you give feedback.)

Lessons 96–100 Planning Page

	Lesson 96	Lesson 97	Lesson 98	Lesson 99	Lesson 100
Student Learning Objectives	**Exercises** 1. Count by twos 2. Say addition and subtraction facts 3. Identify 2-dimensional faces on 3-dimensional objects 4. Order objects by length or height 5. Find and write missing numbers in number families with 2-digit or 3-digit numbers 6. Solve subtraction facts **7. Write digital time with 1-digit minutes** 8. Solve comparison word problems 9. Solve addition facts 10. Complete work independently	**Exercises** 1. Find missing numbers in number families **2. Learn the word _add_** 3. Count forward or backward from a given number 4. Count by twos 5. Order objects by length or height **6. Solve money addition problems with carrying** 7. Find and write missing numbers in number families with 2-digit or 3-digit numbers 8. Write digital time 9. Solve subtraction facts 10. Identify 2-dimensional faces on 3-dimensional objects 11. Complete work independently	**Exercises** 1. Count by twos 2. Say addition and subtraction facts 3. Learn the word _add_ 4. Count forward or backward from a given number 5. Find and write missing numbers in number families with 2-digit or 3-digit numbers 6. Write digital time 7. Write the symbols for word problems in columns and solve 8. Solve subtraction facts 9. Identify 2-dimensional faces on 3-dimensional objects 10. Complete work independently	**Exercises** 1. Count forward or backward from a given number 2. Say addition and subtraction facts with the same small number; find missing numbers in number families 3. Count by twos **4. Use tens and hundreds charts to clarify place value** **5. Learn the words _add_ and _subtract_** 6. Tell time from a digital display 7. Solve addition and subtraction facts 8. Identify 2-dimensional faces on 3-dimensional objects 9. Find and write missing numbers in number families with 2-digit or 3-digit numbers 10. Complete work independently	**Exercises** 1. Count by twos 2. Say addition and subtraction facts with the same small number; find missing numbers in number families 3. Tell time from a digital display 4. Learn the words _add_ and _subtract_ 5. Use tens and hundreds charts to clarify place value 6. Find and write missing numbers in number families with 2-digit or 3-digit numbers **7. Identify 3-dimensional objects (pyramids)** 8. Solve subtraction facts **9. Solve addition problems with carrying multiple digits** 10. Solve addition and subtraction facts 11. Complete work independently

Common Core State Standards for Mathematics					
1.OA 5		✔		✔	✔
1.OA 6	✔	✔	✔	✔	✔
1.OA 8	✔	✔	✔	✔	✔
1.NBT 1		✔	✔	✔	✔
1.NBT 2				✔	✔
1.NBT 4	✔	✔	✔	✔	✔
1.MD 1	✔	✔			
1.MD 3	✔	✔	✔	✔	✔
1.G 1	✔	✔	✔	✔	✔
1.G 2	✔	✔	✔	✔	✔
Teacher Materials	Presentation Book 3, Board Displays CD or chalkboard				
Student Materials	Workbook 2, Pencil				
Additional Practice	Student Practice Software: Block 4: Activity 1 (1.G.1), Activity 2 (1.NBT.1), Activity 3 (1.NBT.1), Activity 4 (1.NBT.3), Activity 5 (1.NBT.1), Activity 6 (1.NBT.1)				
Mastery Test					Student Assessment Book (Present Mastery Test 10 following Lesson 100.)

Lesson 96

EXERCISE 1: COUNT BY TWOS

a. What's the first number you say when you count by twos? (Signal.) *2.*
• Count by twos to 10. Get ready. (Signal.) *2, 4, 6, 8, 10.*
(Repeat until firm.)

b. Start at 45 and count by fives to 80. Get ready. (Signal.) *45, 50, 55, 60, 65, 70, 75, 80.*
• Start at 70 and count by tens to 110. Get 70 going. *Seventyyy.* Count. (Signal.) *80, 90, 100, 110.*
• Start at 600 and count by 25s to 700. Get 600 going. *Six huuundred.* Count. (Signal.) *625, 650, 675, 700.*
• Start at 2 and count by twos to 10. Get ready. (Signal.) *2, 4, 6, 8, 10.*
(Repeat step b until firm.)

c. You have 83. When you plus tens, what's the next number? (Signal.) *93.*
• You have 83. When you minus tens, what's the next number? (Signal.) *73.*
• You have 83. When you minus ones, what's the next number? (Signal.) *82.*
• You have 83. When you plus twos, what's the next number? (Signal.) *85.*

EXERCISE 2: FACTS
ADDITION/SUBTRACTION

a. (Display:) [96:2A]

8 – 3	2 + 9	3 + 4
3 + 6	11 – 6	8 – 4
10 – 4	12 – 6	3 + 3
		10 – 5

For each problem, you'll tell me if the answer is the big number or a small number. Then you'll say the fact.
• (Point to **8 – 3.**) Is the answer the big number or a small number? (Signal.) *A small number.*
• Say the fact. (Signal.) *8 – 3 = 5.*
b. (Point to **3 + 6.**) Is the answer the big number or a small number? (Signal.) *The big number.*
• Say the fact. (Signal.) *3 + 6 = 9.*

c. (Repeat the following tasks for the remaining problems:)

(Point to __.) Is the answer the big number or a small number?		Say the fact.
10 – 4	A small number.	10 – 4 = 6
2 + 9	The big number.	2 + 9 = 11
11 – 6	A small number.	11 – 6 = 5
12 – 6	A small number.	12 – 6 = 6
3 + 4	The big number.	3 + 4 = 7
8 – 4	A small number.	8 – 4 = 4
3 + 3	The big number.	3 + 3 = 6
10 – 5	A small number.	10 – 5 = 5

(Repeat problems that were not firm.)

EXERCISE 3: 3-D OBJECTS WITH 2-D FACES

a. (Display:) [96:3A]

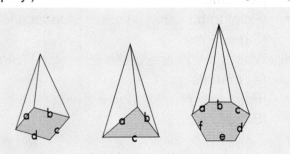

• (Point to ▲.) Here's a picture of an object.
• (Point to bottom.) The bottom of the object is shaded. Raise your hand when you know if the bottom face is a triangle, a rectangle, or a hexagon. ✔
• What is the bottom face of this object? (Touch.) *(A) rectangle.*
• If the bottom face is a rectangle, how many sides does the bottom face have? (Signal.) *4.*

• (Point to ▲.) I'll touch, you'll count the sides of the bottom face. (Touch sides a–d as children count:) *1, 2, 3, 4.*
• How many sides does the bottom face have? (Signal.) *4.*
Yes, the sides of the bottom face have the letters A, B, C, and D.

b. Here's the rule for the objects we'll work with: The number of sides for the bottom face tells the number of side faces.

- (Point to **a**.) Can you see a side face for A? (Signal.) *Yes.*
- (Point to **b**.) Can you see a side face for B? (Signal.) *Yes.*
- (Point to **c**.) Can you see a side face for C? (Signal.) *No.*
- (Point to **d**.) Can you see a side face for D? (Signal.) *No.*
  You can see two side faces.

c. (Point to △.) If the bottom face is a triangle, how many sides does the bottom face have? (Signal.) *3.*

- So how many side faces from the bottom face are there? (Signal.) *3.*
- (Point to **a**.) Can you see a side face for A? (Signal.) *Yes.*
- What's the shape of the side face? (Signal.) *(A) triangle.*
- (Point to **b**.) Can you see a side face for B? (Signal.) *Yes.*
- What's the shape of the side face? (Signal.) *(A) triangle.*
- (Point to **c**.) Can you see a side face for C? (Signal.) *No.*
- You can't see the side face for C, but what shape is it? (Signal.) *(A) triangle.*

d. (Point to ⬡.) The bottom face of this object is shaded. Raise your hand when you know what shape the bottom face is. ✔

- What's the bottom face? (Signal.) *(A) hexagon.*
- If the bottom face is a hexagon, how many sides does it have? (Signal.) *6.*
- So how many side faces are there? (Signal.) *6.*
- (Point to **a**.) Can you see a side for face A? (Signal.) *Yes.*
- What's the shape of the side face? (Signal.) *(A) triangle.*
- (Point to **b**.) Can you see a side face for B? (Signal.) *Yes.*
- What's the shape of the side face? (Signal.) *(A) triangle.*
- (Point to **c**.) Can you see a side face for C? (Signal.) *Yes.*
- What's the shape of the side face? (Signal.) *(A) triangle.*

e. You can't see three side faces. Raise your hand when you know the letters for those faces. ✔

- (Call on a child:) What are the letters of the faces you can't see? *D, E, and F.*
  Yes, you can't see side faces, D, E, and F, but you know what shape they are.
- What shape are the side faces for D, E, and F? (Signal.) *(A) triangle(s).*

f. Remember, if the bottom side of the object is a rectangle, how many side faces are there? (Signal.) *4.*

- If the bottom side of the object is a triangle, how many side faces are there? (Signal.) *3.*
- If the bottom side of the object is a hexagon, how many side faces are there? (Signal.) *6.*

EXERCISE 4: RELATIVE SIZE
ORDERING BY LENGTH

a. (Display:) [96:4A]

This picture shows a ribbon and a rope.

- Which is longer—the ribbon or the rope? (Signal.) *The ribbon.*

b. (Display:) [96:4B]

Here's the rope and a string.

- Which is longer—the rope or the string? (Signal.) *The rope.*
- So if the ribbon were in this picture, which would be longer—the ribbon or the string? (Signal.) *The ribbon.*
  Yes, the string is shorter than the rope. The rope is shorter than the ribbon. So the string is shorter than the ribbon, which means the ribbon is longer than the string.

c. Which is the longest—the string, the rope, or the ribbon? (Signal.) *The ribbon.*
• Which is the next longest—the string or the rope? (Signal.) *The rope.*
• Which is the shortest? (Signal.) *The string.*
(Repeat step c until firm.)

d. (Display:) [96:4C]

Here's the ribbon, the rope, and the string.
• Which is the longest? (Signal.) *The ribbon.*
• Which is the next longest? (Signal.) *The rope.*
• Which is the shortest? (Signal.) *The string.*
Good ordering objects by their lengths.

e. (Display:) [96:4D]

This picture shows a worm and a stick.
• Which is longer—the worm or the stick? (Signal.) *The worm.*

f. (Display:) [96:4E]

Here's the worm and a snake.
• Which is longer—the worm or the snake? (Signal.) *The snake.*
• So if the stick were in this picture, which would be longer—the snake or the stick? (Signal.) *The snake.*
Yes, the snake is longer than the worm. The worm is longer than the stick. So the snake is longer than the stick.

g. Which is the longest—the worm, the stick, or the snake? (Signal.) *The snake.*
• Which is the next longest—the worm or the stick? (Signal.) *The worm.*
• Which is the shortest? (Signal.) *The stick.*
(Repeat step g until firm.)

h. (Display:) [96:4F]

Here's the worm, the stick, and the snake.
• Which is the longest? (Signal.) *The snake.*
• Which is the next longest? (Signal.) *The worm.*
• Which is the shortest? (Signal.) *The stick.*
Good ordering objects by their lengths.

EXERCISE 5: NUMBER FAMILIES
2-DIGIT NUMBERS WITH MISSING NUMBER REMEDY

a. (Distribute unopened workbooks to children.)
• Open your workbook to Lesson 96 and find part 1.
(Observe children and give feedback.)
(Teacher reference:) R Part E

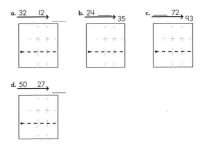

Each number family has 2-digit numbers and a missing number. You're going to write the problem for each number family.
• Touch family A. ✔
• What's missing in this family, the big number or a small number? (Signal.) *The big number.*
• Say the problem for the missing number. (Signal.) *32 plus 12.*

b. What's missing in family B, the big number or a small number? (Signal.) *A small number.*
• Say the problem. (Signal.) *35 minus 24.*

c. What's missing in family C, the big number or a small number? (Signal.) *A small number.*
• Say the problem. (Signal.) *93 minus 72.*

d. What's missing in family D, the big number or a small number? (Signal.) *The big number.*
• Say the problem. (Signal.) *50 plus 27.*
(Repeat until firm.)

e. Go back to family A. Write the column problem for that family. Don't work the problem, just write it.
(Observe children and give feedback.)
(Display:) [96:5A]

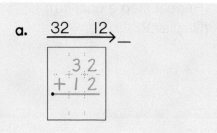

Here's what you should have.

f. Write the column problem for family B.
(Observe children and give feedback.)
(Display:) [96:5B]

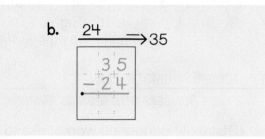

Here's what you should have.

g. Write the column problem for family C.
(Observe children and give feedback.)
(Display:) [96:5C]

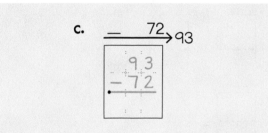

Here's what you should have.

h. Write the column problem for family D.
(Observe children and give feedback.)
(Display:) [96:5D]

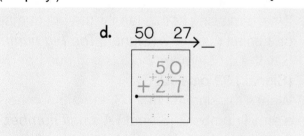

Here's what you should have.

i. Work the problems in part 1. Put your pencil down when you know the missing number for each family.
(Observe children and give feedback.)

j. Check your work. For each family, you'll read the problem and answer.
• Problem A. (Signal.) *32 + 12 = 44.*
• Problem B. (Signal.) *35 – 24 = 11.*
• Problem C. (Signal.) *93 – 72 = 21.*
• Problem D. (Signal.) *50 + 27 = 77.*

k. You know the missing number for each family. That's the answer to each problem.
• Problem A. What's the answer? (Signal.) *44.*
• So what's the missing number in the family? (Signal.) *44.*
• Write that number in the family. ✔

l. Problem B. What's the missing number in the family? (Signal.) *11.*
• Write that number in family B. Then write the missing numbers in the rest of the families.
(Observe children and give feedback.)

EXERCISE 6: FACTS
SUBTRACTION

a. Find part 2 on worksheet 96. ✔
(Teacher reference:)

a. 8 – 4   f. 10 – 5
b. 12 – 10   g. 8 – 5
c. 9 – 6   h. 7 – 6
d. 10 – 2   i. 6 – 3
e. 7 – 3   j. 8 – 2

These are minus problems so the answer is a small number in the family. I'll read each problem and you'll tell me the answer. Then you'll complete each fact.
• Problem A. 8 minus 4. What's the answer? (Signal.) *4.*
• (Repeat the following tasks for remaining problems:)

Problem __. __.		What's the answer?
B	12 – 10	2
C	9 – 6	3
D	10 – 2	8
E	7 – 3	4
F	10 – 5	5
G	8 – 5	3
H	7 – 6	1
I	6 – 3	3
J	8 – 2	6

(Repeat problems that were not firm.)

b. Complete the equations in part 2. Put your pencil down when you're finished.
(Observe children and give feedback.)

c. Check your work. You're going to touch and read the fact for each problem.
- Problem A. (Signal.) *8 – 4 = 4.*
- (Repeat for:) B, *12 – 10 = 2;* C, *9 – 6 = 3;* D, *10 – 2 = 8;* E, *7 – 3 = 4;* F, *10 – 5 = 5;* G, *8 – 5 = 3;* H, *7 – 6 = 1;* I, *6 – 3 = 3;* J, *8 – 2 = 6.*

EXERCISE 7: TELLING TIME
WRITING 1-DIGIT NUMBERS

REMEDY

a. Find part 3. ✔
(Teacher reference:)

a. ___:___
b. ___:___
c. ___:___
d. ___:___
e. ___:___
f. ___:___

I'll say a time for each space. You'll write the time. The dots for each time are already shown.
- If I say o'clock, what will you write after the dots? (Signal.) *Zero, zero.*

b. Time A is 6 o'clock. What's time A? (Signal.) *6 o'clock.*
- What do you write after the dots for 6 o'clock? (Signal.) *Zero, zero.*
- Write 6 o'clock in space A.
(Observe children and give feedback.)
(Display:)   [96:7A]

> a. **6:00**

Here's what you should have written for 6 o'clock.

c. Time B is 5 oh 8. What's time B? (Signal.) *5 oh 8.*
- What's the hour number for 5 oh 8? (Signal.) *5.*
- The minutes are **oh 8,** so you write zero, 8 after the dots. What do you write after the dots? (Signal.) *Zero, 8.*
- Write the time for 5 oh 8 in space B.
(Observe children and give feedback.)
(Display:)   [96:7B]

> b. **5:08**

Here's what you should have written for 5 oh 8.

d. Time C is 12 oh 4. What's time C? (Signal.) *12 oh 4.*
- What do you write after the dots for 12 oh 4? (Signal.) *Zero, 4.*
- Write the time for 12 oh 4 in space C.
(Observe children and give feedback.)
(Display:)   [96:7C]

> c. **12:04**

Here's what you should have written for 12 oh 4.

e. Time D is 5 o'clock. What's time D? (Signal.) *5 o'clock.*
- Write 5 o'clock in space D.
(Observe children and give feedback.)
(Display:)   [96:7D]

> d. **5:00**

Here's what you should have written for 5 o'clock.

f. Time E is 10 oh 1. What's time E? (Signal.) *10 oh 1.*
- Write 10 oh 1 in space E.
(Observe children and give feedback.)
(Display:)   [96:7E]

> e. **10:01**

Here's what you should have written for 10 oh 1.

g. Time F is 11 thirty. What's time F? (Signal.) *11 thirty.*
- Write 11 thirty in space F.
(Observe children and give feedback.)
(Display:)   [96:7F]

> f. **11:30**

Here's what you should have written for 11 thirty.

h. Now you'll read the times you wrote for part 3.
- Read time A. (Signal.) *6 o'clock.*
- (Repeat for:) B, *5 oh 8;* C, *12 oh 4;* D, *5 o'clock;* E, *10 oh 1;* F, *11 thirty.*

EXERCISE 8: WORD PROBLEMS
COMPARISONS

a. Find part 4. ✔
(Teacher reference:)

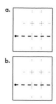

For problem A, I'm going to tell you how long two objects are. You're going to write the problem and figure out how much longer one object is than the other object.

- Listen to problem A: A truck was 48 feet long. A boat was 27 feet long.
- Listen to problem A again and answer some questions. A truck was 48 feet long. How long was the truck? (Signal.) *48 feet.*
- Listen: A boat was 27 feet long. How long was the boat? (Signal.) *27 feet.*
- Which was longer, the truck or the boat? (Signal.) *The truck.*
- Say the problem for figuring out how much longer the truck was than the boat. (Signal.) *48 minus 27.*
- Write the problem for A and work it. Put your pencil down when you know how many feet longer the truck was than the boat. (Observe children and give feedback.)

b. Check your work.
- Read the problem and the answer for A. Get ready. (Signal.) *48 – 27 = 21.*
- The truck was how much longer than the boat? (Signal.) *21 feet.*

c. Touch where you'll write the column problem for B. ✔
You're going to write the problem and figure out how much shorter one tree is than another tree.
- Listen to problem B: The young tree was 25 feet tall. The old tree was 137 feet tall. How much shorter was the young tree than the old tree?
- Listen to problem B again and answer some questions. The young tree was 25 feet tall. How tall was the young tree? (Signal.) *25 feet.*
- Listen: The old tree was 137 feet tall. How tall was the old tree? (Signal.) *137 feet.*
- Say the problem for figuring out how much shorter the young tree was than the old tree. (Signal.) *137 minus 25.*

- Write the problem for B and work it. Put your pencil down when you know how much shorter the young tree was than the old tree. (Observe children and give feedback.)
d. Check your work.
- Read the problem and the answer for B. Get ready. (Signal.) *137 – 25 = 112.*
- How much shorter was the young tree than the old tree? (Signal.) *112 feet.*

EXERCISE 9: FACTS
ADDITION

a. Find part 5. ✔
(Teacher reference:)

a. 4 + 6   f. 5 + 5
b. 5 + 3   g. 4 + 3
c. 6 + 5   h. 6 + 6
d. 3 + 3   i. 2 + 5
e. 6 + 10   j. 4 + 4

These are plus problems so the answer is the big number in the family. I'll read each problem and you'll tell me the answer. Then you'll complete each fact.
- Problem A. 4 plus 6. What's the answer? (Signal.) *10.*
b. (Repeat the following tasks for remaining problems:)

Problem __. __.		What's the answer?
B	5 + 3	8
C	6 + 5	11
D	3 + 3	6
E	6 + 10	16
F	5 + 5	10
G	4 + 3	7
H	6 + 6	12
I	2 + 5	7
J	4 + 4	8

(Repeat problems that were not firm.)
c. Complete the equations in part 5. Put your pencil down when you're finished. (Observe children and give feedback.)
d. Check your work. You're going to touch and read the fact for each problem.
- Problem A. (Signal.) *4 + 6 = 10.*
- (Repeat for:) B, *5 + 3 = 8;* C, *6 + 5 = 11;* D, *3 + 3 = 6;* E, *6 + 10 = 16;* F, *5 + 5 = 10;* G, *4 + 3 = 7;* H, *6 + 6 = 12;* I, *2 + 5 = 7;* J, *4 + 4 = 8.*

EXERCISE 10: INDEPENDENT WORK

a. Turn to the other side of worksheet 96 and find part 6. ✔
(Teacher reference:)

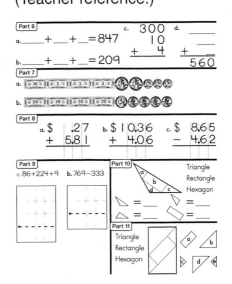

You'll complete the place-value equations.
In part 7, you'll write the dollars and cents amount for each group of bills and coins.
In part 8, you'll work the money problems.
In part 9, you'll write each problem in a column and work it.
In part 10, you'll circle the name for the whole shape. Then you'll complete each equation by writing the letter to show which part it is.
In part 11, you'll circle the name for the whole shape. Then you'll write the letter for each part.

b. Complete worksheet 96.
(Observe children and mark incorrect responses on children's worksheets as you give feedback.)

Lesson

EXERCISE 1: NUMBER FAMILIES
MISSING NUMBER IN FAMILY

a. (Display:) [97:1A]

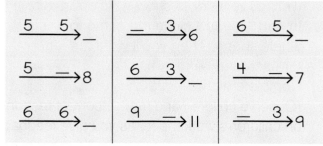

You're going to say the problem for the missing number in each family.

b. (Point to $\xrightarrow{5\quad 5}\_$.) Say the problem for the missing number. Get ready. (Signal.) *5 plus 5.*
• What's 5 plus 5? (Signal.) *10.*
c. (Point to $\xrightarrow{5}8$.) Say the problem for the missing number. (Signal.) *8 minus 5.*
• What's 8 minus 5? (Signal.) *3.*
d. (Repeat the following tasks for the remaining families:)

(Point to __.) Say the problem for the missing number.		What's __?	
$\xrightarrow{6\quad 6}\_$	6 + 6	6 + 6	12
$\xrightarrow{\quad 3}6$	6 − 3	6 − 3	3
$\xrightarrow{6\quad 3}\_$	6 + 3	6 + 3	9
$\xrightarrow{9}11$	11 − 9	11 − 9	2
$\xrightarrow{6\quad 5}\_$	6 + 5	6 + 5	11
$\xrightarrow{4}7$	7 − 4	7 − 4	3
$\xrightarrow{\quad 3}9$	9 − 3	9 − 3	6

(Repeat for families that were not firm.)

EXERCISE 2: ADD
NOMENCLATURE

a. (Display:) [97:2A]

> add

(Point to **add**.) This word is add. Say **add.** (Signal.) *Add.*
b. Add is another word for plus. When you plus 2, you add 2.
• What do you do when you plus 2? (Signal.) *Add 2.*
• What do you do when you plus 4? (Signal.) *Add 4.*
• What do you do when you add 10? (Signal.) *Plus 10.*
• What do you do when you add 7? (Signal.) *Plus 7.*
• What do you do when you plus 15? (Signal.) *Add 15.*
(Repeat until firm.)
c. If you add 10 do you end up with more than 10 or less than 10? (Signal.) *More than 10.* Remember, when you plus you add.

EXERCISE 3: MIXED COUNTING

a. Let's do some counting.
• Count by tens to 100. Get ready. (Tap.) *10, 20, 30, 40, 50, 60, 70, 80, 90, 100.*
• Start with 100 and count backward by tens. Get 100 going. **One huuundred.** Count backward. (Tap.) *90, 80, 70, 60, 50, 40, 30, 20, 10.*
• Start with 100 and count backward by ones to 90. Get 100 going. **One huuundred.** Count backward. (Tap.) *99, 98, 97, 96, 95, 94, 93, 92, 91, 90.*
• Count by fives to 50. Get 5 going. **Fiiive.** Count. (Tap.) *10, 15, 20, 25, 30, 35, 40, 45, 50.*
b. Start with 200 and count by 25s to 300. Get 200 going. **Two huuundred.** Count. (Tap.) *225, 250, 275, 300.*

c. You have 47. When you plus tens, what's the next number? (Signal.) *57.*
• You have 47. When you minus tens, what's the next number? (Signal.) *37.*
• You have 47. When you plus ones, what's the next number? (Signal.) *48.*
• You have 47. When you plus twos, what's the next number? (Signal.) *49.*

EXERCISE 4: COUNT BY TWOS

a. You're going to count by twos to 10. Get ready. (Signal.) *2, 4, 6, 8, 10.*
(Repeat until firm.)
(Display:) [97:4A]

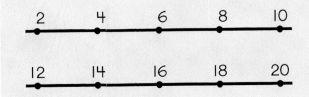

Here are the numbers that come after 10.
• (Point to **12**.) The last part of this number is 2.
• (Point to **14**.) The last part of this number is 4.
• (Point to **16**.) What's the last part of this number? (Signal.) *6.*
• (Point to **18**.) What's the last part of this number? (Signal.) *8.*
• (Point to **20**.) What's the last part of this number? (Signal.) *Zero.*

b. My turn to start with 12 and count by twos. (Touch.) 12, 14, 16, 18, 20.
• (Point to **12**.) Say those numbers with me. Get ready. (Touch as you and children say:) *12, 14, 16, 18, 20.*
(Repeat until firm.)
• Your turn: Say those numbers. Get ready. (Touch.) *12, 14, 16, 18, 20.*

c. My turn to say all the numbers. (Touch numbers as you say:) 2, 4, 6, 8, 10, 12, 14, 16, 18, 20.
• Your turn: Say all the numbers. (Touch as children say:) *2, 4, 6, 8, 10, 12, 14, 16, 18, 20.*
(Repeat until firm.)

d. (Do not show display.)
My turn to say those numbers without looking at them. 12, 14, 16, 18, 20.
• Your turn: Say the numbers. (Signal.) *12, 14, 16, 18, 20.*
(Repeat until firm.)

EXERCISE 5: RELATIVE SIZE
ORDERING BY LENGTH

a. (Display:) [97:5A]

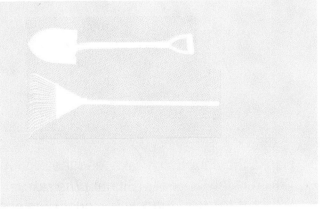

This picture shows a shovel and a rake.
• Which is longer—the shovel or the rake? (Signal.) *The rake.*

b. (Display:) [97:5B]

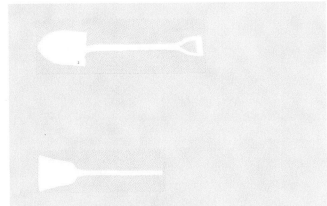

Here's the shovel and a broom.
• Which is longer—the shovel or the broom? (Signal.) *The shovel.*
• So if the rake were in this picture, which would be longer—the broom or the rake? (Signal.) *The rake.*
Yes, the rake is longer than the shovel. The shovel is longer than the broom. So the rake is longer than the broom.

c. Which is the longest—the broom, the shovel, or the rake? (Signal.) *The rake.*
• Which is the next longest—the broom or the shovel? (Signal.) *The shovel.*
• Which is the shortest? (Signal.) *The broom.*
(Repeat step c until firm.)

d. (Display:) [97:5C]

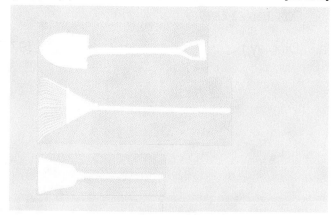

Here's the shovel, the rake, and the broom.
- Which is the longest? (Signal.) *The rake.*
- Which is the next longest? (Signal.) *The shovel.*
- Which is the shortest? (Signal.) *The broom.*
Good ordering objects by their length.

e. (Display:) [97:5D]

This picture shows an airplane and a bus.
- Which is longer—the airplane or the bus? (Signal.) *The airplane.*

f. (Display:) [97:5E]

Here's the airplane and a train.
- Which is longer—the airplane or the train? (Signal.) *The train.*

- So if the bus were in this picture, which would be longer—the bus or the train? (Signal.) *The train.*
  Yes, the train is longer than the airplane. The airplane is longer than the bus. So the train is longer than the bus.
g. Which is the longest—the airplane, the bus, or the train? (Signal.) *The train.*
- Which is the next longest—the airplane or the bus? (Signal.) *The airplane.*
- Which is the shortest? (Signal.) *The bus.*
  (Repeat step g until firm.)

h. (Display:) [97:5F]

Here's the airplane, the bus, and the train.
- Which is the longest? (Signal.) *The train.*
- Which is the next longest? (Signal.) *The airplane.*
- Which is the shortest? (Signal.) *The bus.*
Good ordering objects by their length.

EXERCISE 6: MONEY
DOLLAR PROBLEMS WITH CARRYING

a. (Distribute unopened workbooks to children.)
- Open your workbook to Lesson 97 and find part 1.
  (Observe children and give feedback.)
  (Teacher reference:)

- Touch and read the problem. (Signal.) *8 dollars and 39 cents plus 11 dollars and 62 cents.*
  You have to carry more than twice to work this problem.
- Say the problem for the cent ones column. (Signal.) *9 plus 2.*
- Write the answer. Then stop. ✔
  (Display:) W̄   [97:6A]

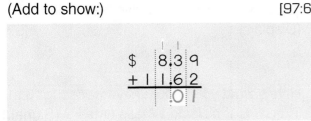

  Here's what you should have.
b. Say the problem for the cent tens column. (Signal.) *1 plus 3 plus 6.*
- Write the answer. Then stop. ✔
  (Add to show:)   [97:6B]

  Here's what you should have.
c. Say the problem for the dollar ones column. (Signal.) *1 plus 8 plus 1.*
- Write the answer for that column. Then stop. ✔
  (Add to show:)   [97:6C]

$ 8.39
+11.62
0.01

  Here's what you should have.

d. Say the problem for the dollar tens column. (Signal.) *1 plus 1.*
- Write the answer. Remember the sign. ✔
  (Add to show:)   [97:6D]

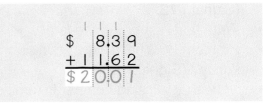

  Here's what you should have.
- Everybody, read the problem we started with and the answer. (Signal.) *8 dollars and 39 cents plus 11 dollars and 62 cents equals 20 dollars and 1 cent.*

EXERCISE 7: NUMBER FAMILIES
MULTIPLE DIGIT NUMBERS

a. Find part 2 on worksheet 97. ✔
  (Teacher reference:)

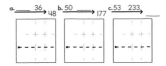

  You're going to say the problem for each number family. Then you're going to write column problems and work them.
- Touch family A. ✔
- What's missing in this family? (Signal.) *A small number.*
- Say the problem. (Signal.) *48 minus 36.*
b. Touch family B. ✔
- What's missing in this family? (Signal.) *A small number.*
- Say the problem. (Signal.) *177 minus 50.*
c. Touch family C. ✔
- What's missing in this family? (Signal.) *The big number.*
- Say the problem. (Signal.) *53 plus 233.*
d. Write all the problems and work them. Raise your hand when you have answers to all the problems. ✔

e. You know the missing number for each family. That's the answer to each problem.
• Problem A. What's the answer? (Signal.) *12.*
• So what's the missing number in that family? (Signal.) *12.*
• Write that number in the family. ✔
• Problem B. What's the missing number in that family? (Signal.) *127.*
• Write that missing number in family B. Then write the missing number in family C. (Observe children and give feedback.)
f. Check your work.
• Problem A. What's the missing number? (Signal.) *12.*
• Problem B. What's the missing number? (Signal.) *127.*
• Problem C. What's the missing number? (Signal.) *286.*

EXERCISE 8: TELLING TIME
WRITING 1-DIGIT NUMBERS

REMEDY

a. Find part 3. ✔
(Teacher reference:)

R Part C

```
a. ___ : ___
b. ___ : ___
c. ___ : ___
d. ___ : ___
e. ___ : ___
f. ___ : ___
```

I'll say a time for each space. You'll write the time. The dots for each time are already shown.
• If I say o'clock, what will you write after the dots? (Signal.) *Zero, zero.*
b. Time A is 4 oh 9. What's time A? (Signal.) *4 oh 9.*
• What do you write after the dots for 4 **oh 9**? (Signal.) *Zero, 9.*
• Write 4 oh 9 in space A. (Observe children and give feedback.) (Display:) [97:8A]

a. __4:09__

Here's what you should have written for 4 oh 9.

c. Time B is 12 fifty-nine. What's time B? (Signal.) *12 fifty-nine.*
• Write 12 fifty-nine in space B. (Observe children and give feedback.) (Display: [97:8B]

b. __12:59__

Here's what you should have written for 12 fifty-nine.
d. Time C is 10 o'clock. What's time C? (Signal.) *10 o'clock.*
• Write the time for 10 o'clock in space C. (Observe children and give feedback.) (Display:) [97:8C]

c. __10:00__

Here's what you should have written for 10 o'clock.
e. Time D is 11 oh 6. What's time D? (Signal.) *11 oh 6.*
• Write 11 oh 6 in space D. (Observe children and give feedback.) (Display:) [97:8D]

d. __11:06__

Here's what you should have written for 11 oh 6.
f. Time E is 3 thirty. What's time E? (Signal.) *3 thirty.*
• Write 3 thirty in space E. (Observe children and give feedback.) (Display:) [97:8E]

e. __3:30__

Here's what you should have written for 3 thirty.
g. Time F is 2 o'clock. What's time F? (Signal.) *2 o'clock.*
• Write 2 o'clock in space F. (Observe children and give feedback.) (Display:) [97:8F]

f. __2:00__

Here's what you should have written for 2 o'clock.
h. Now you'll read the times you wrote for part 3.
• Read time A. (Signal.) *4 oh 9.*
• (Repeat for:) B, *12 fifty-nine;* C, *10 o'clock;* D, *11 oh 6;* E, *3 thirty;* F, *2 o'clock.*

EXERCISE 9: FACTS
SUBTRACTION

a. Find part 4.
(Teacher reference:)

a. $9-3$	f. $7-3$
b. $10-6$	g. $11-9$
c. $11-5$	h. $8-4$
d. $6-4$	i. $7-5$
e. $12-6$	j. $10-5$

These are minus problems you'll work. Think of the number family for each problem and write the answers. Put your pencil down when you've completed the equations in part 4.
(Observe children and give feedback.)

b. Check your work. You'll read the equation for each problem.
- Problem A. (Signal.) $9 - 3 = 6$.
- (Repeat for:) B, $10 - 6 = 4$; C, $11 - 5 = 6$; D, $6 - 4 = 2$; E, $12 - 6 = 6$; F, $7 - 3 = 4$; G, $11 - 9 = 2$; H, $8 - 4 = 4$; I, $7 - 5 = 2$; J, $10 - 5 = 5$.

EXERCISE 10: 3-D OBJECTS WITH 2-D FACES
REMEDY

a. (Display:)   [97:10A]

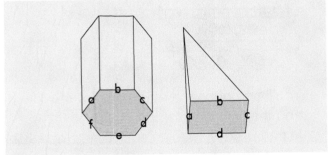

You can see the bottom face and some of the side faces of these objects.

- (Point to .) The bottom face of this object is shaded. Raise your hand when you know what shape the bottom face is. ✔
- Everybody, what shape? (Signal.) *(A) hexagon.*
- How many sides does the bottom face have? (Signal.) *6.*
- So how many side faces come from the bottom face? (Signal.) *6.*
- Can you see all the side faces? (Signal.) *No.*

b. Raise your hand when you know how many side faces you can see. ✔
- How many side faces can you see? (Signal.) *3.*
- Tell me the letters of the side faces you can see. Get ready. (Signal.) *A, B, C.*
- What shape are these side faces? (Signal.) *Rectangles.*

c. You cannot see 3 side faces.
- Raise your hand when you know the letters of the side faces you can't see. ✔
- What are the letters of the side faces you can't see? (Signal.) *D, E, and F.*
  Yes, you can't see the side faces of D, E, and F, but you know the shape they are.
- What shape are side faces D, E, and F? (Signal.) *Rectangles.*

d. (Point to .) The bottom face is shaded. Raise your hand when you know what shape the bottom face is. ✔
- Everybody, what's the shape of the bottom face? (Signal.) *(A) rectangle.*
- How many sides does the bottom face have? (Signal.) *4.*
- So how many side faces come from the bottom face? (Signal.) *4.*
- Can you see all the side faces? (Signal.) *No.*

e. Raise your hand when you know how many side faces you can see. ✔
- How many side faces can you see? (Signal.) *2.*
- Tell me the letters of the side faces you can see. Get ready. (Signal.) *A and B.*
- What shape are those side faces? (Signal.) *Triangles.*

f. You cannot see 2 side faces. Raise your hand when you know the letter of the side faces you can't see. ✔
- What are the letters of the side faces you can't see? (Signal.) *C and D.*
  Yes, you can't see the side faces of C and D, but you know the shape they are.
- What shape are side faces C and D? (Signal.) *Triangles.*

g. Find part 5 on worksheet 97. ✔
(Teacher reference:)

 R Part G

You're going to write a letter for faces of this object to show if they are triangles, rectangles, or hexagons.
You'll write T for each face if it is a triangle, R if it's a rectangle, or H if it's a hexagon.
• What letter do you write for faces that are triangles? (Signal.) *T.*
• What letter do you write for faces that are rectangles? (Signal.) *R.*
• What letter do you write for faces that are hexagons? (Signal.) *H.*
(Repeat until firm.)
h. Touch the bottom face. ✔
• Write the letter for the bottom face. ✔
• Everybody, what letter did you write? (Signal.) *T.*
Yes, the bottom is a triangle.
i. Touch side face A. ✔
• Write T on that face if the face is a triangle, R if it's a rectangle, or H if it's a hexagon. ✔
• Everybody, what letter did you write? (Signal.) *R.*
j. Write the letter for side face B. ✔
• Everybody, what letter did you write? (Signal.) *R.*
k. Does this object have a top face? (Signal.) *Yes.*
• Can you see the top face? (Signal.) *No.*
• Raise your hand when you know what shape the top face is. ✔
• What's the top face? (Signal.) *(A) triangle.*
• Write the letter T on the line above the object. ✔
(Display:)   [97:10B]

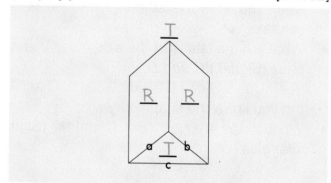

Here's what you should have.

EXERCISE 11: INDEPENDENT WORK

a. Find part 6. ✔
(Teacher reference:)

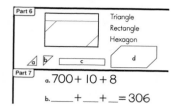

You'll circle the name for the whole shape.
Then you'll write the letter for each part.
In part 7, you'll complete the place-value equations.
b. Turn to the other side of your worksheet and find part 8. ✔
(Teacher reference:)

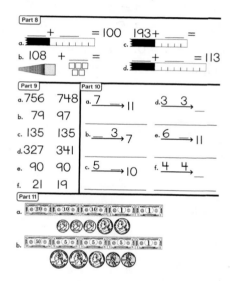

You'll complete the equations for the rulers and the equation for the boxes.
In part 9, you'll write an equals sign, or a sign that has a bigger end next to one of the numbers.
In part 10, below each family, you'll write the fact for the missing number and you'll complete the family.
In part 11, you'll write the dollars and cents amount for each group of bills and coins.
c. Complete worksheet 97. Remember to work parts 6 and 7 on the other side of your worksheet.
(Observe children and mark incorrect responses on children's worksheets as you give feedback.)

Lesson 98

EXERCISE 1: COUNT BY TWOS

a. You're going to count by twos to 10. Get ready.
(Signal.) *2, 4, 6, 8, 10.*
(Repeat until firm.)
(Display:)   [98:1A]

| 2 | 4 | 6 | 8 | 10 |
| 12 | 14 | 16 | 18 | 20 |

b. My turn to start with 12 and count by twos.
(Touch numbers as you say:) 12, 14, 16, 18, 20.
• (Point to **12**.) Say those numbers with me.
Get ready. (Touch as you and children say:)
12, 14, 16, 18, 20.
(Repeat until firm.)
c. Your turn: Say those numbers. Get ready.
(Touch.) *12, 14, 16, 18, 20.*
My turn to say all the numbers. (Touch numbers
as you say:) 2, 4, 6, 8, 10, 12, 14, 16, 18, 20.
• Your turn: Say all the numbers. (Touch as
children say:) *2, 4, 6, 8, 10, 12, 14, 16, 18, 20.*
(Repeat until firm.)

=== INDIVIDUAL TURNS ===

(Call on different students to perform the
following task.)

• Start with 12 and count by twos to 20. (Call on
a student.) *12, 14, 16, 18, 20.*

EXERCISE 2: FACTS
ADDITION/SUBTRACTION

a. (Display:)   [98:2A]

4 + 3	3 + 3	12 − 6
10 − 5	11 − 5	5 + 3
9 − 3	8 − 4	10 − 6

• (Point to **4 + 3**.) Listen: 4 + 3. Is the answer
the big number or a small number? (Signal.)
The big number.
• Say the fact. (Signal.) *4 + 3 = 7.*

b. (Point to **10 − 5**.) 10 − 5. Is the answer the big
number or a small number? (Signal.) *A small
number.*
• Say the fact. (Signal.) *10 − 5 = 5.*
c. (Repeat the following tasks for the remaining
problems:)

(Point to __.) __. Is the answer the big number or a small number?			Say the fact.
9 − 3	9 − 3	A small number.	9 − 3 = 6
3 + 3	3 + 3	The big number.	3 + 3 = 6
11 − 5	11 − 5	A small number.	11 − 5 = 6
8 − 4	8 − 4	A small number.	8 − 4 = 4
12 − 6	12 − 6	A small number.	12 − 6 = 6
5 + 3	5 + 3	The big number.	5 + 3 = 8
10 − 6	10 − 6	A small number.	10 − 6 = 4

(Repeat problems that were not firm.)

EXERCISE 3: ADD
NOMENCLATURE

a. (Display:)   [98:3A]

4 5 + 1 7	163 + 105	5 7 + 8 4
	314 + 93	

• (Point to **45**.) Read this problem. (Signal.)
45 plus 17.
• How many does this problem plus? (Signal.) *17.*
• So this problem adds 17. How many does this
problem add? (Signal.) *17.*
b. (Point to **163**.) Read this problem. (Signal.)
163 plus 105.
• How many does this problem add? (Signal.) *105.*
c. (Point to **314**.) Read this problem. (Signal.)
314 plus 93.
• How many does this problem add? (Signal.) *93.*
d. (Point to **57**.) Read this problem. (Signal.)
57 plus 84.
• How many does this problem add?
(Signal.) *84.*
(Repeat steps that were not firm.)

EXERCISE 4: MIXED COUNTING

a. Let's do some counting.
- Start with 35 and count by fives to 75. Get 35 going. *Thirty fiiive.* Count. (Tap.) *40, 45, 50, 55, 60, 65, 70, 75.*
- Count by hundreds to 1000. Get 100 going. *One huuundred.* Count. (Tap.) *200, 300, 400, 500, 600, 700, 800, 900, 1000.*

b. Start with 80 and count backward by tens. Get 80 going. *Eightyyy.* Count backward. (Tap.) *70, 60, 50, 40, 30, 20, 10.*

c. Start with 700 and count by 25s to 800. Get 700 going. *Seven huuundred.* Count. (Tap.) *725, 750, 775, 800.*
- Count by 25s to 100. Get ready. (Tap.) *25, 50, 75, 100.*
- Start with 125 and plus tens to 175. Get 125 going. *One hundred twenty-fiiive.* Plus tens. (Tap.) *135, 145, 155, 165, 175.*

d. Start with 28 and count backward to 18. Get 28 going. *Twenty-eieieight.* Count backward. (Tap.) *27, 26, 25, 24, 23, 22, 21, 20, 19, 18.* (Repeat until firm.)

e. You have 50. When you plus tens, what's the next number? (Signal.) *60.*
- You have 50. When you minus tens, what's the next number? (Signal.) *40.*
- You have 50. When you plus 25s, what's the next number? (Signal.) *75.*
- You have 50. When you plus fives, what's the next number? (Signal.) *55.*
- You have 50. When you plus ones, what's the next number? (Signal.) *51.*
- You have 50. When you minus ones, what's the next number? (Signal.) *49.*

EXERCISE 5: NUMBER FAMILIES
MULTIPLE DIGIT NUMBERS

a. (Distribute unopened workbooks to children.)
- Open your workbook to Lesson 98 and find part 1.
  (Observe children and give feedback.)
  (Teacher reference:)

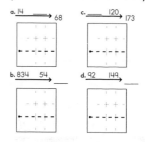

You're going to say the problem for each number family. Then you're going to write column problems and work them.
- Touch family A. ✔
- What's missing in this family? (Signal.) *A small number.*
- Say the problem. (Signal.) *68 minus 14.*

b. Touch family B. ✔
- What's missing in this family? (Signal.) *The big number.*
- Say the problem. (Signal.) *834 plus 54.*

c. Touch family C. ✔
- What's missing in this family? (Signal.) *A small number.*
- Say the problem. (Signal.) *173 minus 120.*

d. Touch family D. ✔
- What's missing in this family? (Signal.) *The big number.*
- Say the problem. (Signal.) *92 plus 149.*

e. Write all the problems and work them. Raise your hand when you have answers to all the problems. ✔

f. Check your work.
- Problem A. 68 minus 14. What's the answer? (Signal.) *54.*
- What's the missing number in that family? (Signal.) *54.*
- Write that number in the family. ✔
- Problem B. What's the missing number in that family? (Signal.) *888.*
- Write that number in family B. Then write the missing numbers in the rest of the families.
  (Observe children and give feedback.)

g. Check your work.
- Family A. What's the missing number? (Signal.) *54.*
- Family B. What's the missing number? (Signal.) *888.*
- Family C. What's the missing number? (Signal.) *53.*
- Family D. What's the missing number? (Signal.) *241.*

EXERCISE 6: TELLING TIME
WRITING 1-DIGIT NUMBERS

a. Find part 2 on worksheet 98. ✔
(Teacher reference:)

a. ___:___
b. ___:___
c. ___:___
d. ___:___
e. ___:___
f. ___:___

I'll say a time for each space. You'll write the time. The dots for each time are already shown.
b. Time A is 4 o'clock. What's time A? (Signal.) *4 o'clock.*
- Write 4 o'clock in space A. ✔
c. Time B is 10 oh 7. What's time B? (Signal.) *10 oh 7.*
- Write 10 oh 7 in space B. ✔
d. Time C is 11 thirty. What's time C? (Signal.) *11 thirty.*
- Write 11 thirty in space C. ✔
e. Time D is 2 oh 5. What's time D? (Signal.) *2 oh 5.*
- Write 2 oh 5 in space D. ✔
f. Time E is 12 o'clock. What's time E? (Signal.) *12 o'clock.*
- Write 12 o'clock in space E. ✔
g. Time F is 2 o'clock. What's time F? (Signal.) *2 o'clock.*
- Write 2 o'clock in space F. ✔

h. Check your work.
(Display:)   [98:6A]

a.	4:00
b.	10:07
c.	11:30
d.	2:05
e.	12:00
f.	2:00

Here are the times you should have written for part 2. You're going to read the times you wrote, so check them carefully and fix up any mistakes.
(Observe children and give feedback.)
i. Now you'll read the times you wrote for part 2.
- Read time A. (Signal.) *4 o'clock.*
- (Repeat for:) B, *10 oh 7;* C, *11 thirty;* D, *2 oh 5;* E, *12 o'clock;* F, *2 o'clock.*

EXERCISE 7: WORD PROBLEMS (COLUMNS)

a. Find part 3. ✔
(Teacher reference:)

You're going to write the symbols for word problems in columns and work them.
- Listen to problem A: A dog had 63 bones. Then the dog found 34 more bones. How many bones did the dog end up with?
- Listen again and write the symbols for the whole problem: A dog had 63 bones. Then the dog found 34 more bones.
(Observe children and give feedback.)
- Everybody, touch and read the problem for A. Get ready. (Signal.) *63 plus 34.*

b. Work problem A. Put your pencil down when you know how many bones the dog ended up with.
(Observe children and give feedback.)
(Teacher reference:)

- Read the problem and the answer you wrote for A. Get ready. (Signal.) *63 + 34 = 97.*
- How many bones did the dog end up with? (Signal.) *97.*

c. Touch where you'll write the symbols for problem B. ✔
Listen to problem B: There were 176 chickens on a farm. Then the farmer sold 45 of those chickens. How many chickens were still on the farm?

- Listen again and write the symbols for the whole problem: There were 176 chickens on a farm. Then the farmer sold 45 of those chickens.
(Observe children and give feedback.)
- Everybody, touch and read the problem for B. Get ready. (Signal.) *176 minus 45.*

d. Work problem B. Put your pencil down when you know how many chickens were still on the farm.
(Observe children and give feedback.)
(Teacher reference:)

- Read the problem and the answer you wrote for B. Get ready. (Signal.) *176 – 45 = 131.*
- How many chickens were still on the farm? (Signal.) *131.*

EXERCISE 8: FACTS
SUBTRACTION

a. Find part 4.
(Teacher reference:)

a. 10−4	f. 12−6
b. 6−3	g. 11−9
c. 11−6	h. 8−4
d. 9−6	i. 6−4
e. 7−3	j. 10−5

I'll read each problem. You'll say the fact.

- Problem A. 10 minus 4. Say the fact. (Signal.) *10 – 4 = 6.*
- Problem B. 6 minus 3. Say the fact. (Signal.) *6 – 3 = 3.*
- Problem C. 11 minus 6. Say the fact. (Signal.) *11 – 6 = 5.*
- Problem D. 9 minus 6. Say the fact. (Signal.) *9 – 6 = 3.*
- Problem E. 7 minus 3. Say the fact. (Signal.) *7 – 3 = 4.*
- Problem F. 12 minus 6. Say the fact. (Signal.) *12 – 6 = 6.*
- Problem G. 11 minus 9. Say the fact. (Signal.) *11 – 9 = 2.*
- Problem H. 8 minus 4. Say the fact. (Signal.) *8 – 4 = 4.*
- Problem I. 6 minus 4. Say the fact. (Signal.) *6 – 4 = 2.*
- Problem J. 10 minus 5. Say the fact. (Signal.) *10 – 5 = 5.*
(Repeat problems that were not firm.)

b. Complete the equations for part 4. Put your pencil down when you're finished.
(Observe children and give feedback.)

c. Check your work. You'll touch and read the equation for each problem.
- Problem A. (Signal.) *10 – 4 = 6.*
- (Repeat for:) B, *6 – 3 = 3;* C, *11 – 6 = 5;* D, *9 – 6 = 3;* E, *7 – 3 = 4;* F, *12 – 6 = 6;* G, *11 – 9 = 2;* H, *8 – 4 = 4;* I, *6 – 4 = 2;* J, *10 – 5 = 5.*

EXERCISE 9: 3-D OBJECTS WITH 2-D FACES

a. Find part 5 on worksheet 98. ✔
(Teacher reference:)

- Touch object 1. ✔
You're going to write a letter on each face you can see. The letter will tell the shape of the face.
- What letter will you write to show a triangle? (Signal.) *T.*
- What letter will you write to show a rectangle? (Signal.) *R.*
- What letter will you write to show a hexagon? (Signal.) *H.*

b. Touch the bottom face of object 1. ✔
   Write the letter to show if the face is a triangle,
   a rectangle, or a hexagon. ✔
- Everybody, what letter did you write?
  (Signal.) *H.*
c. Touch side face A. ✔
- Write the letter to show what shape it is. ✔
- Everybody, what letter did you write?
  (Signal.) *R.*
- Write the letter for the other side faces.
  (Observe children and give feedback.)
- How many side faces have the letter R?
  (Signal.) *3.*
- You can't see some side faces. How many
  side faces can't you see? (Signal.) *3.*
- What shape are the side faces you can't see?
  (Signal.) *Rectangles.*
d. Can you see the top face of the object?
  (Signal.) *No.*
  But you can figure out what shape it is.
- Write the letter for the top face on the line
  above the object. ✔
- Everybody, what letter did you write? (Signal.) *H.*
e. Touch the bottom face of object 2. ✔
- Write the letter to show what shape it is. ✔
- Everybody, what letter did you write?
  (Signal.) *T.*
f. Touch a side face. ✔
- Write the letter to show what shape it is. ✔
- Everybody, what letter did you write?
  (Signal.) *T.*
- Write a letter on the other side face you can see.
  (Observe children and give feedback.)
- How many side faces have the letter T?
  (Signal.) *2.*
- You can't see some side faces. How many
  side faces can't you see? (Signal.) *1.*
- What shape is the side face you can't see?
  (Signal.) *(A) triangle.*
g. Does this object have a top face? (Signal.) *No.*
  So you don't write a letter for the top face.

EXERCISE 10: INDEPENDENT WORK

a. Turn to the other side of your worksheet and
   find part 6. ✔
   (Teacher reference:)

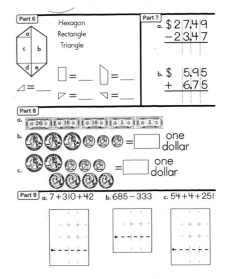

You'll circle the name for the whole shape.
Then you'll complete each equation by writing
the letter to show which part it is.
In part 7, you'll work the money problems.
In part 8, you'll write an equals and the
number of dollars to show what the group of
bills is worth. You'll circle the words one dollar
or write the cents for each group of coins.
In part 9, you'll write each problem in a column
and work it.

b. Complete worksheet 98.
   (Observe children and mark incorrect responses
   on children's worksheets as you give feedback.)

Lesson 99

EXERCISE 1: MIXED COUNTING

a. Let's do some counting.
- Count by hundreds to 1000. Get ready. (Tap.) *100, 200, 300, 400, 500, 600, 700, 800, 900, 1000.*
- Start with 25 and count by fives to 75. Get 25 going. *Twenty-fiiive.* Count. (Tap.) *30, 35, 40, 45, 50, 55, 60, 65, 70, 75.*
- Count by 25s to 100. Get ready. (Tap.) *25, 50, 75, 100.*
- Start with 500 and count by 25s to 600. Get 500 going. *Five huuundred.* Count. (Tap.) *525, 550, 575, 600.*

b. Start with 24 and count backward to 10. Get 24 going. *Twenty-fouuur.* Count backward. (Tap.) *23, 22, 21, 20, 19, 18, 17, 16, 15, 14, 13, 12, 11, 10.*

c. Start with 26 and plus tens to 76. Get 26 going. *Twenty-siiix.* Plus tens. (Tap.) *36, 46, 56, 66, 76.*

d. Start with 100 and count backward by tens. Get 100 going. *One huuundred.* (Tap.) *90, 80, 70, 60, 50, 40, 30, 20, 10.*

e. Start with 324 and plus tens to 374. Get 324 going. *Three hundred twenty-fouuur.* Plus tens. (Tap.) *334, 344, 354, 364, 374.* (Repeat until firm.)

f. You have 76. When you plus ones, what's the next number? (Signal.) *77.*
- You have 76. When you plus tens, what's the next number? (Signal.) *86.*
- You have 76. When you minus ones, what's the next number? (Signal.) *75.*
- You have 76. When you plus hundreds, what's the next number? (Signal.) *176.*

EXERCISE 2: NUMBER FAMILIES
SMALL NUMBER OF 7, 8   REMEDY

a. (Display:)   [99:2A]

The big number is missing in these families. These are new families that have small numbers of 7 and small numbers of 8.

- (Point to $7 \longrightarrow 7$.) What are the small numbers in this family? (Signal.) *7 and 7.*
- The big number is 14. What's the big number? (Signal.) *14.*
- (Point to $8 \longrightarrow 8$.) What are the small numbers in this family? (Signal.) *8 and 8.*
- The big number is 16. What's the big number? (Signal.) *16.*

b. (Point to $7 \longrightarrow 7$.) What are the small numbers in this family? (Signal.) *7 and 7.*
- What's the big number? (Signal.) *14.*

c. (Point to $8 \longrightarrow 8$.) What are the small numbers in this family? (Signal.) *8 and 8.*
- What's the big number? (Signal.) *16.* (Repeat steps b and c until firm.)

d. There's only one plus fact and one minus fact for each family.
- (Point to $7 \longrightarrow 7$.) Say the plus fact. (Signal.) *7 + 7 = 14.*
- Say the minus fact. (Signal.) *14 – 7 = 7.*

e. (Point to $8 \longrightarrow 8$.) Say the plus fact. (Signal.) *8 + 8 = 16.*
- Say the minus fact. (Signal.) *16 – 8 = 8.* (Repeat until firm.)

f. (Display:)   [99:2B]

Each family has a missing number. You'll say the problem for each family. Then you'll say the answer.

g. (Point to $7 \longrightarrow 14$.) Say the problem for the missing number. Get ready. (Signal.) *14 minus 7.*
- What's 14 minus 7? (Signal.) *7.*

h. (Point to $8 \longrightarrow 8 \_$.) Say the problem for the missing number. (Signal.) *8 plus 8.*
- What's 8 plus 8? (Signal.) *16.*

i. (Repeat the following tasks for the remaining families:)

(Point to __.) Say the problem for the missing number.		What's __?	
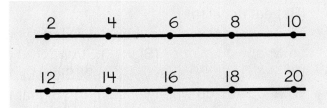 =→6 ³	6 − 3	6 − 3	3
6 3→_	6 + 3	6 + 3	9
10 5→_	10 + 5	10 + 5	15
5→10	10 − 5	10 − 5	5
=→7 ³	7 − 3	7 − 3	4
6 6→_	6 + 6	6 + 6	12
=→16 ⁸	16 − 8	16 − 8	8

(Repeat for families that were not firm.)

EXERCISE 3: COUNT BY TWOS

a. (Display:)   [99:3A]

2   4   6   8   10
12   14   16   18   20

My turn to count by twos to 20. 2, 4, 6, 8, 10, *(pause)* 12, 14, 16, 18, 20.
• Your turn: Count by twos to 20. (Touch.) *2, 4, 6, 8, 10, (pause) 12, 14, 16, 18, 20.*
(Repeat until firm.)

b. (Do not show display.)
My turn to say the numbers without looking. 2, 4, 6, 8, 10 *(pause)* 12, 14, 16, 18, 20.
• Your turn: Count by twos to 20. Get ready. (Tap.) *2, 4, 6, 8, 10, (pause) 12, 14, 16, 18, 20.*
(Repeat until firm.)

═══════════ INDIVIDUAL TURNS ═══════════

(Call on different students to perform the following task.)

• Count by twos to 20. (Call on a student.) *2, 4, 6, 8, 10, 12, 14, 16, 18, 20.*

EXERCISE 4: PLACE VALUE
PICTURES   REMEDY

a. (Display:)   [99:4A]

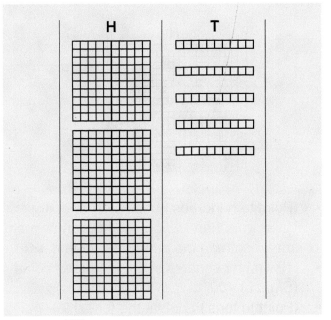

(Point to ▦.) This is a picture of 100 squares.
• How many squares? (Signal.) *100.*
I'll count the hundreds. (Touch each box as you count:) 100, 200, 300.
• How many squares are there in this column? (Signal.) *300.*
• (Point to ▦.) Your turn to count the hundreds. (Touch each box as children count:) *100, 200, 300.*

b. (Point to ▭.) Each row in this column has ten squares. How many squares? (Signal.) *10.*
• Your turn to count the tens in this column. Get ready. (Touch tens.) *10, 20, 30, 40, 50.*
• How many squares are there in this column? (Signal.) *50.*
Yes, there are 50 squares in this column.

c. My turn to say the plus problem for all the squares. (Point to each column as you say:) 300 plus 50.
• Your turn to say the problem. (Touch columns.) *300 plus 50.*
(Repeat step c until firm.)

d. What does 300 plus 50 equal? (Signal.) *350.*
• So how many squares are there? (Signal.) *350.*
  (Repeat step d until firm.)
e. (Display:) [99:4B]

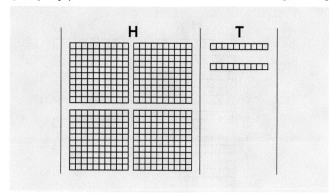

• (Point to hundreds.) Look at the hundreds. Raise your hand when you know how many squares are in the hundreds column. ✔
• How many squares in the hundreds column? (Signal.) *400.*
• (Point to tens.) Look at the tens. How many squares are in the tens column? (Signal.) *20.*
  (Repeat until firm.)
f. My turn to say the plus problem for all the squares. (Touch columns.) *400 plus 20.*
• Say the problem. (Touch columns.) *400 plus 20.*
  (Repeat step f until firm.)
g. What does 400 plus 20 equal? (Signal.) *420.*
• So how many squares are there? (Signal.) *420.*

EXERCISE 5: ADD/SUBTRACT
NOMENCLATURE

a. (Display:) [99:5A]

3 0 7	7 4	5 1 9 + 8 0 0
+ 6 0	+ 2 4 3	
		6 4 + 9 3

• (Point to **307.**) Read this problem. (Signal.) *307 plus 60.*
• How many does this problem plus? (Signal.) *60.*
• So how many does this problem add? (Signal.) *60.*

b. (Point to **74.**) Read this problem. (Signal.) *74 plus 243.*
• How many does this problem add? (Signal.) *243.*
c. (Point to **519.**) Read this problem. (Signal.) *519 plus 800.*
• How many does this problem add? (Signal.) *800.*
d. (Point to **64.**) Read this problem. (Signal.) *64 plus 93.*
• How many does this problem add? (Signal.) *93.*
  (Repeat problems that were not firm.)
e. (Display:) [99:5B]

subtract

• This word is subtract. Say **subtract.** (Signal.) *Subtract.*
  (Repeat until firm.)
f. When you minus you subtract. What do you do when you minus? (Signal.) *Subtract.*
• What do you do when you plus? (Signal.) *Add.*
• What do you do when you minus? (Signal.) *Subtract.*
  (Repeat step f until firm.)
g. If you minus 5, how many do you subtract? (Signal.) *5.*
• If you minus 1, how many do you subtract? (Signal.) *1.*
  Remember, when you minus, you subtract.
h. (Display:) [99:5C]

307 – 60

669 – 120

• (Point to **307.**) Read this problem. (Signal.) *307 minus 60.*
• How many does this problem subtract? (Signal.) *60.*
• (Point to **669.**) Read this problem. (Signal.) *669 minus 120.*
• How many does this problem subtract? (Signal.) *120.*

EXERCISE 6: TELLING TIME
READING DIGITAL DISPLAYS

a. (Display:)   [99:6A]

You're going to read the times shown on these clocks.
- Some of these clocks show a 1-digit number of minutes. When you say the minutes for those clocks, you say oh and the number. What do you say for those clocks? (Signal.) *Oh and the number.* (Repeat until firm.)
- If there are zero minutes shown, what do you say for the minutes? (Signal.) *O'clock.*

b. (Point to **9:02**.) This clock shows a 1-digit number of minutes. So you say oh and the number when you say the minutes.
- Look at the number after the dots. What will you say for the minutes? (Signal.) *Oh 2.*
- Read the time for this clock. (Signal.) *9 oh 2.*

c. (Point to **9:20**.) Does this clock show a 1-digit number of minutes? (Signal.) *No.*
- What will you say for the minutes? (Signal.) *20.*
- Read the time for this clock. (Signal.) *9 twenty.*

d. (Repeat the following tasks for remaining clocks:)

(Point to __.) Does this clock show a 1-digit number of minutes?	So what will you say for the minutes?	Read the time for this clock.
11:03	Yes	Oh 3	11 oh 3
11:30	No	30	11 thirty
6:00	No	O'clock	6 o'clock
6:01	Yes	Oh 1	6 oh 1
10:00	No	O'clock	10 o'clock

(Repeat clocks that were not firm.)

e. Read the time for these clocks again.
- (Point to **9:02**.) What time? (Touch.) *9 oh 2.*
- (Repeat for remaining clocks: 9 twenty, 11 oh 3, 11 thirty, 6 o'clock, 6 oh 1, 10 o'clock.) (Repeat clocks that were not firm.)

INDIVIDUAL TURNS

I'll call on individuals to read one or two clocks.
- (Point to **11:30**.) What's the time for this clock? (Call on a student.) *11 thirty.*
- (Point to **11:03**.) What's the time for this clock? (Call on a student.) *11 oh 3.*
- (Point to **6:00**.) What's the time for this clock? (Call on a student.) *6 o'clock.*
- (Point to **9:02**.) What's the time for this clock? (Call on a student.) *9 oh 2.*

EXERCISE 7: FACTS
PLUS/MINUS MIX   REMEDY

a. (Distribute unopened workbooks to children.)
- Open your workbook to Lesson 99 and find part 1.
  (Observe children and give feedback.)
  (Teacher reference:)   R Test II: Part 0

a. $10-4$	e. $8+8$	i. $6-3$
b. $4+4$	f. $10-5$	j. $6+3$
c. $12-6$	g. $7+7$	k. $10-6$
d. $8-4$	h. $14-7$	

- Touch and read problem A. Get ready. (Signal.) *10 minus 4.*
- Is the answer the big number or a small number? (Signal.) *A small number.*
- What's 10 minus 4? (Signal.) *6.*

b. (Repeat the following tasks for problems B through K:)

Touch and read problem __.	Is the answer the big number or a small number?	What's __?	
B	The big number.	4 + 4	8
C	A small number.	12 − 6	6
D	A small number.	8 − 4	4
E	The big number.	8 + 8	16
F	A small number.	10 − 5	5
G	The big number.	7 + 7	14
H	A small number.	14 − 7	7
I	A small number.	6 − 3	3
J	The big number.	6 + 3	9
K	A small number.	10 − 6	4

(Repeat problems that were not firm.)

c. Write the answer to each fact. Put your pencil down when you've completed the equations for part 1.
  (Observe children and give feedback.)

d. Check your work.
- Touch and read fact A. (Signal.) *10 – 4 = 6.*
- (Repeat for:) B, *4 + 4 = 8;* C, *12 – 6 = 6;* D, *8 – 4 = 4;* E, *8 + 8 = 16;* F, *10 – 5 = 5;* G, *7 + 7 = 14;* H, *14 – 7 = 7;* I, *6 – 3 = 3;* J, *6 + 3 = 9;* K, *10 – 6 = 4.*

EXERCISE 8: 3-D OBJECTS WITH 2-D FACES REMEDY

a. Find part 2 on worksheet 99. ✔
(Teacher reference:)

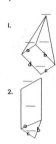

- Touch object 1. ✔
You're going to write a letter on each face you can see. The letter will tell the shape of the face.
- What letter will you write to show a triangle? (Signal.) *T.*
- What letter will you write to show a rectangle? (Signal.) *R.*
- What letter will you write to show a hexagon? (Signal.) *H.*
b. Touch the bottom face of object 1. ✔
- Write the letter to show what shape it is. ✔
- Everybody, what letter did you write? (Signal.) *R.*
c. Touch a side face. ✔
- Write the letter to show what shape it is. ✔
- Everybody, what letter did you write? (Signal.) *T.*
- Write the letter on the other side face you can see.
(Observe children and give feedback.)
- How many side faces have the letter T? (Signal.) *2.*
- You can't see some side faces. How many side faces can't you see? (Signal.) *2.*
- What shape are the side faces you can't see? (Signal.) *Triangles.*
d. Does this object have a top face? (Signal.) *No.*
So you don't write a letter for the top face.

e. Touch the bottom face of object 2. ✔
- Write the letter to show if the face is a triangle, a rectangle, or a hexagon. ✔
- Everybody, what letter did you write? (Signal.) *T.*
f. Touch side face A. ✔
- Write the letter to show what shape it is. ✔
- Everybody, what letter did you write? (Signal.) *R.*
- Can you see any other side faces? (Signal.) *No.*
- How many side faces have the letter R? (Signal.) *1.*
- You can't see some side faces. How many side faces can't you see? (Signal.) *2.*
- What shape are the side faces you can't see? (Signal.) *Rectangles.*
g. Can you see the top face of the object? (Signal.) *No.*
But you can figure out what shape it is.
- Write the letter for the top face on the line above the object. ✔
- Everybody, what letter did you write? (Signal.) *T.*

EXERCISE 9: NUMBER FAMILIES
MULTIPLE DIGIT NUMBERS   REMEDY

a. Find part 3 on worksheet 99. ✔
(Teacher reference:)

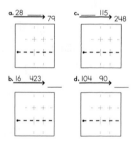

You're going to say the problem for each number family. Then you're going to write column problems and work them.
- Touch family A. ✔
- Is a small number or the big number missing in this family? (Signal.) *A small number.*
- Say the problem for figuring out the missing number. (Signal.) *79 minus 28.*

b. Touch family B. ✔
• Is a small number or the big number missing in this family? (Signal.) *The big number.*
• Say the problem for figuring out the missing number. (Signal.) *16 plus 423.*
c. Touch family C. ✔
• Is a small number or the big number missing in this family? (Signal.) *A small number.*
• Say the problem for figuring out the missing number. (Signal.) *248 minus 115.*
d. Touch family D. ✔
• Is a small number or the big number missing in this family? (Signal.) *The big number.*
• Say the problem for figuring out the missing number. (Signal.) *104 plus 90.*
(Repeat families that were not firm.)
e. Write all the problems and work them. Then write the missing numbers in the families. Raise your hand when you've completed the families in part 3.
(Observe children and give feedback.)
f. Check your work.
• Read problem A and the answer. (Signal.) *79 – 28 = 51.*
• What's the missing number in family A? (Signal.) *51.*
g. Read problem B and the answer. (Signal.) *16 + 423 = 439.*
• What's the missing number in family B? (Signal.) *439.*
h. Read problem C and the answer. (Signal.) *248 – 115 = 133.*
• What's the missing number in family C? (Signal.) *133.*
i. Read problem D and the answer. (Signal.) *104 + 90 = 194.*
• What's the missing number in family D? (Signal.) *194.*

EXERCISE 10: INDEPENDENT WORK

a. Find part 4. ✔
(Teacher reference:)

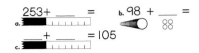

You'll complete the equations for the rulers and the equation for the circles.
b. Turn to the other side of your worksheet and find part 5. ✔
(Teacher reference:)

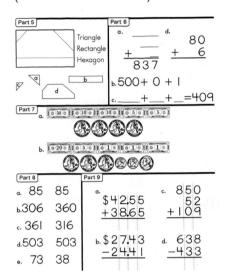

You'll circle the name for the whole shape. Then you'll write the letter for each part.
In part 6, you'll complete the place-value equations.
In part 7, you'll write the dollars and cents amount for each group of bills and coins.
In part 8, you'll write an equals sign, or a sign that has a bigger end next to one of the numbers.
In part 9, you'll work the money problems and the other column problems.
c. Complete worksheet 99. Remember to work part 4 on the other side of your worksheet.
(Observe children and mark incorrect responses on children's worksheets as you give feedback.)

Lesson 100

EXERCISE 1: COUNT BY TWOS

a. (Display:) [100:1A]

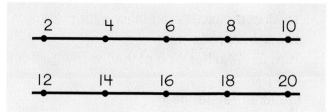

My turn to count by twos to 20. (Touch numbers as you say:) 2, 4, 6, 8, 10, (pause) 12, 14, 16, 18, 20.
- Your turn: Count by twos to 20. (Touch as children say:) *2, 4, 6, 8, 10, (pause) 12, 14, 16, 18, 20.*
  (Repeat until firm.)
b. (Do not show display.)
My turn to say the numbers without looking. 2, 4, 6, 8, 10 *(pause)* 12, 14, 16, 18, 20.
- Your turn: Count by twos to 20. Get ready.
  (Tap.) *2, 4, 6, 8, 10, (pause) 12, 14, 16, 18, 20.*
  (Repeat until firm.)
c. You have 98. When you minus tens, what's the next number? (Signal.) *88.*
- You have 98. When you plus ones, what's the next number? (Signal.) *99.*
- You have 98. When you plus twos, what's the next number? (Signal.) *100.*
- You have 98. When you plus tens, what's the next number? (Signal.) *108.*

=== INDIVIDUAL TURNS ===

(Call on different students to perform the following task.)

- Count by twos to 20. (Call on a student.) *2, 4, 6, 8, 10, 12, 14, 16, 18, 20.*

EXERCISE 2: NUMBER FAMILIES
MISSING NUMBER IN FAMILY

a. (Display:) [100:2A]

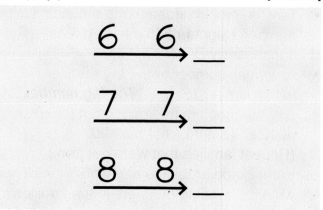

The big number is missing in these families. These families have small numbers of 6, 7, and 8.

- (Point to 6 6→.) What are the small numbers in this family? (Signal.) *6 and 6.*
- The big number is 12. What's the big number? (Signal.) *12.*
- (Point to 7 7→.) What are the small numbers in this family? (Signal.) *7 and 7.*
- The big number is 14. What's the big number? (Signal.) *14.*
- (Point to 8 8→.) What are the small numbers in this family? (Signal.) *8 and 8.*
- The big number is 16. What's the big number? (Signal.) *16.*

b. (Point to 6 6→.) Once more: What are the small numbers in this family? (Signal.) *6 and 6.*
- What's the big number? (Signal.) *12.*
c. (Point to 7 7→.) What are the small numbers in this family? (Signal.) *7 and 7.*
- What's the big number? (Signal.) *14.*
d. (Point to 8 8→.) What are the small numbers in this family? (Signal.) *8 and 8.*
- What's the big number? (Signal.) *16.*
(Repeat steps b through d until firm.)

e. There's only one plus fact and one minus fact for each family.
- (Point to $\overset{6}{\longrightarrow}\overset{6}{\longrightarrow}$.) Say the plus fact. (Signal.)
  6 + 6 = 12.
- Say the minus fact. (Signal.) 12 – 6 = 6.
  (Repeat until firm.)

f. (Point to $\overset{7}{\longrightarrow}\overset{7}{\longrightarrow}$.) Say the plus fact. (Signal.)
   7 + 7 = 14.
- Say the minus fact. (Signal.) 14 – 7 = 7.
  (Repeat step f until firm.)

g. (Point to $\overset{8}{\longrightarrow}\overset{8}{\longrightarrow}$.) Say the plus fact. (Signal.)
   8 + 8 = 16.
- Say the minus fact. (Signal.) 16 – 8 = 8.
  (Repeat step g until firm.)

h. (Display:)   [100:2B]

$\xrightarrow{\ \ 4\ \ }$ 14	8　8 $\longrightarrow$ _	6　6 $\longrightarrow$ _
7 $\longrightarrow$ 14	6　3 $\longrightarrow$ _	5　5 $\longrightarrow$ _
4 $\longrightarrow$ 7	3 $\longrightarrow$ 6	$\xrightarrow{\ \ 4\ \ }$ 8

You're going to say the problem for the missing number in each family. Then you'll say the answer.

i. (Point to $\xrightarrow{\ 4\ }$14.) Say the problem for the missing number. Get ready. (Signal.)
   14 minus 4.
- What's 14 minus 4? (Signal.) 10.

j. (Point to $\overset{7}{\longrightarrow}$14.) Say the problem for the missing number. (Signal.) 14 minus 7.
- What's 14 minus 7? (Signal.) 7.

k. (Repeat the following tasks for the remaining families:)

(Point to __.) Say the problem for the missing number.	What's __?	
$\overset{4}{\longrightarrow}$7	7 – 4	7 – 4	3
8　8 $\longrightarrow$ _	8 + 8	8 + 8	16
6　3 $\longrightarrow$ _	6 + 3	6 + 3	9
3 $\longrightarrow$6	6 – 3	6 – 3	3
6　6 $\longrightarrow$ _	6 + 6	6 + 6	12
5　5 $\longrightarrow$ _	5 + 5	5 + 5	10
$\overset{4}{\longrightarrow}$8	8 – 4	8 – 4	4

(Repeat for families that were not firm.)

EXERCISE 3: TELLING TIME
READING DIGITAL DISPLAYS

a. (Display:)   [100:3A]

You're going to read the times shown on these clocks.
- Some of these clocks show a 1-digit number of minutes. When you say the minutes for those clocks, you say oh and the number. What do you say for those clocks? (Signal.) *Oh and the number.*
  (Repeat until firm.)
- What do you say for the minutes if there are zero minutes shown? (Signal.) *O'clock.*

b. (Point to 8:03.) Does this clock show a 1-digit number of minutes? (Signal.) *Yes.*
- What will you say for the minutes? (Signal.) *Oh 3.*
- Read the time for this clock. (Signal.) *8 oh 3.*

c. (Repeat the following tasks for remaining clocks:)

(Point to __.) Does this clock show a 1-digit number of minutes?		So what will you say for the minutes?	Read the time for this clock.
8:30	No	30	8 thirty
3:08	Yes	Oh 8	3 oh 8
12:00	No	O'clock	12 o'clock
5:12	No	12	5 twelve
12:05	Yes	Oh 5	12 oh 5
12:50	No	50	12 fifty

(Repeat clocks that were not firm.)

d. Read the times for these clocks again.

└─• (Point to **8:03**.) What time? (Touch.) *8 oh 3.*
 • (Repeat for remaining clocks: 8 thirty, 3 oh 8, 12 o'clock, 5 twelve, 12 oh 5, 12 fifty.)

━━━━━━━ **INDIVIDUAL TURNS** ━━━━━━━

I'll call on individual students to read one or two clocks.

• (Point to **12:05**.) What's the time for this clock? (Call on a student.) *12 oh 5.*
• (Point to **5:12**.) What's the time for this clock? (Call on a student.) *5 twelve.*
• (Point to **12:00**.) What's the time for this clock? (Call on a student.) *12 o'clock.*
• (Point to **8:03**.) What's the time for this clock? (Call on a student.) *8 oh 3.*

EXERCISE 4: ADD/SUBTRACT
NOMENCLATURE

a. (Display:) [100:4A]

> add
>
> subtract

You learned that when you plus, you add. When you minus, you subtract.

┌─• What do you do when you plus? (Signal.) *Add.*
│ • What do you do when you minus? (Signal.) *Subtract.*
└─ (Repeat until firm.)

b. (Display:) [100:4B]

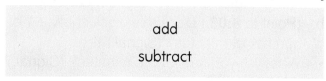

• (Point to **15**.) Read this problem. (Signal.) *15 minus 7.*
• Does this problem add or subtract? (Signal.) *Subtract.*
• How many does this problem subtract? (Signal.) *7.*
c. (Point to **12**.) Read this problem. (Signal.) *12 plus 28.*
• Does this problem add or subtract? (Signal.) *Add.*
• How many does this problem add? (Signal.) *28.*
d. (Point to **340**.) Read this problem. (Signal.) *340 plus 160.*
• Does this problem add or subtract? (Signal.) *Add.*
• How many does this problem add? (Signal.) *160.*

e. (Point to **550**.) Read this problem. (Signal.) *550 minus 100.*
• Does this problem add or subtract? (Signal.) *Subtract.*
• How many does this problem subtract? (Signal.) *100.*
(Repeat steps b through e that were not firm.)

EXERCISE 5: PLACE VALUE
PICTURES

a. (Display:) [100:5A]

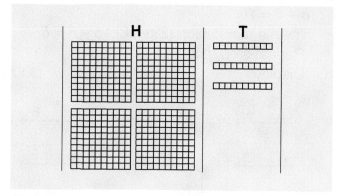

┌─• (Point to hundreds column.) How many squares are in each box? (Signal.) *100.*
│ • (Point to tens column.) How many squares are in each row? (Signal.) *10.*
└─ (Repeat until firm.)
┌─b. (Point to hundreds column.) Look at the hundreds. How many boxes are there in the hundreds column? (Signal.) *4.*
│ • So how many squares are in the hundreds column? (Signal.) *400.*
│ c. (Point to tens.) How many rows are in the tens column? (Signal.) *3.*
│ • So how many squares are in the tens column? (Signal.) *30.*
└─ (Repeat steps b and c until firm.)
d. My turn to say the plus problem for all of the squares. (Touch columns.) *400 plus 30.*
┌─• Your turn: Say the plus problem for all of the squares. (Touch columns.) *400 plus 30.*
└─ (Repeat until firm.)
• What does 400 plus 30 equal? (Signal.) *430.*
• So how many squares are there? (Signal.) *430.* Yes, this picture shows 430 squares.

e. (Display:)   [100:5B]

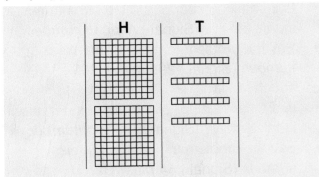

- (Point to hundreds.) Raise your hand when you know the number of squares in the hundreds boxes. ✔
- How many squares are in the hundreds boxes? (Signal.) *200.*
- (Point to tens.) Look at the tens rows. How many squares are in the tens rows? (Signal.) *50.*
  (Repeat until firm.)

f. Say the plus problem for all of the squares. Get ready. (Signal.) *200 plus 50.*
  (Repeat until firm.)
- What does 200 plus 50 equal? (Signal.) *250.*
- (Point to hundreds and tens.) So how many squares are there? (Signal.) *250.*

g. (Display:) W   [100:5C]

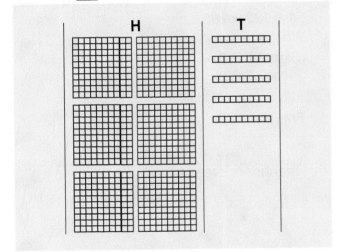

- (Point to hundreds.) Raise your hand when you know the number of squares in the hundreds boxes. ✔
- How many squares are in the hundreds boxes? (Signal.) *600.*
- Look at the tens rows. Raise your hand when you know the number of squares in the tens rows. ✔
- How many squares are in the tens rows? (Signal.) *50.*
  (Repeat until firm.)

h. Say the plus problem for all the squares. Get ready. (Signal.) *600 plus 50.*
  (Repeat until firm.)
- What does 600 plus 50 equal? (Signal.) *650.*
- (Point to hundreds and tens.) So how many squares are there? (Signal.) *650.*

i. (Add to show:)   [100:5D]

$$\underline{\quad} + \underline{\quad} = \underline{\quad}$$

I'm going to write the equation for these squares.
- What's the number for the hundreds boxes? (Signal.) *600.*
  (Add to show:)   [100:5E]

$$\underline{600} + \underline{\quad} = \underline{\quad}$$

- What's the number for the tens rows? (Signal.) *50.*
  (Add to show:)   [100:5F]

$$\underline{600} + \underline{50} = \underline{\quad}$$

- What's the number for all the squares? (Signal.) *650.*
  (Add to show:)   [100:5G]

$$\underline{600} + \underline{50} = \underline{650}$$

j. (Distribute unopened workbooks to children.)
- Open your workbook to Lesson 100 and find part 1.
  (Observe children and give feedback.)
  (Teacher reference:)

$$\underline{\quad} + \underline{\quad} = \underline{\quad}$$

k. (Display:) W̅   [100:5H]

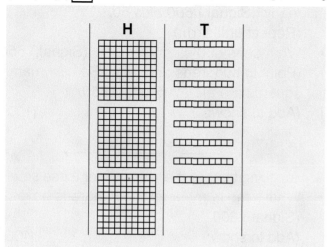

This is the picture for problem A. You'll complete the place-value equation for this picture.
- Write the number of squares for the hundreds boxes. Then stop. ✔
- Everybody, what did you write for hundreds? (Signal.) *300.*
- Write the number of squares for the tens rows. Then complete the equation.
(Observe children and give feedback.)
(Teacher reference:)

$$\underline{300} + \underline{80} = \underline{380}$$

l. Check your work.
- Touch and read the place-value equation for A. Get ready. (Signal.) *300 plus 80 equals 380.*
- (Point to picture.) Everybody, how many squares are there in this picture? (Signal.) *380.*

EXERCISE 6: NUMBER FAMILIES
MULTIPLE DIGIT NUMBERS

a. Find part 2 on worksheet 100.
(Teacher reference:)

You're going to say the problem for figuring out the missing number in each number family. Then you're going to write the problems in columns and work them.
- Touch family A. ✔
- Is a small number or the big number missing in this family? (Signal.) *A small number.*
- Say the problem for finding the missing number. (Signal.) *475 minus 71.*

b. Touch family B. ✔
- Is a small number or the big number missing in this family? (Signal.) *A small number.*
- Say the problem for finding the missing number. (Signal.) *295 minus 263.*
c. Touch family C. ✔
- Is a small number or the big number missing in this family? (Signal.) *The big number.*
- Say the problem for finding the missing number. (Signal.) *54 plus 132.*
(Repeat steps a through c that were not firm.)
d. Write all the problems and work them. Then write the missing numbers in the families. Raise your hand when you've completed the families in part 2.
(Observe children and give feedback.)
e. Check your work.
- Read problem A and the answer. (Signal.) *475 – 71 = 404.*
- What's the missing number in family A? (Signal.) *404.*
f. Read problem B and the answer. (Signal.) *295 – 263 = 32.*
- What's the missing number in family B? (Signal.) *32.*
g. Read problem C and the answer. (Signal.) *54 + 132 = 186.*
- What's the missing number in family C? (Signal.) *186.*

EXERCISE 7: 3-D OBJECTS
PYRAMIDS

a. Find part 3 on worksheet 100. ✔
(Teacher reference:)

- The objects in part 3 are pyramids. Say **pyramid.** (Signal.) *Pyramid.*
- The side faces of pyramids are always triangles. What shape are the side faces? (Signal.) *Triangles.*
- But the bottom faces of pyramids can be a triangle, a rectangle, or a hexagon. Do the bottom faces of pyramids have to be triangles? (Signal.) *No.*
- Think about the top of pyramids. Do pyramids have a top face? (Signal.) *No.*

b. You're going to write the letters on all faces you can see.
- Touch the bottom face of pyramid 1. ✔
- Is the bottom face a triangle, a rectangle, or a hexagon? (Signal.) *(A) hexagon.*
- How many sides does the bottom face have? (Signal.) *6.*
- So how many side faces are there? (Signal.) *6.*
- How many side faces can you see? (Signal.) *3.*
- Are the side faces, triangles, rectangles, or hexagons? (Signal.) *Triangles.*
- Write the letter T, R, or H on each face you can see.
  (Observe children and give feedback.)
  (Display:) W   [100:7A]

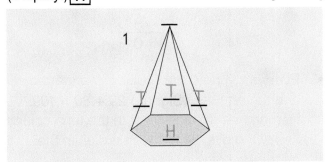

Here's what you should have.
- Is there a top face for object 1? (Signal.) *No.*
- So you don't write any letter in the space above object 1.
c. Touch the bottom of pyramid 2. ✔
- Everybody, what shape is the bottom face? (Signal.) *(A) triangle.*
- What shape are the side faces? (Signal.) *Triangles.*
- How many side faces can you see? (Signal.) *2.*
  Write T, R, or H on each face you can see.
  (Observe children and give feedback.)
  (Display:) W   [100:7B]

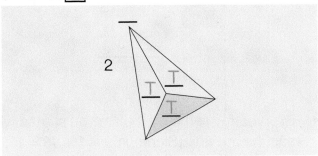

Here's what you should have.
- Everybody, what shape is the bottom face? (Signal.) *(A) triangle.*
- How many side faces are there? (Signal.) *3.*
- What shape are the side faces? (Signal.) *Triangles.*
- Is there a top face for object 2? (Signal.) *No.*

- So what do you write in the space above object 2? (Call on a child. Idea: *Nothing.*)
  Yes, don't write anything above object 2.
d. Touch the bottom of the object in part 4. ✔
  (Teacher reference:)

- Is the object in part 4 a pyramid? (Signal.) *No.*
  (To correct:)
  - Are the sides of the object triangles? (Signal.) *No.*
  - So is the object in part 4 a pyramid? (Signal.) *No.*
- Write the letters T, R, or H on the bottom face and the side faces.
  (Observe children and give feedback.)
e. Check your work.
- What shape is the bottom face of the object? (Signal.) *(A) triangle.*
- What shape are the side faces? (Signal.) *Rectangles.*
f. Is there a top face for the object in part 4? (Signal.) *Yes.*
- Write the letter for the top face above the object. ✔
- What is the shape of the top face? (Signal.) *(A) triangle.*
  (Display:) W   [100:7C]

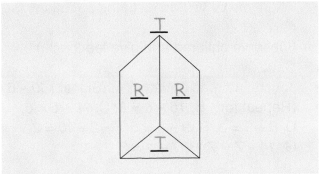

Here's what you should have for the object in part 4.

EXERCISE 8: FACTS
SUBTRACTION

a. Find part 5.
   (Teacher reference:)

   a. 10−6   e. 9−3
   b. 16−8   f. 12−10
   c. 11−5   g. 14−7
   d. 6−3   h. 7−4

   You'll read each problem and say the answer.
- Touch and read problem A. (Signal.) *10 minus 6.*
- What's 10 minus 6? (Signal.) *4.*
b. Touch and read problem B. (Signal.) *16 minus 8.*
- What's 16 minus 8? (Signal.) *8.*
c. Touch and read problem C. (Signal.) *11 minus 5.*
- What's 11 minus 5? (Signal.) *6.*
d. Touch and read problem D. (Signal.) *6 minus 3.*
- What's 6 minus 3? (Signal.) *3.*
e. Touch and read problem E. (Signal.) *9 minus 3.*
- What's 9 minus 3? (Signal.) *6.*
f. Touch and read problem F. (Signal.)
   12 minus 10.
- What's 12 minus 10? (Signal.) *2.*
g. Touch and read problem G. (Signal.)
   14 minus 7.
- What's 14 minus 7? (Signal.) *7.*
h. Touch and read problem H. (Signal.) *7 minus 4.*
- What's 7 minus 4? (Signal.) *3.*
   (Repeat problems that were not firm.)
i. Complete each fact. Put your pencil down
   when you've written all of the equations for
   part 5.
   (Observe children and give feedback.)
j. Check your work.
- Touch and read equation A. (Signal.) *10 − 6 = 4.*
- (Repeat for:) B, *16 − 8 = 8;* C, *11 − 5 = 6;*
   D, *6 − 3 = 3;* E, *9 − 3 = 6;* F, *12 − 10 = 2;*
   G, *14 − 7 = 7;* H, *7 − 4 = 3.*

EXERCISE 9: COLUMN ADDITION
CARRYING (MULTIPLE DIGITS)

a. (Display:) W   [100:9A]

   ```
     2 9
   + 8 0
   ```

   The answer to this problem is more than 100.
   I'll show you how to write the answer.
- (Point to **29**.) Everybody, read the problem.
   (Signal.) *29 plus 80.*

- Say the problem for the ones. (Signal.)
   9 plus zero.
- What's the answer? (Signal.) *9.*
   (Add to show:)   [100:9B]

   ```
     2 9
   + 8 0
       9
   ```

- Say the problem for the tens. (Signal.) *2 plus 8.*
- What's the answer? (Signal.) *10.*
   The answer has more than one digit, so I write
   that digit in the hundreds column.
   (Add to show:)   [100:9C]

   ```
     2 9
   + 8 0
   1 0 9
   ```

- Everybody, read the problem we started with
   and the answer. (Signal.) *29 + 80 = 109.*
   Remember, if the answer in the tens column
   has two digits, write the first digit in the
   hundreds column.
b. Find part 6. ✔
   (Teacher reference:)

   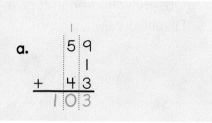

   These problems have answers that are more
   than 100.
- Work problem A. Then stop.
   (Observe children and give feedback.)
   (Display:) W   [100:9D]

   ```
            1
   a.    5 9
            1
      +  4 3
      1 0 3
   ```

   Here's what you should have.
- Everybody, read the problem you started with
   and the answer. (Signal.) *59 + 1 + 43 = 103.*

c. Work problem B. Then stop.
(Observe children and give feedback.)
(Display:) [W]   [100:9E]

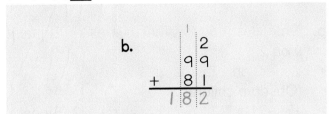

Here's what you should have.
- Everybody, read the problem you started with and the answer. (Signal.) *2 + 99 + 81 = 182.*

EXERCISE 10: FACTS
PLUS/MINUS MIX

a. Turn to the other side of your worksheet and find part 7. ✔
(Teacher reference:)

a. $5 - 3$	c. $4 - 1$ f. $7 + 7$
b. $5 + 5$	d. $4 + 4$ g. $11 - 6$
	e. $7 - 2$ h. $3 + 4$

- Touch and read problem A. Get ready. (Signal.) *5 minus 3.*
- Is the answer the big number or a small number? (Signal.) *A small number.*
- What's 5 minus 3? (Signal.) *2.*

b. (Repeat the following tasks for problems B through H:)

Touch and read problem __.	Is the answer the big number or a small number?	What's __?	
B	*The big number.*	$5 + 5$	*10*
C	*A small number.*	$4 - 1$	*3*
D	*The big number.*	$4 + 4$	*8*
E	*A small number.*	$7 - 2$	*5*
F	*The big number.*	$7 + 7$	*14*
G	*A small number.*	$11 - 6$	*5*
H	*The big number.*	$3 + 4$	*7*

(Repeat problems that were not firm.)

c. Write the answer to each fact. Put your pencil down when you've completed the equations for part 7.
(Observe children and give feedback.)

d. Check your work.
- Touch and read problem A. (Signal.) *5 – 3 = 2.*
- (Repeat for remaining problems:) B, *5 + 5 = 10;* C, *4 – 1 = 3;* D, *4 + 4 = 8;* E, *7 – 2 = 5;* F, *7 + 7 = 14;* G, *11 – 6 = 5;* H, *3 + 4 = 7.*

EXERCISE 11: INDEPENDENT WORK

a. Find part 8. ✔
(Teacher reference:)

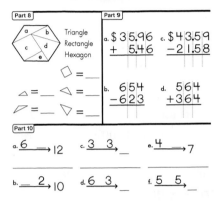

You'll circle the name for the whole shape. Then you'll complete each equation by writing the letter to show which part it is.
In part 9, you'll work the money problems and the other column problems.
In part 10, below each family, you'll write the fact for the missing number and you'll complete the family.

b. Complete worksheet 100.
(Observe children and mark incorrect responses on children's worksheets as you give feedback.)

Mastery Test 10

Teacher Presentation

a. Find Test 10 in your test booklet. ✔
• Touch part 1. ✔
  (Teacher reference:)

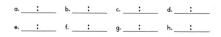

I'll say a time for each space. You'll write the time. The dots for each time are already shown.

• Time A is 12 forty-eight. What's time A? (Signal.) *12 forty-eight.*
• Write 12 forty-eight in space A. (Observe children.)

b. Time B is 2 fifteen. What's time B? (Signal.) *2 fifteen.*
• Write 2 fifteen in space B. (Observe children.)

c. Time C is 6 o'clock. What's time C? (Signal.) *6 o'clock.*
• Write 6 o'clock in space C. (Observe children.)

d. Time D is 11 oh 3. What's time D? (Signal.) *11 oh 3.*
• Write 11 oh 3 in space D. (Observe children.)

e. Time E is 1 thirty. What's time E? (Signal.) *1 thirty.*
• Write 1 thirty in space E. (Observe children.)

f. Time F is 10 o'clock. What's time F? (Signal.) *10 o'clock.*
• Write 10 o'clock in space F. (Observe children.)

g. Time G is 9 oh 2. What's time G? (Signal.) *9 oh 2.*
• Write 9 oh 2 in space G. (Observe children.)

h. Time H is 7 thirty. What's time H? (Signal.) *7 thirty.*
• Write 7 thirty in space H. (Observe children.)

i. Touch part 2 on test sheet 10. ✔
  (Teacher reference:)

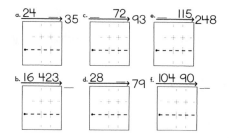

Each of these families has a missing number.
• Write and work the problems to find the missing numbers. Then write the missing numbers in the families. Put your pencil down when you've completed the families in part 2.

j. Touch part 3 on test sheet 10. ✔
  (Teacher reference:)

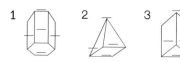

For objects 1, 2, and 3, you'll write a letter for the faces you can see. If the figure has a top face, you'll write the letter for the top face too.
• What letter will you write to show a triangle? (Signal.) *T.*
• What letter will you write to show a rectangle? (Signal.) *R.*
• What letter will you write to show a hexagon? (Signal.) *H.*
  (Repeat until firm.)
• Write the letters for the objects in part 2. (Observe children.)

k. Turn to the other side of test sheet 10 and touch part 4. ✔
(Teacher reference:)

a. $61 + 🪙🪙🪙🪙

b. $33 + 🪙🪙🪙🪙🪙

c. 💵💵🪙🪙🪙🪙

d. 💵💵💵💵🪙🪙

Some of these problems show a dollar amount and coins. Some of the problems show bills and coins.

• Figure out how much money there is for each problem. Write an equals and the amount. Remember the dollar sign.
(Observe children.)

l. Touch part 5 on test sheet 10. ✔
(Teacher reference:)

d. $10 - 5$ h. $8 - 4$
a. $6 - 3$ e. $5 - 5$ i. $3 + 3$
b. $6 + 3$ f. $5 + 5$ j. $8 - 3$
c. $10 + 5$ g. $10 - 4$

All of the problems in part 5 are from families you know.
You'll complete equations in part 5.

m. Touch part 6. ✔
(Teacher reference:)

a.
$$742 + 168$$
b.
$$31 \\ 349 \\ + 4$$
c.
$$261 \\ 6 \\ + 62$$
d.
$$358 + 442$$

You'll work column problems in part 6.
• Your turn: Work the rest of the problems on Test 10.
(Direct students where to put their assessment books when they are finished.)

Scoring Notes

a. Collect test booklets. Use the Answer Key and Passing Criteria Table to score the tests.

Passing Criteria Table — Mastery Test 10			
Part	Score	Possible Score	Passing Score
1	2 for each time	16	14
2	4 for each problem	24	16
3	1 for bottom and side faces	16	14
	2 for each top face		
4	3 for each problem	12	9
5	2 for each problem	20	18
6	3 for each column problem	12	9
	Total	100	

b. Complete the Mastery Test 10 Remedy Summary Sheet to determine whether group remedies are needed. Reproducible Remedy Summary Sheets are at the back of the Answer Key and at the back of the *Teacher's Guide*.

• If ¼ or more of the students did not pass a test part, present the remedy for that part before beginning Lesson 101. The Remedy Table follows and is also at the end of the Mastery Test 10 Answer Key. Remedies worksheets follow Mastery Test 10 in the *Student Assessment Book*.

Remedy Table — Mastery Test 10				
Part	Test Items	Remedy		Student Material Remedies Worksheet
		Lesson	Exercise	
1	Telling Time (Writing Digital Hours)	91	2	—
		94	8	Part A
		96	7	Part B
		97	8	Part C
2	Number Families with 2-Digit and 3-Digit Numbers (Missing Numbers)	92	3	—
		93	6	Part D
		96	5	Part E
		99	9	Part F
3	3D Objects (Faces)	95	2	—
		97	10	Part G
		99	8	Part H
4	Money (Bills and Coins)	88	3	—
		89	4	—
		91	8	Part I
		93	7	Part J
5	Facts (Mix)	89	1	—
		92	5	Part K
		93	5	Part L
6	Column Addition (Carrying)	90	2	—
		91	5	Part M
		93	8	Part N

Retest
Retest individual students on any part failed.

Lessons 101–105 Planning Page

	Lesson 101	Lesson 102	Lesson 103	Lesson 104	Lesson 105
Student Learning Objectives	**Exercises** 1. Say addition and subtraction facts with the same small number; find missing numbers in number families 2. Learn the words *add* and *subtract* 3. **Tell time on an analog clock** 4. **Count by a variety of numbers** 5. **Identify numbers in multi-digit number families** 6. **Use ones, tens, and hundreds charts to clarify place value** 7. Solve addition and subtraction facts 8. Solve money addition problems with carrying 9. Identify 2-dimensional faces on 3-dimensional objects 10. Complete work independently	**Exercises** 1. Say addition and subtraction facts with the same small number; find missing numbers in number families 2. Count by a variety of numbers 3. Tell time on an analog clock 4. **Learn the words *addition* and *subtraction*** 5. Solve addition and subtraction facts 6. Use ones, tens, and hundreds charts to clarify place value 7. Solve money addition problems with carrying 8. Identify 2-dimensional faces on 3-dimensional objects 9. Identify numbers in multi-digit number families 10. Solve subtraction facts 11. Complete work independently	**Exercises** 1. Say addition and subtraction facts with the same small number; find missing numbers in number families 2. Tell time on an analog clock 3. Count by a variety of numbers 4. Solve addition and subtraction facts 5. Learn the words *addition* and *subtraction* 6. Identify numbers in multi-digit number families 7. **Identify 3-dimensional objects with pentagons** 8. Use ones, tens, and hundreds charts to clarify place value 9. Solve subtraction facts 10. Complete work independently	**Exercises** 1. Say addition and subtraction facts 2. Count by a variety of numbers 3. Tell time on an analog clock 4. Solve addition facts 5. Use ones, tens, and hundreds charts to clarify place value 6. **Solve multi-digit problems and write the number families** 7. Identify 3-dimensional objects with pentagons 8. Solve subtraction facts 9. Complete work independently	**Exercises** 1. Say addition and subtraction facts 2. Count by a variety of numbers 3. **Decompose 3-dimenstional objects** 4. Solve addition facts 5. Tell time on an analog clock 6. Write the symbols for word problems in columns and solve; solve comparison word problems 7. Use ones, tens, and hundreds charts to clarify place value 8. Solve multi-digit problems and write the number families 9. Solve subtraction facts 10. Complete work independently
Common Core State Standards for Mathematics					
1.OA 5	✔		✔		✔
1.OA 6	✔	✔	✔	✔	✔
1.OA 8	✔	✔	✔	✔	✔
1.NBT 1	✔	✔	✔	✔	
1.NBT 2	✔			✔	
1.NBT 4	✔	✔	✔	✔	✔
1.MD 3	✔	✔	✔	✔	✔
1.G 1	✔	✔	✔	✔	✔
1.G 2	✔	✔	✔	✔	✔
Teacher Materials	Presentation Book 3, Board Displays CD or chalkboard				
Student Materials	Workbook 2, Pencil				
Additional Practice	• Student Practice Software: Block 5: Activity 1 (1.NBT.1), Activity 2 (1.NBT.1), Activity 3 (1.NBT.4 and 1.OA.6), Activity 4 (1.NBT.1), Activity 5 (1.NBT.1), Activity 6 (1.NBT.3) • Math Fact Worksheets 57–58 (After Lesson 101), 59 (After Lesson 103), 60–64 (After Lesson 104)				
Mastery Test					

Lesson 101

EXERCISE 1: NUMBER FAMILIES

MISSING NUMBER IN FAMILY

a. (Display:)   [101:1A]

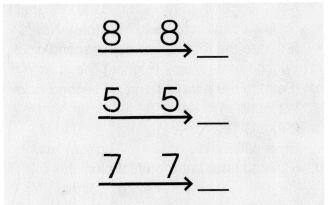

The big number is missing in these families.

- (Point to 8 8→.) What are the small numbers in this family? (Signal.) *8 and 8.*
- What's the big number? (Signal.) *16.*
- (Point to 5 5→.) What are the small numbers in this family? (Signal.) *5 and 5.*
- What's the big number? (Signal.) *10.*
- (Point to 7 7→.) What are the small numbers in this family? (Signal.) *7 and 7.*
- What's the big number? (Signal.) *14.*
(Repeat until firm.)

b. There's only one plus fact and one minus fact for each family.
- (Point to 8 8→.) What are the small numbers in this family again? (Signal.) *8 and 8.*
What's the big number? (Signal.) *16.*
- Say the plus fact. (Signal.) *8 + 8 = 16.*
- Say the minus fact. (Signal.) *16 – 8 = 8.*
c. (Point to 5 5→.) What are the small numbers in this family again? (Signal.) *5 and 5.*
What's the big number? (Signal.) *10.*
- Say the plus fact. (Signal.) *5 + 5 = 10.*
- Say the minus fact. (Signal.) *10 – 5 = 5.*
d. (Point to 7 7→.) What are the small numbers in this family again? (Signal.) *7 and 7.*
What's the big number? (Signal.) *14.*
- Say the plus fact. (Signal.) *7 + 7 = 14.*
- Say the minus fact. (Signal.) *14 – 7 = 7.*
(Repeat steps b though d until firm.)

e. (Display:)   [101:1B]

＿ 8 →16	4 3 →＿	＿ 5 →11
6 3 →＿	5 5 →＿	4 4 →＿
＿ 3 →6	＿ 7 →14	6 →12

You're going to say the problem for the missing number in each family. Then you'll say the answer.

f. (Point to →8 16 .) Say the problem for the missing number. Get ready. (Signal.)
16 minus 8.
- What's 16 minus 8? (Signal.) *8.*
g. (Point to 6 3→＿.) Say the problem for the missing number. (Signal.) *6 plus 3.*
- What's 6 plus 3? (Signal.) *9.*
h. (Repeat the following tasks for the remaining families:)

(Point to __.) Say the problem for the missing number.	What's __?	
＿ 3 →6	6 – 3	6 – 3	3
4 3 →＿	4 + 3	4 + 3	7
5 5 →＿	5 + 5	5 + 5	10
＿ →14	14 – 7	14 – 7	7
＿ →11	11 – 5	11 – 5	6
4 4 →＿	4 + 4	4 + 4	8
6 →12	12 – 6	12 – 6	6

(Repeat for families that were not firm.)

EXERCISE 2: ADD/SUBTRACT
NOMENCLATURE

a. (Display:) [101:2A]

314 − 200	45 − 45 ───── 26 + 75	88 + 107

- (Point to **314.**) Read this problem. (Signal.) *314 minus 200.*
- Does this problem add or subtract? (Signal.) *Subtract.*
- How many does this problem subtract? (Signal.) *200.*

b. (Point to **45.**) Read this problem. (Signal.) *45 minus 45.*
- Does this problem add or subtract? (Signal.) *Subtract.*
- How many does this problem subtract? (Signal.) *45.*

c. (Point to **26.**) Read this problem. (Signal.) *26 plus 75.*
- Does this problem add or subtract? (Signal.) *Add.*
- How many does this problem add? (Signal.) *75.*

d. (Point to **88.**) Read this problem. (Signal.) *88 plus 107.*
- Does this problem add or subtract? (Signal.) *Add.*
- How many does this problem add? (Signal.) *107.*
(Repeat steps a through d that were not firm.)

EXERCISE 3: TELLING TIME
ANALOG CLOCK HANDS   `REMEDY`

a. (Display:) [101:3A]

You've read the times on clocks that show numbers for hours and minutes. These are clocks too. But to figure out the time these clocks show, you have to look at the hands on the clock.

- (Point to the minute hand on 1st clock.) This hand is the minute hand. Which hand? (Touch.) *The minute hand.*
- The minute hand is longer than the other hand. Which hand is longer? (Signal.) *The minute hand.*
(Repeat until firm.)

b. (Point to hour hand on second clock.) Is this hand longer than the other hand? (Signal.) *No.*
- So is it the minute hand? (Signal.) *No.*
- (Point to the minute hand on second clock.) Which hand is this? (Signal.) *The minute hand.*

c. (Point to the minute hand on second clock.) Is this hand longer than the other hand? (Signal.) *Yes.*
- So is it the minute hand? (Signal.) *Yes.*

d. (Point to hour hand on first clock.) Is this hand longer than the other hand? (Signal.) *No.*
- So is it the minute hand? (Signal.) *No.*
(Repeat steps b through d until firm.)

e. Before you can tell time on a clock with hands, you have to know which direction the hands move.
(Display:) [101:3B]

The arrows show the direction that the hands move.

- (Point in the same direction the minute hand arrow on the first clock is pointing.) The minute hand on this clock is moving this way. (Touch 3 on first clock.) This is the next number the minute hand will point to. What's the next number the minute hand will point to? (Signal.) *3.*
- (Point to **2** on first clock.) This is the last number the minute hand pointed to. What's the last number the minute hand pointed to? (Signal.) *2.*

f. (Point to minute hand on second clock.) Is this hand longer than the other hand? (Signal.) *Yes.*
- So which hand is this? (Signal.) *The minute hand.*
- Look at the arrow and get ready to tell me the next number this minute hand will point to.
- Everybody, what's the next number this minute hand will point to? (Signal.) *4.*
- What's the last number this minute hand pointed to? (Signal.) *3.*

g. (Point to minute hand on third clock.) Is this hand longer than the other hand? (Signal.) *Yes.*
- So which hand is this? (Signal.) *The minute hand.*
- Look at the arrow and get ready to tell me the next number this minute hand will point to.
- Everybody, what's the next number this minute hand will point to? (Signal.) *12.*
- What's the last number this minute hand pointed to? Get ready. (Signal.) *11.*

h. (Point to the minute hand on first clock.) Is this hand longer than the other hand? (Signal.) *Yes.*
- So which hand is this? (Signal.) *The minute hand.*
- Look at the arrow and get ready to tell me the next number this minute hand will point to.
- Everybody, what's the next number this minute hand will point to? (Signal.) *3.*
- What's the last number this minute hand pointed to? Get ready. (Signal.) *2.*
  Remember the direction the minute hand moves.

EXERCISE 4: MIXED COUNTING

a. My turn to count by twos to 20: 2, 4, 6, 8, 10, 12, 14, 16, 18, 20.
- Your turn: Count by twos to 20. Get ready. (Signal.) *2, 4, 6, 8, 10, 12, 14, 16, 18, 20.* (Repeat until firm.)

b. Start with 10 and count by twos to 20. Get 10 going. *Tennn.* Count. (Tap.) *12, 14, 16, 18, 20.*

c. Start with 10 and count by fives to 50. Get 10 going. *Tennn.* Count. (Tap.) *15, 20, 25, 30, 35, 40, 45, 50.*

d. Start with 10 and count by tens to 100. Get ready. (Tap.) *10, 20, 30, 40, 50, 60, 70, 80, 90, 100.*

e. Start with 10 and count backward to 1. Get 10 going. *Tennn.* Count backward. (Tap.) *9, 8, 7, 6, 5, 4, 3, 2, 1.*

f. Listen: You have 10. When you count by fives, what's the next number? (Signal.) *15.*
- You have 10. When you count by ones what's the next number? (Signal.) *11.*
- You have 10. When you count backward, what's the next number? (Signal.) *9.*
- You have 10. When you count by tens, what's the next number? (Signal.) *20.*
- You have 10. When you count by twos, what's the next number? (Signal.) *12.*
  (Repeat step f until firm.)

g. Start with 75 and count by 25s to 175. Get 75 going. *Seventy-fiiive.* Count. (Tap.) *100, 125, 150, 175.*

h. Start with 75 and plus tens to 105. Get 75 going. *Seventy-fiiive.* Plus tens. (Tap.) *85, 95, 105.*

EXERCISE 5: NUMBER FAMILIES FROM EQUATIONS
MULTIPLE-DIGIT NUMBERS   REMEDY

a. (Display:) W   [101:5A]

$$
\begin{array}{cccc}
63 & 187 & 305 & 135 \\
+12 & -135 & -201 & +73 \\
\hline
75 & 52 & 104 & 208
\end{array}
$$

Here are equations. Some add and some subtract. You're going to tell me the big number in each equation.

- (Point to **63.**) Read the equation. (Touch.) *63 + 12 = 75.*
- Does the equation add? (Signal.) *Yes.*
- So is the big number the answer? (Signal.) *Yes.*
- What's the big number? (Signal.) *75.* (Repeat until firm.)

b. (Repeat the following tasks for remaining problems:)

	(Point to __.) Read the equation.	Does the equation add?	So is the big number the answer?	What's the big number?
187	187 − 135 = 52	No	No	187
305	305 − 201 = 104	No	No	305
135	135 + 73 = 208	Yes	Yes	208

(Repeat problems that were not firm.)

c. We're going to write number families for these equations.
- (Point to **63.**) Read the equation. (Touch.) *63 + 12 = 75.*
- Does the equation add or subtract? (Signal.) *Add.*
- So what's the big number? (Signal.) *75.*
- 75 is the big number. What are the small numbers? (Signal.) *63 and 12.*
  I write the big number first and then the small numbers.
  (Add to show:) [101:5B]

$$63$$
$$+\ 1\ 2$$
$$\overline{7\ 5}$$

$$\underline{63\quad 12} \rightarrow 75$$

Here's the family.
d. (Point to **187.**) Read the equation. (Touch.) *187 − 135 = 52.*
- Does the equation add or subtract? (Signal.) *Subtract.*
- So what's the big number? (Signal.) *187.*
- 187 is the big number. What are the small numbers? (Signal.) *135 (and) 52.*
  (Add to show:) [101:5C]

$$1\ 8\ 7$$
$$-1\ 3\ 5$$
$$\overline{5\ 2}$$

$$\underline{135\quad 52} \rightarrow 187$$

e. (Point to **305.**) Read the equation. (Touch.) *305 − 201 = 104.*
- Does the equation add or subtract? (Signal.) *Subtract.*
- So what's the big number? (Signal.) *305.*
- 305 is the big number. What are the small numbers? (Signal.) *201 (and) 104.*
  (Add to show:) [101:5D]

$$3\ 0\ 5$$
$$-2\ 0\ 1$$
$$\overline{1\ 0\ 4}$$

$$\underline{201\quad 104} \rightarrow 305$$

f. (Point to **135.**) Read the equation. (Touch.) *135 + 73 = 208.*
- Does the equation add or subtract? (Signal.) *Add.*
- So what's the big number? (Signal.) *208.*
- 208 is the big number. What are the small numbers? (Signal.) *135 (and) 73.*
  (Add to show:) [101:5E]

$$1\ 3\ 5$$
$$+\ 7\ 3$$
$$\overline{2\ 0\ 8}$$

$$\underline{135\quad 73} \rightarrow 208$$

(Repeat steps c through f that were not firm.)

EXERCISE 6: PLACE VALUE
PICTURES   REMEDY

a. (Display:) [101:6A]

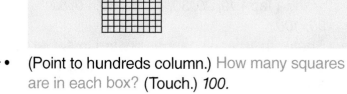

- (Point to hundreds column.) How many squares are in each box? (Touch.) *100.*
- (Point to tens column.) How many squares are in each row? (Touch.) *10.*
- (Point to single squares.) How many squares are in each square? (Touch.) *1.*
  (Repeat until firm.)
b. (Point to hundreds boxes.) Tell me how many squares are in the hundreds boxes. Get ready. (Signal.) *400.*
c. I'll touch each tens row. Count the squares for the tens to yourself. (Touch each tens row.) How many squares are in the tens rows? (Signal.) *60.*

d. I'll touch each single square. Count the squares for the ones to yourself. (Touch each single square.) How many single squares are there? (Signal.) *5.*

e. I'll say the problem for figuring out how many squares are in the picture. (Touch columns as you say:) *400 plus 60 plus 5.*

- Your turn: Say the problem for figuring out the number of squares in the picture. Get ready. (Touch columns as students say:) *400 plus 60 plus 5.*

- What's the answer? (Signal.) *465.*
(Add to show:)   [101:6B]

$$\underline{400} + \underline{60} + \underline{5} = \underline{465}$$

Here's the equation for the picture.

f. (Display:)   [101:6C]

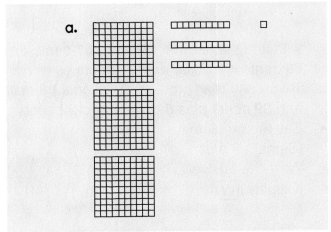

This is the picture for A. You're going to write the equation for how many squares are in picture A.

- Look at the hundreds boxes for this problem. Everybody, how many squares are in the hundreds boxes? (Signal.) *300.*

- Look at the tens rows. How many squares are in the tens rows? (Signal.) *30.*

- How many single squares are there? (Signal.) *1.*

- What does 300 plus 30 plus 1 equal? (Signal.) *331.*

- Say the equation for the squares in picture A. Get ready. (Touch.) *300 + 30 + 1 = 331.* (Repeat until firm.)

g. (Distribute unopened workbooks to children.)
- Open your workbook to Lesson 101 and find part 1.
(Observe children and give feedback.)
(Teacher reference:)   **R Part L**

- Touch the space where you'll write the numbers for picture A. ✔

- Say the equation for picture A. Get ready. (Signal.) *300 + 30 + 1 = 331.*

- Write the equation for picture A. ✔

h. (Display:)   [101:6D]

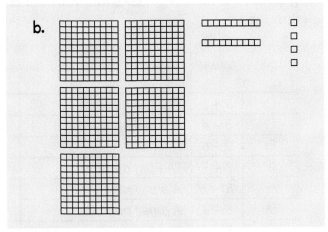

This is the picture for B.
- I'll touch each hundreds box. Count the squares for the hundreds to yourself. (Touch each hundreds box.) How many squares are in the hundreds boxes? (Signal.) *500.*

- Look at the tens rows. How many squares are in the tens rows? (Signal.) *20.*

- Look at the single squares. How many single squares are there? (Signal.) *4.*

- What does 500 plus 20 plus 4 equal? (Signal.) *524.*

- Say the equation for the squares in picture B. Get ready. (Touch.) *500 + 20 + 4 = 524.* (Repeat until firm.)

- Write the equation for B. ✔ (Answer key:)

a. $\underline{300} + \underline{30} + \underline{1} = \underline{331}$ b. $\underline{500} + \underline{20} + \underline{4} = \underline{524}$

EXERCISE 7: FACTS
PLUS/MINUS MIX

a. Find part 2 on worksheet 101. ✔
   (Teacher reference:)

   a. $7 - 4$ c. $8 - 4$ f. $10 - 5$
   b. $3 + 3$ d. $9 - 6$ g. $6 - 4$
             e. $7 + 7$ h. $8 + 8$

- Touch and read problem A. Get ready.
  (Signal.) *7 minus 4.*
- Is the answer the big number or a small
  number? (Signal.) *A small number.*
- What's 7 minus 4? (Signal.) *3.*

b. (Repeat the following tasks for problems B
   through H:)

Touch and read problem __.		Is the answer the big number or a small number?	What's __?	
B	3 + 3	The big number.	3 + 3	6
C	8 – 4	A small number.	8 – 4	4
D	9 – 6	A small number.	9 – 6	3
E	7 + 7	The big number.	7 + 7	14
F	10 – 5	A small number.	10 – 5	5
G	6 – 4	A small number.	6 – 4	2
H	8 + 8	The big number.	8 + 8	16

(Repeat problems that were not firm.)

c. Write the answer to each problem. Put your
   pencil down when you've completed the
   equations for part 2.
   (Observe children and give feedback.)

d. Check your work.
- Touch and read fact A. (Signal.) *7 – 4 = 3.*
- (Repeat for:) B, *3 + 3 = 6;* C, *8 – 4 = 4;*
  D, *9 – 6 = 3;* E, *7 + 7 = 14;* F, *10 – 5 = 5;*
  G, *6 – 4 = 2;* H, *8 + 8 = 16.*

EXERCISE 8: MONEY
DOLLAR PROBLEMS WITH CARRYING

a. Find part 3 on worksheet 101. ✔
   (Teacher reference:)

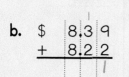

- Touch and read problem A. (Signal.) *51 cents
  plus 69 cents.*
- Say the problem for the cent ones column.
  (Signal.) *1 plus 9.*

- Write the answer. Then stop. ✔
  (Display:) W   [101:8A]

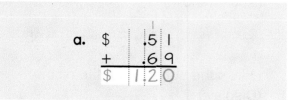

  Here's what you should have.

b. Say the problem for the cent tens column.
   (Signal.) *1 plus 5 plus 6.*
- Write the answer. Then stop. ✔
  (Add to show:)   [101:8B]

  a. $ .5 1
      + .6 9
      $ 1.2 0

  Here's what you should have.

- Everybody, read the problem we started
  with and the answer. (Signal.) *51 cents +
  69 cents = 1 dollar and 20 cents.*

c. Touch and read problem B. (Signal.) *8 dollars
   and 39 cents plus 8 dollars and 22 cents.*
- Say the problem for the cent ones column.
  (Signal.) *9 plus 2.*
- Write the answer. Then stop. ✔
  (Display:) W   [101:8C]

  b. $ 8.3 9
      + 8.2 2
             1

  Here's what you should have.

d. Say the problem for the cent tens column.
   (Signal.) *1 plus 3 plus 2.*
- Write the answer. Then stop. ✔
  (Add to show:)   [101:8D]

  b. $ 8.3 9
      + 8.2 2
           6 1

  Here's what you should have.

e. Say the problem for the dollar ones column. (Signal.) *8 plus 8.*
• Write the answer. Then stop. ✔
(Add to show:) [101:8E]

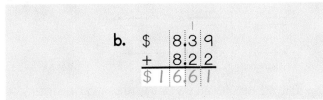

b. $ 8,39
+ 8,22
$16,61

Here's what you should have.
• Everybody, read the problem we started with and the answer. (Signal.) *8 dollars and 39 cents + 8 dollars and 22 cents = 16 dollars and 61 cents.*

f. Problem C is 86 plus 5 plus 79.
• Raise your hand when you can read the problem for the ones column and say the answer. ✔
• Read the problem for the ones column and say the answer. (Signal.) *6 + 5 + 9 = 20.*
• 20 is a two-digit number. What are the two digits of 20? (Signal.) *2 (and) zero.*
• What do you write in the tens column? (Signal.) *2.*
• Write the two digits of 20 and complete problem C.
• Touch and read problem C and the answer. (Signal.) *86 + 5 + 79 = 170.*
(Display:) [101:8F]

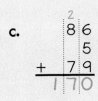

c. 86
5
+ 79
170

Here's what you should have.

EXERCISE 9: 3-D OBJECTS
PYRAMIDS

a. Find part 4 on your worksheet. ✔
(Teacher reference:)

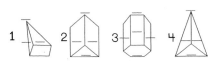

Some of these objects are pyramids.
• Touch object 1. ✔

• Are the side faces of object 1 triangles? (Signal.) *Yes.*
• So is this object a pyramid? (Signal.) *Yes.*
• Touch object 2. ✔
• Are the side faces of object 2 triangles? (Signal.) *No.*
• So is this object a pyramid? (Signal.) *No.*
• Touch object 3. ✔
• Are the side faces of object 3 triangles? (Signal.) *No.*
• So is this object a pyramid? (Signal.) *No.*
• Touch object 4. ✔
• Are the side faces of object 4 triangles? (Signal.) *Yes.*
• So is this object a pyramid? (Signal.) *Yes.*

b. Raise your hand when you know how many objects are pyramids. ✔
• How many objects are pyramids? (Signal.) *2.*

c. Touch the bottom face of object 1. ✔
Write T, R, or H for the bottom face and the side faces you can see. If there's a top face write the letter for it above object 1. Don't write anything in the space for the top if there isn't a top face. Put your pencil down when you've labeled the faces for object 1.
(Observe children and give feedback.)

d. (Display:) [101:9A]

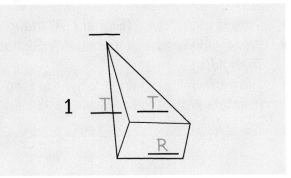

Here's what you should have.
• What's the shape of the bottom face? (Signal.) *Rectangle.*
• What's the shape of the side faces? (Signal.) *Triangle(s).*
• Did you write anything for the top face? (Signal.) *No.*

e. Write T, R, and H on all the faces for objects
    2, 3, and 4. For each object, if there's a top
    face, write the letter. Don't write anything if
    there's not a top face. Put your pencil down
    when you're finished.
    (Observe children and give feedback.)
    (Teacher reference:)

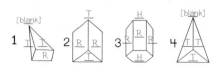

f. Touch the bottom face of object 2.
•   What's the shape? (Signal.) *(A) triangle.*
•   What's the shape of the sides? (Signal.)
    Rectangle(s).
•   Did you write anything for the top face?
    (Signal.) *Yes.*
•   What's the shape of the top face? (Signal.)
    (A) triangle.
g. Touch the bottom face of object 3. ✔
•   What's the shape? (Signal.) *(A) hexagon.*
•   What's the shape of the sides? (Signal.)
    Rectangle(s).
•   Did you write anything for the top face?
    (Signal.) *Yes.*
•   What's the shape for the top face? (Signal.)
    (A) hexagon.
h. Touch the bottom face of object 4. ✔
•   What's the shape? (Signal.) *(A) triangle.*
•   What's the shape of the sides? (Signal.)
    Triangle(s).
•   Did you write anything for the top face?
    (Signal.) *No.*

EXERCISE 10: INDEPENDENT WORK

a. Find part 5. ✔
    (Teacher reference:)

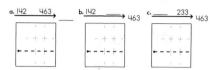

These are families with missing numbers.
You'll write a column problem to figure out the
missing number in each family. Then you'll
complete the family.

b. Turn to the other side of your worksheet and
    find part 6. ✔
    (Teacher reference:)

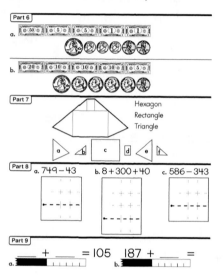

You'll write the dollars and cents amount for
each group of bills and coins.
In part 7, you'll circle the name for the whole
shape. Then you'll write the letter for each part.
In part 8, you'll write each problem in a column
and work it.
In part 9, you'll complete the equations for the
rulers.

c. Complete worksheet 101. Remember to work
    part 5 on the other side of your worksheet.
    **(Observe children and mark incorrect responses
    on children's worksheets as you give feedback.)**

Lesson 102

EXERCISE 1: NUMBER FAMILIES
MISSING NUMBER IN FAMILY

a. (Display:) [102:1A]

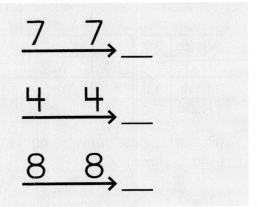

The big number is missing in these families. These families have small numbers of 4, 7, and 8.

- (Point to 7 ⟶ 7.) What are the small numbers in this family? (Signal.) *7 and 7.*
- The big number is 14. What's the big number? (Signal.) *14.*
- (Point to 4 ⟶ 4.) What are the small numbers in this family? (Signal.) *4 and 4.*
- The big number is 8. What's the big number? (Signal.) *8.*
- (Point to 8 ⟶ 8.) What are the small numbers in this family? (Signal.) *8 and 8.*
- The big number is 16. What's the big number? (Signal.) *16.*

b. (Point to 7 ⟶ 7.) Once more: What are the small numbers in this family? (Signal.) *7 and 7.*
- What's the big number? (Signal.) *14.*

c. (Point to 4 ⟶ 4.) What are the small numbers in this family? (Signal.) *4 and 4.*
- What's the big number? (Signal.) *8.*

d. (Point to 8 ⟶ 8.) What are the small numbers in this family? (Signal.) *8 and 8.*
- What's the big number? (Signal.) *16.*
(Repeat steps b through d until firm.)

e. There's only one plus fact and one minus fact for each family.
- (Point to 7 ⟶ 7.) Say the plus fact. (Signal.) *7 + 7 = 14.*
- Say the minus fact. (Signal.) *14 − 7 = 7.*
(Repeat until firm.)

f. (Point to 4 ⟶ 4.) Say the plus fact. (Signal.) *4 + 4 = 8.*
- Say the minus fact. (Signal.) *8 − 4 = 4.*
(Repeat step f until firm.)

g. (Point to 8 ⟶ 8.) Say the plus fact. (Signal.) *8 + 8 = 16.*
- Say the minus fact. (Signal.) *16 − 8 = 8.*
(Repeat step g until firm.)

h. (Display:) [102:1B]

4 ⟶ 8	⟶ 3 ⟶ 7	⟶ 8 ⟶ 16
⟶ 3 ⟶ 9	5 ⟶ 10	6 ⟶ 11
7 7 ⟶ _	6 4 ⟶ _	3 3 ⟶ _

You're going to say the problem for the missing number in each family. Then you'll say the answer.

i. (Point to 4 ⟶ 8.) Say the problem for the missing number. Get ready. (Signal.) *8 minus 4.*
- What's 8 minus 4? (Signal.) *4.*

j. (Point to ⟶ 3 ⟶ 9.) Say the problem for the missing number. (Signal.) *9 minus 3.*
- What's 9 minus 3? (Signal.) *6.*

k. (Repeat the following tasks for remaining families:)

(Point to __.) Say the problem for the missing number.		What's __?	
7 7 ⟶ _	7 + 7	7 + 7	14
⟶ 3 ⟶ 7	7 − 3	7 − 3	4
5 ⟶ 10	10 − 5	10 − 5	5
6 4 ⟶ _	6 + 4	6 + 4	10
⟶ 8 ⟶ 16	16 − 8	16 − 8	8
6 ⟶ 11	11 − 6	11 − 6	5
3 3 ⟶ _	3 + 3	3 + 3	6

(Repeat for families that were not firm.)

EXERCISE 2: MIXED COUNTING

a. Your turn to count by twos to 20. Get ready. (Tap.) *2, 4, 6, 8, 10, 12, 14, 16, 18, 20.* (Repeat until firm.)

b. Your turn to start with 12 and count by twos to 20.

- What will you start with? (Signal.) *12.*
- Get 12 going. *Tweeelve*. Count. (Tap.) *14, 16, 18, 20.*

c. Listen: Start with 75 and count by 25s to 175. Get 75 going. *Seventy-fiiive*. Count. (Tap.) *100, 125, 150, 175.*

d. Start with 75 and count by fives to 100. Get 75 going. *Seventy-fiiive*. Count. (Tap.) *80, 85, 90, 95, 100.*

e. Start with 75 and count by ones to 85. Get 75 going. *Seventy-fiiive*. Count. (Tap.) *76, 77, 78, 79, 80, 81, 82, 83, 84, 85.*

f. Start with 75 and plus tens to 105. Get 75 going. *Seventy-fiiive*. Plus tens. (Tap.) *85, 95, 105.*

g. Listen: Start with 75 and count backward by ones to 65. Get 75 going. *Seventy-fiiive*. Count backward. (Tap.) *74, 73, 72, 71, 70, 69, 68, 67, 66, 65.*
(Repeat steps c though g until firm.)

h. Listen: You have 75. When you count by fives, what's the next number? (Signal.) *80.*

- You have 75. When you county by 25s what's the next number? (Signal.) *100.*
- You have 75. When you count by ones, what's the next number? (Signal.) *76.*
- You have 75. When you count backward by ones, what's the next number? (Signal.) *74.*
(Repeat step h until firm.)

EXERCISE 3: TELLING TIME
ANALOG CLOCK HANDS

a. (Display:) [102:3A]

These are clocks with hands. Last time you learned about the minute hand, and you learned which direction the hands move.

- (Point to hour hand on 1st clock.) Is this hand longer than the other hand? (Touch.) *No.*
- So is this the minute hand? (Signal.) *No.*

b. (Point to minute hand on 2nd clock.) Is this hand longer than the other hand? (Touch.) *Yes.*

- So is it the minute hand? (Signal.) *Yes.*
- Which hand is this? (Signal.) *The minute hand.*

c. (Repeat the following tasks for the specified clock hands:)

(Point to __ hand on __ clock.) Is this hand longer than the other hand?		So is it the minute hand?	Which hand is this?
Minute	3rd	Yes	Yes	The minute hand.
Hour	2nd	No	No	
Minute	1st	Yes	Yes	The minute hand.
Hour	3rd	No	No	

(Repeat hands that were not firm.)

d. (Add to show:) [102:3B]

The arrows show the direction that the hands move.

- (Point in the same direction the minute hand arrow on 1st clock is pointing.) The minute hand on this clock is moving this way. (Touch 8 on first clock.) This is the next number the minute hand will point to. What's the next number the minute hand will point to? (Signal.) *8.*
- (Point to 7 on 1st clock.) What's the last number the minute hand pointed to? (Touch.) *7.*

e. (Point to minute hand on 2nd clock.) Is this hand longer than the other hand? (Signal.) *Yes.*

- So which hand is this? (Signal.) *The minute hand.*
- Look at the arrow and get ready to tell me the next number this minute hand will point to.
- Everybody, what's the next number this minute hand will point to? (Signal.) *1.*
- What's the last number this minute hand pointed to? (Signal.) *12.*

f. (Point to minute hand on 3rd clock.) Is this hand longer than the other hand? (Signal.) *Yes.*

- So which hand is this? (Signal.) *The minute hand.*
- Look at the arrow and get ready to tell me the next number this minute hand will point to.
- Everybody, what's the next number this minute hand will point to? (Signal.) *5.*
- What's the last number this minute hand pointed to? (Signal.) *4.*

g. (Point to minute hand on 1st clock.) Is this hand longer than the other hand? (Signal.) *Yes.*
- So which hand is this? (Signal.) *The minute hand.*
- Look at the arrow and get ready to tell me the next number this minute hand will point to.
- Everybody, what's the next number this minute hand will point to? (Signal.) *8.*
- What's the last number this minute hand pointed to? (Signal.) *7.*

h. I'm going to show you some clocks.
(Display:) [102:3C]

- You can see a finger pointing. Is that finger pointing the direction the clock hand is moving? (Signal.) *No.*
(Display:) [102:3D]

- Is the finger pointing the direction the clock hand is moving, now? (Signal.) *Yes.*
(Display:) [102:3E]

- Is that finger pointing the direction the clock hand is moving? (Signal.) *No.*
(Display:) [102:3F]

- Is the finger pointing the direction the clock hand is moving, now? (Signal.) *Yes.*
(Repeat until firm.)

i. Now I'll make the arrow to show the direction for the minute hand.
(Display:) [102:3G]

- (Point to hour hand.) Is this the minute hand? (Signal.) *No.*
So we'll make the arrow for the other hand.
(Display:) [102:3H]

- Is the finger pointing the direction the clock hand is moving? (Signal.) *No.*
(Display:) [102:3I]

- Is the finger pointing the direction the clock hand is moving, now? (Signal.) *Yes.*
(Repeat until firm.)
(Display:) [102:3J]

Here's the arrow to show the direction the minute hand is moving.

j. New clock.
(Display:) [102:3K]

• You can see a finger pointing. Is that finger pointing the direction the clock hand is moving? (Signal.) *Yes.*
(Display:) [102:3L]

• Is that finger pointing the direction the clock hand is moving? (Signal.) *No.*
(Display:) [102:3M]

• Is the finger pointing the direction the clock hand is moving? (Signal.) *No.*
(Display:) [102:3N]

• Is the finger pointing the direction the clock hand is moving? (Signal.) *Yes.*
(Repeat step j until firm.)

k. This time I'll make the arrow to show the direction for the minute hand.
(Display:) [102:3O]

• (Point to minute hand.) Is this the minute hand? (Touch.) *Yes.*
(Display:) [102:3P]

• Is the finger pointing the direction the clock hand is moving? (Signal.) *Yes.*
(Display:) [102:3Q]

Here's the arrow to show the direction the minute hand is moving.

l. Now you're going to tell me the next number the minute hand will point to.
• Look at the minute hand. Everybody, what's the next number the minute hand will point to? (Signal.) *6.*
• What's the last number the minute hand pointed to? (Signal.) *5.*
(Repeat until firm.)

m. The other hand doesn't have an arrow on it, but you can figure out the numbers for that hand.
• Look at it and get ready to tell me the next number it will point to. Everybody, what's the next number the other hand will point to? (Signal.) *10.*
• What's the last number the other hand pointed to? (Signal.) *9.*
(Repeat until firm.)

EXERCISE 4: ADD/SUBTRACT

NOMENCLATURE

a. (Display:)   [102:4A]

$$53 + 81$$

$$76 - 32$$

- (Point to **53**.) Read this problem. (Signal.)
  53 plus 81.
- Does this problem add or subtract?
  (Signal.) *Add.*

b. (Point to **76**.) Read this problem. (Signal.)
  76 minus 32.
- Does this problem add or subtract? (Signal.)
  Subtract.

c. Listen: Problems that add are called **addition problems.**
- What do we call problems that add? (Signal.)
  Addition problems.

d. Problems that subtract are called **subtraction problems.**
- What do we call problems that subtract?
  (Signal.) *Subtraction problems.*
  (Repeat steps c and d until firm.)

e. (Distribute unopened workbooks to children.)
- Open your workbook to Lesson 102 and find
  part 1.
  (Observe children and give feedback.)
  (Teacher reference:)

a. $\begin{array}{r} 67 \\ +102 \end{array}$ b. $\begin{array}{r} 137 \\ -27 \end{array}$ c. $\begin{array}{r} 25 \\ +35 \end{array}$ d. $\begin{array}{r} 87 \\ -55 \end{array}$

Some of these problems are addition
problems. Some are subtraction problems.
- Circle the subtraction problems.
  (Observe children and give feedback.)

f. Check your work.
- Touch and read the first subtraction problem.
  (Signal.) *137 minus 27.*
- Touch and read the other subtraction problem.
  (Signal.) *87 minus 55.*

g. Touch problem A. ✔
- What kind of problem is A? (Signal.) *(An)
  addition (problem).*

h. (Repeat the following tasks for remaining
  problems:)

Touch problem __.	What kind of problem is __?	
B	B	*(A) subtraction (problem).*
C	C	*(An) addition (problem).*
D	D	*(A) subtraction (problem).*

(Repeat steps g and h until firm.)
You'll work these problems later. Remember,
addition problems add. Subtraction problems
subtract.

EXERCISE 5: FACTS

PLUS/MINUS MIX

a. Find part 2 on worksheet 102. ✔
  (Teacher reference:)

a. $3 + 6$ d. $12 - 6$ h. $14 - 7$
b. $6 - 3$ e. $8 + 8$ i. $5 + 5$
c. $6 + 5$ f. $8 - 3$ j. $7 - 4$
            g. $10 - 4$ k. $7 + 2$

- Touch and read problem A. Get ready.
  (Signal.) *3 plus 6.*
- Is the answer the big number or a small
  number? (Signal.) *The big number.*
- What's 3 plus 6? (Signal.) *9.*

b. (Repeat the following tasks for the remaining
  problems:)

Touch and read problem __.		Is the answer the big number or a small number?	What's __?	
B	$6 - 3$	*A small number.*	$6 - 3$	*3*
C	$6 + 5$	*The big number.*	$6 + 5$	*11*
D	$12 - 6$	*A small number.*	$12 - 6$	*6*
E	$8 + 8$	*The big number.*	$8 + 8$	*16*
F	$8 - 3$	*A small number.*	$8 - 3$	*5*
G	$10 - 4$	*A small number.*	$10 - 4$	*6*
H	$14 - 7$	*A small number.*	$14 - 7$	*7*
I	$5 + 5$	*The big number.*	$5 + 5$	*10*
J	$7 - 4$	*A small number.*	$7 - 4$	*3*
K	$7 + 2$	*The big number.*	$7 + 2$	*9*

(Repeat problems that were not firm.)

c. Complete each of the facts. Put your pencil down when you've written all of the equations for part 1.
(Observe children and give feedback.)
d. Check your work.
• Touch and read the equation for problem A. (Signal.) *3 + 6 = 9.*
• (Repeat for:) B, *6 – 3 = 3;* C, *6 + 5 = 11;* D, *12 – 6 = 6;* E, *8 + 8 = 16;* F, *8 – 3 = 5;* G, *10 – 4 = 6;* H, *14 – 7 = 7;* I, *5 + 5 = 10;* J, *7 – 4 = 3;* K, *7 + 2 = 9.*

EXERCISE 6: PLACE VALUE
PICTURES

a. (Display:)   [102:6A]

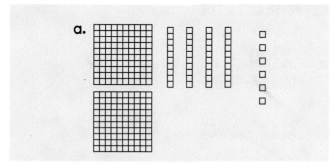

• (Point to hundreds column.) How many squares are in each box? (Touch.) *100.*
• (Point to tens column.) The tens rows are not shown in rows for this picture. They are shown in columns. How many squares are in each column? (Touch.) *10.*
• (Point to single squares.) How many squares are in each square? (Touch.) *1.*
(Repeat until firm.)
b. (Point to hundreds boxes.) Tell me how many squares are in the hundreds boxes. Get ready. (Signal.) *200.*
• (Point to tens columns.) Tell me how many squares are in the tens columns. Get ready. (Signal.) *40.*
• (Point to single squares.) I'll touch each single square. Count them to yourself. (Touch each single square.)
• How many single squares are there? (Signal.) *6.*
• What does 200 plus 40 plus 6 equal? (Signal.) *246.*
• Say the equation for the number of squares in picture A. Get ready. (Touch columns.) *200 + 40 + 6 = 246.*
(Repeat step b until firm.)

c. Find part 3 on worksheet 102. ✔
(Teacher reference:)

• Write the equation for picture A. ✔
d. (Display:)   [102:6B]

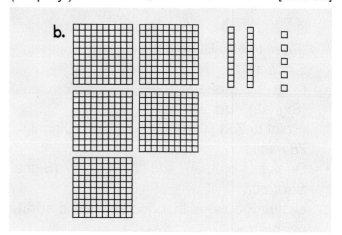

This is the picture for B.
• (Point to hundreds column.) I'll touch each box. Count the squares to yourself. (Touch boxes.)
• Everybody, how many squares are in the hundreds boxes? (Signal.) *500.*
• (Point to tens column.) I'll touch each tens column. Count the tens to yourself. (Touch each tens column.)
• Everybody, how many squares are in the tens columns? (Signal.) *20.*
• (Point to single squares.) I'll touch each single square. Count the squares for the ones to yourself. (Touch each single square.)
• Everybody, how many single squares are there? (Signal.) *5.*
• What does 500 plus 20 plus 5 equal? (Signal.) *525.*
• Say the equation for the squares in picture B. Get ready. (Touch.) *500 + 20 + 5 = 525.*
(Repeat until firm.)
• Write the equation for B. ✔
(Answer key:)

a. 200+40+6=246   b. 500+20+5=525

EXERCISE 7: MONEY
DOLLAR PROBLEMS WITH CARRYING

a. Find part 4 on worksheet 102. ✔
(Teacher reference:)

```
a. $  .95
      .43
   +  .68

b. $  4.95
      .65
   +  6.42
```

- Problem A is 95 cents plus 43 cents plus
  68 cents. The answer is more than one dollar.
  Work problem A. Remember to write the signs,
  the dot, and the numbers in the right columns.
  (Observe children and give feedback.)
  (Display:) W   [102:7A]

```
a.
          1
    $   .9 5
        .4 3
    +   .6 8
    $  2.0 6
```

- Touch and read the problem and the answer
  for A. Get ready. (Signal.) 95 cents + 43 cents +
  68 cents = 2 dollars and 6 cents.

b. Problem B is 4 dollars and 95 cents plus 65
  cents plus 6 dollars and 42 cents. The answer
  is more than 10 dollars. Work problem B.
  Remember to write the signs, the dot, and the
  numbers in the right columns.
  (Observe children and give feedback.)
  (Display:) W   [102:7B]

```
b.         2 1
    $   4.9 5
        .6 5
    +  6.4 2
    $ 12.0 2
```

- Touch and read the problem and the answer for
  B. Get ready. (Signal.) 4 dollars and 95 cents +
  65 cents + 6 dollars and 42 cents = 12 dollars
  and 2 cents.

EXERCISE 8: 3-D OBJECTS
PYRAMIDS

a. Find part 5 on worksheet 102. ✔
(Teacher reference:)

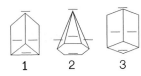

1   2   3

Some of these objects are pyramids.
- Raise your hand when you know how many
  objects are pyramids. ✔
- How many objects are pyramids? (Signal.) 1.
b. Touch object 1. ✔
- Are the side faces of this object triangles?
  (Signal.) No.
- So is this object a pyramid? (Signal.) No.
- Touch object 2. ✔
- Are the side faces of this object triangles?
  (Signal.) Yes.
- So is this object a pyramid? (Signal.) Yes.
- Touch object 3. ✔
- Are the side faces of this object triangles?
  (Signal.) No.
- So is this object a pyramid? (Signal.) No.
c. Touch the bottom face of object 1. ✔
- Write the letter T, R, or H on the bottom face.
  Then write T, R, or H on the other faces you
  can see. Put your pencil down when you've
  labeled the faces for object 1.
  (Observe children and give feedback.)
d. (Display:)   [102:8A]

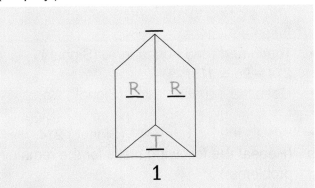

1

Here's what you should have.
- What's the shape of the bottom face? (Signal.)
  (A) triangle.
- What's the shape of the side faces? (Signal.)
  Rectangles.
- Is there a top face? (Signal.) Yes.
  You can't see it but you can figure out what
  shape it is.
- Write the letter for the top face above object 1. ✔
- Everybody, what's the shape of the top face?
  (Signal.) (A) triangle.

e. Write T, R, and H on all the faces for objects 2 and 3. Write the letter for the top face if there is a top face. If there isn't a top face leave the space blank.
(Observe children and give feedback.)

f. Touch the bottom face of object 2. ✔
- What's the shape? (Signal.) *(A) hexagon.*
- What's the shape of the sides? (Signal.) *Triangles.*
- Did you write a letter for the top face? (Signal.) *No.*

g. Touch the bottom face of object 3. ✔
- What's the shape? (Signal.) *(A) rectangle.*
- What's the shape of the sides? (Signal.) *Rectangles.*
- Did you write a letter for the top face? (Signal.) *Yes.*
- What's the top face? (Signal.) *(A) rectangle.*

EXERCISE 9: NUMBER FAMILIES FROM EQUATIONS
MULTIPLE-DIGIT NUMBERS
REMEDY

a. Turn to the other side of worksheet 102 and find part 6. ✔
(Teacher reference:) R Part J

$$
\begin{array}{cc}
\text{a.}\ \ 214 & \text{c.}\ \ 385 \\
-\ \ 96 & -214 \\
\hline
118 & 171
\end{array}
$$
$$
\begin{array}{cc}
\text{b.}\ \ 137 & \text{d.}\ \ 171 \\
+385 & +\ \ 96 \\
\hline
522 & 267
\end{array}
$$

You're going to make number families for these equations.

- Touch and read equation A. (Signal.) *214 – 96 = 118.*
- Does the equation add? (Signal.) *No.*
- So is the big number the answer? (Signal.) *No.*
- What's the big number? (Signal.) *214.*

b. (Repeat the following tasks for the remaining problems:)

Touch and read equation __.	Does the equation add?	So is the big number the answer?	What's the big number?
B	137 + 385 = 522	Yes	Yes	522
C	385 – 214 = 171	No	No	385
D	171 + 96 = 267	Yes	Yes	267

(Repeat problems that were not firm.)

c. Go back to problem A. Write the numbers in family A.
(Observe children and give feedback.)
- Check your work. What's the big number for family A? (Signal.) *214.*
- What are the small numbers? (Call on a child.) *96 (and) 118.*

d. (Repeat the following tasks for the remaining problems:)

Write the numbers in family __.	What's the big number for family __?	What are the small numbers? (Call on a child.)
B	B	522	137 (and) 385
C	C	385	214 (and) 171
D	D	267	171 (and) 96

EXERCISE 10: FACTS
SUBTRACTION

a. Find part 7 on worksheet 102. ✔
(Teacher reference:)

a. 8 – 4	c. 16 – 8	f. 10 – 6
b. 10 – 8	d. 8 – 5	g. 7 – 4
	e. 6 – 4	h. 10 – 5

- Touch and read problem A. Get ready. (Signal.) *8 minus 4.*
- What's 8 minus 4? (Signal.) *4.*

b. (Repeat the following tasks for problems B through H:)

Touch and read problem __.		What's ___?	
B	10 – 8	10 – 8	2
C	16 – 8	16 – 8	8
D	8 – 5	8 – 5	3
E	6 – 4	6 – 4	2
F	10 – 6	10 – 6	4
G	7 – 4	7 – 4	3
H	10 – 5	10 – 5	5

(Repeat problems that were not firm.)

c. Complete each of the facts. Put your pencil down when you've written all of the equations for part 7.
(Observe children and give feedback.)

d. Check your work.
- Touch and read the equation for problem A. (Signal.) *8 – 4 = 4.*
- (Repeat for:) B, 10 – 8 = 2; C, 16 – 8 = 8; D, 8 – 5 = 3; E, 6 – 4 = 2; F, 10 – 6 = 4; G, 7 – 4 = 3; H, 10 – 5 = 5.

EXERCISE 11: INDEPENDENT WORK

a. Find part 8. ✔
    (Teacher reference:)

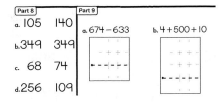

You'll write an equals sign, or a sign that has a bigger end next to one of the numbers.
In part 9, you'll write each problem in a column and work it.

b. Complete worksheet 102. Remember to work the addition and subtraction problems in part 1 on the other side of your worksheet.
    (Observe children and mark incorrect responses on children's worksheets as you give feedback.)

Lesson 103

MISSING NUMBER IN FAMILY

REMEDY

a. (Display:) [103:1A]

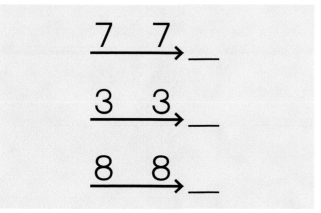

The big number is missing in these families. These families have small numbers of 7, 3, and 8.

- (Point to 7 →7.) What are the small numbers in this family? (Signal.) *7 and 7.*
- The big number is 14. What's the big number? (Signal.) *14.*
- (Point to 3 →3.) What are the small numbers in this family? (Signal.) *3 and 3.*
- The big number is 6. What's the big number? (Signal.) *6.*
- (Point to 8 →8.) What are the small numbers in this family? (Signal.) *8 and 8.*
- The big number is 16. What's the big number? (Signal.) *16.*

b. (Point to 7 →7.) What are the small numbers in this family? (Signal.) *7 and 7.*
- What's the big number? (Signal.) *14.*

c. (Point to 3 →3.) What are the small numbers in this family? (Signal.) *3 and 3.*
- What's the big number? (Signal.) *6.*

d. (Point to 8 →8.) What are the small numbers in this family? (Signal.) *8 and 8.*
- What's the big number? (Signal.) *16.*
(Repeat steps b through d until firm.)

e. There's only one plus fact and one minus fact for each family.
- (Point to 7 →7.) Say the plus fact. (Signal.) *7 + 7 = 14.*
- Say the minus fact. (Signal.) *14 − 7 = 7.*

f. (Point to 3 →3.) Say the plus fact. (Signal.) *3 + 3 = 6.*
- Say the minus fact. (Signal.) *6 − 3 = 3.*

g. (Point to 8 →8.) Say the plus fact. (Signal.) *8 + 8 = 16.*
- Say the minus fact. (Signal.) *16 − 8 = 8.*
(Repeat steps e through g until firm.)

h. (Display:) [103:1B]

6 →12	5 5 →_	7 2 →_
— 3→6	— 5→11	— 3→7
6 3 →_	— 7→14	8 →16

You're going to say the problem for the missing number in each family. Then you'll say the answer.

i. (Point to 6 →12.) Say the problem for the missing number. Get ready. (Signal.) *12 minus 6.*
- What's 12 minus 6? (Signal.) *6.*

j. (Point to —3→6.) Say the problem for the missing number. (Signal.) *6 minus 3.*
- What's 6 minus 3? (Signal.) *3.*

k. (Repeat the following tasks for remaining families:)

(Point to __.) Say the problem for the missing number.		What's __?	
6 3 →_	6 + 3	6 + 3	9
5 5 →_	5 + 5	5 + 5	10
— 5→11	11 − 5	11 − 5	6
— 7→14	14 − 7	14 − 7	7
7 2 →_	7 + 2	7 + 2	9
— 3→7	7 − 3	7 − 3	4
8 →16	16 − 8	16 − 8	8

(Repeat for families that were not firm.)

Exercise 2: Telling Time
Analog Clock Hands

a. (Display:) [103:2A]

These are clocks with hands. First you're going to tell me which hand is the minute hand for each clock.

- (Point to minute hand on 1st clock.) Is this hand longer than the other hand? (Touch.) *Yes.*
- So is this the minute hand? (Signal.) *Yes.*
- Which hand is this? (Signal.) *The minute hand.*

b. (Point to hour hand on 2nd clock.) Is this hand longer than the other hand? (Touch.) *No.*
- So is this the minute hand? (Signal.) *No.*

c. (Point to hour hand on 3rd clock.) Is this hand longer than the other hand? (Touch.) *No.*
- So is this the minute hand? (Signal.) *No.*

d. (Point to minute hand on 3rd clock.) Is this hand longer than the other hand? (Touch.) *Yes.*
- So is this the minute hand? (Signal.) *Yes.*
- Which hand is this? (Signal.) *The minute hand.*

e. (Point to minute hand on 2nd clock.) Is this hand longer than the other hand? (Touch.) *Yes.*
- So is this the minute hand? (Signal.) *Yes.*
- Which hand is this? (Signal.) *The minute hand.*

f. (Point to hour hand on 1st clock.) Is this hand longer than the other hand? (Touch.) *No.*
- So is this the minute hand? (Signal.) *No.*
(Repeat steps c through f that were not firm.)

g. I'm going to show you two of those clocks again.
(Display:) [103:2B]

- You can see a finger pointing. Is that finger pointing the direction the clock hand is moving? (Signal.) *No.*
(Display:) [103:2C]

- Is that finger pointing the direction the clock hand is moving? (Signal.) *Yes.*
(Display:) [103:2D]

- Is that finger pointing the direction the clock hand is moving? (Signal.) *Yes.*
(Display:) [103:2E]

- Is that finger pointing the direction the clock hand is moving? (Signal.) *No.*
(Repeat until firm.)

h. This time I'll make the arrow to show the direction for the minute hand.

• The hands for this clock are pointing towards the numbers 2 and 6. Which number is the minute hand pointing towards? (Signal.) *6.*
(Display:) [103:2F]

• Is the finger pointing the direction the clock hand is moving? (Signal.) *No.*
(Display:) [103:2G]

• Is the finger pointing the direction the clock hand is moving, now? (Signal.) *Yes.*
(Display:) [103:2H]

Here's the arrow to show the direction the minute hand is moving.

i. Now you're going to tell me the next number the minute hand will point to.

• Look at the minute hand. Everybody, what's the next number the minute hand will point to? (Signal.) *7*

• What's the last number the minute hand pointed to? (Signal.) *6.*
(Repeat steps h and i until firm.)

j. The other hand doesn't have an arrow on it, but you can figure out the numbers for that hand.

• Look at it and get ready to tell me the next number it will point to. Everybody, what's the next number the other hand will point to? (Signal.) *3.*

• What's the last number the other hand pointed to? (Signal.) *2.*
(Repeat step j until firm.)

k. New clock.
(Display:) [103:2I]

• You can see a finger pointing. Is that finger pointing the direction the clock hand is moving? (Signal.) *Yes.*
(Display:) [103:2J]

• Is that finger pointing the direction the clock hand is moving? (Signal.) *No.*
(Display:) [103:2K]

• Is the finger pointing the direction the clock hand is moving? (Signal.) *Yes.*
(Display:) [103:2L]

• Is the finger pointing the direction the clock hand is moving? (Signal.) *No.*
(Repeat step k until firm.)

l. This time I'll make the arrow to show the direction for the minute hand.
• The hands for this clock are pointing towards the numbers 3 and 7. Which number is the minute hand pointing towards? (Signal.) *3.*
(Display:) [103:2M]

• Is the finger pointing the direction the clock hand is moving? (Signal.) *Yes.*
(Display:) [103:2N]

Here's the arrow to show the direction the minute hand is moving.
m. Now you're going to tell me the next number the minute hand will point to.
• Look at the minute hand. Everybody, what's the next number the minute hand will point to? (Signal.) *3.*
• What's the last number the minute hand pointed to? (Signal.) *2.*
(Repeat steps l and m until firm.)
n. The other hand doesn't have an arrow on it, but you can figure out the numbers for that hand.
• Look at it and get ready to tell me the next number it will point to. Everybody, what's the next number the other hand will point to? (Signal.) *8.*
• What's the last number the other hand pointed to? (Signal.) *7.*
(Repeat step n until firm.)

EXERCISE 3: MIXED COUNTING

a. Your turn: Count by twos to 20. Get ready. (Tap.) *2, 4, 6, 8, 10, 12, 14, 16, 18, 20.*
(Repeat until firm.)

b. My turn to start with 20 and count by twos to 30: Twentyyy, 22, 24, 26, 28, 30.
• Once more: 20, 22, 24, 26, 28, 30.
• Your turn to start with 20 and count by twos to 30. Get 20 going. *Twentyyy.* Count. (Tap.) *22, 24, 26, 28, 30.*
(Repeat until firm.)
c. Start with 20 and plus tens to 100. Get 20 going. *Twentyyy.* Plus tens. (Tap.) *30, 40, 50, 60, 70, 80, 90, 100.*
d. Start with 20 and count by fives to 50. Get 20 going. *Twentyyy.* Count. (Tap.) *25, 30, 35, 40, 45, 50.*
e. Start with 20 and count by ones to 30. Get 20 going. *Twentyyy.* Count. (Tap.) *21, 22, 23, 24, 25, 26, 27, 28, 29, 30.*
f. Start with 20 and count backward by ones to 10. Get 20 going. *Twentyyy.* Count backward. (Tap.) *19, 18, 17, 16, 15, 14, 13, 12, 11, 10.*
g. Listen: You have 20. When you count by twos, what's the next number? (Signal.) *22.*
• You have 20. When you count backward by ones, what's the next number? (Signal.) *19.*
• You have 20. When you count by fives, what's the next number? (Signal.) *25.*
• You have 20. When you plus tens, what's the next number? (Signal.) *30.*
• You have 20. When you count by ones, what's the next number? (Signal.) *21.*
h. You have 82. When you count by twos, what's the next number? (Signal.) *84.*
• You have 82. When you count backward by ones, what's the next number? (Signal.) *81.*
• You have 82. When you plus tens, what's the next number? (Signal.) *92.*
(Repeat steps c though h until firm.)

EXERCISE 4: FACTS
PLUS/MINUS MIX · [REMEDY]

a. Distribute unopened workbooks to children.)
• Open your workbook to Lesson 103 and find part 1.
(Observe children and give feedback.)
(Teacher reference:) [R | Part P]

	c. $6 + 3$	f. $10 - 5$
a. $6 + 4$	d. $11 - 6$	g. $7 + 7$
b. $6 - 2$	e. $3 + 4$	h. $8 - 4$

Think of the number families for each problem and write the answers. Put your pencil down when you've completed the equations in part 1.
(Observe children and give feedback.)

b. Check your work. You'll touch and read the equation for each problem.
- Problem A. (Signal.) *6 + 4 = 10.*
- (Repeat for:) B, *6 – 2 = 4;* C, *6 + 3 = 9;* D, *11 – 6 = 5;* E, *3 + 4 = 7;* F, *10 – 5 = 5;* G, *7 + 7 = 14;* H, *8 – 4 = 4.*

EXERCISE 5: ADDITION/SUBTRACTION
NOMENCLATURE

a. (Display:) [103:5A]

- (Point to –.) Is this sign for addition or subtraction? (Touch.) *Subtraction.*
- (Point to +.) What is this sign for? (Touch.) *Addition.*
- (Point to +.) What is this sign for? (Touch.) *Addition.*
- (Point to –.) What is this sign for? (Touch.) *Subtraction.*
  (Repeat until firm.)
b. (Point to –.) If a problem has this sign, is it an addition or a subtraction problem? (Signal.) *(A) subtraction problem.*
- (Point to +.) What kind of problem has this sign? (Signal.) *(An) addition problem.*
- (Point to –.) What kind of problem has this sign? (Signal.) *(A) subtraction problem.*
c. Find part 2 on worksheet 103. ✔
  (Teacher reference:)

a. 58   b. 314   c. 75   d. 539
 – 46   +304   + 26   –316

Some of these problems are addition problems. Some are subtraction problems.
- Circle the subtraction problems.
  (Observe children and give feedback.)
d. Touch and read the first subtraction problem. (Signal.) *58 – 46.*
- Touch and read the other subtraction problem. (Signal.) *539 – 316.*
e. Touch problem A. ✔
- What kind of problem is that? (Signal.) *(A) subtraction problem.*

f. (Repeat the following tasks for the remaining problems:)

Touch problem __.	What kind of problem is that?
B	*(An) addition problem.*
C	*(An) addition problem.*
D	*(A) subtraction problem.*

(Repeat problems that were not firm.)
You'll work these problems later. Remember, addition problems add, subtraction problems subtract.

EXERCISE 6: NUMBER FAMILIES FROM EQUATIONS
MULTIPLE-DIGIT NUMBERS   REMEDY

a. Find part 3 on worksheet 103. ✔
  (Teacher reference:)   R Part K

a. 142   b. 181   c. 220
  +181   –142   –142
   323   39   78

d. 220
  +181
   401

You're going to make number families for these equations.
- Touch and read equation A. (Signal.) *142 + 181 = 323.*
- Does the equation add? (Signal.) *Yes.*
- So is the big number the answer? (Signal.) *Yes.*
- What's the big number? (Signal.) *323.*
b. (Repeat the following tasks for the remaining problems:)

	Touch and read equation __.	Does the equation add?	So is the big number the answer?	What's the big number?
B	*181 – 142 = 39*	*No*	*No*	*181*
C	*220 – 142 = 78*	*No*	*No*	*220*
D	*220 + 181 = 401*	*Yes*	*Yes*	*401*

(Repeat problems that were not firm.)

c. Write the number families for all of the equations in part 3. Put your pencil down when you've completed all of the families. (Observe children and give feedback.)

d. Check your work.

• What's the big number for family A? (Signal.) *323.*

• What are the small numbers? (Call on a child.) *142 (and) 181.*

e. (Repeat the following tasks for the remaining problems:)

What's the big number for family __?		What are the small numbers? (Call on a child.)
B	181	142 (and) 39
C	220	142 (and) 78
D	401	220 (and) 181

EXERCISE 7: 3-D OBJECTS
PENTAGONS

a. (Display:) [103:7A]

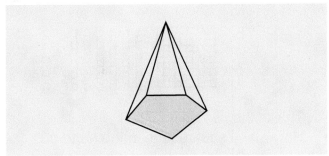

The bottom face of this object is not a hexagon. I'll touch each side. Count the sides of the bottom face to yourself. (Touch each side.)

• Everybody, how many sides does the bottom face have? (Signal.) *5.*

• A hexagon doesn't have 5 sides. How many sides does a hexagon have? (Signal.) *6.*

b. Listen: A face with 5 sides is called a pentagon. Say **pentagon.** (Signal.) *Pentagon.*

• What is a face with 5 sides? (Signal.) *(A) pentagon.*

• What is a shape with 6 sides? (Signal.) *(A) hexagon.*

c. (Point to △.) How many sides does this face have? (Signal.) *5.*

• What shape is the bottom face? (Signal.) *(A) pentagon.*

d. Are all of the side faces of this object triangles? (Signal.) *Yes.*

• So is this a pyramid? (Signal.) *Yes.*

• What kind of object is this? (Signal.) *(A) pyramid.*

e. Find part 4 on worksheet 103. ✔ (Teacher reference:)

1   2

You're going to write letters on all the faces of the objects. You'll write P on a face that is a pentagon.

• What letter will you write on a face that is a pentagon? (Signal.) *P.*

f. What letter will you write on a face that is a triangle? (Signal.) *T.*

• What letter will you write on a face that is a rectangle? (Signal.) *R.*

• What letter will you write on a face that is a hexagon? (Signal.) *H.* (Repeat step f until firm.)

g. For the objects in part 4, write letters for the faces that you can see. Then write the letter for the top face if there is a top face. Don't write anything for the top face if there isn't a top face. (Observe children and give feedback.)

h. Check your work.

• Object 1. What letter did you write on the bottom face? (Signal.) *H.*

• What letter did you write on the side faces? (Signal.) *R.*

• Did you write a letter for the top face? (Signal.) *Yes.*

• What letter? (Signal.) *H.*

i. Object 2. What letter did you write on the bottom face? (Signal.) *P.*

• What letter did you write on the side faces? (Signal.) *R.*

• Did you write a letter for the top face? (Signal.) *Yes.*

• What letter? (Signal.) *P.*

j. (Display:) [103:7B]

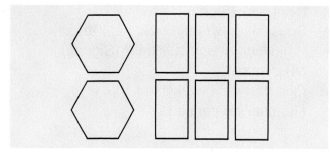

Here are the faces of one of the objects in part 4. You're going to figure out which object.

- (Point to hexagons.) These are the bottom and top faces. What shape are these faces? (Touch.) *Hexagon.*
- (Point to rectangles.) These are the side faces. What shape? (Touch.) *Rectangles.*
- Raise your hand when you know how many side faces there are. ✔
- How many side faces? (Touch.) *6.*
- Look at the objects in part 4 carefully and circle the object that has these faces. (Observe children and give feedback.)
- Everybody, which object did you circle? (Signal.) *(Object) 1.*

EXERCISE 8: PLACE VALUE
PICTURES

a. (Display:) [103:8A]

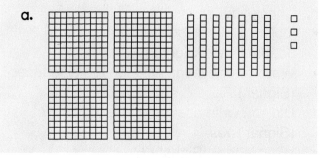

This is the picture for A. You're going to write the equation for the number of squares in picture A.

- (Point to a hundreds box.) How many squares are in each box? (Touch.) *100.*
- (Point to a tens column.) How many squares are in each column? (Touch.) *10.*
- (Point to a single square.) How many squares are in each square? (Touch.) *1.*
- (Repeat until firm.)

b. (Point to hundreds boxes.) Tell me how many squares are in the hundreds boxes. Get ready. (Signal.) *400.*
- (Point to tens columns.) I'll touch each tens column. Count for the squares to yourself. (Touch each tens column.) Tell me how many squares are in the tens columns. Get ready. (Signal.) *70.*
- (Point to single squares.) Tell me how many single squares there are. (Signal.) *3.*
- What does 400 plus 70 plus 3 equal? (Signal.) *473.*
- Say the equation for the number of squares in picture A. Get ready. (Touch columns.) *400 + 70 + 3 = 473.* (Repeat until firm.)

c. Find part 5 on worksheet 103. ✔ (Teacher reference:)

- Write the equation for picture A. ✔

d. (Display:) [103:8B]

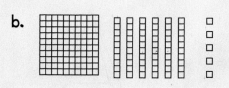

This is the picture for B.
- How many squares are in the hundreds boxes? (Signal.) *100.*
- I'll touch each tens column. Count for the squares to yourself. (Touch each tens column.) How many squares are in the tens columns? (Signal.) *60.*
- I'll touch each single square. Count the squares for the ones to yourself. (Touch each single square.) How many single squares are there? (Signal.) *5.*
- What does 100 plus 60 plus 5 equal? (Signal.) *165.*
- Say the equation for the number of squares in picture B. Get ready. (Touch columns.) *100 + 60 + 5 = 165.* (Repeat until firm.)
- Write the equation for B. ✔ (Answer key:)

a. 400+70+3=473   b. 100+60+5=165

EXERCISE 9: FACTS

SUBTRACTION

a. Find part 6 on worksheet 103. ✔
(Teacher reference:)

a. 16 − 8
b. 9 − 3
c. 10 − 2
d. 7 − 3
e. 10 − 4
f. 14 − 7
g. 9 − 7
h. 8 − 3

- Touch and read problem A. Get ready.
(Signal.) *16 minus 8.*
- What's 16 minus 8? (Signal.) *8.*

b. (Repeat the following tasks for problems B through H:)

Touch and read problem __.	What's __?	
B	9 − 3	9 − 3	6
C	10 − 2	10 − 2	8
D	7 − 3	7 − 3	4
E	10 − 4	10 − 4	6
F	14 − 7	14 − 7	7
G	9 − 7	9 − 7	2
H	8 − 3	8 − 3	5

(Repeat problems that were not firm.)

c. Complete each of the facts. Put your pencil down when you've written all of the equations for part 6.
(Observe children and give feedback.)

d. Check your work.
- Touch and read the equation for problem A.
(Signal.) *16 − 8 = 8.*
- (Repeat for:) B, *9 − 3 = 6;* C, *10 − 2 = 8;* D, *7 − 3 = 4;* E, *10 − 4 = 6;* F, *14 − 7 = 7;* G, *9 − 7 = 2;* H, *8 − 3 = 5.*

EXERCISE 10: INDEPENDENT WORK

a. Find part 7. ✔
(Teacher reference:)

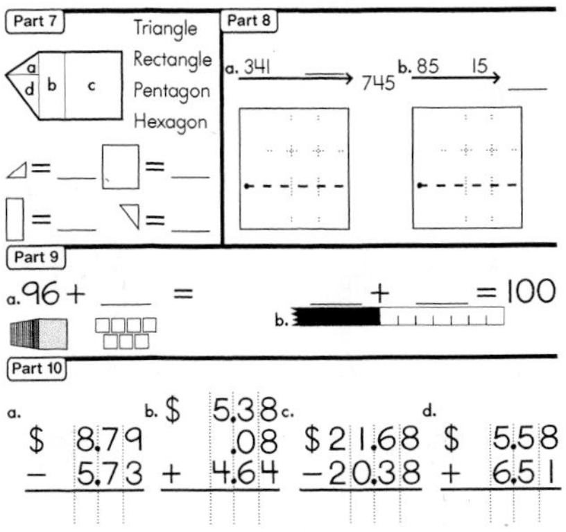

You'll circle the name for the whole figure. The names are triangle, rectangle, pentagon, and hexagon. Then you'll write the letter for each part.
In part 8, you'll write a column problem to figure out the missing number in each family. Then you'll complete the family.
In part 9, you'll complete the equation for the squares and the equation for the ruler.
In part 10, you'll work the money problems.

b. Complete worksheet 103. Remember to work the addition and subtraction problems in part 2 on the other side of your worksheet.
(Observe children and mark incorrect responses on children's worksheets as you give feedback.)

Lesson

EXERCISE 1: FACTS
ADDITION/SUBTRACTION

a. (Display:) [104:1A]

8 – 5	7 – 4	9 – 3
8 + 8	14 – 7	6 + 6
11 – 5	3 + 3	10 – 6

- (Point to **8 – 5**.) Listen: 8 – 5. Is the answer the big number or a small number? (Signal.) *A small number.*
- Say the fact. (Signal.) *8 – 5 = 3.*

b. (Point to **8 + 8**.) 8 + 8. Is the answer the big number or a small number? (Signal.) *The big number.*
- Say the fact. (Signal.) *8 + 8 = 16.*

c. (Repeat the following tasks for the remaining problems:)

(Point to __.) __. Is the answer the big number or a small number?			Say the fact.
11 – 5	11 – 5	*A small number.*	*11 – 5 = 6*
7 – 4	7 – 4	*A small number.*	*7 – 4 = 3*
14 – 7	14 – 7	*A small number.*	*14 – 7 = 7*
3 + 3	3 + 3	*The big number.*	*3 + 3 = 6*
9 – 3	9 – 3	*A small number.*	*9 – 3 = 6*
6 + 6	6 + 6	*The big number.*	*6 + 6 = 12*
10 – 6	10 – 6	*A small number.*	*10 – 6 = 4*

(Repeat problems that were not firm.)

EXERCISE 2: MIXED COUNTING

a. Your turn: Count by twos to 20. Get ready. (Tap.) *2, 4, 6, 8, 10, 12, 14, 16, 18, 20.* (Repeat until firm.)

b. My turn to start with 20 and count by twos to 30. 20, 22, 24, 26, 28, 30.
- Once more: 20, 22, 24, 26, 28, 30.
- Your turn to start with 20 and count by twos to 30. Get 20 going. *Twentyyy.* Count. (Tap.) *22, 24, 26, 28, 30.* (Repeat until firm.)

c. My turn to start with 30 and count by twos to 40. 30, 32, 34, 36, 38, 40.
- Your turn: Start with 30 and count by twos to 40. Get 30 going. *Thirtyyy.* Count. (Tap.) *32, 34, 36, 38, 40.* (Repeat until firm.)

d. Start with 50 and plus tens to 150. Get 50 going. *Fiftyyy.* Plus tens. (Tap.) *60, 70, 80, 90, 100, 110, 120, 130, 140, 150.*

e. Start with 50 and count by 25s to 150. Get 50 going. *Fiftyyy.* Count. (Tap.) *75, 100, 125, 150.*

f. Start with 50 and count by fives to 100. Get 50 going. *Fiftyyy.* Count. (Tap.) *55, 60, 65, 70, 75, 80, 85, 90, 95, 100.*

g. Start with 50 and count by ones to 60. Get 50 going. *Fiftyyy.* Count. (Tap.) *51, 52, 53, 54, 55, 56, 57, 58, 59, 60.*

h. Start with 50 and count backward by ones to 40. Get 50 going. *Fiftyyy.* Count backward. (Tap.) *49, 48, 47, 46, 45, 44, 43, 42, 41, 40.* (Repeat steps d through h until firm.)

i. Listen: You have 50. When you count by 25s, what's the next number? (Signal.) *75.*
- You have 50. When you count backward by tens, what's the next number? (Signal.) *40.*
- You have 50. When you count by ones, what's the next number? (Signal.) *51.*
- You have 50. When you count by fives, what's the next number? (Signal.) *55.* (Repeat step i until firm.)

EXERCISE 3: TELLING TIME
ANALOG CLOCK HANDS `REMEDY`

a. I'm going to show you clocks, and we're going to make the arrows to show which direction the minute hand moves.
(Display:) [104:3A]

- The hands for this clock are pointing towards the numbers 4 and 8. Which number is the minute hand pointing towards? (Signal.) *8.*
(Display:) [104:3B]

- Is that finger pointing the direction the clock hand is moving? (Signal.) *No.*
(Display:) [104:3C]

- Is that finger pointing the direction the clock hand is moving? (Signal.) *Yes.*
(Display:) [104:3D]

Here's the arrow to show the direction the minute hand is moving.

b. Now you're going to tell me the next number the minute hand will point to.
- Look at the minute hand. Everybody, what's the next number the minute hand will point to? (Signal.) *8.*
- What's the last number the minute hand pointed to? (Signal.) *7.*
(Repeat step b until firm.)

c. Look at the other hand and get ready to tell me the next number it will point to.
- Everybody, what's the next number the other hand will point to? (Signal.) *4.*
- What's the last number the other hand pointed to? (Signal.) *3.*
(Repeat step c until firm.)

d. (Display:) [104:3E]

Let's make the arrow to show the direction the minute hand is moving for this clock.
- The hands for this clock are pointing towards the numbers 4 and 6. Which number is the minute hand pointing towards? (Signal.) *4.*
(Display:) [104:3F]

- Is that finger pointing the direction the clock hand is moving? (Signal.) *No.*
(Display:) [104:3G]

- Is that finger pointing the direction the clock hand is moving? (Signal.) *Yes.*
(Display:) [104:3H]

Here's the arrow to show the direction the minute hand is moving.

e. Now you're going to tell me the next number the minute hand will point to.
- Look at the minute hand. Everybody, what's the next number the minute hand will point to? (Signal.) *4.*
- What's the last number the minute hand pointed to? (Signal.) *3.*
(Repeat step e until firm.)

f. Look at the other hand and get ready to tell me the next number it will point to.
- Everybody, what's the next number the other hand will point to? (Signal.) *7.*
- What's the last number the other hand pointed to? (Signal.) *6.*
(Repeat step f until firm.)

g. (Display:) [104:3I]

Let's make the arrow to show the direction the minute hand is moving for this clock.
- The hands for this clock are pointing towards the numbers 8 and 12. Which number is the minute hand pointing towards? (Signal.) *12.*
(Display:) [104:3J]

- Is that finger pointing the direction the clock hand is moving? (Signal.) *No.*
(Display:) [104:3K]

- Is that finger pointing the direction the clock hand is moving? (Signal.) *Yes.*
(Display:) [104:3L]

Here's the arrow to show the direction the minute hand is moving.

h. Now you're going to tell me the next number the minute hand will point to.
- Look at the minute hand. Everybody, what's the next number the minute hand will point to? (Signal.) *1.*
- What's the last number the minute hand pointed to? (Signal.) *12.*
(Repeat step h until firm.)

i. Look at the other hand and get ready to tell me the next number it will point to. Everybody, what's the next number the other hand will point to? (Signal.) *9.*
- What's the last number the other hand pointed to? (Signal.) *8.*
(Repeat step i until firm.)

j. (Distribute unopened workbooks to children.)
- Open your workbook to Lesson 104 and find part 1.
(Observe children and give feedback.)
(Teacher reference:) [R] **Part G**

You're going to make the arrow to show the direction the minute hand is moving.
- Touch the minute hand for clock A.
(Observe children and give feedback.)
- Point the direction the minute hand is moving.
(Observe children and give feedback.)
- Make an arrow to show the direction the minute hand is moving.
(Observe children and give feedback.)
(Display:) [104:3M]

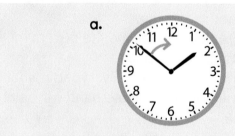

Here's what you should have for clock A.

k. Touch the minute hand for clock B. ✔
• Point the direction the minute hand is moving. (Observe children and give feedback.)
• Make an arrow to show the direction the minute hand is moving. ✔
(Display:)   [104:3N]

Here's what you should have for clock B.
l. Touch the minute hand for clock C. ✔
• Point the direction the minute hand is moving. ✔
• Make an arrow to show the direction the minute hand is moving. ✔
(Display:)   [104:3O]

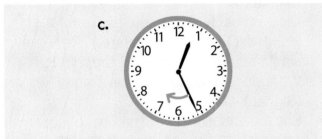

Here's what you should have for clock C.
m. Touch the minute hand for clock A again and get ready to tell me the next number.
• Everybody, what's the next number the minute hand will point to? (Signal.) *11.*
• What's the last number the minute hand pointed to? (Signal.) *10.*
(Repeat step m until firm.)
n. The other hand is the hour hand. What's the other hand? (Signal.) *The hour hand.*
• Touch the hour hand. ✔
• Everybody, what's the next number the hour hand will point to? (Signal.) *2.*
• What's the last number the hour hand pointed to? (Signal.) *1.*
(Repeat step n until firm.)

o. Touch the minute hand for clock B and get ready to tell me the next number.
• Everybody, what's the next number the minute hand will point to? (Signal.) *12.*
• What's the last number the minute hand pointed to? (Signal.) *11.*
p. Touch the hour hand for clock B. ✔
• Everybody, what's the next number the hour hand will point to? (Signal.) *8.*
• What's the last number the hour hand pointed to? (Signal.) *7.*
(Repeat steps o and p until firm.)
q. Touch the minute hand for clock C and get ready to tell me the next number.
• Everybody, what's the next number the minute hand will point to? (Signal.) *6.*
• What's the last number the minute hand pointed to? (Signal.) *5.*
r. Touch the hour hand for clock C. ✔
• Everybody, what's the next number the hour hand will point to? (Signal.) *1.*
• What's the last number the hour hand pointed to? (Signal.) *12.*
(Repeat steps q and r until firm.)

EXERCISE 4: FACTS
ADDITION

a. Find part 2 on worksheet 104. ✔
(Teacher reference:)

a. $3 + 5$ c. $4 + 3$ f. $5 + 6$
b. $7 + 7$ d. $5 + 5$ g. $4 + 4$
            e. $3 + 6$ h. $4 + 6$

These are plus problems so the answer is the big number in the family. I'll read each problem and you'll tell me the answer. Then you'll complete each fact.
• Problem A. 3 plus 5. What's the answer? (Signal.) *8.*
• (Repeat the following tasks for remaining problems:)

Problem __. __. What's the answer?

B	$7 + 7$	*14*
C	$4 + 3$	*7*
D	$5 + 5$	*10*
E	$3 + 6$	*9*
F	$5 + 6$	*11*
G	$4 + 4$	*8*
H	$4 + 6$	*10*

(Repeat problems that were not firm.)

b. Complete the equations in part 2. Put your pencil down when you're finished.
   (Observe children and give feedback.)
c. Check your work. You're going to touch and read the fact for each problem.
   - Problem A. (Signal.) *3 + 5 = 8.*
   - (Repeat for:) B, *7 + 7 = 14;* C, *4 + 3 = 7;* D, *5 + 5 = 10;* E, *3 + 6 = 9;* F, *5 + 6 = 11;* G, *4 + 4 = 8;* H, *4 + 6 = 10.*

EXERCISE 5: PLACE VALUE
PICTURES

a. (Display:)   [104:5A]

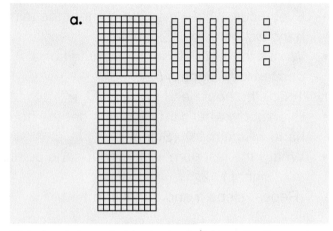

This is picture A. You're going to write the equation for the number of squares in picture A.
   - (Point to a hundreds box.) How many squares are in each box? (Touch.) *100.*
   - (Point to a tens column.) How many squares are in each column? (Touch.) *10.*
   - (Point to a single square.) How many squares are in each square? (Touch.) *1.*
   (Repeat until firm.)
b. (Point to hundreds boxes.) Tell me how many squares are in the hundreds boxes. Get ready. (Signal.) *300.*
   - (Point to tens columns.) I'll touch each tens column. Count for the squares to yourself. (Touch each tens column.) Tell me how many squares are in the tens columns. Get ready. (Signal.) *60.*
   - (Point to single squares.) Tell me how many single squares there are. (Signal.) *4.*
   - What does 300 plus 60 plus 4 equal? (Signal.) *364.*
   - Say the equation for the number of squares in picture A. Get ready. (Touch columns.) *300 + 60 + 4 = 364.*
   (Repeat until firm.)

c. Find part 3 on worksheet 104. ✔
   (Teacher reference:)

   - Write the equation for picture A. ✔
d. (Display:)   [104:5B]

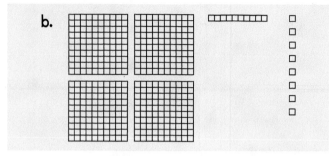

This is picture B.
   - (Point to hundreds column.) How many squares are in the hundreds boxes? (Signal.) *400.*
   - (Point to tens row.) The tens are shown in a row. How many squares are in the tens row? (Signal.) *10.*
   - I'll touch each single square. Count them to yourself. (Touch each square.) How many single squares are there? *(Signal.) 8.*
   - What does 400 plus 10 plus 8 equal? (Signal.) *418.*
   - Say the equation for the number of squares in picture B. (Touch columns.) *400 + 10 + 8 = 418.* (Repeat until firm.)
   - Write the equation for B. ✔
   (Answer key:)

   a.
   300+60+4=364   400+10+8=418
                   b.

EXERCISE 6: NUMBER FAMILIES FROM PROBLEMS
MULTIPLE-DIGIT NUMBERS

a. Find part 4. ✔
(Teacher reference:)

a.   386   b.   96   c.   298
     − 73   +156   − 96
   ═══⟶_   ═══⟶_   ═══⟶_

You're going to make number families for these problems, but first you'll solve each problem.

- Touch and read problem A. (Signal.) *386 − 73.*
- Does the problem add? (Signal.) *No.*
- So is the answer the big number? (Signal.) *No.*
- What's the big number? (Signal.) *386.*

b. (Repeat the following tasks for the remaining problems:)

Touch and read problem __.	Does the problem add?	So is the answer the big number?	What's the big number?
B	96 + 156	Yes	Yes	
C	298 − 96	No	No	298

(Repeat problems that were not firm.)

c. Go back to problem A and work it. Stop when you've completed equation A.
(Observe children and give feedback.)

d. Check your work.

- Read equation A. (Signal.) *386 − 73 = 313.*
(Display:)   [104:6A]

```
        a.    3 8 6
            −   7 3
              3 1 3
```

- Think of the numbers for the family. What's the big number for family A? (Signal.) *386.*
- What are the small numbers? (Signal.) *73 (and) 313.*
- Write the numbers in family A.
(Observe children and give feedback.)

e. Work problems B and C. Stop when you've completed the rest of the equations in part 4.
(Observe children and give feedback.)

f. Check your work.

- Read equation B. (Signal.) *96 + 156 = 252.*
- Read equation C. (Signal.) *298 − 96 = 202.*
(Display:)   [104:6B]

```
    b.      9 6         c.    2 9 8
          +1 5 6            −   9 6
            2 5 2            2 0 2
```

Here's what you should have for equations B and C.

g. Write the numbers in families B and C. Put your pencil down when you've completed the families in part 4.
(Observe children and give feedback.)

h. Check your work.

- What's the big number for family B? (Signal.) *252.*
- What are the small numbers? (Call on a child.) *96 (and) 156.*

i. What's the big number for family C? (Signal.) *298.*

- What are the small numbers? (Call on a child.) *96 (and) 202.*
(Display:)   [104:6C]

```
a. ─73──313─⟶ 386   b. ─96──156─⟶ 252   c. ─96──202─⟶ 298
```

Here are the families for part 4.

EXERCISE 7: 3-D OBJECTS
PENTAGON   REMEDY

a. In the last lesson you learned how many sides a pentagon has. How many sides does a pentagon have? (Signal.) *5.*

- What is a shape with 5 sides? (Signal.) *(A) pentagon.*
- What is a shape with 6 sides? (Signal.) *(A) hexagon.*
- What is a shape with 4 sides? (Signal.) *(A) rectangle.*

b. (Display:) [104:7A]

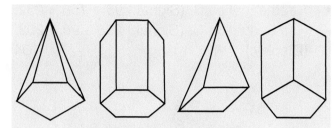

- (Point to pentagonal pyramid) What's the bottom face of this object? (Signal.) *(A) pentagon.*
- Are the side faces triangles? (Signal.) *Yes.*
- So is this object a pyramid? (Signal.) *Yes.*

c. (Repeat the following tasks for the remaining problems:)

(Point to __.) What's the bottom face of this object?		Are the side faces triangles?	So is this object a pyramid?
Hexagonal prism	*(A) hexagon*	*No*	*No*
Rectangular pyramid	*(A) rectangle*	*Yes*	*Yes*
Pentagonal prism	*(A) pentagon*	*No*	*No*

(Repeat objects that were not firm.)

d. Find part 5 on worksheet 104. ✔
(Teacher reference:) R Part A

 1   2

You're going to write letters on all the faces of the objects. You'll write P on a face that is a pentagon.

- What letter will you write on a face that is a pentagon? (Signal.) *P.*
- What letter will you write on a face that is a triangle? (Signal.) *T.*
- What letter will you write on a face that is a rectangle? (Signal.) *R.*
- What letter will you write on a face that is a hexagon? (Signal.) *H.*
(Repeat until firm.)

e. For the objects in part 5, write letters for the faces that you can see. Then write the letter for the top face if there is a top face. Don't write anything for the top face if there isn't a top face.
(Observe children and give feedback.)

f. Check your work.
- Object 1, what letter did you write for the bottom face? (Signal.) *P.*
- What letter did you write for the side faces? (Signal.) *R.*
- Did you write a letter for the top face? (Signal.) *Yes.*
- What letter? (Signal.) *P.*
- Object 2, what letter did you write for the bottom face? (Signal.) *T.*
- What letter did you write for the side faces? (Signal.) *R.*
- Did you write a letter for the top face? (Signal.) *Yes.*
- What letter? (Signal.) *T.*

g. I'm going to show you faces of one of the objects in part 5. You're going to figure out if it is object 1 or 2.
(Display:) [104:7B]

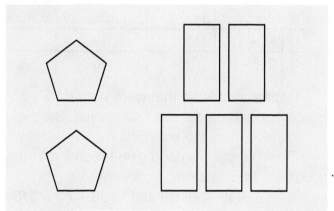

Here are the faces.

(Point to bottom ⬠.) This is the bottom face.

(Point to top ⬠.) This is the top face.
(Point to rectangles.) These are the sides.

- (Point to top and bottom ⬠.) What shape are the top and bottom? (Signal.) *Pentagons.*
Yes, the top and the bottom faces are pentagons.
- (Point to sides.) What shape are the sides? (Signal.) *Rectangles.*
Look at objects 1 and 2.
- Pentagons are the top face and the bottom face of which object—object 1 or object 2? (Signal.) *Object 1.*
- Circle object 1. It has the right faces. ✔

EXERCISE 8: FACTS
SUBTRACTION

a. Turn to the other side of your worksheet and find part 6. ✔
(Teacher reference:)

a. $12-6$ c. $16-8$ f. $10-5$
b. $12-10$ d. $9-6$ g. $8-2$
e. $6-4$ h. $8-4$

• Touch and read problem A. Get ready.
(Signal.) *12 minus 6.*
• What's 12 minus 6? (Signal.) *6.*

b. (Repeat the following tasks for problems B through H:)

Touch and read problem __.		What's __?	
B	12 – 10	12 – 10	2
C	16 – 8	16 – 8	8
D	9 – 6	9 – 6	3
E	6 – 4	6 – 4	2
F	10 – 5	10 – 5	5
G	8 – 2	8 – 2	6
H	8 – 4	8 – 4	4

(Repeat problems that were not firm.)

c. Complete each of the facts. Put your pencil down when you've written all of the equations for part 6.
(Observe children and give feedback.)

d. Check your work.
• Touch and read the equation for problem A.
(Signal.) *12 – 6 = 6.*
• (Repeat for:) B, *12 – 10 = 2;* C, *16 – 8 = 8;*
D, *9 – 6 = 3;* E, *6 – 4 = 2;* F, *10 – 5 = 5;*
G, *8 – 2 = 6;* H, *8 – 4 = 4.*

EXERCISE 9: INDEPENDENT WORK

a. Find part 7. ✔
(Teacher reference:)

You'll write the dollars and cents amount for each group of bills and coins.
In part 8, you'll circle the name for the whole shape. Then you'll write the letter for each part.
In part 9, below each family, you'll write the fact for the missing number and you'll complete the family.
In part 10, you'll write each problem in a column and work it.

b. Complete worksheet 104.
(Observe children and mark incorrect responses on children's worksheets as you give feedback.)

Lesson 105

EXERCISE 1: FACTS
ADDITION/SUBTRACTION

a. (Display:) [105:1A]

16 − 8	5 + 3	7 + 7
16 − 10	5 + 6	7 − 3
10 − 5	6 − 3	10 − 4

- (Point to **16 − 8**.) Listen: 16 minus 8. Is the answer the big number or a small number? (Signal.) *A small number.*
- Say the fact. (Signal.) *16 − 8 = 8.*

b. (Point to **16 − 10**.) 16 minus 10. Is the answer the big number or a small number? (Signal.) *A small number.*
- Say the fact. (Signal.) *16 − 10 = 6.*

c. (Repeat the following tasks for the remaining problems:)

(Point to __.) __. Is the answer the big number or a small number?			Say the fact.
10 − 5	10 − 5	A small number.	10 − 5 = 5
5 + 3	5 + 3	The big number.	5 + 3 = 8
5 + 6	5 + 6	The big number.	5 + 6 = 11
6 − 3	6 − 3	A small number.	6 − 3 = 3
7 + 7	7 + 7	The big number.	7 + 7 = 14
7 − 3	7 − 3	A small number.	7 − 3 = 4
10 − 4	10 − 4	A small number.	10 − 4 = 6

(Repeat problems that were not firm.)

EXERCISE 2: MIXED COUNTING

a. Your turn: Count by twos to 20. Get ready. (Tap.) *2, 4, 6, 8, 10, 12, 14, 16, 18, 20.*
(Repeat step a until firm.)

b. Your turn to start with 20 and count by twos to 30. Get 20 going. *Twentyyy.* Count. (Tap.) *22, 24, 26, 28, 30.*
(Repeat step b until firm.)

c. My turn to start with 40 and count by twos to 50. 40, 42, 44, 46, 48, 50.
- Your turn: Start with 40 and count by twos to 50. Get 40 going. *Fortyyy.* Count. (Tap.) *42, 44, 46, 48, 50.*
(Repeat until firm.)

d. Start with 100 and plus tens to 200. Get 100 going. *One huuundred.* Plus tens. (Tap.) *110, 120, 130, 140, 150, 160, 170, 180, 190, 200.*

e. Start with 100 and count by 25s to 200. Get 100 going. *One huuundred.* Count. (Tap.) *125, 150, 175, 200.*

f. Start with 100 and count by ones to 110. Get 100 going. *One huuundred.* Count. (Tap.) *101, 102, 103, 104, 105, 106, 107, 108, 109, 110.*

g. Start with 100 and count by 100s to 1000. Get 100 going. *One huuundred.* Count. (Tap.) *200, 300, 400, 500, 600, 700, 800, 900, 1000.*

h. Start with 100 and count backward by ones to 90. Get 100 going. *One huuundred.* Count backward. (Tap.) *99, 98, 97, 96, 95, 94, 93, 92, 91, 90.*

i. Start with 100 and count backward by tens. Get 100 going. *One huuundred.* Count backward. (Tap.) *90, 80, 70, 60, 50, 40, 30, 20, 10.*
(Repeat steps d through i that were not firm.)

j. Listen: You have 100. When you count by 25s, what's the next number? (Signal.) *125.*
- You have 100. When you count backward by tens, what's the next number? (Signal.) *90.*
- You have 100. When you count backward by ones, what's the next number? (Signal.) *99.*
- You have 100. When you count by 100s, what's the next number? (Signal.) *200.*
(Repeat step j until firm.)

EXERCISE 3: 3-D OBJECTS
DECOMPOSITION

a. (Display:) [105:3A]

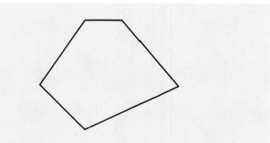

- (Point to pentagon.) I'll touch each side of this shape. Count the sides to yourself. (Touch each side.) Everybody, how many sides does this shape have? (Touch.) *5.*

- Is this figure a hexagon or a pentagon? (Touch.) *(A) pentagon.*
  Yes, a pentagon.
b. Tell me what letter you write for a pentagon. Get ready. (Signal.) *P.*
- Tell me what letter you write for a rectangle. Get ready. (Signal.) *R.*
- Tell me what letter you write for a hexagon. Get ready. (Signal.) *H.*
- Tell me what letter you write for a triangle. Get ready. (Signal.) *T.*
  (Repeat step b until firm.)
c. (Distribute unopened workbooks to children.)
- Open your workbook to Lesson 105 and find part 1.
  (Observe children and give feedback.)
  (Teacher reference:)

- Write letters for the faces of object 1 and object 2 that you can see. Then write the letter for the top face if there is a top face.
  (Observe children and give feedback.)
d. What letter did you write on the bottom face of object 1? (Signal.) *R.*
- What letter did you write on the side faces? (Signal.) *R.*
- Did you write a letter for the top face? (Signal.) *Yes.*
- What letter? (Signal.) *R.*
- Object 2. What letter did you write on the bottom face? (Signal.) *P.*
- What letter did you write on the side faces? (Signal.) *R.*
- Did you write a letter for the top face? (Signal.) *Yes.*
- What letter? (Signal.) *P.*
e. (Display:)   [105:3B]

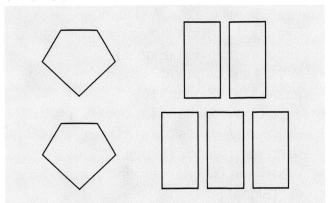

Here are the faces of one of the objects in part 1. You're going to figure out which object.

- (Point to top and bottom faces.) Here are the top and bottom faces. What shape? (Signal.) *(A) pentagon.*
- (Point to side faces.) Here are the side faces. What shape? (Signal.) *(A) rectangle.*
- I'll touch each face. Count them to yourself. (Touch each rectangle.) How many side faces are there? *(Signal.) 5.*
- Look at the faces carefully and circle the object in part 1 that has these faces. (Observe children and give feedback.)
- Everybody, which object did you circle? (Signal.) *(Object) 2.*

EXERCISE 4: FACTS
ADDITION

a. Find part 2 on worksheet 105. ✔
  (Teacher reference:)

a.	3 + 3	f.	6 + 6
b.	3 + 5	g.	4 + 6
c.	5 + 5	h.	4 + 4
d.	5 + 2	i.	7 + 7
e.	6 + 3	j.	8 + 8

These are addition problems so the answer is the big number in the family. I'll read each problem and you'll tell me the answer. Then you'll complete each fact.

- Problem A. 3 plus 3. What's the answer? (Signal.) *6.*
- (Repeat the following tasks for remaining problems:)

Problem __. __. What's the answer?

B	3 + 5	8
C	5 + 5	10
D	5 + 2	7
E	6 + 3	9
F	6 + 6	12
G	4 + 6	10
H	4 + 4	8
I	7 + 7	14
J	8 + 8	16

(Repeat problems that were not firm.)

b. Complete the equations in part 2. Put your pencil down when you're finished.
  (Observe children and give feedback.)
c. Check your work. You're going to touch and read the fact for each problem.
- Problem A. (Signal.) *3 + 3 = 6.*
- (Repeat for:) B, *3 + 5 = 8;* C, *5 + 5 = 10;* D, *5 + 2 = 7;* E, *6 + 3 = 9;* F, *6 + 6 = 12;* G, *4 + 6 = 10;* H, *4 + 4 = 8;* I, *7 + 7 = 14;* J, *8 + 8 = 16.*

EXERCISE 5: TELLING TIME
ANALOG CLOCK HANDS

a. Find part 3 on your worksheet. ✔
(Teacher reference:)

You're going to make the arrow to show the direction the minute hand is moving.
- Touch the minute hand for clock A.
(Observe children and give feedback.)
- Point the direction the minute hand is moving.
(Observe children and give feedback.)
- Make an arrow to show the direction the minute hand is moving.
(Observe children and give feedback.)
(Display:)   [105:5A]

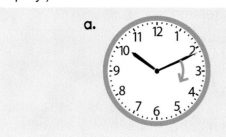

Here's what you should have for clock A.
b. Touch the minute hand for clock B. ✔
- Point the direction the minute hand is moving.
(Observe children and give feedback.)
- Make an arrow to show the direction the minute hand is moving. ✔
(Display:)   [105:5B]

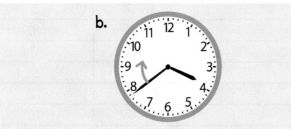

Here's what you should have for clock B.
c. Touch the minute hand for clock C. ✔
- Point the direction the minute hand is moving. ✔
- Make an arrow to show the direction the minute hand is moving. ✔
(Display:)   [105:5C]

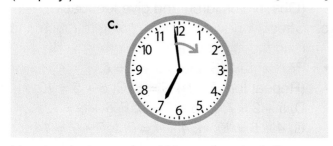

Here's what you should have for clock C.

d. Make an arrow to show the direction the minute hand is moving for clock D. ✔
(Display:)   [105:5D]

Here's what you should have for clock D.
e. Touch the minute hand for clock A again and get ready to tell me the next number.
- Everybody, what's the next number the minute hand will point to? (Signal.) *3.*
- What's the last number the minute hand pointed to? (Signal.) *2.*
f. The other hand is not called the minute hand. What's the other hand called? (Signal.) *The hour hand.*
- Touch the hour hand. ✔
- Everybody, what's the next number the hour hand will point to? (Signal.) *11.*
- What's the last number the hour hand pointed to? (Signal.) *10.*
(Repeat steps e and f until firm.)
g. Touch the minute hand for clock B and get ready to tell me the next number.
- Everybody, what's the next number the minute hand will point to? (Signal.) *8.*
- What's the last number the minute hand pointed to? (Signal.) *7.*
h. Touch the hour hand for clock B. ✔
- Everybody, what's the next number the hour hand will point to? (Signal.) *4.*
- What's the last number the hour hand pointed to? (Signal.) *3.*
(Repeat steps g and h until firm.)
i. Touch the minute hand for clock C and get ready to tell me the next number.
- Everybody, what's the next number the minute hand will point to? (Signal.) *12.*
- What's the last number the minute hand pointed to? (Signal.) *11.*
j. Touch the hour hand for clock C. ✔
- Everybody, what's the next number the hour hand will point to? (Signal.) *7.*
- What's the last number the hour hand pointed to? (Signal.) *6.*
(Repeat steps i and j until firm.)

k. Touch the minute hand for clock D and get ready to tell me the next number.
• Everybody, what's the next number the minute hand will point to? (Signal.) *5.*
• What's the last number the minute hand pointed to? (Signal.) *4.*
l. Touch the hour hand for clock D. ✔
• Everybody, what's the next number the hour hand will point to? (Signal.) *12.*
• What's the last number the hour hand pointed to? (Signal.) *11.*
(Repeat steps k and l until firm.)

EXERCISE 6: WORD PROBLEMS (COLUMNS)
CONVENTIONAL AND COMPARISON

a. Find part 4. ✔
(Teacher reference:)

You're going to write the symbols for word problems in columns and work them.
• Listen to problem A: A library case had 96 books in it. 874 more books were put in the case. How many books ended up in the case?
• Listen again and write the symbols for the whole problem: A library case had 96 books in it. 874 more books were put in the case.
(Observe children and give feedback.)
• Everybody, touch and read the problem for A. Get ready. (Signal.) *96 plus 874.*
b. Work problem A. Put your pencil down when you know how many books ended up in the case.
(Observe children and give feedback.)
(Teacher reference:)

• Read the addition problem and the answer you wrote for A. Get ready. (Signal.) *96 + 874 = 970.*
• How many books ended up in the case? (Signal.) *970.*

c. Touch where you'll write the symbols for problem B. ✔
Listen to problem B: Jenny had 375 cookies. People ate 312 of those cookies. How many cookies did Jenny end up with?
• Listen again and write the symbols for the whole problem: Jenny had 375 cookies. People ate 312 of those cookies.
(Observe children and give feedback.)
• Everybody, touch and read the problem for B. Get ready. (Signal.) *375 minus 312.*
d. Work problem B. Put your pencil down when you know how many cookies Jenny ended up with.
(Observe children and give feedback.)
(Teacher reference:)

• Read the subtraction problem and the answer you wrote for B. Get ready. (Signal.) *375 – 312 = 63.*
• How many cookies did Jenny end up with? (Signal.) *63.*
e. Touch where you'll write the symbols for problem C. ✔
You're going to figure out how much heavier one animal is than another animal.
• Listen to problem C: A dog weighs 67 pounds. A goat weighs 188 pounds. How much heavier is the goat than the dog?
• Listen to problem C again and write the problem for figuring out how much heavier the goat is. A dog weighs 67 pounds. (Pause.) A goat weighs 188 pounds. (Pause.)
• Read the problem you wrote for figuring out how much more the goat is than the dog. (Signal.) *188 minus 67.*
• Work the problem. Put your pencil down when you know how many pounds heavier the goat is.
(Observe children and give feedback.)
(Teacher reference:)

f. Check your work.
• Read the subtraction problem and the answer for C. Get ready. (Signal.) *188 – 67 = 121.*
• How much heavier is the goat than the dog? (Signal.) *121 pounds.*

EXERCISE 7: PLACE VALUE
PICTURES

a. (Display:) [105:7A]

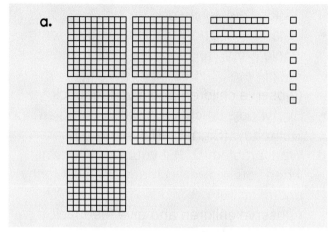

This is the picture A. You're going to write the equation for the number of squares in picture A.
- (Point to a hundreds box.) How many squares are in each box? (Touch.) *100.*
- (Point to a tens row.) How many squares are in each row? (Touch.) *10.*
- (Point to a single square.) How many squares are in each square? (Touch.) *1.*
  (Repeat until firm.)

b. (Point to hundreds boxes.) I'll touch each hundreds box. Count for the squares to yourself. (Touch each hundreds square.) Tell me how many squares are in the hundreds boxes. Get ready. (Touch.) *500.*
- Tell me how many squares are in the tens rows. Get ready. (Signal.) *30.*
- (Point to single squares.) I'll touch for the single squares. Count to yourself. (Touch each single square.) Tell me how many single squares there are. Get ready. (Signal.) *7.*
- What does 500 plus 30 plus 7 equal? (Signal.) *537.*
- Say the equation for the number of squares in picture A. Get ready. (Touch columns.) *500 + 30 + 7 = 537.*
  (Repeat until firm.)

c. Find part 5. ✔
(Teacher reference:)

a.
___+__+_=___

b.
___+__+_=___

- Write the equation for picture A. ✔
  (Observe children and give feedback.)

d. (Display:) [105:7B]

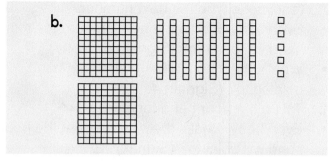

This is the picture for B.
- How many squares are in the hundreds boxes? (Signal.) *200.*
- (Point to tens columns.) I'll touch each tens column. Count for the squares to yourself. (Touch each tens column.) How many squares are in the tens columns? (Signal.) *80.*
- I'll touch each single square. Count them to yourself. (Touch each single square.) How many single squares are there? (Signal.) *5.*
- What does 200 plus 80 plus 5 equal? (Signal.) *285.*
- Say the equation for the number of squares in picture B. (Touch columns.) *200 + 80 + 5 = 285.*
  (Repeat until firm.)

e. Write the equation for picture B. ✔
(Observe children and give feedback.)
(Answer key:)

a.
500+30+7=537

b.
200+80+5=285

EXERCISE 8: NUMBER FAMILIES FROM PROBLEMS
MULTIPLE-DIGIT NUMBERS

a. Find part 6. ✔
(Teacher reference:)

a. 295   b. 415   c. 327
 −163   + 36   −122

You're going to make number families for these problems, but first you'll solve each problem.
- Touch and read problem A. (Signal.) *295 minus 163.*
- Does the problem add? (Signal.) *No.*
- So is the answer the big number? (Signal.) *No.*
- What's the big number? (Signal.) *295.*

b. (Repeat the following tasks for the remaining problems:)

Touch and read problem __.		Does the problem add?	So is the answer the big number?	What's the big number?
B	415 + 36	Yes	Yes	
C	327 – 122	No	No	327

(Repeat problems that were not firm.)

c. Go back to problem A and work it. Stop when you've completed equation A.
(Observe children and give feedback.)

d. Check your work.
- Read equation A. (Signal.) *295 – 163 = 132.*
(Display:) [105:8A]

$$a. \quad \begin{array}{r} 2\,9\,5 \\ -\,1\,6\,3 \\ \hline 1\,3\,2 \end{array}$$

- Think of the numbers for the family. What's the big number for family A? (Signal.) *295.*
- What are the small numbers? (Signal.) *163 (and) 132.*
- Write the numbers in family A.
(Observe children and give feedback.)

e. Work problems B and C. Stop when you've completed the rest of the equations in part 6.
(Observe children and give feedback.)

f. Check your work.
- Read equation B. (Signal.) *415 + 36 = 451.*
- Read equation C. (Signal.) *327 – 122 = 205.*
(Display:) [105:8B]

$$b. \quad \begin{array}{r} 4\,1\,5 \\ +\ \ 3\,6 \\ \hline 4\,5\,1 \end{array} \qquad c. \quad \begin{array}{r} 3\,2\,7 \\ -\,1\,2\,2 \\ \hline 2\,0\,5 \end{array}$$

Here's what you should have for equations B and C.

g. Write the numbers in families B and C. Put your pencil down when you've completed the families in part 6.
(Observe children and give feedback.)

h. Check your work.
- What's the big number for family B? (Signal.) *451.*
- What are the small numbers? (Call on a child.) *415 (and) 36.*

i. What's the big number for family C? (Signal.) *327.*
- What are the small numbers? (Call on a child.) *122 (and) 205.*
(Display:) [105:8C]

$$a. \xrightarrow[\ \ \ \ \ \ \ ]{163 \qquad 132} 295 \qquad b. \xrightarrow[\ \ \ \ \ \ \ ]{415 \qquad 36} 451 \qquad c. \xrightarrow[\ \ \ \ \ \ \ ]{122 \qquad 205} 327$$

Here are the families for part 6.

EXERCISE 9: FACTS
SUBTRACTION

a. Turn to the other side of your worksheet and find part 7. ✔
(Teacher reference:)

a. 8 – 4	c. 8 – 5	f. 10 – 4
b. 8 – 2	d. 14 – 7	g. 11 – 6
	e. 7 – 4	h. 11 – 9

- These are subtraction problems you'll work. Think of the number families for each problem and write the answers. Put your pencil down when you've completed the equations in part 7.
(Observe children and give feedback.)

b. Check your work. You'll read the equation for each problem.
- Problem A. (Signal.) *8 – 4 = 4.*
- (Repeat for:) B, *8 – 2 = 6;* C, *8 – 5 = 3;* D, *14 – 7 = 7;* E, *7 – 4 = 3;* F, *10 – 4 = 6;* G, *11 – 6 = 5;* H, *11 – 9 = 2.*

a. Find part 8. ✔
(Teacher reference:)

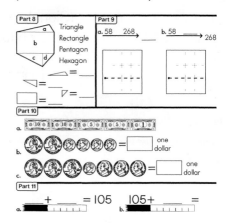

You'll circle the name for the whole figure. The names are triangle, rectangle, pentagon, and hexagon. Then you'll write the letter for each part.

In part 9, you'll write a column problem to figure out the missing number in each family. Then you'll complete the family.

In part 10, you'll write an equals and the number of dollars to show what the group of bills is worth. You'll circle the words one dollar or write the cents for each group of coins.

In part 11, you'll complete the equations for the rulers.

b. Complete worksheet 105.

(Observe children and mark incorrect responses on children's worksheets as you give feedback.)

Lessons 106-110 Planning Page

	Lesson 106	Lesson 107	Lesson 108	Lesson 109	Lesson 110
Student Learning Objectives	**Exercises** 1. Say addition and subtraction facts 2. **Solve problems from multi-digit number families** 3. Count by a variety of numbers 4. **Say facts for plus 3** 5. Tell time from an analog clock 6. Use ones, tens, and hundreds charts to clarify place value 7. Decompose a 3-dimensional object 8. Solve subtraction facts 9. Solve multi-digit problems and write the number families 10. Complete work independently	**Exercises** 1. Say facts for plus 3 2. Solve problems from multi-digit number families 3. Find missing numbers in number families 4. Count by a variety of numbers 5. **Write the time from an analog clock (hours)** 6. Write the symbols for word problems in columns and solve; solve comparison word problems 7. Use ones, tens, and hundreds charts to clarify place value 8. Decompose a 3-dimensional object 9. Solve subtraction facts 10. Complete work independently	**Exercises** 1. Say addition and subtraction facts 2. Solve problems from multi-digit number families 3. Say facts for plus 3 4. Decompose a 3-dimensional object 5. **Write the time from an analog clock (hours and minutes)** 6. Use ones, tens, and hundreds charts to clarify place value 7. Solve subtraction facts 8. Write the symbols for word problems in columns and solve; solve comparison word problems 9. Complete work independently	**Exercises** 1. Say facts for plus 3 2. **Say two addition and two subtraction facts with a small number of 3; find missing numbers in number families** 3. Count by a variety of numbers 4. **Identify the missing value (big number, small number)** 5. Identify 2-dimensional faces on 3-dimensional objects 6. Solve addition and subtraction facts 7. Write the time from an analog clock (hours and minutes) 8. Use ones, tens, and hundreds charts to clarify place value 9. Solve problems from multi-digit number families 10. Complete work independently	**Exercises** 1. Say facts for plus 3 2. Identify the missing value (big number, small number) 3. Say two addition and two subtraction facts with a small number of 3; find missing numbers in number families 4. Identify 2-dimensional faces on 3-dimensional objects 5. Solve problems from multi-digit number families 6. Write the time from an analog clock (hours and minutes) 7. Solve addition and subtraction facts 8. Write the symbols for word problems in columns and solve; solve comparison word problems 9. Use ones, tens, and hundreds charts to clarify place value 10. Complete work independently

Common Core State Standards for Mathematics					
1.OA 3				✔	✔
1.OA 4				✔	✔
1.OA 5			✔	✔	✔
1.OA 6 and 8	✔	✔	✔	✔	✔
1.NBT 1	✔	✔	✔	✔	✔
1.NBT 2			✔		
1.NBT 4	✔	✔	✔	✔	✔
1.NBT 6				✔	
1.MD 3	✔	✔	✔	✔	✔
1.G 1–2	✔	✔	✔	✔	✔
Teacher Materials	Presentation Book 3, Board Displays CD or chalkboard				
Student Materials	Workbook 2, Pencil				
Additional Practice	Student Practice Software: Block 5: Activity 1 (1.NBT.1), Activity 2 (1.NBT.1), Activity 3 (1.NBT.4 and 1.OA.6), Activity 4 (1.NBT.1), Activity 5 (1.NBT.1), Activity 6 (1.NBT.3)				
Mastery Test					Student Assessment Book (Present Mastery Test 11 following Lesson 110.)

Lesson  106

EXERCISE 1: FACTS
ADDITION/SUBTRACTION

a. (Display:) [106:1A]

9 – 6	11 – 6	3 + 4
6 + 4	8 + 8	14 – 7
8 – 4	8 – 3	5 + 5

- (Point to **9 – 6.**) Listen: 9 minus 6. Is the answer the big number or a small number? (Signal.) *A small number.*
- Say the fact. (Signal.) *9 – 6 = 3.*
b. (Point to **6 + 4.**) 6 plus 4. Is the answer the big number or a small number? (Signal.) *The big number.*
- Say the fact. (Signal.) *6 + 4 = 10.*
c. (Repeat the following tasks for the remaining problems:)

(Point to __.) __. Is the answer the big number or a small number?			Say the fact.
8 – 4	8 – 4	A small number.	8 – 4 = 4
11 – 6	11 – 6	A small number.	11 – 6 = 5
8 + 8	8 + 8	The big number.	8 + 8 = 16
8 – 3	8 – 3	A small number.	8 – 3 = 5
3 + 4	3 + 4	The big number.	3 + 4 = 7
14 – 7	14 – 7	A small number.	14 – 7 = 7
5 + 5	5 + 5	The big number.	5 + 5 = 10

(Repeat problems that were not firm.)

EXERCISE 2: SOLVING PROBLEMS FROM NUMBER FAMILIES
MULTI-DIGIT NUMBERS

a. (Display:) [106:2A]

$$\xrightarrow[\text{61} \quad \text{114}]{} 175 \qquad \xrightarrow[\text{96} \quad \text{156}]{} 252$$

252	175	114	156
– 96	– 114	+ 61	+ 96

Here are two families and four problems. Each problem comes from one of the families. You're going to figure out the family that goes with each problem and say the answer.

- (Point to **252 – 96.**) Read the problem. (Signal.) *252 minus 96.*
- Does the problem add? (Signal.) *No.*
- So is the answer the big number? (Signal.) *No.* The answer is not the big number. So this problem shows the big number.
- What's the big number? (Signal.) *252.* The big number for this family is 252 and 96 is a small number.
b. (Point to $\xrightarrow[\text{61} \quad \text{114}]{} 175$.) Does this family have a big number of 252 and a small number of 96? (Signal.) *No.*
- So this is not the family for 252 minus 96.
- (Point to $\xrightarrow[\text{96} \quad \text{156}]{} 252$.) Does this family have a big number of 252 and a small number of 96? (Signal.) *Yes.*
- So this is the family that shows the answer to 252 minus 96.
- Look at this family. It shows the missing number. Tell me what 252 minus 96 equals. (Signal.) *156.* (Add to show:) [106:2B]

$$\begin{array}{r} 252 \\ - \ 96 \\ \hline 156 \end{array}$$

c. (Point to **175 – 114.**) Read the problem. (Signal.) *175 minus 114.*
- Does the problem add? (Signal.) *No.*
- So is the answer the big number? (Signal.) *No.*
- What's the big number? (Signal.) *175.*
- What is a small number? (Signal.) *114.*
d. (Point to $\xrightarrow[\text{61} \quad \text{114}]{} 175$.) Does this family have a big number of 175 and a small number of 114? (Signal.) *Yes.*
- So is this the family for 171 minus 114? (Signal.) *Yes.*
- Look at this family. Tell me what 175 minus 114 equals? (Signal.) *61.* (Add to show:) [106:2C]

$$\begin{array}{r} 175 \\ - \ 114 \\ \hline 61 \end{array}$$

e. (Point to **114 + 61.**) Read the problem. (Signal.) *114 plus 61.*
- Does the problem add? (Signal.) *Yes.*
- So is the answer the big number? (Signal.) *Yes.*
- Does this problem show the big number for the family? (Signal.) *No.*
- What are the small numbers? (Signal.) *114 (and) 61.*

f. Raise your hand when you know which family has those two numbers. ✔
- (Point to ⁶¹ ¹¹⁴→175.) Is this the family for 114 plus 61? (Signal.) *Yes.*
- Tell me what 114 plus 61 equals. (Signal.) *175.* (Add to show:)   [106:2D]

$$\begin{array}{r} 114 \\ + \ 61 \\ \hline 175 \end{array}$$

g. (Point to **156 + 96.**) Read the problem. (Signal.) *156 plus 96.*
- Does the problem add? (Signal.) *Yes.*
- So is the answer the big number? (Signal.) *Yes.*
- Does this problem show the big number for the family? (Signal.) *No.*
- What are the small numbers? (Signal.) *156 (and) 96.*

h. Raise your hand when you know which family has those two numbers. ✔
- (Point to ⁹⁶ ¹⁵⁶→252.) Is this the family for 156 plus 96? (Signal.) *Yes.*
- Tell me what 156 plus 96 equals. (Signal.) *252.* (Add to show:)   [106:2E]

$$\begin{array}{r} 156 \\ + \ 96 \\ \hline 252 \end{array}$$

EXERCISE 3: MIXED COUNTING

a. Your turn: Count by twos to 20. Get ready. (Tap.) *2, 4, 6, 8, 10, 12, 14, 16, 18, 20.* (Repeat until firm.)

b. My turn to start with 50 and count by twos to 60. 50, 52, 54, 56, 58, 60.
- Your turn: Start with 50 and count by twos to 60. Get 50 going. *Fiftyyy.* Count. (Tap.) *52, 54, 56, 58, 60.* (Repeat step b until firm.)

c. Start with 50 and count by ones to 60. Get 50 going. *Fiftyyy.* Count. (Tap.) *51, 52, 53, 54, 55, 56, 57, 58, 59, 60.*

d. Start with 50 and count backward by ones to 40. Get 50 going. *Fiftyyy.* Count backward. (Tap.) *49, 48, 47, 46, 45, 44, 43, 42, 41, 40.*

e. Start with 50 and count backward by tens. Get 50 going. *Fiftyyy.* Count backward. (Tap.) *40, 30, 20, 10.*

f. Start with 50 and count by fives to 100. Get 50 going. *Fiftyyy.* Count. (Tap.) *55, 60, 65, 70, 75, 80, 85, 90, 95, 100.*

g. Start with 50 and count by 25s to 150. Get 50 going. *Fiftyyy.* Count. (Tap.) *75, 100, 125, 150.*

h. Start with 50 and plus tens to 150. Get 50 going. *Fiftyyy.* Plus tens. (Tap.) *60, 70, 80, 90, 100, 110, 120, 130, 140, 150.* (Repeat steps c through h that were not firm.)

i. Listen: You have 50. When you count by 25s, what number comes next? (Signal.) *75.*
- You have 50. When you count by fives, what number comes next? (Signal.) *55.*
- You have 50. When you count by tens, what number comes next? (Signal.) *60.*
- You have 50. When you count by twos, what number comes next? (Signal.) *52.*
- You have 50. When you count by ones, what number comes next? (Signal.) *51.*
- You have 50. When you count backward by ones, what number comes next? (Signal.) *49.*
- You have 50. When you count backward by tens, what number comes next? (Signal.) *40.* (Repeat tasks in step i that were not firm.)

EXERCISE 4: FACT RELATIONSHIPS
PLUS 3 | REMEDY |

a. (Display:)   [106:4A]

1 + 3	4 + 3	7 + 3
2 + 3	5 + 3	8 + 3
3 + 3	6 + 3	9 + 3

These are problems that plus 3. You know the answer to some of the problems. My turn to read the plus-3 problems and say the answers. Say the answers with me if you can keep up.
1 plus 3 equals **4.**
2 plus 3 equals **5.**
- (Repeat for 3 + 3 = **6**, 4 + 3 = **7**, 5 + 3 = **8**, 6 + 3 = **9**, 7 + 3 = **10**, 8 + 3 = **11**, 9 + 3 = **12**.)

b. Now you'll read each problem and say the answer.
- (Point to **1 + 3.**) Read the problem. (Touch.) *1 plus 3.*
- What's the answer? (Signal.) *4.*

c. (Point to **2 + 3.**) Read the problem. (Touch.) *2 plus 3.*
- What's the answer? (Signal.) *5.*

d. (Repeat the following tasks for remaining problems:)

(Point to __.) Read the problem.	What's the answer?
3 + 3	6
4 + 3	7
5 + 3	8
6 + 3	9
7 + 3	10
8 + 3	11
9 + 3	12

e. (Do not show display.) This time I'll say the problems, and you'll say the whole facts.
- Listen: 1 plus 3. Say the fact. (Signal.) *1 + 3 = 4.*
- Listen: 2 plus 3. Say the fact. (Signal.) *2 + 3 = 5.*
- Say the fact for 3 plus 3. (Signal.) *3 + 3 = 6.*
- Say the fact for 4 plus 3. (Signal.) *4 + 3 = 7.*
- Say the fact for 5 plus 3. (Signal.) *5 + 3 = 8.*
- Say the fact for 6 plus 3. (Signal.) *6 + 3 = 9.*
- Say the fact for 7 plus 3. (Signal.) *7 + 3 = 10.*
- Say the fact for 8 plus 3. (Signal.) *8 + 3 = 11.*
- Say the fact for 9 plus 3. (Signal.) *9 + 3 = 12.*
  (Repeat step e until firm.)

EXERCISE 5: TELLING TIME
ANALOG CLOCK HANDS
REMEDY

a. (Distribute unopened workbooks to children.)
- Open your workbook to Lesson 106 and find part 1.
  (Observe children and give feedback.)
  (Teacher reference:)
  R Part H

For these clocks you're going to make an arrow to show the direction the hour hand is moving. The minute hand for these clocks is pointing to a number.

- Is the minute hand the longer hand or the shorter hand? (Signal.) *The longer hand.*
- Touch the minute hand for clock A. ✔
- Touch the hour hand for clock A. ✔
  (Repeat until firm.)
- Point the direction the hour hand is moving. ✔
- Make an arrow to show the direction the hour hand is moving.
  (Observe children and give feedback.)
  (Display:) [106:5A]

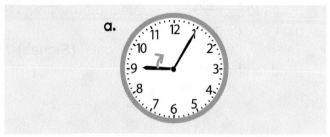

Here's what you should have for clock A.

b. Touch the minute hand for clock B. ✔
- Touch the hour hand for clock B. ✔
- Make an arrow to show the direction the hour hand is moving for clock B. ✔
  (Display:) [106:5B]

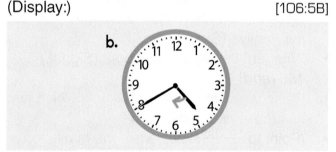

Here's what you should have for clock B.

c. Touch the minute hand for clock C. ✔
- Touch the hour hand for clock C. ✔
- Make an arrow to show the direction the hour hand is moving for clock C. ✔
  (Display:) [106:5C]

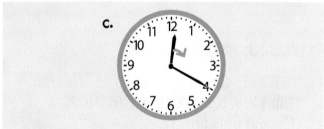

Here's what you should have for clock C.

d. Touch the minute hand for clock D. ✔
- Touch the hour hand for clock D. ✔
- Make an arrow to show the direction the hour hand is moving for clock D. ✔
(Display:) [106:5D]

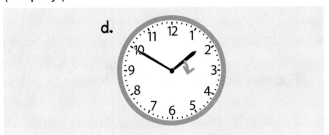

Here's what you should have for clock D.

e. Touch the hour hand for clock A again and get ready to tell me the next number.
- Everybody, what's the next number the hour hand will point to? (Signal.) 10.
- What's the last number the hour hand pointed to? (Signal.) 9.
- The hour number for the time clock A shows is 9. What's the hour number for clock A? (Signal.) 9.
- Look at the minute hand. What number is the minute hand pointing to? (Signal.) 1.

f. Touch the hour hand for clock B and get ready to tell me the next number.
- Everybody, what's the next number the hour hand will point to? (Signal.) 5.
- What's the last number the hour hand pointed to? (Signal.) 4.
- So what's the hour number for clock B? (Signal.) 4.
- Look at the minute hand. What number is the minute hand pointing to? (Signal.) 8.

g. Touch the hour hand for clock C and get ready to tell me the next number.
- Everybody, what's the next number the hour hand will point to? (Signal.) 1.
- What's the last number the hour hand pointed to? (Signal.) 12.
- So what's the hour number for clock C? (Signal.) 12.
- Look at the minute hand. What number is the minute hand pointing to? (Signal.) 4.

h. Touch the hour hand for clock D and get ready to tell me the next number.
- Everybody, what's the next number the hour hand will point to? (Signal.) 2.
- What's the last number the hour hand pointed to? (Signal.) 1.

- So what's the hour number for clock D? (Signal.) 1.
- Look at the minute hand. What number is the minute hand pointing to? (Signal.) 10.

EXERCISE 6: PLACE VALUE
PICTURES

a. (Display:) [106:6A]

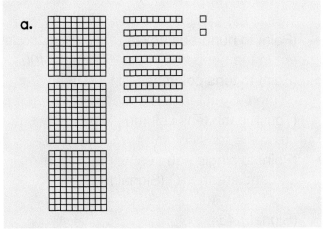

This is picture A. You're going to write the equation for the number of squares in picture A.
- (Point to a hundreds box.) How many squares are in each box? (Touch.) 100.
- (Point to a tens row.) How many squares are in each row? (Touch.) 10.
- (Point to a single square.) How many squares are in each square? (Touch.) 1.
(Repeat until firm.)

b. (Point to hundreds boxes.) Tell me how many squares are in the hundreds boxes. Get ready. (Signal.) 300.
- (Point to tens rows.) I'll touch each tens row. Count for the squares to yourself. (Touch each tens row.) How many squares are in the tens rows? (Signal.) 70.
- (Point to single squares.) Tell me how many single squares there are. Get ready. (Signal.) 2.
- What does 300 plus 70 plus 2 equal? (Signal.) 372.
- Say the equation for the number of squares in picture A. Get ready. (Touch columns.) 300 + 70 + 2 = 372.
(Repeat until firm.)

c. Find part 2 on worksheet 106. ✔
(Teacher reference:)

- Write the equation for picture A. ✔
(Observe children and give feedback.)

d. (Display:)   [106:6B]

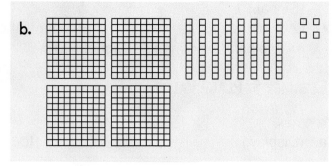

This is the picture for B.

- (Point to hundreds boxes.) How many squares are in the hundreds boxes? (Signal.) *400.*
- (Point to tens columns.) I'll touch each tens column. Count for the squares to yourself. (Touch each tens column.) How many squares are in the tens columns? *(Signal.) 80.*
- (Point to single squares.) How many single squares are there? (Signal.) *4.*
- What does 400 plus 80 plus 4 equal? (Signal.) *484.*
- Say the equation for the number of squares in picture B. Get ready. (Touch columns.) *400 + 80 + 4 = 484.* (Repeat until firm.)
- Write the equation for picture B. ✔ (Observe children and give feedback.) (Answer key:)

a.
300+70+2=372

b.
400+80+4=484

EXERCISE 7: 3-D OBJECTS
DECOMPOSITION   REMEDY

a. I'm going to ask you a question about a pentagon. Does a pentagon have 6 or 5 sides? (Signal.) *5.*
- What shape has 6 sides? (Signal.) *(A) hexagon.* Yes, a hexagon.
- What shape has 5 sides? (Signal.) *(A) pentagon.*
- What shape has 4 sides? (Signal.) *(A) rectangle.*
- What shape has 3 sides? (Signal.) *(A) triangle.*
b. Tell me what letter you write on a pentagon. Get ready. (Signal.) *P.*
- Tell me what letter you write on a triangle. Get ready. (Signal.) *T.*
- Tell me what letter you write on a rectangle. Get ready. (Signal.) *R.*
- Tell me what letter you write on a hexagon. Get ready. (Signal.) *H.*
- (Repeat step b until firm.)

c. Find part 3 on worksheet 106. ✔ (Teacher reference:)   R Part B

1   2

- Write letters for all of the faces for object 1 and object 2. (Observe children and give feedback.)
d. Check your work.
- What letter did you write on the bottom face of object 1? (Signal.) *P.*
- What letter did you write on the side faces? (Signal.) *R.*
- Did you write a letter for the top face? (Signal.) *Yes.*
- What letter? (Signal.) *P.*
e. Object 2. What letter did you write on the bottom face? (Signal.) *P.*
- What letter did you write on the side faces? (Signal.) *T.*
- Did you write a letter for the top face? (Signal.) *No.*
f. (Display:)   [106:7A]

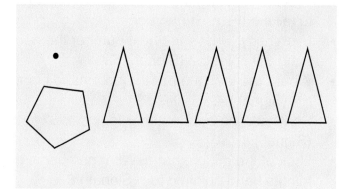

Here are the bottom, top, and sides of one of the objects. You're going to figure out which object. (Point to bottom face.) Here's the bottom face. (Point to top.) This dot is the top of the object. The top is a point not a face.

- (Point to side faces.) Here are the side faces. What shape are the side faces? (Touch.) *Triangles.*

g. Now you're going to circle the object that has this top, bottom, and sides.
- Touch object 1. ✔
- Is the top of object 1 a point, and are the sides triangles? (Signal.) *No.*
- Touch object 2. ✔
- Is the top of object 2 a point, and are the sides triangles? (Signal.) *Yes.*
- Circle object 2.
(Observe children and give feedback.)

h. Find part 4 on worksheet 106. ✔
(Teacher reference:)

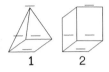

1   2

- Write letters for all of the faces for objects 1 and 2.
(Observe children and give feedback.)
i. Check your work.
- What letter did you write for the bottom face of object 1? (Signal.) *R.*
- What letter did you write for the side faces? (Signal.) *T.*
- Did you write a letter for the top face? (Signal.) *No.*
j. Object 2. What letter did you write for the bottom face? (Signal.) *R.*
- What letter did you write for the side faces? (Signal.) *R.*
- Did you write a letter for the top face? (Signal.) *Yes.*
- What letter? (Signal.) *R.*
k. (Display:)   [106:7B]

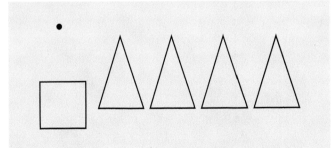

Here are the bottom, top, and sides of one of the objects. You're going to figure out which object.
(Point to bottom face.) Here's the bottom face.
(Point to top.) This point is the top of the object.
- (Point to side faces.) Here are the side faces. What shape are the side faces? (Touch.) *Triangles.*

l. Look at the objects 1 and 2 and circle the object that has this top, bottom, and sides.
(Observe children and give feedback.)
- Everybody, which object did you circle? (Signal.) *(Object) 1.*
m. Look at the objects you circled in parts 3 and 4. ✔
- The side faces of both objects are triangles. So are both of the objects you circled pyramids? (Signal.) *Yes.*
- Everybody, what do you call both of the objects you circled? (Signal.) *Pyramids.*

EXERCISE 8: FACTS
SUBTRACTION

a. Find part 5. ✔
(Teacher reference:)

	c. 10 − 6	f. 10 − 2
a. 12 − 6	d. 6 − 3	g. 16 − 8
b. 14 − 7	e. 11 − 5	h. 10 − 5

- Touch and read problem A. Get ready. (Signal.) *12 minus 6.*
- What's 12 minus 6? (Signal.) *6.*
b. (Repeat the following tasks for problems B through H:)

Touch and read problem __.		What's __?	
B	14 − 7	14 − 7	7
C	10 − 6	10 − 6	4
D	6 − 3	6 − 3	3
E	11 − 5	11 − 5	6
F	10 − 2	10 − 2	8
G	16 − 8	16 − 8	8
H	10 − 5	10 − 5	5

(Repeat problems that were not firm.)
c. Complete each of the facts. Put your pencil down when you've written all of the equations for part 5.
(Observe children and give feedback.)
d. Check your work.
- Touch and read the equation for problem A. (Signal.) *12 − 6 = 6.*
- (Repeat for:) B, *14 − 7 = 7;* C, *10 − 6 = 4;* D, *6 − 3 = 3;* E, *11 − 5 = 6;* F, *10 − 2 = 8;* G, *16 − 8 = 8;* H, *10 − 5 = 5.*

EXERCISE 9: NUMBER FAMILIES FROM PROBLEMS
MULTIPLE-DIGIT NUMBERS

a. Turn to the other side of your worksheet and
find part 6. ✔
(Teacher reference:)

$$
\begin{array}{c}
\text{a.} \;\; 740 \\
-540 \\
\hline
\end{array}
\qquad
\begin{array}{c}
\text{b.} \;\; 617 \\
+ \;\; 92 \\
\hline
\end{array}
\qquad
\begin{array}{c}
\text{c.} \;\; 389 \\
-136 \\
\hline
\end{array}
$$

⟶ ___   ⟶ ___   ⟶ ___

You're going to solve each problem and then
write a number family for each equation.

• Work problem A. Put your pencil down when
you have a number family with three numbers
for family A.
(Observe children and give feedback.)

b. Check your work.

• Read problem A and the answer. (Signal.)
740 – 540 = 200.

• What's the big number for family A?
(Signal.) *740.*
(Display:)   [106:9A]

$$
\begin{array}{c}
\text{a.} \;\; 740 \\
-540 \\
\hline
200 \\
\end{array}
$$

$$
\underset{}{540 \quad 200} \longrightarrow 740
$$

Here's what you should have.
You'll work the rest of the problems and write
the families as part of your independent work.

EXERCISE 10: INDEPENDENT WORK

a. Find part 7 on worksheet 106. ✔
(Teacher reference:)

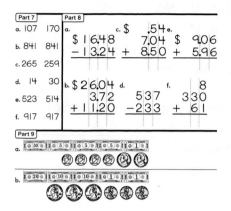

You'll write an equals sign or a sign that has a
bigger end next to one of the numbers.
In part 8, you'll work the money problems and
the column problems.
In part 9, you'll write the dollars and cents
amount for each group of bills and coins.

b. Complete worksheet 106. Remember to work
the problems and write the families in part 6.
(Observe children and mark incorrect responses
on children's worksheets as you give feedback.)

Lesson 107

EXERCISE 1: FACT RELATIONSHIPS
PLUS 3

a. (Display:) [107:1A]

1 + 3	4 + 3	7 + 3
2 + 3	5 + 3	8 + 3
3 + 3	6 + 3	9 + 3

These are problems that plus 3. You know the answer to some of the problems. You're going to read each problem and say the answer.

- (Point to **1 + 3**.) Read the problem. (Touch.) *1 plus 3.*
- What's the answer? (Signal.) *4.*
- (Point to **2 + 3**.) Read the problem. (Touch.) *2 plus 3.*
- What's the answer? (Signal.) *5.*

b. (Repeat the following tasks for remaining problems:)

(Point to __.) Read the problem.	What's the answer?	
3 + 3	3 + 3	6
4 + 3	4 + 3	7
5 + 3	5 + 3	8
6 + 3	6 + 3	9
7 + 3	7 + 3	10
8 + 3	8 + 3	11
9 + 3	9 + 3	12

c. (Do not show display.) This time I'll say the problems and you'll say the whole facts.
- Say the fact for 1 plus 3. (Signal.) *1 + 3 = 4.*
- Say the fact for 2 plus 3. (Signal.) *2 + 3 = 5.*
- 3 plus 3. (Signal.) *3 + 3 = 6.*
- 4 plus 3. (Signal.) *4 + 3 = 7.*
- 5 plus 3. (Signal.) *5 + 3 = 8.*
- 6 plus 3. (Signal.) *6 + 3 = 9.*
- 7 plus 3. (Signal.) *7 + 3 = 10.*
- 8 plus 3. (Signal.) *8 + 3 = 11.*
- 9 plus 3. (Signal.) *9 + 3 = 12.*
(Repeat step c until firm.)

d. One of the facts is new. You'll start with 6 plus 3 and say the facts in order.
- Say the fact for 6 plus 3. (Signal.) *6 + 3 = 9.*
- 7 plus 3. (Signal.) *7 + 3 = 10.*
- 8 plus 3. (Signal.) *8 + 3 = 11.*
- 9 plus 3. (Signal.) *9 + 3 = 12.*
- 10 plus 3. (Signal.) *10 + 3 = 13.*
(Repeat step d until firm.)

EXERCISE 2: SOLVING PROBLEMS FROM NUMBER FAMILIES
MULTI-DIGIT NUMBERS

a. (Display:) [107:2A]

$$\underrightarrow{278 \quad 94}372 \qquad \underrightarrow{184 \quad 94}278$$

94	372	278	94
+ 184	− 94	− 94	+ 278

Here are two families and four problems. Each problem comes from one of the families. You're going to figure out the family that goes with each problem and say the answer.

- (Point to **94 + 184**.) Read the problem. (Signal.) *94 plus 184.*
- Does the problem add? (Signal.) *Yes.*
- So is the answer the big number? (Signal.) *Yes.* The answer is the big number. So this problem doesn't show the big number.
- What are the small numbers? (Signal.) *94 (and) 184.*

b. (Point to $\underrightarrow{278 \quad 94}372$.) Does this family have small numbers of 94 and 184? (Signal.) *No.*
- So this is not the family for 94 plus 184.
- (Point to $\underrightarrow{184 \quad 94}278$.) Does this family have small numbers of 94 and 184? (Signal.) *Yes.*
- So this is the family that shows the answer to 94 plus 184.
- Look at this family. It shows the missing number. Tell me what 94 plus 184 equals. (Signal.) *278.*
(Add to show:) [107:2B]

$$\begin{array}{r} 94 \\ + 184 \\ \hline 278 \end{array}$$

c. (Point to **372 – 94.**) Read the problem. (Signal.) *372 minus 94.*
- Does this problem add? (Signal.) *No.*
- So is the answer the big number? (Signal.) *No.*
- What's the big number for the family? (Signal.) *372.*
- What's a small number? (Signal.) *94.*

d. Raise your hand when you know which family has a big number of 372 and a small number of 94. ✔
- (Point to $\xrightarrow[\text{278 \quad 94}]{}$**372.**) Is this the family for 372 minus 94? (Signal.) *Yes.*
- Tell me what 372 minus 94 equals. (Signal.) *278.* (Add to show:) [107:2C]

$$\begin{array}{r} 372 \\ -\ 94 \\ \hline 278 \end{array}$$

e. (Point to **278 – 94.**) Read the problem. (Signal.) *278 minus 94.*
- Does this problem add? (Signal.) *No.*
- So is the answer the big number? (Signal.) *No.*
- What's the big number for the family? (Signal.) *278.*
- What's a small number? (Signal.) *94.*

f. Raise your hand when you know which family has a big number of 278 and a small number of 94. ✔
- (Point to $\xrightarrow[\text{184 \quad 94}]{}$**278.**) Is this the family for 278 minus 94? (Signal.) *Yes.*
- Tell me what 278 minus 94 equals. (Signal.) *184.* (Add to show:) [107:2D]

$$\begin{array}{r} 278 \\ -\ 94 \\ \hline 184 \end{array}$$

g. (Point to **94 + 278.**) Read the problem. (Signal.) *94 plus 278.*
- Does the problem add? (Signal.) *Yes.*
- So is the answer the big number? (Signal.) *Yes.* The answer is the big number. So this problem doesn't show the big number.
- What are the small numbers? (Signal.) *94 (and) 278.*

h. Raise your hand when you know which family has small numbers of 94 and 278. ✔
- (Point to $\xrightarrow[\text{278 \quad 94}]{}$**372.**) Is this the family for 94 plus 278? (Signal.) *Yes.*
- Tell me what 94 plus 278 equals. (Signal.) *372.* (Add to show:) [107:2E]

$$\begin{array}{r} 94 \\ +\ 278 \\ \hline 372 \end{array}$$

EXERCISE 3: NUMBER FAMILIES
MISSING NUMBER IN FAMILY

a. (Display:) [107:3A]

You're going to say the problem for the missing number in each family.
- (Point to $\xrightarrow{6}$12.) Say the problem for the missing number. Get ready. (Touch.) *12 minus 6.*
- What's 12 minus 6? (Signal.) *6.*

b. (Point to $\xrightarrow{3}$6.) Say the problem for the missing number. (Touch.) *6 minus 3.*
- What's 6 minus 3? (Signal.) *3.*

c. (Repeat the following tasks for remaining families:)

(Point to __.) Say the problem for the missing number.	What's __?	
$\xrightarrow{3}$9	9 – 3	9 – 3	6
$\xrightarrow{4\quad4}$	4 + 4	4 + 4	8
$\xrightarrow{6\quad5}$	6 + 5	6 + 5	11
$\xrightarrow{8}$16	16 – 8	16 – 8	8
$\xrightarrow{4}$7	7 – 4	7 – 4	3
$\xrightarrow{4}$10	10 – 4	10 – 4	6
$\xrightarrow{5\quad5}$	5 + 5	5 + 5	10

(Repeat for families that were not firm.)

EXERCISE 4: MIXED COUNTING

a. Your turn: Count by twos to 20. Get ready.
(Tap.) *2, 4, 6, 8, 10, 12, 14, 16, 18, 20.*
- Everybody, start with 60 and count by twos
to 70. Get 60 going. *Sixtyyy.* Count. (Tap.)
62, 64, 66, 68, 70.
(Repeat step a until firm.)

b. I'm going to count by twos and stop. When I
point to you, say the next number.
- Listen: 10, 12, fourteeen. (Point.) *16.*
- Listen: 22, 24, twenty-siiix. (Point.) *28.*
- Listen: 6, 8, tennn. (Point.) *12.*
- Listen: 34, 36, thirty-eieieight. (Point.) *40.*
(Repeat tasks in step b that were not firm.)

c. Start with 300 and count by hundreds to 1000.
Get 300 going. *Three huuundred.* Count.
(Tap.) *400, 500, 600, 700, 800, 900, 1000.*

d. Start with 300 and count by 25s to 400. Get
300 going. *Three huuundred.* Count. (Tap.)
325, 350, 375, 400.

e. Start with 300 and plus tens to 400. Get 300
going. *Three huuundred.* Plus tens. (Tap.)
310, 320, 330, 340, 350, 360, 370, 380, 390, 400.

f. Start with 300 and count by fives to 350. Get
300 going. *Three huuundred.* Count. (Tap.)
305, 310, 315, 320, 325, 330, 335, 340, 345, 350.

g. Start with 300 and count by ones to 310. Get
300 going. *Three huuundred.* Count. (Tap.)
301, 302, 303, 304, 305, 306, 307, 308, 309, 310.
(Repeat steps c through g that were not firm.)

h. Listen: You have 70. When you count
backward by tens, what number comes next?
(Signal.) *60.*
- You have 70. When you count backward by
ones, what number comes next? (Signal.) *69.*
- You have 70. When you count by twos, what
number comes next? (Signal.) *72.*
- You have 70. When you count by fives, what
number comes next? (Signal.) *75.*
(Repeat tasks in step h that were not firm.)

INDIVIDUAL TURNS

(Call on individual students to perform one of
the following tasks.)

- Start with 60 and count by twos to 70. (Call on
a student.) *Sixtyyy, 62, 64, 66, 68, 70.*
- Start with 40 and count by twos to 50. (Call on
a student.) *Fortyyy, 42, 44, 46, 48, 50.*

EXERCISE 5: TELLING TIME
ANALOG HOURS   REMEDY

a. (Distribute unopened workbooks to children.)
- Open your workbook to Lesson 107 and find
part 1.
(Observe children and give feedback.)
(Teacher reference:)   R Part I

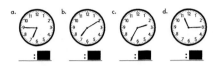

For these clocks you're going to make an arrow
to show the direction the hour hand is moving.
- Is the minute hand the longer hand or the
shorter hand? (Signal.) *The longer hand.*
- Touch the minute hand for clock A. ✔
- Touch the hour hand for clock A. ✔
(Repeat until firm.)
- Point the direction the hour hand is moving. ✔
- Make an arrow to show the direction the hour
hand is moving.
(Observe children and give feedback.)
(Display:)   [107:5A]

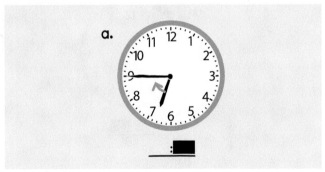

Here's what you should have for clock A.

b. Touch the minute hand for clock B. ✔
- Touch the hour hand for clock B. ✔
- Make an arrow to show the direction the hour
hand is moving for clock B.
(Observe children and give feedback.)
(Display:)   [107:5B]

Here's what you should have for clock B.

c. Make arrows to show the direction the hour hand is moving for clocks C and D.
(Observe children and give feedback.)
(Display:) [107:5C]

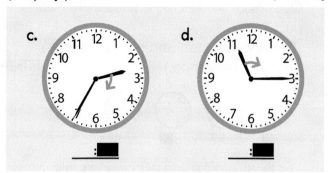

Here's what you should have for clocks C and D.

d. Touch the hour hand for clock A again and get ready to tell me the next number.
• Everybody, what's the next number the hour hand will point to? (Signal.) 7.
• What's the last number the hour hand pointed to? (Signal.) 6.
• The hour for clock A is 6. What's the hour for clock A? (Signal.) 6.
(Display:) [107:5D]

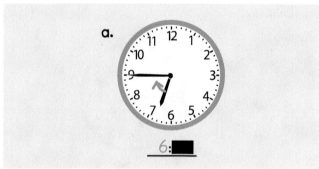

Here's the hour for clock A. The minute number is blacked out.
• Write the hour number for clock A in the space below.
e. Touch the minute hand for clock A. ✔
• What number is the minute hand pointing to? (Signal.) 9.
f. Touch the hour hand for clock B and get ready to tell me the next number.
• Everybody, what's the next number the hour hand will point to? (Signal.) 8.

• What's the last number the hour hand pointed to? (Signal.) 7.
• So what's the hour number for clock B? (Signal.) 7.
• Write the hour number for clock B in the space below.
(Observe children and give feedback.)
(Display:) [107:5E]

Here's what you should have for clock B.

g. Touch the minute hand for clock B. ✔
• What number is the minute hand pointing to? (Signal.) 2.
h. Touch the hour hand for clock C. ✔
• Figure out the last number the hour hand pointed to and get ready to tell me the hour number. ✔
• What's the hour number for clock C? (Signal.) 2.
• Write the hour number for clock C in the space below.
(Observe children and give feedback.)
(Display:) [107:5F]

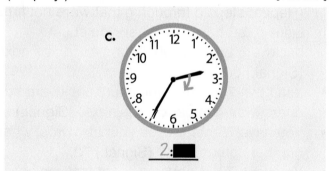

Here's what you should have for clock C.

i. Touch the minute hand for clock C. ✔
• What number is the minute hand pointing to? (Signal.) 7.

j. Touch the hour hand for clock D. ✔
• Figure out the last number the hour hand pointed to and get ready to tell me the hour number. ✔
• What's the hour number for clock D? (Signal.) *11.*
• Write the hour number for clock D in the space below.
(Observe children and give feedback.)
(Display:)   [107:5G]

Here's what you should have for clock D.
k. Touch the minute hand for clock D. ✔
• What number is the minute hand pointing to? (Signal.) *3.*

EXERCISE 6: WORD PROBLEMS (COLUMNS)
CONVENTIONAL AND COMPARISON   REMEDY

a. Find part 2 on worksheet 107. ✔
(Teacher reference:)   R Part D

You're going to write the symbols for word problems in columns and work them.
For problem A, you're going to tell me how much deeper the north end of a lake is.
• Listen to problem A: The south end of the lake is 43 feet deep. The north end of the lake is 68 feet deep. How many feet deeper is the north end of the lake?
• Listen to problem A again and write the problem for figuring out how much deeper the north end of the lake is: The south end of the lake is 43 feet deep. (Pause.) The north end of the lake is 68 feet deep.
(Observe children and give feedback.)

• Read the problem you wrote for figuring out how much deeper the north end of the lake is. (Signal.) *68 minus 43.*
• Work the problem. Put your pencil down when you know how many feet deeper the north end of the lake is.
(Observe children and give feedback.)
(Teacher reference:)

b. Check your work.
• Read the subtraction problem and the answer for A. Get ready. (Signal.) *68 – 43 = 25.*
• How many feet deeper is the north end of the lake? (Signal.) *25 feet.*
c. Touch where you'll write the symbols for problem B. ✔
• Listen to problem B: Jane had 496 dollars in her bank account. Jane put 274 more dollars into her account. How many dollars ended up in Jane's bank account?
• Listen again and write the symbols for the whole problem: Jane had 496 dollars in her bank account. Jane put 274 more dollars into her account.
(Observe children and give feedback.)
• Everybody, touch and read the problem for B. Get ready. (Signal.) *496 plus 274.*
d. Work problem B. Put your pencil down when you know how many dollars ended up in Jane's bank account.
(Observe children and give feedback.)
(Teacher reference:)

• Read the addition problem and the answer you wrote for B. Get ready. (Signal.) *496 + 274 = 770.*
• How many dollars ended up in Jane's bank account? (Signal.) *770.*

e. Touch where you'll write the symbols for problem C. ✔
Listen to problem C: 685 students were in a school. 613 of those students left the school. How many students ended up in the school?

• Listen again and write the symbols for the whole problem: 685 students were in a school. 613 of those students left the school.
(Observe children and give feedback.)

• Everybody, touch and read the problem for C. Get ready. (Signal.) *685 minus 613.*

f. Work problem C. Put your pencil down when you know how many students ended up in the school.
(Observe children and give feedback.)
(Teacher reference:)

c.
```
  685
- 613
  ̄ ̄ ̄
   72
```

• Read the subtraction problem and the answer you wrote for C. Get ready.
(Signal.) *685 – 613 = 72.*

• How many students ended up in the school? (Signal.) *72.*

EXERCISE 7: PLACE VALUE
PICTURES

a. (Display:)   [107:7A]

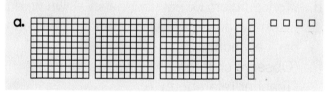

This is picture A. You're going to write the equation for the number of squares in picture A.

⌐ • (Point to a hundreds box.) Look at the picture. Count the hundreds, tens, and single squares to yourself. Raise your hand when you can say the equation for the number of the squares.
(Observe children and give feedback.)

• Everybody, say the equation for the number of squares in picture A. Get ready. (Touch columns.) *300 + 20 + 4 = 324.*
└ (Repeat until firm.)

b. Find part 3 on worksheet 107. ✔
(Teacher reference:)

a.   b.

___+__+_=___   ___+__+_=___

• Write the equation for picture A. ✔
(Observe children and give feedback.)

c. (Display:)   [107:7B]

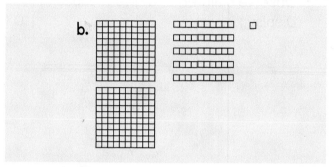

This is picture B.

• (Point to a hundreds box.) Look at the picture. Count the hundreds, tens, and single squares to yourself. Raise your hand when you can say the equation for the number of the squares in picture B.
(Observe children and give feedback.)

⌐ • Everybody, say the equation for the number of squares in picture B. Get ready. (Touch columns.) *200 + 50 + 1 = 251.*
└ (Repeat until firm.)

• Write the equation for picture B. ✔
(Observe children and give feedback.)
(Answer key:)

a.   b.

300+20+4 = 324   200+50+1 = 251

EXERCISE 8: 3-D OBJECTS
DECOMPOSITION   REMEDY

a. You're going to answer questions about hexagons, pentagons, rectangles, and triangles.

⌐ • What shape has 4 sides? (Signal.) *(A) rectangle.*
• What shape has 6 sides? (Signal.) *(A) hexagon.*
• What shape has 3 sides? (Signal.) *(A) triangle.*
• What shape has 5 sides? (Signal.) *(A) pentagon.*
└ (Repeat until firm.)

b. Find part 4 on worksheet 107. ✔
(Teacher reference:)

1   2   3

Write letters for all of the faces for object 1, object 2, and object 3. Put your pencil down when you've labeled all of the faces with the letter for its shape.
(Observe children and give feedback.)

c. Check your work.
- Look at object 1. What letter did you write on the bottom face? (Signal.) *R.*
- What letter did you write on the side faces? (Signal.) *T.*
- Did you write a letter for the top face? (Signal.) *No.*
- Object 2: What letter did you write on the bottom face? (Signal.) *T.*
- What letter did you write on the side faces? (Signal.) *R.*
- Did you write a letter for the top face? (Signal.) *Yes.*
- What letter? (Signal.) *T.*
- Object 3: What letter did you write on the bottom face? (Signal.) *R.*
- What letter did you write on the side faces? (Signal.) *R.*
- Did you write a letter for the top face? (Signal.) *Yes.*
- What letter? (Signal.) *R.*

d. Look at object 1. What do you call that object? (Call on a student.) *(A) pyramid.*
- Yes, object 1 is a pyramid. Everybody, what do you call object 1? (Signal.) *(A) pyramid.*

e. (Display:) [107:8A]

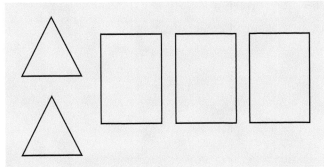

Here are the bottom, top, and sides of one of the objects. You're going to figure out which object.

- (Point to bottom face.) Here's the bottom face.
- Everybody, what is the shape of the bottom face? (Signal.) *(A) triangle.*
- (Point to top face.) Everybody, what is the shape of the top face? (Signal.) *(A) triangle.*
- (Point to side faces.) Here are the side faces. What shape are the side faces? (Touch.) *Rectangle(s).*

f. Circle the object that has the right top, bottom, and sides.
(Observe children and give feedback.)
- Everybody, which object did you circle? (Signal.) *(Object) 2.*

EXERCISE 9: FACTS
SUBTRACTION

a. Turn to the other side of your worksheet and find part 5. ✔
(Teacher reference:)

	c. $8-5$	f. $8-4$
a. $14-4$	d. $7-3$	g. $16-8$
b. $14-7$	e. $11-5$	h. $10-5$

- Touch and read problem A. Get ready. (Signal.) *14 minus 4.*
- What's 14 minus 4? (Signal.) *10.*

b. (Repeat the following tasks for problems B through H:)

Touch and read problem __.		What's __?	
B	$14-7$	$14-7$	7
C	$8-5$	$8-5$	3
D	$7-3$	$7-3$	4
E	$11-5$	$11-5$	6
F	$8-4$	$8-4$	4
G	$16-8$	$16-8$	8
H	$10-5$	$10-5$	5

(Repeat problems that were not firm.)

c. Complete each of the facts. Put your pencil down when you've written all of the equations for part 5.
(Observe children and give feedback.)

d. Check your work.
- Touch and read the equation for problem A. (Signal.) *14 − 4 = 10.*
- (Repeat for:) B, *14 − 7 = 7;* C, *8 − 5 = 3;* D, *7 − 3 = 4;* E, *11 − 5 = 6;* F, *8 − 4 = 4;* G, *16 − 8 = 8;* H, *10 − 5 = 5.*

EXERCISE 10: INDEPENDENT WORK

a. Find part 6. ✔
   (Teacher reference:)

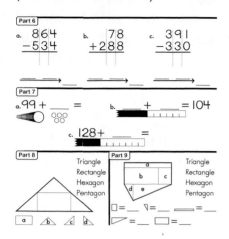

You'll work the problems and write the numbers in the family for each problem.
In part 7, you'll complete the equations for the circles and the rulers.
In part 8, you'll circle the name for the whole shape. The names are triangle, rectangle, hexagon, and pentagon. Then you'll write the letter for each part.
In part 9, you'll circle the name for the whole figure. Then you'll complete each equation by writing the letter to show which part it is.

b. Complete worksheet 107.
   (Observe children and mark incorrect responses on children's worksheets as you give feedback.)

Lesson 108

EXERCISE 1: FACTS
ADDITION/SUBTRACTION

a. (Display:)   [108:1A]

$10 - 6$	$8 + 8$	$7 + 7$
$3 + 4$	$8 - 5$	$9 - 6$
$8 - 4$	$11 - 6$	$10 - 5$

• (Point to **10 – 6.**) Listen: 10 minus 6. Is the answer the big number or a small number? (Signal.) *A small number.*
• Say the fact. (Signal.) *10 – 6 = 4.*
b. (Point to **3 + 4.**) 3 plus 4. Is the answer the big number or a small number? (Signal.) *The big number.*
• Say the fact. (Signal.) *3 + 4 = 7.*
c. (Repeat the following tasks for the remaining problems:)

(Point to __.) __. Is the answer the big number or a small number?		Say the fact.
$8 - 4$	$8 - 4$ *A small number.*	$8 - 4 = 4$
$8 + 8$	$8 + 8$ *The big number.*	$8 + 8 = 16$
$8 - 5$	$8 - 5$ *A small number.*	$8 - 5 = 3$
$11 - 6$	$11 - 6$ *A small number.*	$11 - 6 = 5$
$7 + 7$	$7 + 7$ *The big number.*	$7 + 7 = 14$
$9 - 6$	$9 - 6$ *A small number.*	$9 - 6 = 3$
$10 - 5$	$10 - 5$ *A small number.*	$10 - 5 = 5$

(Repeat problems that were not firm.)

EXERCISE 2: SOLVING PROBLEMS FROM NUMBER FAMILIES
MULTI-DIGIT NUMBERS

a. (Display:)   [108:2A]

$$\xrightarrow[\quad 94 \quad\quad 188 \quad]{}282 \qquad \xrightarrow[\quad 94 \quad\quad 282 \quad]{}376$$

282	282	376	188
+ 94	– 188	– 282	+ 94

Here are two families and four problems. Each problem comes from one of the families. You're going to figure out the family that goes with each problem. But you have to be really careful because the families have two numbers that are the same.

So you have to make sure that you know if each number is the big number in the family or a small number.
• (Point to $\xrightarrow[\text{94 \quad 188}]{}$282.) What are the small numbers in this family? (Signal.) *94 (and) 188.*
• What's the big number? (Signal.) *282.*
• (Point to $\xrightarrow[\text{94 \quad 282}]{}$376.) What are the small numbers in this family? (Signal.) *94 (and) 282.*
• What's the big number? (Signal.) *376.*
b. (Point to **282 + 94.**) Read this problem. (Signal.) *282 plus 94.*
• Does this problem show the big number? (Signal.) *No.*
• What are the small numbers? (Signal.) *282 (and) 94.*
• Raise your hand when you know which family has the right numbers.
c. (Point to $\xrightarrow[\text{94 \quad 188}]{}$282.) Does this family have the right numbers? (Signal.) *No.*
  This family does not have a small number of 282.
• (Point to $\xrightarrow[\text{94 \quad 282}]{}$376.) Does this family have the right numbers? (Signal.) *Yes.*
  Yes, but they are not in the same order as they are in the problem.
• Raise your hand when you know what 282 plus 94 equals. ✔
• What's does it equal? (Signal.) *376.*
  (Add to show:)   [108:2B]

282
+ 94
376

d. (Point to **282 – 188.**) Read this problem. (Signal.) *282 minus 188.*
• Does this problem show the big number? (Signal.) *Yes.*
• What's the big number? (Signal.) *282.*
• What's the small number shown? (Signal.) *188.*
• Raise your hand when you know which family has the right numbers. ✔

e. (Point to $\overset{94\ 188}{\longrightarrow}$282.) Does this family have the
    right numbers? (Signal.) *Yes.*
    Yes, it has a big number of 282 and a small
    number of 188.
* Raise your hand when you know what 282
  minus 188 equals. ✔
* What does it equal? (Signal.) *94.*
  (Add to show:)   [108:2C]

$$282$$
$$-\ 188$$
$$\overline{\quad 94}$$

f. (Point to **376 − 282.**) Read this problem.
    (Signal.) *376 minus 282.*
* Does this problem show the big number?
  (Signal.) *Yes.*
* What's the big number? (Signal.) *376.*
* What's the small number shown?
  (Signal.) *282.*
* Raise your hand when you know which family
  has the right numbers. ✔
g. (Point to $\overset{94\ 188}{\longrightarrow}$282.) Does this family have the
    right numbers? (Signal.) *No.*
* (Point to $\overset{94\ 282}{\longrightarrow}$376.) Does this family have the
  right numbers. (Signal.) *Yes.*
* Raise your hand when you know what 376
  minus 282 equals. ✔
* What does it equal? (Signal.) *94.*
  (Add to show:)   [108:2D]

$$376$$
$$-\ 282$$
$$\overline{\quad 94}$$

h. (Point to **188 + 94.**) Read this problem.
    (Signal.) *188 plus 94.*
* Does this problem show the big number?
  (Signal.) *No.*
* What are the small numbers? (Signal.) *188*
  (and) 94.
* Raise your hand when you know which family
  has the right numbers. ✔
i. (Point to $\overset{94\ 188}{\longrightarrow}$282.) Does this family have the
    right numbers? (Signal.) *Yes.*
* Raise your hand when you know what 188
  plus 94 equals. ✔

* What's does it equal? (Signal.) *282.*
  (Add to show:)   [108:2E]

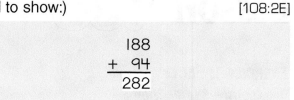

$$188$$
$$+\ \ 94$$
$$\overline{\quad 282}$$

EXERCISE 3: FACT RELATIONSHIPS
PLUS 3

a. (Display:)   [108:3A]

2 + 3	5 + 3	8 + 3
3 + 3	6 + 3	9 + 3
4 + 3	7 + 3	10 + 3

These are problems that plus 3. You know the
answer to some of the problems. You're going
to read each problem and say the answer.
* (Point to **2 + 3.**) Read the problem. (Touch.)
  2 plus 3.
* What's the answer? (Signal.) *5.*
b. (Repeat the following tasks for remaining
    problems:)

(Point to __.) Read the problem.	What's the answer?
3 + 3	6
4 + 3	7
5 + 3	8
6 + 3	9
7 + 3	10
8 + 3	11
9 + 3	12
10 + 3	13

c. (Do not show display.) This time I'll say the
    problems and you'll say the whole facts.
* Say the fact for 2 plus 3. (Signal.) *2 + 3 = 5.*
* 3 plus 3. (Signal.) *3 + 3 = 6.*
* 4 plus 3. (Signal.) *4 + 3 = 7.*
* 5 plus 3. (Signal.) *5 + 3 = 8.*
* 6 plus 3. (Signal.) *6 + 3 = 9.*
* 7 plus 3. (Signal.) *7 + 3 = 10.*
* 8 plus 3. (Signal.) *8 + 3 = 11.*
* 9 plus 3. (Signal.) *9 + 3 = 12.*
* 10 plus 3. (Signal.) *10 + 3 = 13.*
  (Repeat step c until firm.)

d. You'll start with 6 plus 3 and say the facts in order.
- Say the fact for 6 plus 3. (Signal.) *6 + 3 = 9.*
- 7 plus 3. (Signal.) *7 + 3 = 10.*
- 8 plus 3. (Signal.) *8 + 3 = 11.*
- 9 plus 3. (Signal.) *9 + 3 = 12.*
- 10 plus 3. (Signal.) *10 + 3 = 13.*
  (Repeat step d until firm.)

EXERCISE 4: 3-D OBJECTS
DECOMPOSITION

a. You're going to answer questions about hexagons, pentagons, rectangles, and triangles.
- What shape has 3 sides? (Signal.) *(A) triangle.*
- What shape has 4 sides? (Signal.) *(A) rectangle.*
- What shape has 5 sides? (Signal.) *(A) pentagon.*
- What shape has 6 sides? (Signal.) *(A) hexagon.*

b. (Display:) [108:4A]

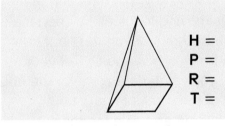

Here's a pyramid.
- (Point to bottom face.) What's the shape of the bottom face? (Touch.) *(A) rectangle.*
- How many sides does the bottom face have? (Signal.) *4.*
- So how many side faces does this pyramid have? (Signal.) *4.*

c. (Point to **H**.) These letters stand for shapes.
- What shape does H stand for? (Signal.) *(A) hexagon.*
- (Point to **P**.) What shape does P stand for? (Signal.) *(A) pentagon.*
- (Point to **R**.) What shape does R stand for? (Signal.) *(A) rectangle.*
- (Point to **T**.) What shape does T stand for? (Signal.) *(A) triangle.*
  (Repeat step c until firm.)

d. (Point to pyramid.) You're going to tell me the number of faces that are hexagons, pentagons, rectangles, and triangles for the pyramid.
- How many faces of this pyramid are hexagons? (Touch.) *Zero.*

(Add to show:) [108:4B]

H = 0

- (Point to pyramid.) How many faces of this pyramid are pentagons? (Touch.) *Zero.*
(Add to show:) [108:4C]

H = 0
P = 0

- (Point to pyramid.) How many faces of this pyramid are rectangles? (Touch.) *1.*
(Add to show:) [108:4D]

H = 0
P = 0
R = 1

- (Point to pyramid.) How many faces of this pyramid are triangles? (Touch.) *4.*
(Add to show:) [108:4E]

H = 0
P = 0
R = 1
T = 4

(Point to pyramid.) This pyramid has 1 face that's a rectangle and 4 faces that are triangles.

e. (Display:) [108:4F]

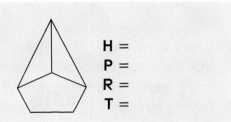

- (Point to △.) Are the side faces for this object triangles? (Signal.) *Yes.*
- So is this object a pyramid? (Signal.) *Yes.*
  You're going to tell me the number of faces that are hexagons, pentagons, rectangles, and triangles for the pyramid.
- I'll touch the sides for the bottom face. Count them to yourself and figure out the shape of the bottom face and the number of side faces. (Touch sides of bottom face.)
- Raise your hand if you know the shape of the bottom face. ✔
- Raise your hand if you know how many faces this pyramid has. ✔

f. Now we'll write the numbers for each shape.
* (Point to pyramid.) How many faces of this pyramid are hexagons? (Touch.) *Zero.*
  (Add to show:)   [108:4G]

> **H** = 0

* (Point to pyramid.) How many faces are pentagons? (Touch.) *1.*
  (Add to show:)   [108:4H]

> **H** = 0
> **P** = |

Yes, the bottom face is a pentagon.
* (Point to pyramid.) How many faces are rectangles? (Touch.) *Zero.*
  (Add to show:)   [108:4I]

> **H** = 0
> **P** = |
> **R** = 0

* (Point to pyramid.) How many faces are triangles? (Touch.) *5.*
  (Add to show:)   [108:4J]

> **H** = 0
> **P** = |
> **R** = 0
> **T** = 5

(Point to pyramid.) This pyramid has 1 face that's a pentagon and 5 faces that are triangles. Good figuring out the shapes of faces for objects.

EXERCISE 5: TELLING TIME
ANALOG HOURS   REMEDY

a. (Distribute unopened workbooks to children.)
* Open your workbook to Lesson 108 and find part 1.
  (Observe children and give feedback.)
  (Teacher reference:)   R Test 12: Part M

For these clocks you're going to make an arrow to show the direction the hour hand is moving, and write the time for each clock.
* Is the minute hand the longer hand or the shorter hand? (Signal.) *The longer hand.*

* • Touch the minute hand for clock A. ✔
* • Touch the hour hand for clock A. ✔
  (Repeat until firm.)
* Point the direction the hour hand is moving. ✔
* Make an arrow to show the direction the hour hand is moving.
  (Observe children and give feedback.)
  (Display:)   [108:5A]

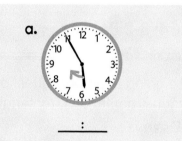

Here's what you should have for clock A.
b. Touch the minute hand for clock B. ✔
* • Touch the hour hand for clock B. ✔
  (Repeat until firm.)
* Make an arrow to show the direction the hour hand is moving for clock B. Then make arrows to show the direction the hour hands are moving for clocks C and D. Put your pencil down when you've made arrows for the hour hand for all of the clocks in part 1.
  (Observe children and give feedback.)
  (Display:)   [108:5B]

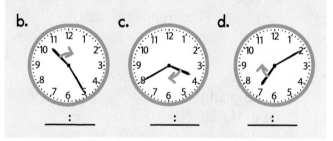

Here's what you should have for clocks B, C, and D.
* Make sure the arrows you made are going the right direction.
  (Observe children and give feedback.)
c. Touch the hour hand for clock A. ✔
* Figure out the last number the hour hand pointed to and get ready to tell me the hour number. ✔
* What's the hour number for clock A?
  (Signal.) *5.*

d. Touch the hour hand for clock B. ✔
- Figure out the last number the hour hand pointed to and get ready to tell me the hour number. ✔
- What's the hour number for clock B? (Signal.) *10.*
e. Touch the hour hand for clock C and get ready to tell me the hour number. ✔
- What's the hour number for clock C? (Signal.) *3.*
f. Touch the hour hand for clock D and get ready to tell me the hour number. ✔
- What's the hour number for clock D? (Signal.) *7.*
  (Repeat hour numbers that were not firm.)
g. Write the hour number for each clock in the space below. Do not write anything for the minutes. Just the hours numbers. Put your pencil down when you've done that much. (Observe children and give feedback.)
h. Check your work. You'll read the hour number you wrote for each clock.
- Clock A? (Signal.) *5.*
- (Repeat for:) B, *10;* C, *3;* D, *7.*
i. (Display:)   [108:5C]

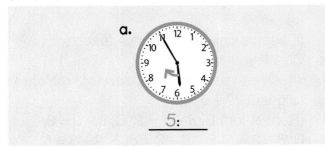

a.

5:___

Here's what you should have for clock A. I'll show you how to count minutes.
- (Point to minute marks between numbers.) It takes 1 minute for the minute hand to go from one of these marks to the next mark. It takes 5 minutes for the minute hand to go from one number to the next number.
- (Touch 12.) How long does it take the minute hand to go from here to (touch 1) here? (Signal.) *5 minutes.*
- (Touch 1.) How long does it take the minute hand to go from here to (touch 2) here? (Signal.) *5 minutes.*
  (Repeat for next numbers until firm.)

j. I'll count the minutes for clock A the fast way. After I get to the minute hand, tell me to stop. (Touch 1 through 11 as you count.) 5, 10, 15, 20, 25, 30, 35, 40, 45, 50, 55. *Stop.*
- Your turn: Touch 12 on clock A. ✔
- You'll touch the numbers and count for the minutes. Stop after you get to the minute hand. Get ready. (Tap 11.) *5, 10, 15, 20, 25, 30, 35, 40, 45, 50, 55.*
  (Repeat until firm.)
- What's the minute number for clock A? (Signal.) *55.*
k. Touch 12 on clock B. Touch the numbers and count for the minutes. Stop after you get to the minute hand. Get ready. (Tap 5.) *5, 10, 15, 20, 25.*
  (Repeat until firm.)
- What's the minute number for clock B? (Signal.) *25.*
l. Touch 12 on clock C. Touch the numbers and count for the minutes. Stop after you get to the minute hand. Get ready. (Tap 8.) *5, 10, 15, 20, 25, 30, 35, 40.*
  (Repeat until firm.)
- What's the minute number for clock C? (Signal.) *40.*
m. Touch 12 on clock D. Touch the numbers and count for the minutes. Stop after you get to the minute hand. Get ready. (Tap 2.) *5, 10.*
  (Repeat until firm.)
- What's the minute number for clock D? (Signal.) *10.*
n. Touch 12 on clock A again. Touch the numbers and count for the minutes. Stop after you get to the minute hand. Get ready. (Tap 11.) *5, 10, 15, 20, 25, 30, 35, 40, 45, 50, 55.*
- What's the minute number for clock A? (Signal.) *55.*
  (Repeat steps j through n that were not firm.)
o. Write the minutes for clock A. (Observe children and give feedback.)
- Read the time you wrote for clock A. Get ready. (Signal.) *Five 55.*
  (Display:)   [108:5D]

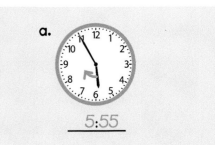

a.

5:55

Here's what you should have for clock A.

p. Write the minutes for clock B.
(Observe children and give feedback.)

• Read the time you wrote for clock B. Get
ready. (Signal.) *Ten 25.*

q. Write the minutes for clock C.
(Observe children and give feedback.)

• Read the time you wrote for clock C. Get
ready. (Signal.) *Three 40.*

r. Write the minutes for clock D.
(Observe children and give feedback.)

• Read the time you wrote for clock D. Get
ready. (Signal.) *Seven 10.*

EXERCISE 6: PLACE VALUE
PICTURES

REMEDY

a. Find part 2 on worksheet 108. ✔
(Teacher reference:)

R Part M

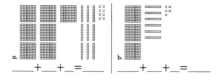

You're going to write the equation for each
picture.

• Count the squares for picture A. Raise your
hand when you can say the whole equation. ✔

• Everybody, say the equation for the number
of squares in picture A. Get ready. (Signal.)
700 + 90 + 8 = 798.
(Repeat until firm.)

b. Count the squares for picture B. Raise your
hand when you can say the whole equation.
(Observe children and give feedback.)

• Everybody, say the equation for the number
of squares in picture B. Get ready. (Signal.)
300 + 60 + 5 = 365.
(Repeat until firm.)

c. Write the equations for picture A and picture B.
(Observe children and give feedback.)
(Answer key:)

a.
$\underline{700} + \underline{90} + \underline{8} = \underline{798}$
b.
$\underline{300} + \underline{60} + \underline{5} = \underline{365}$

EXERCISE 7: FACTS
SUBTRACTION

a. Find part 3 on worksheet 108. ✔
(Teacher reference:)

a. 12 – 6   c. 6 – 3   f. 9 – 3
b. 8 – 6   d. 8 – 3   g. 9 – 7
              e. 10 – 2   h. 10 – 4

• Touch and read problem A. Get ready.
(Signal.) *12 minus 6.*

• What's 12 minus 6? (Signal.) *6.*

b. (Repeat the following tasks for problems B
through H:)

Touch and read problem __.		What's __?	
B	8 – 6	8 – 6	2
C	6 – 3	6 – 3	3
D	8 – 3	8 – 3	5
E	10 – 2	10 – 2	8
F	9 – 3	9 – 3	6
G	9 – 7	9 – 7	2
H	10 – 4	10 – 4	6

(Repeat problems that were not firm.)

c. Complete each of the facts. Put your pencil
down when you've written all of the equations
for part 3.
(Observe children and give feedback.)

d. Check your work.

• Touch and read the equation for problem A.
(Signal.) *12 – 6 = 6.*

• (Repeat for:) B, *8 – 6 = 2;* C, *6 – 3 = 3;*
D, *8 – 3 = 5;* E, *10 – 2 = 8;* F, *9 – 3 = 6;*
G, *9 – 7 = 2;* H, *10 – 4 = 6.*

EXERCISE 8: WORD PROBLEMS (COLUMNS)
CONVENTIONAL AND COMPARISON

REMEDY

a. Find part 4 on worksheet 108. ✔
(Teacher reference:)

R **Part E**

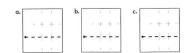

You're going to write the symbols for word problems in columns and work them. For problem A, you're going to tell me how much shorter one of the boats was.

- Listen to problem A: The blue boat was 83 feet long. The red boat was 42 feet long. How many feet shorter was the red boat?
- Listen to problem A again and write the problem for figuring out how much shorter the red boat was: The blue boat was 83 feet long. (Pause.) The red boat was 42 feet long. (Observe children and give feedback.)
- Read the problem you wrote for figuring out how much shorter the red boat was. (Signal.) *83 minus 42.*
- Work the problem. Put your pencil down when you know how many feet shorter the red boat was.
(Observe children and give feedback.)
(Teacher reference:)

b. Check your work.
- Read the subtraction problem and the answer for A. Get ready. (Signal.) *83 – 42 = 41.*
- How many feet shorter was the red boat? (Signal.) *41 feet.*
c. Touch where you'll write the symbols for problem B. ✔
- Listen to problem B: There were 84 apples in a crate. 70 of those apples were removed from the crate. How many apples ended up in the crate?
- Listen again and write the symbols for the whole problem: There were 84 apples in a crate. 70 of those apples were removed from the crate.
(Observe children and give feedback.)
- Everybody, touch and read the problem for B. Get ready. (Signal.) *84 minus 70.*

d. Work problem B. Put your pencil down when you know how many apples ended up in the crate.
(Observe children and give feedback.)
(Teacher reference:)

- Read the subtraction problem and the answer you wrote for B. Get ready. (Signal.) *84 – 70 = 14.*
- How many apples ended up in the crate? (Signal.) *14.*
e. Touch where you'll write the symbols for problem C. ✔
- Listen to problem C: A rancher had 74 cows in a field. The rancher moved 94 more cows into the field. How many cows ended up in the field?
- Listen again and write the symbols for the whole problem: A rancher had 74 cows in a field. The rancher moved 94 more cows into the field.
(Observe children and give feedback.)
- Everybody, touch and read the problem for C. Get ready. (Signal.) *74 plus 94.*
f. Work problem C. Put your pencil down when you know how many cows ended up in the field.
(Observe children and give feedback.)
(Teacher reference:)

- Read the addition problem and the answer you wrote for C. Get ready. (Signal.) *74 + 94 = 168.*
- How many cows ended up in the field? (Signal.) *168.*

EXERCISE 9: INDEPENDENT WORK

a. Find part 5. ✔
(Teacher reference:)

You'll complete the equations for the rulers.

b. Turn to the other side of your worksheet and find part 6. ✔
(Teacher reference:)

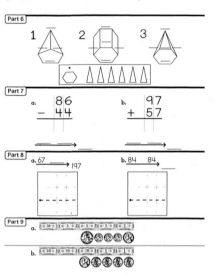

- Touch object 1. ✔
- Touch object 2. ✔
- Touch object 3. ✔
  You're going to write letters for all of the faces for objects 1, 2, and 3.
- Touch the picture of the faces for the bottom, top, and sides. ✔
  After you label the faces, you'll circle the object that has the right bottom, top, and sides.

c. In part 7, you'll work the problems and write the numbers in the family for each problem.
  In part 8, you'll write a column problem to figure out the missing number in each family. Then you'll complete the family.
  In part 9, you'll write the dollars and cents amount for each group of bills and coins.

d. Complete worksheet 108.
  (Observe children and mark incorrect responses on children's worksheets as you give feedback.)

Lesson 109

EXERCISE 1: FACT RELATIONSHIPS
PLUS 3 [REMEDY]

a. I'll say problems that plus 3. You'll say the whole facts.
- Listen: 1 plus 3. Say the fact for 1 plus 3. (Signal.) *1 + 3 = 4.*
- Say the fact for 2 plus 3. (Signal.) *2 + 3 = 5.*

b. (Repeat the following task for numbers 3 through 10:)
- Say the fact for __ plus 3.

c. You'll start with 6 plus 3 and you'll say those facts again.
- Listen: 6 plus 3. Say the fact for 6 plus 3. (Signal.) *6 + 3 = 9.*
- Say the fact for 7 plus 3. (Signal.) *7 + 3 = 10.*
- Say the fact for 8 plus 3. (Signal.) *8 + 3 = 11.*
- Say the fact for 9 plus 3. (Signal.) *9 + 3 = 12.*
- Say the fact for 10 plus 3. (Signal.) *10 + 3 = 13.*
(Repeat step c until firm.)

d. Now I'll mix them up.
- Listen: 10 plus 3. Say the fact for 10 plus 3. (Signal.) *10 + 3 = 13.*
- Listen: 9 plus 3. Say the fact for 9 plus 3. (Signal.) *9 + 3 = 12.*
- Listen: 8 plus 3. Say the fact for 8 plus 3. (Signal.) *8 + 3 = 11.*
- Listen: 6 plus 3. Say the fact for 6 plus 3. (Signal.) *6 + 3 = 9.*
- Listen: 7 plus 3. Say the fact for 7 plus 3. (Signal.) *7 + 3 = 10.*

INDIVIDUAL TURNS

(Call on individual students to perform one of the following tasks.)

- Say the fact for 9 plus 3. (Call on a student.) *9 + 3 = 12.*
- Say the fact for 8 plus 3. (Call on a student.) *8 + 3 = 11.*
- Say the fact for 7 plus 3. (Call on a student.) *7 + 3 = 10.*

EXERCISE 2: NUMBER FAMILIES
MISSING NUMBER IN FAMILY [REMEDY]

a. (Display:) [109:2A]

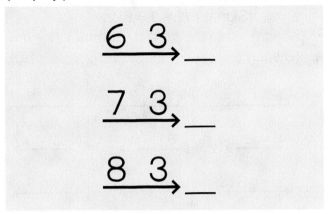

Here are three new families. The big number is missing in these families. All of these families have a small number of 3.
- (Point to 6→3→.) What are the small numbers in this family? (Touch.) *6 and 3.*
  What's the big number? (Signal.) *9.*
- (Point to 7→3→.) What are the small numbers in this family? (Touch.) *7 and 3.*
  What's the big number? (Signal.) *10.*
- (Point to 8→3→.) What are the small numbers in this family? (Touch.) *8 and 3.*
  What's the big number? (Signal.) *11.*
(Repeat until firm.)

b. (Point to 6→3→.) What are the small numbers in this family? (Touch.) *6 and 3.*
- What's the big number? (Signal.) *9.*
- Say the fact that starts with 6. (Signal.) *6 + 3 = 9.*
- Say the fact that starts with the other small number. (Signal.) *3 + 6 = 9.*
- Say the fact that goes backward down the arrow. (Signal.) *9 − 3 = 6.*
- Say the other minus fact. (Signal.) *9 − 6 = 3.*

c. (Point to 7→3→.) What are the small numbers in this family? (Touch.) *7 and 3.*
- What's the big number? (Signal.) *10.*
- Say the fact that starts with 7. (Signal.) *7 + 3 = 10.*
- Say the fact that starts with the other small number. (Signal.) *3 + 7 = 10.*
- Say the fact that goes backward down the arrow. (Signal.) *10 − 3 = 7.*
- Say the other minus fact. (Signal.) *10 − 7 = 3.*

d. (Point to 8 →3 .) What are the small numbers in this family? (Touch.) *8 and 3.*
- What's the big number? (Signal.) *11.*
- Say the fact that starts with 8. (Signal.) *8 + 3 = 11.*
- Say the fact that starts with the other small number. (Signal.) *3 + 8 = 11.*
- Say the fact that goes backward down the arrow. (Signal.) *11 – 3 = 8.*
- Say the other minus fact. (Signal.) *11 – 8 = 3.*

e. (Display:)   [109:2B]

6 3, → _	7 3, → _	4, → 8
6 → 12	8 8, → _	3, → 10
3, → 7	8, → 11	7 7, → _

Each family has a missing number. You'll say the problem for each family. Then you'll say the answer.

f. (Point to 6 →3 _.) Say the problem for the missing number. (Signal.) *6 plus 3.*
- What's 6 plus 3? (Signal.) *9.*

g. (Point to 6 → 12.) Say the problem for the missing number. (Signal.) *12 minus 6.*
- What's 12 minus 6? (Signal.) *6.*

h. (Repeat the following tasks for the remaining families:)

(Point to __.) Say the problem for the missing number.	What's __?	
3, → 7	7 – 3	7 – 3	4
7 3, → _	7 + 3	7 + 3	10
8 8, → _	8 + 8	8 + 8	16
8, → 11	11 – 8	11 – 8	3
4, → 8	8 – 4	8 – 4	4
3, → 10	10 – 3	10 – 3	7
7 7, → _	7 + 7	7 + 7	14

(Repeat for families that were not firm.)

EXERCISE 3: MIXED COUNTING
COUNTING BY TWOS

a. Everybody, count by twos to 20. Get ready. (Tap.) *2, 4, 6, 8, 10, 12, 14, 16, 18, 20.*
- Everybody, start with 50 and count by twos to 60. Get 50 going. *Fiftyyy.* Count. (Tap.) *52, 54, 56, 58, 60.*

b. Start with 50 and count by fives to 75. Get 50 going. *Fiftyyy.* Count. (Tap.) *55, 60, 65, 70, 75.*

c. Start with 50 and count backward by tens. Get 50 going. *Fiftyyy.* Count backward. (Tap.) *40, 30, 20, 10.*

d. Start with 50 and count backward by ones to 40. Get 50 going. *Fiftyyy.* Count backward. (Tap.) *49, 48, 47, 46, 45, 44, 43, 42, 41, 40.*

e. Start with 50 and plus tens to 100. Get 50 going. *Fiftyyy.* Plus tens. (Tap.) *60, 70, 80, 90, 100.*

f. Start with 50 and count by 25s to 150. Get 50 going. *Fiftyyy.* Count. (Tap.) *75, 100, 125, 150.*
(Repeat steps a through f that were not firm.)

g. Listen: You have 50. When you count backward by tens, what number comes next? (Signal.) *40.*
- So what's 50 minus 10? (Signal.) *40.*

h. You have 50. When you count by twos, what number comes next? (Signal.) *52.*
- So what's 50 plus 2? (Signal.) *52.*

i. You have 50. When you count by 25s, what number comes next? (Signal.) *75.*
- So what's 50 plus 25? (Signal.) *75.*

j. You have 50. When you count by ones, what number comes next? (Signal.) *51.*
- So what's 50 plus 1? (Signal.) *51.*

k. You have 50. When you count backward by ones, what number comes next? (Signal.) *49.*
- So what's 50 minus 1? (Signal.) *49.*
(Repeat steps g through k that were not firm.)

EXERCISE 4: MISSING VALUES
BIG NUMBERS OR SMALL NUMBERS

a. (Display:) [109:4A]

47 – ____ = 18	____ – 64 = 217
____ + 145 = 197	305 + 76 = ____
263 – ____ = 132	

These equations have a missing number.
- (Point to 47 – ___ = 18.) I'll read this problem. (Touch symbols.) 47 minus how many equals 18. Read the problem. (Touch.) *47 minus how many equals 18.*
  (Repeat until firm.)
- (Point to ___ + 145 = 197.) I'll read this problem. (Touch symbols.) How many plus 145 equals 197. Read the problem. (Touch.) *How many plus 145 equals 197.*
- (Point to 263 – ___ = 132.) Read the problem. (Touch.) *263 minus how many equals 132.*
- (Point to ___ – 64 = 217.) Read the problem. (Signal.) *How many minus 64 equals 217.*
- (Point to 305 + 76 = ___.) Read the problem. (Signal.) *305 plus 76 equals how many.*
  (Repeat problems that were not firm.)

b. You're going to tell me if the missing number is a small number or the big number.
   Remember, if you add, the answer is the big number. If you subtract, the first number is the big number.
- (Point to 47 – ___ = 18.) Does this problem add or subtract? (Signal.) *Subtract.*
- Raise your hand when you know if the missing number is a small number or the big number. ✔
- Is the missing number a small number or the big number? (Signal.) *A small number.*

c. (Repeat the following tasks for the remaining problems:)

(Point to __.) Does this problem add or subtract?	Raise your hand when you know if the missing number is a small number or the big number. ✔ Is the missing number a small number or the big number?
___ + 145 = 197	Add	A small number.
263 – ___ = 132	Subtract	A small number.
___ – 64 = 217	Subtract	The big number.
305 + 76 = ___	Add	The big number.

(Repeat problems that were not firm.)

EXERCISE 5: 3-D OBJECTS
DECOMPOSITION

a. (Display:) [109:5A]

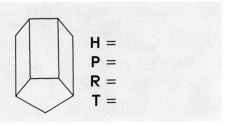

Here's an object.
- (Point to bottom face.) I'll touch the sides of the bottom face. Count them to yourself and figure out the shape of the bottom face and the number of side faces.
- (Touch each side of bottom face.) What's the shape of the bottom face? (Touch.) *(A) pentagon.*
- How many sides does the bottom face have? (Signal.) *5.*
- So how many side faces does this object have? (Signal.) *5.*
- What shape are the side faces? (Signal.) *Rectangle(s).*
- Does this object have a top face? (Signal.) *Yes.*
- What is the shape of the top face? (Signal.) *(A) pentagon.*
  (Repeat until firm.)

b. (Point to **H**.) These letters stand for shapes.
- What shape does H stand for? (Signal.) *(A) hexagon.*
- (Point to **P**.) What shape does P stand for? (Signal.) *(A) pentagon.*
- (Point to **R**.) What shape does R stand for? (Signal.) *(A) rectangle.*
- (Point to **T**.) What shape does T stand for? (Signal.) *(A) triangle.*
  (Repeat step b until firm.)

c. Now we'll write the numbers for each shape.
- (Point to prism.) How many faces of this object are hexagons? (Touch.) *Zero.*
  (Add to show:) [109:5B]

H = 0

d. (Point to prism.) How many faces are pentagons? (Touch.) *2.*
(Add to show:) [109:5C]

> **H** = 0
> **P** = 2

- (Point to prism.) How many faces are rectangles? (Touch.) *5.*
(Add to show:) [109:5D]

> **H** = 0
> **P** = 2
> **R** = 5

Yes, there are 5 side faces that are rectangles.
- (Point to prism.) How many faces are triangles? (Touch.) *Zero.*
(Add to show:) [109:5E]

> **H** = 0
> **P** = 2
> **R** = 5
> **T** = 0

(Point to prism.) This object has 2 faces that are pentagons and 5 faces that are rectangles.
e. (Display:) [109:5F]

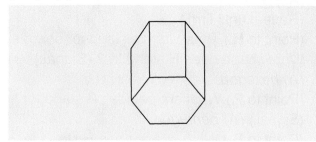

You're going to tell me the number of faces that are hexagons, pentagons, rectangles, and triangles for this object. I'll touch the sides of the bottom face. Count them to yourself and figure out the shape of the bottom face and the number of faces.
- (Touch each side of bottom face.) What's the shape of the bottom face? (Touch.) *(A) hexagon.*
- How many sides does the bottom face have? (Signal.) *6.*
- So how many side faces does this object have? (Signal.) *6.*

- What shape are the side faces? (Signal.) *Rectangle(s).*
- Does this object have a top face? (Signal.) *Yes.*
- What is the shape of the top face? (Signal.) *(A) hexagon.*
f. (Distribute unopened workbooks to children.)
- Open your workbook to Lesson 109 and find part 1.
(Observe children and give feedback.)
(Teacher reference:)

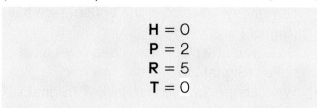

> H =
> P =
> R =
> T =

You're going to complete the equations.
- Touch H equals. ✔
- (Point to prism.) How many faces of this object are hexagons? (Touch.) *2.*
- Write 2. ✔
g. Touch P equals. ✔
- (Point to prism.) How many faces are pentagons? (Touch.) *Zero.*
- Write zero. ✔
h. Touch R equals. ✔
- (Point to prism.) How many faces are rectangles? (Touch.) *6.*
- Write 6. ✔
i. Touch T equals. ✔
- (Point to prism.) How many faces are triangles? (Touch.) *Zero.*
- Write zero. ✔
(Display:) [109:5G]

> **H** = 2
> **P** = O
> **R** = 6
> **T** = O

Here's what you should have.
(Point to prism.) This object has 2 faces that are hexagons and 6 faces that are rectangles.

EXERCISE 6: FACTS
PLUS/MINUS MIX

a. Find part 2 on worksheet 109. ✔
(Teacher reference:)

a. 12 − 6   e. 11 − 5
b. 6 − 3   f. 16 − 8
c. 7 + 3   g. 8 + 3
d. 14 − 7   h. 10 − 5

- Touch and read problem A. Get ready.
(Signal.) *12 minus 6.*
- Is the answer the big number or a small
number? (Signal.) *A small number.*
- What's 12 minus 6? (Signal.) *6.*

b. (Repeat the following tasks for problems B
through H:)

Touch and read problem __.		Is the answer the big number or a small number?	What's __?	
B	6 − 3	A small number.	6 − 3	3
C	7 + 3	The big number.	7 + 3	10
D	14 − 7	A small number.	14 − 7	7
E	11 − 5	A small number.	11 − 5	6
F	16 − 8	A small number.	16 − 8	8
G	8 + 3	The big number.	8 + 3	11
H	10 − 5	A small number.	10 − 5	5

(Repeat problems that were not firm.)

c. Complete each of the facts. Put your pencil
down when you've written all of the equations
for part 2.
(Observe children and give feedback.)

d. Check your work.
- Touch and read the equation for problem A.
(Signal.) *12 − 6 = 6.*
- (Repeat for:) B, *6 − 3 = 3;* C, *7 + 3 = 10;*
D, *14 − 7 = 7;* E, *11 − 5 = 6;* F, *16 − 8 = 8;*
G, *8 + 3 = 11;* H, *10 − 5 = 5.*

EXERCISE 7: TELLING TIME
ANALOG HOURS   REMEDY

a. Find part 3 on your worksheet. ✔
(Teacher reference:)   R Test 12: Part N

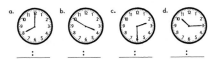

For these clocks you're going to make an
arrow to show the direction the hour hand is
moving, and write the time for each clock.
- Is the minute hand the longer hand or the
shorter hand? (Signal.) *The longer hand.*

- Touch the minute hand for clock A. ✔
- Touch the hour hand for clock A. ✔
(Repeat until firm.)
- Point the direction the hour hand is moving. ✔
- Make an arrow to show the direction the hour
hand is moving for clock A. Then make arrows
to show the direction the hour hands are
moving for the rest of the clocks in part 3. Put
your pencil down when you've made arrows
for all of the hour hands.
(Observe children and give feedback.)
(Display:)   [109:7A]

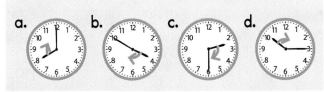

Here's what you should have for the clocks in
part 3.
- Make sure the arrows you made are going the
right direction.
(Observe children and give feedback.)

b. Touch the hour hand for clock A. ✔
- The hour hand is pointing right at a number.
What's the hour number for clock A?
(Signal.) *8.*

c. Touch the hour hand for clock B. ✔
- Figure out the last number the hour hand
pointed to and get ready to tell me the hour
number. ✔
- What's the hour number for clock B?
(Signal.) *3.*

d. Touch the hour hand for clock C and get ready
to tell me the hour number. ✔
- What's the hour number for clock C?
(Signal.) *2.*

e. Touch the hour hand for clock D and get ready
to tell me the hour number. ✔
- What's the hour number for clock D?
(Signal.) *10.*
(Repeat hour numbers that were not firm.)

f. Write the hour number for each clock in the
space below. Do not write anything for the
minutes. Just the hour numbers. Put your
pencil down when you've done that much.
(Observe children and give feedback.)

g. Check your work. You'll read the hour number
for each clock.
- Clock A? (Signal.) *8.*
- (Repeat for:) B, *3;* C, *2;* D, *10.*

h. (Display:) [109:7B]

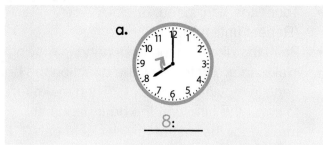

Here's what you should have for clock A. The minute hand is pointing right at 12, so the minute number is zero. What's the minute number? (Signal.) *Zero.*

• Write zero for the minutes below clock A. ✔
• Everybody, read the time for clock A. Get ready. (Signal.) *8 o'clock.*
(Display:) [109:7C]

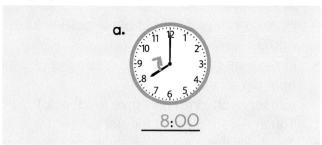

Here's what you should have for clock A. Clock A shows 8 o'clock.

i. (Display:) [109:7D]

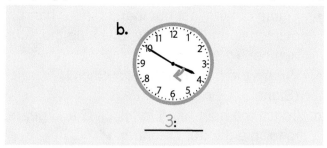

Here's what you should have for clock B. I'll show you how to count minutes.

• It takes 5 minutes for the minute hand to go from one number to the next number. How many minutes do you count for each number? (Signal.) *5.*
I'll count the minutes for clock B the fast way. After I get to the minute hand, tell me to stop. (Touch 1 through 10 as you count.) *5, 10, 15, 20, 25, 30, 35, 40, 45, 50, stop.*
• Your turn: Touch 12 on clock B. Touch the numbers and count for the minutes. Stop after you get to the minute hand. Get ready. (Tap 10.) *5, 10, 15, 20, 25, 30, 35, 40, 45, 50.*
(Repeat until firm.)
• What's the minute number for clock B? (Signal.) *50.*

j. Touch 12 on clock C. Touch the numbers and count for the minutes. Stop after you get to the minute hand. Get ready. (Tap 6.) *5, 10, 15, 20, 25, 30.*
(Repeat until firm.)
• What's the minute number for clock C? (Signal.) *30.*

k. Yes, when the minute hand is straight down, the clock shows 30 minutes. When the minute hand is straight up, the clock shows zero minutes.
• When the minute hand is straight up, how many minutes does the clock show? (Signal.) *Zero.*
• When the minute hand is straight down, how many minutes does the clock show? (Signal.) *30.*
(Repeat until firm.)

l. Touch 12 on clock D. Touch the numbers and count for the minutes. Stop after you get to the minute hand. Get ready. (Tap 3.) *5, 10, 15.*
(Repeat until firm.)
• What's the minute number for clock D? (Signal.) *15.*
(Repeat steps k through l that were not firm.)

m. Write the minutes for clock B. (Observe children and give feedback.)
• Read the time you wrote for clock B. Get ready. (Signal.) *Three 50.*

n. Write the minutes for clock C. (Observe children and give feedback.)
• Read the time you wrote for clock C. Get ready. (Signal.) *Two 30.*

o. Write the minutes for clock D. (Observe children and give feedback.)
• Read the time you wrote for clock D. Get ready. (Signal.) *Ten 15.*
(Display:) [109:7E]

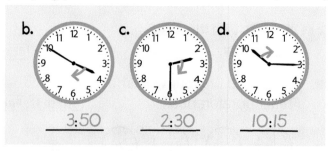

Here's what you should have for clocks B, C, and D.

p. When the minute hand is pointing straight
    up or straight down, think about how many
    minutes the clock shows.
    • When the minute hand is straight up,
      how many minutes does the clock
      show? (Signal.) *Zero.*
    • When the minute hand is straight down,
      how many minutes does the clock show?
      (Signal.) *30.*
      (Repeat until firm.)

EXERCISE 8: PLACE VALUE
PICTURES

a. Find part 4 on worksheet 109. ✔
    (Teacher reference:)

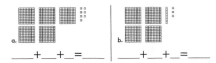

___+__+_=___ | ___+__+_=___

    You're going to write the equation for each
    picture.
    • Count the squares for picture A. There are no
      tens in the picture for A, so remember to say
      zero for the tens. Raise your hand when you
      can say the whole equation. ✔
    • Everybody, say the equation for the number
      of squares for picture A. Get ready. (Touch
      columns.) *500 + 0 + 7 = 507.*
      (Repeat until firm.)
b. Count the squares for picture B. Raise your
    hand when you can say the whole equation. ✔
    • Everybody, say the equation for the number
      of squares for picture B. Get ready. (Touch
      columns.) *400 + 10 + 3 = 413.*
      (Repeat until firm.)
c. Write the equations for picture A and picture B.
    (Observe children and give feedback.)
    (Answer key:)

    a.
    500+ 0 +7=507 | b.
                    400+10+3=413

EXERCISE 9: SOLVING PROBLEMS FROM NUMBER FAMILIES
MULTI-DIGIT NUMBERS

a. Find part 5. ✔
    (Teacher reference:)

    $$\xrightarrow{\underline{77 \quad 99}}176 \qquad \xrightarrow{\underline{99 \quad 176}}275$$

    a.176 − 77   c.275 − 176
    b.176 + 99   d. 99 + 77

    You're going to use number famillies to work
    problems.
    • First family: What are the small numbers?
      (Signal.) *77 and 99.*
    • What's the big number? (Signal.) *176.*
    • Second family: What are the small numbers?
      (Signal.) *99 and 176.*
    • What's the big number? (Signal.) *275.*
b. Touch and read problem A. (Signal.)
    176 minus 77.
    • Does this problem show the big number?
      (Signal.) *Yes.*
    • What's the big number? (Signal.) *176.*
    • What's the small number that is shown?
      (Signal.) *77.*
      (Repeat until firm.)
    • Touch the family that has the right numbers. ✔
      You should be touching the first family.
c. Look at the numbers in the family. Raise
    your hand when you know what 176 minus
    77 equals. ✔
    • What does 176 minus 77 equal? (Signal.) *99.*
    • Complete equation A. ✔
d. Touch and read problem B. (Signal.)
    176 plus 99.
    • Does this problem show the big number?
      (Signal.) *No.*
    • What are the small numbers? (Signal.)
      176 and 99.
    • Touch the family that has the right numbers. ✔
      You should be touching the second family.

e. Look at the number in the family. Raise your hand when you know what 176 plus 99 equals. ✔
• What does 176 plus 99 equal? (Signal.) *275.*
• Complete equation B. ✔
f. Touch and read problem C. (Signal.)
 275 minus 176.
• Does this problem show the big number? (Signal.) *Yes.*
• What's the big number? (Signal.) *275.*
• What's the small number that is shown? (Signal.) *176.*
• Find the family with the right numbers and complete equation C.
 (Observe children and give feedback.)
• Everybody, what does 275 minus 176 equal? (Signal.) *99.*
g. Touch and read problem D. (Signal.) *99 plus 77.*
• Does this problem show the big number? (Signal.) *No.*
• What are the small numbers? (Signal.) *99 and 77.*
• Find the family with the right numbers and complete equation D.
 (Observe children and give feedback.)
• Everybody, what does 99 plus 77 equal? (Signal.) *176.*

EXERCISE 10: INDEPENDENT WORK

a. Turn to the other side of your worksheet and find part 6. ✔
 (Teacher reference:)

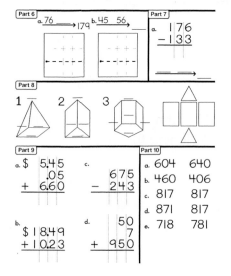

You'll write a column problem to figure out the missing number in each family. Then you'll complete the family.
In part 7, you'll work the problem and write the numbers in the family.
b. Find part 8. ✔
• Touch object 1. ✔
• Touch object 2. ✔
• Touch object 3. ✔
You're going to write letters for all of the faces for objects 1, 2, and 3.
• Touch the picture of the faces for the bottom, top, and sides. ✔
After you label the faces, you'll circle the object that has the right bottom, top, and sides.
In part 9, you'll work the money problems and the column problems.
In part 10, you'll write an equals, or a sign that has a bigger end next to one of the numbers.
c. Complete worksheet 109.
 (Observe children and mark incorrect responses on children's worksheets as you give feedback.)

Lesson 110

EXERCISE 1: FACT RELATIONSHIPS
PLUS 3

a. I'll say problems that plus 3. You'll say the whole facts.
- Listen: 1 plus 3. Say the fact for 1 plus 3. (Signal.) *1 + 3 = 4.*
- Say the fact for 2 plus 3. (Signal.) *2 + 3 = 5.*

b. (Repeat the following task for numbers 3 through 10:)
- Say the fact for __ plus 3.

c. You'll start with 6 plus 3 and you'll say those facts again.
- Listen: 6 plus 3. Say the fact for 6 plus 3. (Signal.) *6 + 3 = 9.*
- Say the fact for 7 plus 3. (Signal.) *7 + 3 = 10.*
- Say the fact for 8 plus 3. (Signal.) *8 + 3 = 11.*
- Say the fact for 9 plus 3. (Signal.) *9 + 3 = 12.*
- Say the fact for 10 plus 3. (Signal.) *10 + 3 = 13.* (Repeat step c until firm.)

d. Now I'll mix some of them up.
- Listen: 10 plus 3. Say the fact for 10 plus 3. (Signal.) *10 + 3 = 13.*
- Listen: 9 plus 3. Say the fact for 9 plus 3. (Signal.) *9 + 3 = 12.*
- Listen: 6 plus 3. Say the fact for 6 plus 3. (Signal.) *6 + 3 = 9.*
- Listen: 7 plus 3. Say the fact for 7 plus 3. (Signal.) *7 + 3 = 10.*
- Listen: 4 plus 3. Say the fact for 4 plus 3. (Signal.) *4 + 3 = 7.*
- Listen: 8 plus 3. Say the fact for 8 plus 3. (Signal.) *8 + 3 = 11.*

=== INDIVIDUAL TURNS ===

(Call on individual students to perform one of the following tasks.)

- Say the fact for 7 plus 3. (Call on a student.) *7 + 3 = 10.*
- Say the fact for 9 plus 3. (Call on a student.) *9 + 3 = 12.*
- Say the fact for 8 plus 3. (Call on a student.) *8 + 3 = 11.*

EXERCISE 2: MISSING VALUES
BIG NUMBERS OR SMALL NUMBERS

a. (Display:) [110:2A]

$$\_\_\_\_ + 463 = 587 \qquad 215 + \_\_\_\_ = 298$$
$$291 - \_\_\_\_ = 184 \qquad \_\_\_\_ - 191 = 84$$
$$36 + 302 = \_\_\_\_ \qquad 289 - 107 = \_\_\_\_$$

You're going to read these problems. Remember, you say **how many** for the missing number.
- (Point to ___ + 463 = 587.) Read the problem. (Touch.) *How many plus 463 equals 587.*
- (Point to 291 – ___ = 184.) Read the problem. (Touch.) *291 minus how many equals 184.*
- (Point to 36 + 302 = ___.) Read the problem. (Touch.) *36 plus 302 equals how many.*
- (Point to 215 + ___ = 298.) Read the problem. (Touch.) *215 plus how many equals 298.*
- (Point to ___ – 191 = 84.) Read the problem. (Touch.) *How many minus 191 equals 84.*
- (Point to 289 – 107 = ___.) Read the problem. (Touch.) *289 minus 107 equals how many.* (Repeat problems that were not firm.)

b. You're going to tell me if the missing number is a small number or the big number. Remember, if you add, the answer is the big number. If you subtract, the first number is the big number.
- (Point to ___ + 463 = 587.) Does this problem add or subtract? (Signal.) *Add.*
- Raise your hand when you know if the missing number is a small number or the big number. ✔
- Is the missing number a small number or the big number? (Signal.) *A small number.*

c. (Repeat the following tasks for the remaining problems:)

(Point to __.) Does this problem add or subtract?	Raise your hand when you know if the missing number is a small number or the big number. ✔ Is the missing number a small number or the big number?
291 – ____ = 184	Subtract	A small number.
36 + 302 = ____	Add	The big number.
215 + ____ = 298	Add	A small number.
____ – 191 = 84	Subtract	The big number.
289 – 107 = ____	Subtract	A small number.

(Repeat problems that were not firm.)

d. Now you're going to tell me about the numbers that are shown.

• (Point to ___ + 463 = 587.) This problem has the numbers 463 and 587.

• Is 463 a small number or the big number? (Signal.) *A small number.*

• What is 587? (Signal.) *The big number.*

e. (Repeat the following tasks for remaining problems:)

(Point to __.) This problem has the numbers __.		Is __ a small number or the big number?		What is __?	
291 – ____ = 184	291 and 184	291	The big number.	184	A small number.
36 + 302 = ____	36 and 302	36	A small number.	302	A small number.
215 + ____ = 298	215 and 298	215	A small number.	298	The big number.
____ – 191 = 84	191 and 84	191	A small number.	84	A small number.
289 – 107 = ____	289 and 107	289	The big number.	107	A small number.

a. (Display:) [110:3A]

All of these families have a small number of 3. The big number is missing in these families.

• (Point to 6 3→.) What are the small numbers in this family? (Touch.) *6 and 3.*

• What's the big number? (Signal.) *9.*

• (Point to 7 3→.) What are the small numbers in this family? (Touch.) *7 and 3.*

• What's the big number? (Signal.) *10.*

• (Point to 8 3→.) What are the small numbers in this family? (Touch.) *8 and 3.*

• What's the big number? (Signal.) *11.*
(Repeat until firm.)

b. (Point to 6 3→.) What are the small numbers in this family? (Touch.) *6 and 3.*

• What's the big number? (Signal.) *9.*

• Say the fact that starts with 6. (Signal.) *6 + 3 = 9.*

• Say the fact that starts with the other small number. (Signal.) *3 + 6 = 9.*

• Say the fact that goes backward down the arrow. (Signal.) *9 – 3 = 6.*

• Say the other minus fact. (Signal.) *9 – 6 = 3.*

c. (Point to 7 3→.) What are the small numbers in this family? (Touch.) *7 and 3.*

• What's the big number? (Signal.) *10.*

• Say the fact that starts with 7. (Signal.) *7 + 3 = 10.*

• Say the fact that starts with the other small number. (Signal.) *3 + 7 = 10.*

• Say the fact that goes backward down the arrow. (Signal.) *10 – 3 = 7.*

• Say the other minus fact. (Signal.) *10 – 7 = 3.*

d. (Point to $\overset{8\quad 3}{\longrightarrow}$.) What are the small numbers in this family? (Touch.) *8 and 3.*
- What's the big number? (Signal.) *11.*
- Say the fact that starts with 8. (Signal.) *8 + 3 = 11.*
- Say the fact that starts with the other small number. (Signal.) *3 + 8 = 11.*
- Say the fact that goes backward down the arrow. (Signal.) *11 − 3 = 8.*
- Say the other minus fact. (Signal.) *11 − 8 = 3.*

e. (Display:)   [110:3B]

$\overset{3}{=\!\!\longrightarrow}8$	$\overset{4}{\longrightarrow}7$	$\overset{8}{\longrightarrow}10$
$\overset{6\quad 5}{\longrightarrow}\_$	$\overset{5\quad 5}{\longrightarrow}\_$	$\overset{3}{=\!\!\longrightarrow}11$
$\overset{7}{\longrightarrow}10$	$\overset{6}{=\!\!\longrightarrow}10$	$\overset{7\quad 3}{\longrightarrow}\_$

Each family has a missing number. You'll say the problem for each family. Then you'll say the answer.

f. (Point to $=\!\!\overset{3}{\longrightarrow}8$.) Say the problem for the missing number. Get ready. (Signal.) *8 minus 3.*
- What's 8 minus 3? (Signal.) *5.*

g. (Point to $\overset{6\quad 5}{\longrightarrow}\_$.) Say the problem for the missing number. (Signal.) *6 plus 5.*
- What's 6 plus 5? (Signal.) *11.*

h. (Repeat the following tasks for the remaining families:)

(Point to __.) Say the problem for the missing number.		What's __?	
$\overset{7}{\longrightarrow}10$	10 − 7	10 − 7	3
$\overset{4}{\longrightarrow}7$	7 − 4	7 − 4	3
$\overset{5\quad 5}{\longrightarrow}\_$	5 + 5	5 + 5	10
$\overset{6}{\longrightarrow}10$	10 − 6	10 − 6	4
$\overset{8}{\longrightarrow}10$	10 − 8	10 − 8	2
$=\!\!\overset{3}{\longrightarrow}11$	11 − 3	11 − 3	8
$\overset{7\quad 3}{\longrightarrow}\_$	7 + 3	7 + 3	10

(Repeat for families that were not firm.)

EXERCISE 4: 3-D OBJECTS
DECOMPOSITION   REMEDY

a. (Display:)   [110:4A]

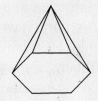

Here's an object.
- (Point to bottom face.) I'll touch the sides of the bottom face. Count them to yourself and figure out the shape of the bottom face and the number of side faces.
- (Touch each side of bottom face.) What's the shape of the bottom face? (Touch.) *(A) hexagon.*
- How many sides does the bottom face have? (Signal.) *6.*
- So how many side faces does this object have? (Signal.) *6.*
- What shape are the side faces? (Signal.) *Triangles(s).*
- Does this object have a top face? (Signal.) *No.* (Repeat until firm.)

b. (Distribute unopened workbooks to children.)
- Open your workbook to Lesson 110 and find part 1.
(Observe children and give feedback.)
(Teacher reference:)   R Test 12: Part G

Part 1	Part 2
H =	H =
P =	P =
R =	R =
T =	T =

The letters in part 1 and part 2 stand for shapes.
- Touch H equals. ✔
What shape does H stand for? (Signal.) *(A) hexagon.*
- Touch P equals. ✔
What shape does P stand for? (Signal.) *(A) pentagon.*
- Touch R equals. ✔
What shape does R stand for? (Signal.) *(A) rectangle.*
- Touch T equals. ✔
What shape does T stand for? (Signal.) *(A) triangle.*
- (Repeat until firm.)

c. (Point to pyramid) Now you're going to complete each equation.
- How many faces of this pyramid are hexagons? (Touch.) *1.*
- Complete the equation for hexagons. (Observe children and give feedback.)
d. (Point to pyramid) How many faces are pentagons? (Touch.) *Zero.*
- How many faces are rectangles? (Touch.) *Zero.*
- Complete the equations for pentagons and rectangles. (Observe children and give feedback.)
e. (Point to pyramid) How many faces are triangles? (Touch.) *6.*
  Yes, there are 6 side faces that are triangles.
- Complete the equation for triangles. (Observe children and give feedback.)
f. (Display:) [110:4B]

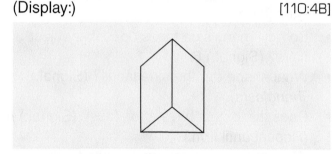

You're going to complete the equations for hexagons, pentagons, rectangles, and triangles for this object in part 2.
- (Touch bottom face.) What's the shape of the bottom face? (Touch.) *(A) triangle.*
- How many sides does the bottom face have? (Signal.) *3.*
- So how many side faces does this object have? (Signal.) *3.*
- What shape are the side faces? (Signal.) *Rectangle(s).*
- Does this object have a top face? (Signal.) *Yes.*
- What is the shape of the top face? (Signal.) *(A) triangle.*
  Yes, the bottom face and the top face are triangles.
g. (Point to prism.) Now you're going to complete each equation.
- How many faces of this object are hexagons? (Touch.) *Zero.*
- How many faces are pentagons? (Touch.) *Zero.*
- Complete the equations for hexagons and pentagons in part 2. (Observe children and give feedback.)

h. (Point to prism.) How many faces are rectangles? (Signal.) *3.*
- Complete the equation for rectangles. (Observe children and give feedback.)
i. (Point to prism.) How many faces are triangles? (Signal.) *2.*
- Complete the equation for triangles. (Observe children and give feedback.) (Answer key:)

Part 1	Part 2
H = *1*	H = *0*
P = *0*	P = *0*
R = *0*	R = *3*
T = *6*	T = *2*

EXERCISE 5: SOLVING PROBLEMS FROM NUMBER FAMILIES
MULTI-DIGIT NUMBERS

a. Find part 3 on your worksheet. ✔ (Teacher reference:)

$\dfrac{79\quad 268}{}$ 347 $\dfrac{268\quad 347}{}$ 615

a. 615 − 268
b. 268 + 79
c. 347 − 268
d. 347 + 268

You're going to use number families to work problems.
- First family: What are the small numbers? (Signal.) *79 and 268.*
- What's the big number? (Signal.) *347.*
- Second family: What are the small numbers? (Signal.) *268 and 347.*
- What's the big number? (Signal.) *615.*
b. Touch and read problem A. (Signal.) *615 minus 268.*
- Does this problem show the big number? (Signal.) *Yes.*
- What's the big number? (Signal.) *615.*
- What's the small number that is shown? (Signal.) *268.* (Repeat until firm.)
- Touch the family that has the right numbers. ✔ You should be touching the second family.

c. Look at the numbers in the family. Raise your hand when you know what 615 minus 268 equals. ✔
- What does 615 minus 268 equal? (Signal.) *347.*
- Complete equation A.

d. Touch and read problem B. (Signal.) *268 plus 79.*
- Does this problem show the big number? (Signal.) *No.*
- What are the small numbers? (Signal.) *268 and 79.*
- Touch the family that has the right numbers. ✔ You should be touching the first family.

e. Look at the number in the family. Raise your hand when you know what 268 plus 79 equals. ✔
- What does 268 plus 79 equal? (Signal.) *347.*
- Complete equation B. ✔

f. Touch and read problem C. (Signal.) *347 minus 268.*
- Does this problem show the big number? (Signal.) *Yes.*
- What's the big number? (Signal.) *347.*
- What's the small number that is shown? (Signal.) *268.*
- Find the family with the right numbers and complete equation C.
  (Observe children and give feedback.)
- Everybody, what does 347 minus 268 equal? (Signal.) *79.*

g. Touch and read problem D. (Signal.) *347 plus 268.*
- Does this problem show the big number? (Signal.) *No.*
- What are the small numbers? (Signal.) *347 and 268.*
- Find the family with the right numbers and complete equation D.
  (Observe children and give feedback.)
- Everybody, what does 347 plus 268 equal? (Signal.) *615.*

EXERCISE 6: TELLING TIME
ANALOG HOURS

a. Find part 4 on your worksheet. ✔
(Teacher reference:)

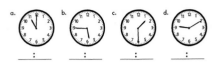

For these clocks you're going to make an arrow to show the direction the hour hand is moving, and write the time for each clock.
- Make the arrows to show the direction the hour hands are moving. Put your pencil down when you've made arrows for all the hour hands.
(Observe children and give feedback.)
(Display:)   [110:6A]

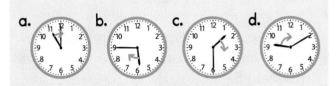

Here's what you should have for the clocks in part 4.
- Make sure the arrows you made are going the right direction.
(Observe children and give feedback.)

b. Now you'll tell me the hour numbers for each clock. Remember, the last number that the hour hand pointed to is the hour number.
- Clock A: The hour hand is pointing right at a number. What's the hour number for clock A? (Signal.) *11.*
- Clock B: What's the hour number for clock B? (Signal.) *5.*
- Clock C: What's the hour number for clock C? (Signal.) *1.*
- Clock D: What's the hour number for clock D? (Signal.) *9.*
(Repeat until firm.)

c. Write the hour number for each clock in the space below. Do not write anything for the minutes. Just write the hour numbers. Put your pencil down when you've done that much.
(Observe children and give feedback.)

d. Check your work. You'll read the hour number for each clock.
- Clock A? (Signal.) *11.*
- (Repeat for:) B, *5;* C, *1;* D, *9.*

e. Look at clock A and get ready to tell me the minute number.
 • Everybody, what's the minute number for clock A? (Signal.) *Zero.*
 • Write the minute number for clock A. ✔
 • Everybody, read the time for clock A. Get ready. (Signal.) *11 o'clock.*
 (Display:)   [110:6B]

a.

11:00

Here's what you should have for clock A. Clock A shows 11 o'clock.

f. It takes 5 minutes for the minute hand to go from one number to the next number. How many minutes do you count for each number? (Signal.) *5.*
 • Your turn: Touch 12 on clock B. ✔
 • Touch each number and count for the minutes. Stop after you get to the minute hand. Get ready. (Tap 9.) *5, 10, 15, 20, 25, 30, 35, 40, 45.* (Repeat until firm.)
 • What's the minute number for clock B? (Signal.) *45.*

g. Touch 12 on clock C. ✔
 • Touch each number and count for the minutes. Stop after you get to the minute hand. Get ready. (Tap 5.) *5, 10, 15, 20, 25, 30.* (Repeat until firm.)
 • What's the minute number for clock C? (Signal.) *30.*

h. Yes, when the minute hand is straight down, the clock shows 30 minutes. When the minute hand is straight up, the clock shows zero minutes.
 • When the minute hand is straight up, how many minutes does the clock show? (Signal.) *Zero.*
 • When the minute hand is straight down, how many minutes does the clock show? (Signal.) *30.* (Repeat until firm.)

i. Touch 12 on clock D. ✔
 • Touch each number and count for the minutes. Stop after you get to the minute hand. Get ready. (Tap 2.) *5, 10.* (Repeat until firm.)
 • What's the minute number for clock D? (Signal.) *10.* (Repeat steps e through i that were not firm.)

j. Write the minutes for clocks B, C, and D. (Observe children and give feedback.)

k. Check your work. You'll read the time you wrote for clocks B, C, and D.
 • Clock B. (Signal.) *Five 45.*
 • (Repeat for:) C, *One 30;* D, *Nine 10.* (Display:)   [110:6C]

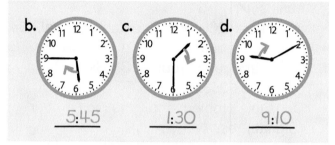

b.   c.   d.
5:45   1:30   9:10

Here's what you should have for clocks B, C, and D.

l. When the minute hand is straight up or straight down, think about how many minutes the clock shows.
 • When the minute hand is straight up, how many minutes does the clock show? (Signal.) *Zero.*
 • When the minute hand is straight down, how many minutes does the clock show? (Signal.) *30.* (Repeat until firm.)

EXERCISE 7: FACTS
PLUS/MINUS MIX

a. Find part 5. ✔
 (Teacher reference:)

	c. $7 + 7$	f. $8 - 5$
a. $3 + 7$	d. $11 - 6$	g. $4 + 4$
b. $7 - 3$	e. $11 - 8$	h. $9 - 3$

 • Touch and read problem A. Get ready. (Signal.) *3 plus 7.*
 • Is the answer the big number or a small number? (Signal.) *The big number.*
 • What's 3 plus 7? (Signal.) *10.*

b. (Repeat the following tasks for problems B through H:)

Touch and read problem __.		Is the answer the big number or a small number?	What's __?	
B	7 – 3	A small number.	7 – 3	4
C	7 + 7	The big number.	7 + 7	14
D	11 – 6	A small number.	11 – 6	5
E	11 – 8	A small number.	11 – 8	3
F	8 – 5	A small number.	8 – 5	3
G	4 + 4	The big number.	4 + 4	8
H	9 – 3	A small number.	9 – 3	6

(Repeat problems that were not firm.)

c. Complete each of the facts. Put your pencil down when you've written all of the equations for part 5.
(Observe children and give feedback.)

d. Check your work.
- Touch and read the equation for problem A. (Signal.) *3 + 7 = 10.*
- (Repeat for:) B, *7 – 3 = 4;* C, *7 + 7 = 14;* D, *11 – 6 = 5;* E, *11 – 8 = 3;* F, *8 – 5 = 3;* G, *4 + 4 = 8;* H, *9 – 3 = 6.*

EXERCISE 8: WORD PROBLEMS (COLUMNS)
CONVENTIONAL AND COMPARISON   REMEDY

a. Find part 6. ✔
(Teacher reference:)   R | Part F

You're going to write the symbols for word problems in columns and work them. For problem A, you're going to tell me how much shorter one building was.
- Listen to problem A: The white building was 173 feet tall. The blue building was 50 feet tall. How many feet shorter was the blue building?
- Listen to problem A again and write the problem for figuring out how much shorter the blue building was: The white building was 173 feet tall. (Pause.) The blue building was 50 feet tall.
(Observe children and give feedback.)

- Read the problem you wrote for figuring out how much shorter the blue building was. (Signal.) *173 minus 50.*
- Work the problem. Put your pencil down when you know how many feet shorter the blue building was.
(Observe children and give feedback.)
(Teacher reference:)

b. Check your work.
- Read the subtraction problem and the answer for A. Get ready. (Signal.) *173 – 50 = 123.*
- How many feet shorter was the blue building? (Signal.) *123 feet.*

c. Touch where you'll write the symbols for problem B. ✔
- Listen to problem B: There were 218 children in the school. 72 more children came to the school. How many children ended up in the school?
- Listen again and write the symbols for the whole problem: There were 218 children in the school. 72 more children came to the school.
(Observe children and give feedback.)
- Everybody, touch and read the problem for B. Get ready. (Signal.) *218 plus 72.*

d. Work problem B. Put your pencil down when you know how many children ended up in the school.
(Observe children and give feedback.)
(Teacher reference:)

- Read the addition problem and the answer you wrote for B. Get ready. (Signal.) *218 + 72 = 290.*
- How many children ended up in the school? (Signal.) *290.*

e. Touch where you'll write the symbols for problem C. ✔
• Listen to problem C: A truck started out with 278 cans in it. Workers took 165 cans from the truck. How many cans were still in the truck?
• Listen again and write the symbols for the whole problem: A truck started out with 278 cans in it. Workers took 165 cans from the truck.
• Everybody, touch and read the problem for C. Get ready. (Signal.) *278 minus 165.*
f. Work problem C. Put your pencil down when you know how many cans were still in the truck.
(Observe children and give feedback.)
(Teacher reference:)

• Read the subtraction problem and the answer you wrote for C. Get ready. (Signal.) *278 – 165 = 113.*
• How many cans were still in the truck? (Signal.) *113.*

EXERCISE 9: PLACE VALUE
PICTURES   REMEDY

a. Turn to the other side of your worksheet and find part 7. ✔
(Teacher reference:)   R Part N

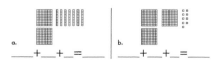

You're going to write the equation for each picture.
• Count the squares for picture A. There are no ones in the picture for A so remember to say zero for the ones. Raise your hand when you can say the whole equation.
(Observe children and give feedback.)
• Everybody, say the equation for the number of squares for picture A. Get ready. (Touch columns.) *200 + 70 + 0 = 270.*
(Repeat until firm.)

b. Count the squares for picture B. There are no tens in the picture for B so remember to say zero for the tens. Raise your hand when you can say the whole equation.
(Observe children and give feedback.)
• Everybody, say the equation for the number of squares for picture B. Get ready. (Touch columns.) *300 + 0 + 9 = 309.*
(Repeat until firm.)
c. Write the equations for picture A and picture B.
(Observe children and give feedback.)
(Answer key:)

EXERCISE 10: INDEPENDENT WORK

a. Find part 8. ✔
(Teacher reference:)

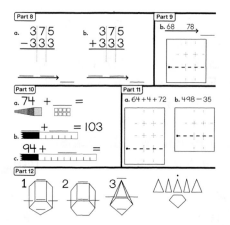

You'll work the problems and write the numbers in the family for each problem.
In part 9, you'll write a column problem to figure out the missing number in the family. Then you'll complete the family.
In part 10, you'll complete the equation for the squares and the equations for the rulers.
In part 11, you'll write each problem in a column and work it.
b. Find part 12. ✔
• Touch object 1. ✔
• Touch object 2. ✔
• Touch object 3. ✔
You're going to write letters for all of the faces for objects 1, 2, and 3.
• Touch the picture of the faces for the bottom, top, and sides. ✔
After you label the faces, you'll circle the object that has the right bottom, top, and sides.
c. Complete worksheet 110.
(Observe children and mark incorrect responses on children's worksheets as you give feedback.)

Mastery Test 11

Teacher Presentation

a. Find Test 11 in your test booklet. ✔
- Touch part 1. ✔
  (Teacher reference:)

You're going to write letters for the faces of objects 1, 2, and 3.
- For faces that are shaped like triangles, what letter will you write? (Signal.) *T.*
- For faces that are shaped like rectangles, what letter will you write? (Signal.) *R.*
- For faces that are shaped like pentagons, what letter will you write? (Signal.) *P.*
- For faces that are shaped like hexagons, what letter will you write? (Signal.) *H.*
  (Repeat until firm.)

b. Write the letters for all of the faces for objects 1, 2, and 3. Remember, if an object doesn't have a top face, leave the space for the top face blank. Put your pencil down when you've labeled all of the faces with the letter for its shape.
(Observe children.)

c. (Display:)   [Test 11]

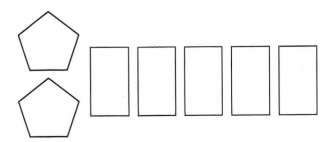

Here are the bottom, top, and side faces of one of the objects. Circle the object that has the correct bottom, top, and sides.
(Observe children.)

d. Touch part 2 on test sheet 11. ✔
(Teacher reference:)

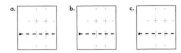

You're going to write the symbols for word problems and work them.
- Touch where you'll write the symbols for problem A. ✔
- Listen to problem A: 685 students were in a school. 613 of those students left the school. How many students ended up in the school?
- Listen again and write the symbols for both parts: 685 students were in a school. 613 of those students left the school.
(Observe children.)
- Work problem A. Put your pencil down when you know how many students ended up in the school.
(Observe children.)

e. For problem B you're going to figure out how much shorter one building was.
- Touch where you'll write problem B. ✔
- Listen to problem B: The white building was 173 feet tall. The blue building was 50 feet tall. How many feet shorter was the blue building?
- Listen to problem B again and write the symbols for the problem as I say it: The white building was 173 feet tall. The blue building was 50 feet tall. How many feet shorter was the blue building?
(Observe children.)
- Work problem B. Put your pencil down when you know how many feet shorter the blue building was.
(Observe children.)

f. Touch where you'll write the symbols for problem C. ✔
- Listen to problem C: There were 218 children in the school. 72 more children came to the school. How many children ended up in the school?
- Listen again and write the symbols for both parts: There were 218 children in the school. 72 more children came to the school.
(Observe children.)
- Work problem C. Put your pencil down when you know how many children ended up in the school.
(Observe children.)
g. Touch part 3 on test sheet 11. ✔
(Teacher reference:)

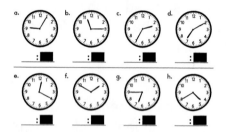

For these clocks you're going to make an arrow to show the direction the hour hand is moving. Then you'll write the hour number for each clock.
- Make an arrow to show the direction the hour hand is moving for each clock. Put your pencil down when you've made the arrow for each hour hand.
(Observe children.)
- Now, write the hour number for each clock. Put your pencil down when you've written the hour number for all of the clocks in part 3.
(Observe children.)

h. Turn to the other side of test sheet 11 and touch part 4. ✔
(Teacher reference:)

a.
```
  220
 -142
   78
```
═══⟶ __

b.
```
  181
 -142
   39
```
═══⟶ __

c.
```
  220
 +181
  401
```
═══⟶ __

d.
```
  142
 +181
  323
```
═══⟶ __

e.
```
  142
 - 39
  103
```
═══⟶ __

You're going to make number families for these equations.
- Write the number families for all of the equations in part 4. Put your pencil down when you've completed all of the families.
(Observe children.)
i. Touch part 5 on test sheet 11. ✔
(Teacher reference:)

a. 1 + 3
b. 2 + 3
c. 3 + 3
d. 4 + 3
e. 5 + 3
f. 6 + 3
g. 7 + 3
h. 8 + 3
i. 9 + 3
j. 10 + 3

You'll complete equations in part 5.
j. Touch part 6. ✔
(Teacher reference:)

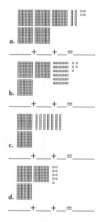

a.
___ + ___ + __ = ____

b.
___ + ___ + __ = ____

c.
___ + ___ + __ = ____

d.
___ + ___ + __ = ____

You'll write place-value equations for each picture in part 6.

k. Touch part 7. ✔
(Teacher reference:)

a. 16 − 8
b. 16 − 10
c. 7 + 2
d. 7 + 7
e. 8 − 4   h. 6 − 3
f. 8 + 8   i. 6 + 6
g. 10 − 5   j. 14 − 7

All of the problems in part 7 are from families you know. You'll complete equations in part 7.

• Your turn: Work the rest of the problems on Test 11.
(Direct students where to put their assessment books when they are finished.)

Scoring Notes

a. Collect test booklets. Use the Answer Key and Passing Criteria Table to score the tests.

Passing Criteria Table — Mastery Test 11				
Part	Score		Possible Score	Passing Score
1	1 for bottom and side faces		19	17
	2 for each top face			
	3 for circling the right object			
2	4 for each problem		12	8
3	2 for each problem		16	12
4	3 for each problem		15	12
5	1 for each problem		10	10
6	2 for each problem		8	6
7	2 for each problem		20	18
		Total	100	

b. Complete the Mastery Test 11 Remedy Summary Sheet to determine whether group remedies are needed. Reproducible Remedy Summary Sheets are at the back of the Answer Key and at the back of the *Teacher's Guide.*

• If ¼ or more of the students did not pass a test part, present the remedy for that part before beginning Lesson 111. The Remedy Table follows and is also at the end of the Mastery Test 11 Answer Key. Remedies worksheets follow Mastery Test 11 in the *Student Assessment Book.*

Remedy Table — Mastery Test 11				
Part	Test Items	Remedy		Student Material Remedies Worksheet
		Lesson	Exercise	
1	Decomposition	104	7	Part A
		106	7	Part B
		107	8	Part C
2	Word Problems (Conventional and Comparison)	107	6	Part D
		108	8	Part E
		110	8	Part F
3	Telling Time (Analog Hours)	101	3	—
		104	3	Part G
		106	5	Part H
		107	5	Part I
4	Number Families from Equations	101	5	—
		102	9	Part J
		103	6	Part K
5	Plus 3 Fact Relationships	106	4	—
		109	1	—
6	Place-Value (Pictures)	99	4	—
		101	6	Part L
		108	6	Part M
		110	9	Part N
7	Facts (Mix)	99	2	—
		99	7	Part O
		103	1	—
		103	4	Part P

Retest
Retest individual students on any part failed.

Lessons 111–115 Planning Page

	Lesson 111	Lesson 112	Lesson 113	Lesson 114	Lesson 115
Student Learning Objectives	**Exercises** 1. **Say facts for 3 plus** 2. Identify the missing value (big number, small number) 3. **Count by a variety of numbers (evens)** 4. Say two addition and two subtraction facts with a small number of 3; find missing numbers in number families 5. Write the time from an analog clock (hours and minutes) 6. Write the symbols for word problems in columns and solve; solve comparison word problems 7. Solve addition and subtraction facts 8. Identify 2-dimensional faces on 3-dimensional objects 9. Solve problems from multi-digit number families 10. Complete work independently	**Exercises** 1. Say facts for 3 plus 2. **Solve problems with a missing value using multi-digit number families** 3. Count by a variety of numbers (evens) 4. Say two addition and two subtraction facts with a small number of 3; find missing numbers in number families 5. Write the time from an analog clock (hours and minutes) 6. Identify 2-dimensional faces on 3-dimensional objects 7. Solve addition and subtraction facts 8. Solve problems from multi-digit number families 9. **Identify and explain the place values of tens and ones in 2-digit numbers with drawings** 10. Complete work independently	**Exercises** 1. **Say facts for minus 3** 2. **Mentally add 10 and subtract 10 and explain reasoning** 3. Say two addition and two subtraction facts with a small number of 3; find missing numbers in number families 4. Identify and explain the place values of tens and ones in 2-digit numbers with drawings 5. Solve problems with a missing value using multi-digit number families 6. **Write the time from an analog clock (half hours)** 7. Solve addition and subtraction facts 8. **Identify 2-dimensional faces on 3-dimensional objects (cube)** 9. Complete work independently	**Exercises** 1. Say facts for minus 3 2. **Identify 3-dimensional objects (prisms and cubes)** 3. Mentally add 10 and subtract 10 and explain reasoning 4. Say addition and subtraction facts 5. **Identify even and odd numbers** 6. Identify and explain the place values of tens and ones in 2-digit numbers with drawings 7. Write the symbols for word problems in columns and solve; solve comparison word problems 8. Write the time from an analog clock (half hours) 9. Solve addition facts 10. Solve problems with a missing value using multi-digit number families 11. Complete work independently	**Exercises** 1. Say facts for plus 3, 3 plus, and minus 3 2. **Identify 3-dimensional objects (prism vs. pyramid)** 3. Mentally add 10 and subtract 10 and explain reasoning 4. **Identify and explain the place values of tens and ones in 2-digit numbers** 5. Say the time from an analog clock (half hours) 6. Identify even and odd numbers 7. Solve addition and subtraction facts 8. Solve problems with a missing value using multi-digit number families 9. Solve addition facts 10. Write the symbols for word problems in columns and solve; solve comparison word problems 11. Complete work independently

Common Core State Standards for Mathematics

	Lesson 111	Lesson 112	Lesson 113	Lesson 114	Lesson 115
1.OA 1				✔	✔
1.OA 3	✔	✔	✔		
1.OA 4–6, 8	✔	✔	✔	✔	✔
1.NBT 1–2	✔	✔	✔	✔	✔
1.NBT 4	✔	✔	✔	✔	✔
1.NBT 5			✔	✔	✔
1.NBT 6		✔			
1.MD 3	✔	✔	✔	✔	✔
1.G 1–2	✔	✔	✔	✔	✔

Teacher Materials	Presentation Book 3, Board Displays CD or chalkboard
Student Materials	Workbook 2, Pencil
Additional Practice	• Student Practice Software: Block 5: Activity 1 (1.NBT.1), Activity 2 (1.NBT.1), Activity 3 (1.NBT.4 and 1.OA.6), Activity 4 (1.NBT.1), Activity 5 (1.NBT.1), Activity 6 (1.NBT.3) • Math Fact Worksheets 65–66 (After Lesson 111), 67 (After Lesson 113), 68–72 (After Lesson 114)
Mastery Test	

Lesson  111

EXERCISE 1: FACT RELATIONSHIPS
3 PLUS   `REMEDY`

a. You've said facts for problems that plus 3. This time you'll say the plus facts that start with 3.
- Listen: 3 plus 1. Say the fact for 3 plus 1. (Signal.) *3 + 1 = 4.*
- Say the fact for 3 plus 2. (Signal.) *3 + 2 = 5.*

b. (Repeat the following task for numbers 3 through 10:)
- Say the fact for 3 plus __.

c. You'll start with 3 plus 6 and you'll say those facts again.
- Say the fact for 3 plus 6. (Signal.) *3 + 6 = 9.*
- Say the fact for 3 plus 7. (Signal.) *3 + 7 = 10.*
- Say the fact for 3 plus 8. (Signal.) *3 + 8 = 11.*
- Say the fact for 3 plus 9. (Signal.) *3 + 9 = 12.*
- Say the fact for 3 plus 10. (Signal.) *3 + 10 = 13.*
  (Repeat step c until firm.)

d. Now I'll mix some of them up.
- Listen: 3 plus 10. Say the fact for 3 plus 10. (Signal.) *3 + 10 = 13.*
- Listen: 3 plus 9. Say the fact for 3 plus 9. (Signal.) *3 + 9 = 12.*
- Listen: 3 plus 6. Say the fact for 3 plus 6. (Signal.) *3 + 6 = 9.*
- Listen: 3 plus 7. Say the fact for 3 plus 7. (Signal.) *3 + 7 = 10.*
- Listen: 3 plus 4. Say the fact for 3 plus 4. (Signal.) *3 + 4 = 7.*
- Listen: 3 plus 8. Say the fact for 3 plus 8. (Signal.) *3 + 8 = 11.*

=========== INDIVIDUAL TURNS ===========

(Call on individual students to perform one of the following tasks.)

- Say the fact for 3 plus 7. (Call on a student.) *3 + 7 = 10.*
- Say the fact for 3 plus 9. (Call on a student.) *3 + 9 = 12.*
- Say the fact for 3 plus 8. (Call on a student.) *3 + 8 = 11.*

EXERCISE 2: MISSING VALUES
BIG NUMBERS OR SMALL NUMBERS

a. (Display:)   [111:2A]

$$614 - \underline{\quad} = 601 \qquad \underline{\quad} + 273 = 491$$

$$\underline{\quad} - 56 = 82 \qquad \underline{\quad} - 108 = 79$$

$$317 + \underline{\quad} = 530 \qquad 95 + 416 = \underline{\quad}$$

These problems have a number missing. You're going to read these problems. You'll say how many for the missing number. Then you'll tell me if the numbers are small numbers or the big number.

- (Point to 614 – ___ = 601.) Read this problem. (Signal.) *614 minus how many equals 601.*
- Is 614 a small number or the big number? (Signal.) *The big number.*
- So what is the missing number? (Signal.) *A small number.*
- What is 601? (Signal.) *A small number.* (Repeat until firm.)

b. (Point to ___ – 56 = 82.) Read this problem. (Signal.) *How many minus 56 equals 82.*
- Is the missing number a small number or the big number? (Signal.) *The big number.*
- What is 56? (Signal.) *A small number.*
- What is 82? (Signal.) *A small number.*

c. (Point to 317 + ___ = 530.) Read this problem. (Signal.) *317 plus how many equals 530.*
- Is the missing number a small number or the big number? (Signal.) *A small number.*
- What's 317? (Signal.) *A small number.*
- What's 530? (Signal.) *The big number.*

d. (Point to ___ + 273 = 491.) Read this problem. (Signal.) *How many plus 273 equals 491.*
- Is the missing number a small number or the big number? (Signal.) *A small number.*
- What's 273? (Signal.) *A small number.*
- What's 491? (Signal.) *The big number.*

e. (Point to ___ – 108 = 79.) Read this problem. (Signal.) *How many minus 108 equals 79.*
- Is the missing number a small number or the big number? (Signal.) *The big number.*
- What's 108? (Signal.) *A small number.*
- What's 79? (Signal.) *A small number.*

f. (Point to 95 + 416 = ___.) Read this problem.
(Signal.) *95 plus 416 equals how many.*
- Is the missing number a small number or the big number? (Signal.) *The big number.*
- What's 95? (Signal.) *A small number.*
- What's 416? (Signal.) *A small number.*
(Repeat problems that were not firm.)

EXERCISE 3: MIXED COUNTING
EVEN NOMENCLATURE   **REMEDY**

a. Everybody, count by twos to 20. Get ready. (Tap.) *2, 4, 6, 8, 10, 12, 14, 16, 18, 20.*
- Everybody, start with 38 and count by twos to 48. Get 38 going. *Thirty-eieieight.* Count. (Tap.) *40, 42, 44, 46, 48.*
- The numbers for counting by twos have a special name. They are called **even numbers.** What's the special name for the numbers you say when you count by twos? (Signal.) *Even numbers.*
b. Count by hundreds to 1000. Get ready. (Tap.) *100, 200, 300, 400, 500, 600, 700, 800, 900, 1000.*
c. Start with 75 and count by ones to 85. Get 75 going. *Seventy-fiiive.* Count. (Tap.) *76, 77, 78, 79, 80, 81, 82, 83, 84, 85.*
d. Start with 75 and count backward by ones to 65. Get 75 going. *Seventy-fiiive.* Count backward. (Tap.) *74, 73, 72, 71, 70, 69, 68, 67, 66, 65.*
e. Start with 75 and count by fives to 100. Get 75 going. *Seventy-fiiive.* Count. (Tap.) *80, 85, 90, 95, 100.*
f. Start with 75 and count by 25s to 175. Get 75 going. *Seventy-fiiive.* Count. (Tap.) *100, 125, 150, 175.*
(Repeat steps a through f that were not firm.)
g. Listen: You have 75. When you count by fives, what number comes next? (Signal.) *80.*
- So what's 75 plus 5? (Signal.) *80.*
h. Listen: You have 75. When you count by twos, what number comes next? (Signal.) *77.*
- So what's 75 plus 2? (Signal.) *77.*
i. Listen: You have 75. When you plus tens, what number comes next? (Signal.) *85.*
- So what's 75 plus 10? (Signal.) *85.*
j. You have 75. When you count by ones, what number comes next? (Signal.) *76.*
- So what's 75 plus 1? (Signal.) *76.*
(Repeat steps g through j that were not firm.)
k. Earlier you learned a name for the numbers you say when you count by twos. What's the name? (Signal.) *Even numbers.*

EXERCISE 4: NUMBER FAMILIES
MISSING NUMBER IN FAMILY

a. (Display:)   [111:4A]

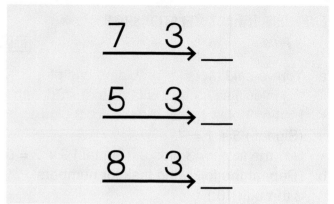

All of these families have a small number of 3. The big number is missing in these families.
- (Point to 7→3.) What are the small numbers in this family? (Touch.) *7 and 3.*
- What's the big number? (Signal.) *10.*
- (Point to 5→3.) What are the small numbers in this family? (Touch.) *5 and 3.*
- What's the big number? (Signal.) *8.*
- (Point to 8→3.) What are the small numbers in this family? (Touch.) *8 and 3.*
- What's the big number? (Signal.) *11.*
(Repeat until firm.)
b. (Point to 7→3.) What are the small numbers in this family? (Touch.) *7 and 3.*
- What's the big number? (Signal.) *10.*
- Say the fact that starts with 7. (Signal.) *7 + 3 = 10.*
- Say the fact that starts with the other small number. (Signal.) *3 + 7 = 10.*
- Say the fact that goes backward down the arrow. (Signal.) *10 – 3 = 7.*
- Say the other minus fact. (Signal.) *10 – 7 = 3.*
c. (Point to 5→3.) What are the small numbers in this family? (Touch.) *5 and 3.*
- What's the big number? (Signal.) *8.*
- Say the fact that starts with 5. (Signal.) *5 + 3 = 8.*
- Say the fact that starts with the other small number. (Signal.) *3 + 5 = 8.*
- Say the fact that goes backward down the arrow. (Signal.) *8 – 3 = 5.*
- Say the other minus fact. (Signal.) *8 – 5 = 3.*

d. (Point to 8 →3.) What are the small numbers in this family? (Touch.) *8 and 3.*
- What's the big number? (Signal.) *11.*
- Say the fact that starts with 8. (Signal.) *8 + 3 = 11.*
- Say the fact that starts with the other small number. (Signal.) *3 + 8 = 11.*
- Say the fact that goes backward down the arrow. (Signal.) *11 – 3 = 8.*
- Say the other minus fact. (Signal.) *11 – 8 = 3.*

e. (Display:) [111:4B]

=—3→8	6—→10	3—→4
8 3,—→_	10 3,—→_	=—5→10
=—3→10	4 3,—→_	4—→8

Each family has a missing number. You'll say the problem for each family. Then you'll say the answer.

f. (Point to =→3 8.) Say the problem for the missing number. Get ready. (Signal.) *8 minus 3.*
- What's 8 minus 3? (Signal.) *5.*

g. (Point to 8 →3 _.) Say the problem for the missing number. (Signal.) *8 plus 3.*
- What's 8 plus 3? (Signal.) *11.*

h. (Repeat the following tasks for the remaining families:)

(Point to __.) Say the problem for the missing number.		What's __?	
=—3→10	10 – 3	10 – 3	7
6—→10	10 – 6	10 – 6	4
10 3,—→_	10 + 3	10 + 3	13
4 3,—→_	4 + 3	4 + 3	7
3—→4	4 – 3	4 – 3	1
=—5→10	10 – 5	10 – 5	5
4—→8	8 – 4	8 – 4	4

(Repeat for families that were not firm.)

EXERCISE 5: TELLING TIME
ANALOG HOURS AND MINUTES

a. (Distribute unopened workbooks to children.)
- Open your workbook to Lesson 111 and find part 1.
(Observe children and give feedback.)
(Teacher reference:)

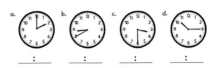

For these clocks you're going to make an arrow to show the direction the hour hand is moving, and write the time for each clock.
- Make the arrows to show the direction the hour hands are moving. Put your pencil down when you've made arrows for all of the hour hands.
(Observe children and give feedback.)
(Display:) [111:5A]

Here's what you should have for the clocks in part 1.
- Make sure the arrows you made are going the right direction.
(Observe children and give feedback.)

b. Now you'll tell me the hour number for each clock. Remember, the last number that the hour hand pointed to is the hour number.
- Clock A: The hour hand is pointing right at a number. What's the hour number for clock A? (Signal.) *2.*
- Clock B: What's the hour number for clock B? (Signal.) *8.*
- Clock C: What's the hour number for clock C? (Signal.) *3.*
- Clock D: What's the hour number for clock D? (Signal.) *10.*
(Repeat until firm.)

c. Write the hour number for each clock in the space below. Do not write anything for the minutes. Just write the hour numbers. Put your pencil down when you've done that much.
(Observe children and give feedback.)

d. Check your work. You'll read the hour number for each clock.
- Clock A. (Signal.) *2.*
- (Repeat for:) B, *8;* C, *3;* D, *10.*

e. Look at clock A and get ready to tell me the minute number.
• Everybody, what's the minute number for clock A? (Signal.) *Zero.*
• Write the minute number for clock A. ✔
• Everybody, read the time for clock A. Get ready. (Signal.) *2 o'clock.*
(Display:) [111:5B]

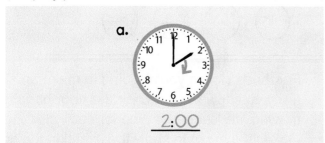

Here's what you should have for clock A. Clock A shows 2 o'clock.

f. It takes 5 minutes for the minute hand to go from one number to the next number. How many minutes do you count for each number? (Signal.) *5.*
• Your turn: Touch 12 on clock B. ✔
• Touch each number and count for the minutes. Stop after you get to the minute hand. Get ready. (Tap 8.) *5, 10, 15, 20, 25, 30, 35, 40.* (Repeat until firm.)
• What's the minute number for clock B? (Signal.) *40.*

g. Touch 12 on clock C. ✔
• Touch each number and count for the minutes. Stop after you get to the minute hand. Get ready. (Tap 6.) *5, 10, 15, 20, 25, 30.* (Repeat until firm.)
• What's the minute number for clock C? (Signal.) *30.*

h. Yes, when the minute hand is straight down, the clock shows 30 minutes. When the minute hand is straight up, the clock shows zero minutes.
• When the minute hand is straight up, how many minutes does the clock show? (Signal.) *Zero.*
• When the minute hand is straight down, how many minutes does the clock show? (Signal.) *30.* (Repeat until firm.)

i. Touch 12 on clock D. ✔
• Touch each number and count for the minutes. Stop after you get to the minute hand. Get ready. (Tap 3.) *5, 10, 15.* (Repeat until firm.)

• What's the minute number for clock D? (Signal.) *15.* (Repeat steps f through i that were not firm.)
j. Write the minutes for clocks B, C, and D. (Observe children and give feedback.)
k. Check your work. You'll read the time you wrote for clocks B, C, and D.
• Clock B? (Signal.) *Eight 40.*
• (Repeat for:) C, *Three 30;* D, *Ten 15.*
(Display:) [111:5C]

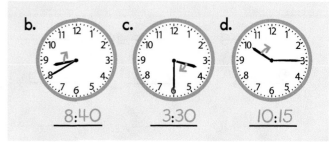

Here's what you should have for clocks B, C, and D.

l. When the minute hand is straight up or straight down, think about how many minutes the clock shows.
• When the minute hand is straight up, how many minutes does the clock show? (Signal.) *Zero.*
• When the minute hand is straight down, how many minutes does the clock show? (Signal.) *30.* (Repeat until firm.)

EXERCISE 6: WORD PROBLEMS (COLUMNS)
CONVENTIONAL AND COMPARISON

a. Find part 2 on your worksheet. ✔ (Teacher reference:)

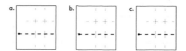

You're going to write the symbols for word problems in columns and work them.
• Touch where you'll write the symbols for problem A. ✔
• Listen to problem A: There were 97 people on a train. 68 more people got on the train. How many people ended up on the train?
• Listen again and write the symbols for the whole problem. There were 97 people on a train. 68 more people got on the train. (Observe and give feedback.)
• Everybody, touch and read the problem for A. Get ready. (Signal.) *97 plus 68.*

b. Work problem A. Put your pencil down when you know how many people ended up on the train.
(Observe children and give feedback.)
(Teacher reference:)

- Read the addition problem and the answer you wrote for A. Get ready. (Signal.)
  97 + 68 = 165.
- How many people ended up on the train? (Signal.) *165.*

c. Touch where you'll write the symbols for problem B. ✔
You're going to tell me how much shorter one house is.

- Listen to problem B: The brown house was 32 feet tall. The white house was 66 feet tall. How many feet shorter was the brown house than the white house?
- Listen to problem B again and write the problem for figuring out how much shorter the brown house was. The brown house was 32 feet tall. (Pause.) The white house was 66 feet tall. (Observe children and give feedback.)
- Read the problem you wrote for figuring out how much shorter the brown house was. (Signal.) *66 minus 32.*
- Work the problem. Put your pencil down when you know how many feet shorter the brown house was.
(Observe children and give feedback.)
(Teacher reference:)

d. Check your work.
- Read the subtraction problem and the answer for B. Get ready. (Signal.) *66 – 32 = 34.*
- How many feet shorter was the brown house? (Signal.) *34 feet.*

e. Touch where you'll write the symbols for problem C. ✔
Listen to problem C: Troy had 375 nails. Troy used 172 of those nails. How many nails did Troy end up with?

- Listen again and write the symbols for the whole problem. Troy had 375 nails. Troy used 172 of those nails.
(Observe children and give feedback.)

- Everybody, touch and read the problem for C. Get ready. (Signal.) *375 minus 172.*

f. Write 375 minus 172 equals and work problem C. Put your pencil down when you know how many nails Troy ended up with.
(Observe children and give feedback.)
(Teacher reference:)

- Read the subtraction problem and the answer you wrote for C. Get ready. (Signal.)
  375 – 172 = 203.
- How many nails did Troy end up with? (Signal.) *203.*

EXERCISE 7: FACTS
PLUS/MINUS MIX

a. Find part 3. ✔
(Teacher reference:)

a. 7 – 4
b. 7 + 3
c. 7 – 2
d. 11 – 3
e. 11 – 9
f. 11 – 5
g. 3 + 6
h. 6 – 3

- Touch and read problem A. Get ready. (Signal.) *7 minus 4.*
- Is the answer the big number or a small number? (Signal.) *A small number.*
- What's 7 minus 4? (Signal.) *3.*

b. (Repeat the following tasks for problems B through H:)

Touch and read problem __.		Is the answer the big number or a small number?	What's __?	
B	7 + 3	The big number.	7 + 3	10
C	7 – 2	A small number.	7 – 2	5
D	11 – 3	A small number.	11 – 3	8
E	11 – 9	A small number.	11 – 9	2
F	11 – 5	A small number.	11 – 5	6
G	3 + 6	The big number.	3 + 6	9
H	6 – 3	A small number.	6 – 3	3

(Repeat problems that were not firm.)

c. Complete each of the facts. Put your pencil down when you've written all of the equations for part 3.
(Observe children and give feedback.)

d. Check your work.
- Touch and read the equation for problem A. (Signal.) *7 – 4 = 3.*
- (Repeat for:) B, *7 + 3 = 10;* C, *7 – 2 = 5;* D, *11 – 3 = 8;* E, *11 – 9 = 2;* F, *11 – 5 = 6;* G, *3 + 6 = 9;* H, *6 – 3 = 3.*

EXERCISE 8: 3-D OBJECTS
DECOMPOSITION

REMEDY

a. You've learned a name for objects with side faces that are triangles, a top that comes to a point, and a bottom face that can have any shape. What's the name for those objects? (Signal.) *Pyramids.*

b. (Display:)   [111:8A]

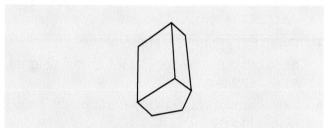

Here's an object.

• Is this object a pyramid? (Signal.) *No.*
You'll learn the name of objects like this one very soon.

c. You're going to figure out the number of faces for each shape.

• (Point to bottom face.) I'll touch the sides for the bottom face. Count them to yourself and figure out the shape of the bottom face and the number of side faces.
• (Touch each side of bottom face.) What's the shape of the bottom face? (Touch.)
(A) pentagon.
• How many sides does the bottom face have? (Signal.) *5.*
• So how many side faces does this object have? (Signal.) *5.*
• What shape are the side faces? (Signal.) *Rectangle(s).*
• Does this object have a top face? (Signal.) *Yes.*
• What shape is the top face? (Signal.)
(A) pentagon.
(Repeat until firm.)

d. Find part 4 on worksheet 111. ✔
(Teacher reference:)

R Part H

Part 4	Part 5
H =	H =
P =	P =
R =	R =
T =	T =

• (Point to prism.) You're going to complete each equation.
• How many faces of this object are hexagons? (Touch.) *Zero.*
• Complete the equation for H.
(Observe children and give feedback.)

e. (Point to prism.) How many faces are pentagons? (Touch.) *2.*
• Complete the equation for P.
(Observe children and give feedback.)
f. (Point to prism.) How many faces are rectangles? (Touch.) *5.*
• Complete the equation for R.
(Observe children and give feedback.)
g. (Point to prism.) How many faces are triangles? (Touch.) *Zero.*
• Complete the equation for triangles.
(Observe children and give feedback.)

h. (Display:)   [111:8B]

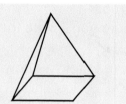

You're going to complete the equations for hexagons, pentagons, rectangles, and triangles for this object in part 5.

• I'll touch the sides for the bottom face. Count them to yourself and figure out the shape of the bottom face and the number of side faces. (Touch each side of bottom face.)
• Look at the object and figure out if there is a top face.

i. Now you'll tell me the number you'll write after each letter.

• (Point to pyramid.) How many faces of this object are hexagons? (Touch.) *Zero.*
• (Point to pyramid.) How many faces are pentagons? (Touch.) *Zero.*
• (Point to pyramid.) How many faces are rectangles? (Signal.) *1.*
• (Point to pyramid.) How many faces are triangles? (Signal.) *4.*

j. Complete the equations for H, P, R, and T in part 5. Put your pencil down when you're finished.
(Observe children and give feedback.)
(Answer key:)

Part 4	Part 5
H = 0	H = 0
P = 2	P = 0
R = 5	R = 1
T = 0	T = 4

EXERCISE 9: SOLVING PROBLEMS FROM NUMBER FAMILIES
MULTI-DIGIT NUMBERS

a. Find part 6. ✔
(Teacher reference:)

$$\underset{a.\ 213+84}{\overset{129\quad 84}{\underline{\qquad}}} \longrightarrow 213 \qquad \underset{}{\overset{84\quad 213}{\underline{\qquad}}} \longrightarrow 297$$
a. 213 + 84   c. 213 − 84
b. 297 − 213   d. 84 + 129

You're going to use number families to work problems.

- Touch and read problem A. (Signal.) *213 plus 84.*
- Does this problem show the big number? (Signal.) *No.*
- What are the small numbers? (Signal.) *213 and 84.*
- Find the family with the right numbers and complete equation A.
  (Observe children and give feedback.)
- Everybody, what does 213 plus 84 equal? (Signal.) *297.*

b. Touch and read problem B. (Signal.) *297 minus 213.*
- Find the family with the right numbers and complete equation B.
  (Observe children and give feedback.)
- Everybody, what does 297 minus 213 equal? (Signal.) *84.*

c. Touch and read problem C. (Signal.) *213 minus 84.*
- Find the family with the right numbers and complete equation C.
  (Observe children and give feedback.)
- Everybody, what does 213 minus 84 equal? (Signal.) *129.*

d. Touch and read problem D. (Signal.) *84 plus 129.*
- Find the family with the right numbers and complete equation D.
  (Observe children and give feedback.)
- Everybody, what does 84 plus 129 equal? (Signal.) *213.*

EXERCISE 10: INDEPENDENT WORK

a. Find part 7. ✔
(Teacher reference:)

You'll complete the place-value equation for each picture.

b. Turn to the other side of your worksheet and find part 8. ✔
(Teacher reference:)

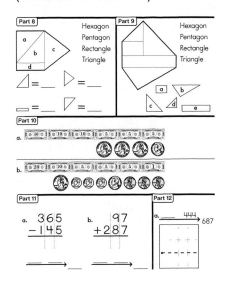

You'll circle the name for the whole figure. The names are hexagon, pentagon, rectangle, and triangle. Then you'll complete each equation by writing the letter to show which part it is.
In part 9, you'll circle the name for the whole shape. Then you'll write the letter for each part.
In part 10, you'll write the dollars and cents amount for each group of bills and coins.
In part 11, you'll work the problems and write the numbers in the family for each problem.
In part 12, you'll write a column problem to figure out the missing number in the family. Then you'll complete the family.

c. Complete worksheet 111.
(Observe children and mark incorrect responses on children's worksheets as you give feedback.)

Lesson 112

EXERCISE 1: FACT RELATIONSHIPS

3 PLUS

a. You've said facts for problems that plus 3. This time you'll say the plus facts that start with 3.
- Listen: 3 plus 1. Say the fact for 3 plus 1. (Signal.) *3 + 1 = 4.*
- Say the fact for 3 plus 2. (Signal.) *3 + 2 = 5.*

b. (Repeat the following task for numbers 3 through 10:)
- Say the fact for 3 plus __.

c. You'll start with 3 plus 6 and you'll say those facts again.
- Say the fact for 3 plus 6. (Signal.) *3 + 6 = 9.*
- Say the fact for 3 plus 7. (Signal.) *3 + 7 = 10.*
- Say the fact for 3 plus 8. (Signal.) *3 + 8 = 11.*
- Say the fact for 3 plus 9. (Signal.) *3 + 9 = 12.*
- Say the fact for 3 plus 10. (Signal.) *3 + 10 = 13.*
(Repeat step c until firm.)

d. Now I'll mix some of them up.
- Listen: 3 plus 6. Say the fact for 3 plus 6. (Signal.) *3 + 6 = 9.*
- Listen: 3 plus 8. Say the fact for 3 plus 8. (Signal.) *3 + 8 = 11.*
- Listen: 3 plus 5. Say the fact for 3 plus 5. (Signal.) *3 + 5 = 8.*
- Listen: 3 plus 9. Say the fact for 3 plus 9. (Signal.) *3 + 9 = 12.*
- Listen: 3 plus 7. Say the fact for 3 plus 7. (Signal.) *3 + 7 = 10.*

══ INDIVIDUAL TURNS ══

(Call on individual students to perform one of the following tasks.)

- Say the fact for 3 plus 8. (Call on a student.) *3 + 8 = 11.*
- Say the fact for 3 plus 5. (Call on a student.) *3 + 5 = 8.*
- Say the fact for 3 plus 7. (Call on a student.) *3 + 7 = 10.*

EXERCISE 2: SOLVING PROBLEMS FROM NUMBER FAMILIES

MISSING ADDEND, MINUEND, OR SUBTRAHEND

a. (Display:) [112:2A]

$$118 + \underline{\quad} = 197 \qquad 315 - \underline{\quad} = 118$$
$$\underline{\quad} - 118 = 197 \qquad \underline{\quad} + 79 = 197$$

These problems have a missing number. You're going to read these problems and say how many for the missing number. Then tell me if each value is a small number or the big number.
- (Point to $118 + \underline{\quad} = 197$.) Read this problem. (Signal.) *118 plus how many equals 197.*
- Is the missing number a small number or the big number? (Signal.) *A small number.*
- What is 118? (Signal.) *A small number.*
- What is 197? (Signal.) *The big number.*

b. (Point to $\underline{\quad} - 118 = 197$.) Read this problem. (Signal.) *How many minus 118 equals 197.*
- Is the missing number a small number or the big number? (Signal.) *The big number.*
- What's 118? (Signal.) *A small number.*
- What's 197? (Signal.) *A small number.*

c. (Point to $315 - \underline{\quad} = 118$.) Read this problem. (Signal.) *315 minus how many equals 118.*
- Is the missing number a small number or the big number? (Signal.) *A small number.*
- What's 315? (Signal.) *The big number.*
- What's 118? (Signal.) *A small number.*

d. (Point to $\underline{\quad} + 79 = 197$.) Read this problem. (Signal.) *How many plus 79 equals 197.*
- Is the missing number a small number or the big number? (Signal.) *A small number.*
- What's 79? (Signal.) *A small number.*
What's 197? (Signal.) *The big number.*
(Repeat problems that were not firm.)

e. (Add to show:) [112:2B]

a. $\underline{118 \quad 197}\rightarrow 315$	b. $\underline{118 \quad 79}\rightarrow 197$
118 + _____ = 197	315 − _____ = 118
_____ − 118 = 197	_____ + 79 = 197

Families A and B show the numbers for these problems. You're going to tell me if each number in the problems is a small number or the big number. Then you're going to find the family and tell me the missing number.

- (Point to 118 + ___ = 197.) This problem is 118 plus how many equals 197. Raise your hand when you can tell me about 118 and 197. ✔
- Is 118 a small number or the big number? (Signal.) *A small number.*
- What's 197? (Signal.) *The big number.*
- Look at families A and B. Find the family with a small number of 118 and the big number of 197. ✔
- What's the letter of the right family? (Signal.) *B.*
- Raise your hand when you know the missing number in the problem. ✔
- 118 plus how many equals 197? (Signal.) *79.* (Add to show:) [112:2C]

a. $\underline{118 \quad 197}\rightarrow 315$	b. $\underline{118 \quad 79}\rightarrow 197$
118 + __79__ = 197	315 − _____ = 118
_____ − 118 = 197	_____ + 79 = 197

f. (Point to ___ − 118 = 197.) This problem is how many minus 118 equals 197. Raise your hand when you can tell me about 118 and 197. ✔
- Is 118 a small number or the big number? (Signal.) *A small number.*
- What's 197? (Signal.) *A small number.*
- Look at families A and B. Find the family with small numbers of 118 and 197. ✔
- What's the letter of the right family? (Signal.) *A.*
- Raise your hand when you know the missing number in the problem. ✔

- How many minus 118 equals 197? (Signal.) *315.* (Add to show:) [112:2D]

a. $\underline{118 \quad 197}\rightarrow 315$	b. $\underline{118 \quad 79}\rightarrow 197$
118 + __79__ = 197	315 − _____ = 118
__315__ − 118 = 197	_____ + 79 = 197

g. (Point to 315 − ___ = 118.) This problem is 315 minus how many equals 118. Raise your hand when you can tell me about 315 and 118. ✔
- Is 315 a small number or the big number? (Signal.) *The big number.*
- What's 118? (Signal.) *A small number.*
- Look at families A and B. Find the family with the big number of 315 and a small number of 118. ✔
- What's the letter of the right family? (Signal.) *A.*
- Raise your hand when you know the missing number in the problem. ✔
- 315 minus how many equals 118? (Signal.) *197.* (Add to show:) [112:2E]

a. $\underline{118 \quad 197}\rightarrow 315$	b. $\underline{118 \quad 79}\rightarrow 197$
118 + __79__ = 197	315 − __197__ = 118
__315__ − 118 = 197	_____ + 79 = 197

h. (Point to ___ + 79 = 197.) This problem is how many plus 79 equals 197. Raise your hand when you can tell me about 79 and 197. ✔
- Is 79 a small number or the big number? (Signal.) *A small number.*
- What's 197? (Signal.) *The big number.*
- Look at families A and B. Find the family with a small number of 79 and the big number of 197. ✔
- What's the letter of the right family? (Signal.) *B.*
- Raise your hand when you know the missing number in the problem. ✔
- How many plus 79 equals 197? (Signal.) *118.* (Add to show:) [112:2F]

a. $\underline{118 \quad 197}\rightarrow 315$	b. $\underline{118 \quad 79}\rightarrow 197$
118 + __79__ = 197	315 − __197__ = 118
__315__ − 118 = 197	__118__ + 79 = 197

EXERCISE 3: MIXED COUNTING
EVEN NOMENCLATURE

a. Last lesson you learned about even numbers. What numbers do you count by to say even numbers? (Signal.) *Twos.*
- Everybody, start with 2 and say even numbers to 20. Get ready. (Tap.) *2, 4, 6, 8, 10, 12, 14, 16, 18, 20.*
- Everybody, start with 38 and say even numbers to 48. Get 38 going. *Thirty-eieieight.* Count. (Tap.) *40, 42, 44, 46, 48.*
b. Count by hundreds to 1000. Get ready. (Tap.) *100, 200, 300, 400, 500, 600, 700, 800, 900, 1000.*
c. Start with 60 and count by ones to 70. Get 60 going. *Sixtyyy.* Count. (Tap.) *61, 62, 63, 64, 65, 66, 67, 68, 69, 70.*
d. Start with 60 and count backward by ones to 50. Get 60 going. *Sixtyyy.* Count backward. (Tap.) *59, 58, 57, 56, 55, 54, 53, 52, 51, 50.*
e. Start with 60 and count by fives to 100. Get 60 going. *Sixtyyy.* Count. (Tap.) *65, 70, 75, 80, 85, 90, 95, 100.*
f. Start with 60 and count by tens to 100. Get 60 going. *Sixtyyy.* Count. (Tap.) *70, 80, 90, 100.*
g. Start with 60 and count backward by tens. Get 60 going. *Sixtyyy.* Count backward. (Tap.) *50, 40, 30, 20, 10.*
(Repeat steps a through g that were not firm.)
h. Listen: You have 60. When you count by tens, what number comes next? (Signal.) *70.*
- So what's 60 plus 10? (Signal.) *70.*
i. Listen: You have 60. When you count by twos, what number comes next? (Signal.) *62.*
- So what's 60 plus 2? (Signal.) *62.*
j. Listen: You have 60. When you count backward by tens, what number comes next? (Signal.) *50.*
- So what's 60 minus 10? (Signal.) *50.*
k. You have 60. When you count backward by ones, what number comes next? (Signal.) *59.*
- So what's 60 minus 1? (Signal.) *59.*
(Repeat steps h through k that were not firm.)

EXERCISE 4: NUMBER FAMILIES
MISSING NUMBER IN FAMILY   [REMEDY]

a. (Display:)   [112:4A]

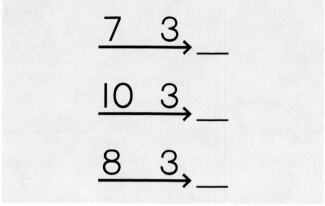

All of these families have a small number of 3. The big number is missing in these families.
- (Point to $\xrightarrow{7\quad 3}$.) What are the small numbers in this family? (Touch.) *7 and 3.*
- What's the big number? (Signal.) *10.*
- (Point to $\xrightarrow{10\quad 3}$.) What are the small numbers in this family? (Touch.) *10 and 3.*
- What's the big number? (Signal.) *13.*
- (Point to $\xrightarrow{8\quad 3}$.) What are the small numbers in this family? (Touch.) *8 and 3.*
- What's the big number? (Signal.) *11.*
(Repeat until firm.)
b. (Point to $\xrightarrow{7\quad 3}$.) What are the small numbers in this family? (Touch.) *7 and 3.*
- What's the big number? (Signal.) *10.*
- Say the fact that starts with 7. (Signal.) *7 + 3 = 10.*
- Say the fact that starts with the other small number. (Signal.) *3 + 7 = 10.*
- Say the fact that goes backward down the arrow. (Signal.) *10 – 3 = 7.*
- Say the other minus fact. (Signal.) *10 – 7 = 3.*
c. (Point to $\xrightarrow{10\quad 3}$.) What are the small numbers in this family? (Touch.) *10 and 3.*
- What's the big number? (Signal.) *13.*
- Say the fact that starts with 10. (Signal.) *10 + 3 = 13.*
- Say the fact that starts with the other small number. (Signal.) *3 + 10 = 13.*
- Say the fact that goes backward down the arrow. (Signal.) *13 – 3 = 10.*
- Say the other minus fact. (Signal.) *13 – 10 = 3.*

d. (Point to $\xrightarrow{8 \quad 3}$.) What are the small numbers in this family? (Touch.) *8 and 3.*
- What's the big number? (Signal.) *11.*
- Say the fact that starts with 8. (Signal.) *8 + 3 = 11.*
- Say the fact that starts with the other small number. (Signal.) *3 + 8 = 11.*
- Say the fact that goes backward down the arrow. (Signal.) *11 − 3 = 8.*
- Say the other minus fact. (Signal.) *11 − 8 = 3.*

e. (Display:) [112:4B]

$\xrightarrow{\quad 4 \quad} 8$	$\xrightarrow{7 \quad} 10$	$\xrightarrow{8 \quad 8} \_$
$\xrightarrow{6 \quad 4} \_$	$\xrightarrow{\quad 3 \quad} 7$	$\xrightarrow{\quad 8 \quad} 11$
$\xrightarrow{5 \quad} 8$	$\xrightarrow{6 \quad} 9$	$\xrightarrow{6 \quad} 11$

Each family has a missing number. You'll say the problem for each family. Then you'll say the answer.

f. (Point to $\xrightarrow{\quad 4 \quad} 8$.) Say the problem for the missing number. Get ready. (Signal.) *8 minus 4.*
- What's 8 minus 4? (Signal.) *4.*

g. (Point to $\xrightarrow{6 \quad 4} \_$.) Say the problem for the missing number. (Signal.) *6 plus 4.*
- What's 6 plus 4? (Signal.) *10.*

h. (Repeat the following tasks for the remaining families:)

(Point to __.) Say the problem for the missing number.	What's __?	
$\xrightarrow{5 \quad} 8$	$8 - 5$	$8 - 5$	3
$\xrightarrow{7 \quad} 10$	$10 - 7$	$10 - 7$	3
$\xrightarrow{\quad 3 \quad} 7$	$7 - 3$	$7 - 3$	4
$\xrightarrow{6 \quad} 9$	$9 - 6$	$9 - 6$	3
$\xrightarrow{8 \quad 8} \_$	$8 + 8$	$8 + 8$	16
$\xrightarrow{\quad 8 \quad} 11$	$11 - 8$	$11 - 8$	3
$\xrightarrow{6 \quad} 11$	$11 - 6$	$11 - 6$	5

(Repeat for families that were not firm.)

EXERCISE 5: TELLING TIME
ANALOG HOURS AND MINUTES

a. (Distribute unopened workbooks to children.)
- Open your workbook to Lesson 112 and find part 1.
  (Observe children and give feedback.)
  (Teacher reference:)

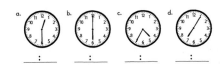

For these clocks you're going to make an arrow to show the direction the hour hand is moving. Then you'll write the time for each clock.
- For each clock, make an arrow to show the direction the hour hand is moving. Then write the hour number for each clock. Put your pencil down when you've made arrows for all of the hour hands and written the hour number for each clock.
  (Observe children and give feedback.)

b. Check your work. You'll read the hour number for each clock.
- Clock A. (Signal.) *12.*
- (Repeat for:) B, *6;* C, *4;* D, *7.*
  (Display:) [112:5A]

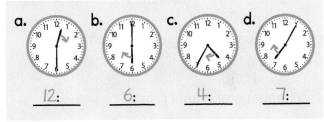

Here's what you should have for the clocks in part 1.

c. Now you'll figure out the minute number for each clock.
- How many minutes do you count for each number on a clock? (Signal.) *5.*
- Touch each number on clock A and count to yourself. Raise your hand when you know the minute number for clock A.
  (Observe children and give feedback.)
- What's the minute number for clock A? (Signal.) *30.*

d. Look at clock B and get ready to tell me the minute number.
- What's the minute number for clock B? (Signal.) *Zero.*

e. Touch each number on clock C and count to yourself. Raise your hand when you know the minute number for clock C.
(Observe children and give feedback.)
• What's the minute number for clock C? (Signal.) *35.*
f. Look at clock D and get ready to tell me the minute number.
• What's the minute number for clock D? (Signal.) *5.*
g. Write the minutes for the clocks in part 1.
(Observe children and give feedback.)
h. Check your work. You'll read the time for each clock.
• Clock A. (Signal.) *Twelve 30.*
• (Repeat for:) B, *6 o'clock;* C, *Four 35;* D, *Seven oh 5.*
(Display:)   [112:5B]

a.	b.	c.	d.
12:30	6:00	4:35	7:05

Here's what you should have for the clocks.
i. Listen: When the minute hand is straight up, how many minutes does the clock show? (Signal.) *Zero.*
• When the minute hand is straight down, how many minutes does the clock show? (Signal.) *30.*
(Repeat step i until firm.)

EXERCISE 6: 3-D OBJECTS
DECOMPOSITION   REMEDY

a. You've learned a name for objects with side faces that are triangles, a top that comes to a point, and a bottom face that can be any shape. What's the name for these objects? (Signal.) *Pyramids.*
(Display:)   [112:6A]

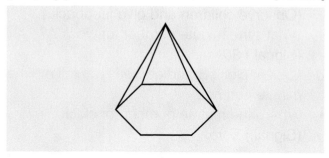

Here's an object.

• Is this object a pyramid? (Signal.) *Yes.* You're going to figure out the number of faces for each shape.
• (Point to bottom face.) I'll touch the sides for the bottom face. Count them to yourself and figure out the shape of the bottom face and the number of side faces.
• (Touch each side of bottom face.) What's the shape of the bottom face? (Touch.) *(A) hexagon.*
• How many sides does the bottom face have? (Signal.) *6.*
• So how many side faces does this object have? (Signal.) *6.*
• What shape are the side faces? (Signal.) *Triangle(s).*
• Does this object have a top face? (Signal.) *No.*
(Repeat until firm.)

b. Find part 2 on worksheet 112. ✔
(Teacher reference:)   R|Part I|

Part 2	Part 3
H =	H =
P =	P =
R =	R =
T =	T =

(Point to pyramid.) You're going to complete each equation.
• How many faces of this pyramid are hexagons? (Touch.) *1.*
• (Point to pyramid.) How many faces are pentagons? (Touch.) *Zero.*
• (Point to pyramid.) How many faces are rectangles? (Touch.) *Zero.*
• (Point to pyramid.) How many faces are triangles? (Touch.) *6.*
• Complete the equations for H, P, R, and T. (Observe children and give feedback.)

c. (Display:) [112:6B]

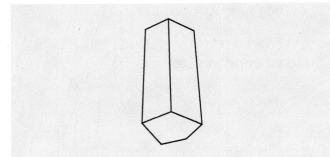

You're going to complete the equations for hexagons, pentagons, rectangles, and triangles for this object in part 3.

• I'll touch the sides for the bottom face. Count them to yourself and figure out the shape of the bottom face and the number of side faces. (Touch each side of bottom face.)

• Look at this object and figure out if there is a top face and what shape it is.

d. Now you'll tell me the number you'll write after each letter.

• (Point to prism.) How many faces of this object are hexagons? (Touch.) *Zero.*

• (Point to prism.) How many faces are pentagons? (Touch.) *2.*

• (Point to prism.) How many faces are rectangles? (Signal.) *5.*

• (Point to prism.) How many faces are triangles? (Signal.) *Zero.*

e. Complete the equations for H, P, R, and T in part 3. Put your pencil down when you're finished.
(Observe children and give feedback.)
(Answer key:)

Part 2	Part 3
H = 1	H = 0
P = 0	P = 2
R = 0	R = 5
T = 6	T = 0

EXERCISE 7: FACTS
PLUS/MINUS MIX

a. Find part 4. ✔
(Teacher reference:)

a. 10 – 4	e. 5 + 6
b. 14 – 7	f. 8 – 6
c. 7 + 3	g. 12 – 6
d. 14 – 10	h. 8 + 3
	i. 8 + 8

• Touch and read problem A. Get ready. (Signal.) *10 minus 4.*

• Is the answer the big number or a small number? (Signal.) *A small number.*

• What's 10 minus 4? (Signal.) *6.*

b. (Repeat the following tasks for problems B through I:)

Touch and read problem __.		Is the answer the big number or a small number?	What's __?	
B	14 – 7	A small number.	14 – 7	7
C	7 + 3	The big number.	7 + 3	10
D	14 – 10	A small number.	14 – 10	4
E	5 + 6	The big number.	5 + 6	11
F	8 – 6	A small number.	8 – 6	2
G	12 – 6	A small number.	12 – 6	6
H	8 + 3	The big number.	8 + 3	11
I	8 + 8	The big number.	8 + 8	16

(Repeat problems that were not firm.)

c. Complete each of the facts. Put your pencil down when you've written all of the equations for part 4.
(Observe children and give feedback.)

d. Check your work.

• Touch and read the equation for problem A. (Signal.) *10 – 4 = 6.*

• (Repeat for:) B, *14 – 7 = 7;* C, *7 + 3 = 10;* D, *14 – 10 = 4;* E, *5 + 6 = 11;* F, *8 – 6 = 2;* G, *12 – 6 = 6;* H, *8 + 3 = 11;* I, *8 + 8 = 16.*

EXERCISE 8: SOLVING PROBLEMS FROM NUMBER FAMILIES
MULTI-DIGIT NUMBERS

a. Find part 5. ✔
(Teacher reference:)

$$\underset{a.\ 387 - 278}{\underset{b.\ 109 + 278}{\overset{278\ \ 109}{\longrightarrow}387}} \qquad \underset{c.\ 387 - 209}{\underset{d.\ 387 - 178}{\overset{178\ \ 209}{\longrightarrow}387}}$$

You're going to use number families to work problems.

- Touch and read problem A. (Signal.)
  387 minus 278.
- Does this problem show the big number?
  (Signal.) *Yes.*
- What's the big number? (Signal.) *387.*
- What is the small number shown? (Signal.) *278.*
- Find the family with the right numbers and complete equation A.
  (Observe children and give feedback.)
- Everybody, what does 387 minus 278 equal?
  (Signal.) *109.*

b. Touch and read problem B. (Signal.)
  109 plus 278.
- Find the family with the right numbers and complete equation B.
  (Observe children and give feedback.)
- Everybody, what does 109 plus 278 equal?
  (Signal.) *387.*

c. Touch and read problem C. (Signal.) *387 minus 209.*
- Find the family with the right numbers and complete equation C.
  (Observe children and give feedback.)
- Everybody, what does 387 minus 209 equal?
  (Signal.) *178.*

d. Touch and read problem D. (Signal.)
  387 minus 178.
- Find the family with the right numbers and complete equation D.
  (Observe children and give feedback.)
- Everybody, what does 387 minus 178 equal?
  (Signal.) *209.*

EXERCISE 9: PLACE VALUE (TWO-DIGIT NUMBERS)
DRAWINGS AND EXPLANATIONS

a. Find part 6. ✔
(Teacher reference:)

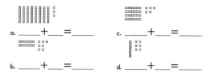

These pictures show tens columns or rows and single squares.

- Write the place-value equation for each picture. Put your pencil down when you've completed the place-value addition for all of the pictures in part 6.
  (Observe children and give feedback.)

b. Check your work. You'll read the place-value equations for each problem.
- Equation A. (Signal.) *80 + 3 = 83.*
- (Repeat for:) B, *20 + 7 = 27;* C, *40 + 5 = 45;* D, *10 + 6 = 16.*

c. Touch the picture for problem A again. ✔
- How many columns of 10 does that picture show? (Signal.) *8.*
- So what number did you write for the tens? (Signal.) *80.*
- How many single squares does that picture show? (Signal.) *3.*
- So what number did you write for the ones? (Signal.) *3.*
- What does 8 tens and 3 ones equal? (Signal.) *83.*
  (Repeat until firm.)

d. Listen: Why did you write 80 plus 3 equals 83 for equation A? (Call on a student. Ideas: *There are 8 tens so I wrote 80. There are 3 single squares so I wrote 3. 80 + 3 = 83, so I wrote 83 as the answer.*)

e. Touch the picture for problem C. ✔
- How many rows of 10 does that picture show? (Signal.) *4.*
- So what number did you write for the tens? (Signal.) *40.*
- How many single squares does that picture show? (Signal.) *5.*
- So what number did you write for the ones? (Signal.) *5.*
- What does 4 tens and 5 ones equal? (Signal.) *45.*
  (Repeat until firm.)

f. Listen: Why did you write 40 plus 5 equals
    45 for equation C? (Call on a student. Ideas:
    *There are 4 tens so I wrote 40. There are
    5 single squares so I wrote 5. 40 + 5 = 45,
    so I wrote 45 as the answer.*)

EXERCISE 10: INDEPENDENT WORK

a. Turn to the other side of your worksheet and
    find part 7. ✔
    (Teacher reference:)

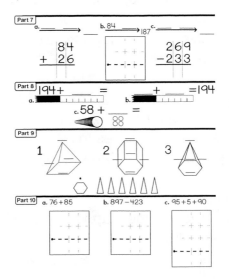

For each problem in part 7, you'll work the
column problem and complete the family.
In part 8, you'll complete the equations for the
rulers and the equation for the circles.

b. Find part 9. ✔
•   Touch object 1. ✔
•   Touch object 2. ✔
•   Touch object 3. ✔
    You're going to write letters for all of the faces
    for objects 1, 2, and 3.
•   Touch the picture of the faces for the bottom,
    top, and sides. ✔
    After you label the faces, you'll circle the object
    that has the right bottom, top, and sides.
    In part 10, you'll write each problem in a
    column and work it.

c. Complete worksheet 112.
    (Observe children and mark incorrect responses
    on children's worksheets as you give feedback.)

Lesson

EXERCISE 1: FACT RELATIONSHIPS

MINUS 3   REMEDY

a. You'll say facts that minus 3.
- Listen: 4 minus 3. Say the fact for 4 minus 3. (Signal.) *4 – 3 = 1.*
- Say the fact for 5 minus 3. (Signal.) *5 – 3 = 2.*
- Say the fact for 6 minus 3. (Signal.) *6 – 3 = 3.*
- Say the fact for 7 minus 3. (Signal.) *7 – 3 = 4.*
- Say the fact for 8 minus 3. (Signal.) *8 – 3 = 5.*
- Say the fact for 9 minus 3. (Signal.) *9 – 3 = 6.*
- Say the fact for 10 minus 3. (Signal.) *10 – 3 = 7.*
- Say the fact for 11 minus 3. (Signal.) *11 – 3 = 8.*
- Say the fact for 12 minus 3. (Signal.) *12 – 3 = 9.*
- Say the fact for 13 minus 3. (Signal.) *13 – 3 = 10.*

b. You'll start with 9 minus 3 and you'll say those facts again.
- Say the fact for 9 minus 3. (Signal.) *9 – 3 = 6.*
- Say the fact for 10 minus 3. (Signal.) *10 – 3 = 7.*
- Say the fact for 11 minus 3. (Signal.) *11 – 3 = 8.*
- Say the fact for 12 minus 3. (Signal.) *12 – 3 = 9.*
- Say the fact for 13 minus 3. (Signal.) *13 – 3 = 10.*
(Repeat step b until firm.)

c. Now I'll mix some of them up.
- Listen: 11 minus 3. Say the fact for 11 minus 3. (Signal.) *11 – 3 = 8.*
- Listen: 12 minus 3. Say the fact for 12 minus 3. (Signal.) *12 – 3 = 9.*
- Listen: 10 minus 3. Say the fact for 10 minus 3. (Signal.) *10 – 3 = 7.*

━━━━━━━ **INDIVIDUAL TURNS** ━━━━━━━

(Call on individual students to perform one of the following tasks.)

- Say the fact for 11 minus 3. (Call on a student.) *11 – 3 = 8.*
- Say the fact for 12 minus 3. (Call on a student.) *12 – 3 = 9.*
- Say the fact for 10 minus 3. (Call on a student.) *10 – 3 = 7.*

EXERCISE 2: MIXED COUNTING

EXPLAIN ADDING OR SUBTRACTING 10

a. Start with 2 and say even numbers to 20. Get ready. (Tap.) *2, 4, 6, 8, 10, 12, 14, 16, 18, 20.*
b. Count by hundreds to 1000. Get ready. (Tap.) *100, 200, 300, 400, 500, 600, 700, 800, 900, 1000.*
c. Start with 25 and count by ones to 35. Get 25 going. *Twenty-fiiive.* Count. (Tap.) *26, 27, 28, 29, 30, 31, 32, 33, 34, 35.*
d. Start with 25 and count backward by ones to 15. Get 25 going. *Twenty-fiiive.* Count backward. (Tap.) *24, 23, 22, 21, 20, 19, 18, 17, 16, 15.*
e. Start with 25 and count by fives to 50. Get 25 going. *Twenty-fiiive.* Count. (Tap.) *30, 35, 40, 45, 50.*
f. Start with 25 and count by 25s to 100. Get 25 going. *Twenty-fiiive.* Count. (Tap.) *50, 75, 100.*
g. Start with 25 and plus tens to 95. Get 25 going. *Twenty-fiiive.* Plus tens. (Tap.) *35, 45, 55, 65, 75, 85, 95.*
(Repeat steps a through g that were not firm.)

h. Listen: You have 25. When you count by tens, what number comes next? (Signal.) *35.*
- So what's 25 plus 10? (Signal.) *35.*
i. Listen: You have 25. When you count by 25s, what number comes next? (Signal.) *50.*
- So what's 25 plus 25? (Signal.) *50.*
j. Listen: You have 25. When you count backward by ones, what number comes next? (Signal.) *24.*
- So what's 25 minus 1? (Signal.) *24.*
k. You have 25. When you count by fives, what number comes next? (Signal.) *30.*
- So what's 25 plus 5? (Signal.) *30.*
(Repeat steps h through k that were not firm.)
l. Listen: If you have 25, how many tens do you have? (Signal.) *2.*
- How many ones do you have? (Signal.) *5.*
- So if you add 10 to 25, how many tens do you end up with? (Signal.) *3.*
- So what's 25 plus 10? (Signal.) *35.*
- How do you know that 25 plus 10 equals 35? (Call on a student. Idea: *25 is 2 tens and 5 ones. When you add a ten, you end up with 3 tens and 5 ones which is 35.*)

m. Listen: If you have 25, and subtract 10, how many tens do you end up with? (Signal.) *1.*
• How many ones do you end up with? (Signal.) *5.*
• So what's 25 minus 10? (Signal.) *15.*
• How do you know that 25 minus 10 equals 15? (Call on a student. Idea: *25 is 2 tens and 5 ones. When you subtract a ten, you end up with 1 ten and 5 ones which is 15.*)

EXERCISE 3: NUMBER FAMILIES
MISSING NUMBER IN FAMILY

a. (Display:) [113:3A]

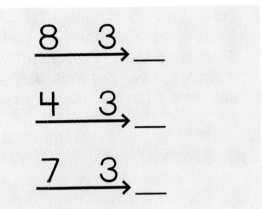

All of these families have a small number of 3. The big number is missing in these families.
• (Point to 8 3→.) What are the small numbers in this family? (Touch.) *8 and 3.*
• What's the big number? (Signal.) *11.*
• (Point to 4 3→.) What are the small numbers in this family? (Touch.) *4 and 3.*
• What's the big number? (Signal.) *7.*
• (Point to 7 3→.) What are the small numbers in this family? (Touch.) *7 and 3.*
• What's the big number? (Signal.) *10.*
  (Repeat until firm.)
b. (Point to 8 3→.) What are the small numbers in this family? (Touch.) *8 and 3.*
• What's the big number? (Signal.) *11.*
• Say the fact that starts with 8. (Signal.) *8 + 3 = 11.*
• Say the fact that starts with the other small number. (Signal.) *3 + 8 = 11.*
• Say the fact that goes backward down the arrow. (Signal.) *11 – 3 = 8.*
• Say the other minus fact. (Signal.) *11 – 8 = 3.*

c. (Point to 4 3→.) What are the small numbers in this family? (Touch.) *4 and 3.*
• What's the big number? (Signal.) *7.*
• Say the fact that starts with 4. (Signal.) *4 + 3 = 7.*
• Say the fact that starts with the other small number. (Signal.) *3 + 4 = 7.*
• Say the fact that goes backward down the arrow. (Signal.) *7 – 3 = 4.*
• Say the other minus fact. (Signal.) *7 – 4 = 3.*
d. (Point to 7 3→.) What are the small numbers in this family? (Touch.) *7 and 3.*
• What's the big number? (Signal.) *10.*
• Say the fact that starts with 7. (Signal.) *7 + 3 = 10.*
• Say the fact that starts with the other small number. (Signal.) *3 + 7 = 10.*
• Say the fact that goes backward down the arrow. (Signal.) *10 – 3 = 7.*
• Say the other minus fact. (Signal.) *10 – 7 = 3.*
e. (Display:) [113:3B]

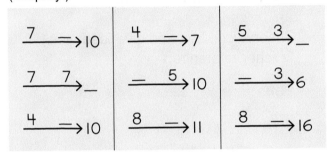

Each family has a missing number. You'll say the problem for each family. Then you'll say the answer.
f. (Point to 7 →10.) Say the problem for the missing number. Get ready. (Signal.) *10 minus 7.*
• What's 10 minus 7? (Signal.) *3.*
g. (Point to 7 7→_.) Say the problem for the missing number. (Signal.) *7 plus 7.*
• What's 7 plus 7? (Signal.) *14.*

h. (Repeat the following tasks for the remaining families:)

(Point to __.) Say the problem for the missing number.		What's __?	
$\xrightarrow{4} 10$	$10 - 4$	$10 - 4$	6
$\xrightarrow{4} 7$	$7 - 4$	$7 - 4$	3
$\xrightarrow{5} 10$	$10 - 5$	$10 - 5$	5
$\xrightarrow{8} 11$	$11 - 8$	$11 - 8$	3
$\xrightarrow{5 \quad 3}$	$5 + 3$	$5 + 3$	8
$\xrightarrow{3} 6$	$6 - 3$	$6 - 3$	3
$\xrightarrow{8} 16$	$16 - 8$	$16 - 8$	8

(Repeat for families that were not firm.)

EXERCISE 4: PLACE VALUE (TWO-DIGIT NUMBERS)
DRAWINGS AND EXPLANATIONS

a. (Distribute unopened workbooks to children.)
- Open your workbook to Lesson 113 and find part 1.
  (Observe children and give feedback.)
  (Teacher reference:)

  a. ___ + ___ = ___   c. ___ + ___ = ___
  b. ___ + ___ = ___   d. ___ + ___ = ___

  These pictures show tens columns or rows and single squares.
- Write the place-value equation for each picture. Put your pencil down when you've completed the place-value addition for all of the pictures in part 1.
  (Observe children and give feedback.)
b. Check your work. You'll read the place-value equation for each problem.
- Equation A. (Signal.) $60 + 8 = 68$.
- (Repeat for:) B, $50 + 4 = 54$; C, $30 + 9 = 39$; D, $80 + 0 = 80$.
c. Touch the picture for problem A again. ✔
- How many rows of 10 does that picture show? (Signal.) 6.
- So what number did you write for the tens? (Signal.) 60.
- How many single squares does that picture show? (Signal.) 8.
- So what number did you write for the ones? (Signal.) 8.
- So what does 6 tens and 8 ones equal? (Signal.) 68.
  (Repeat until firm.)

d. Listen: Why did you write 60 plus 8 equals 68 for equation A? (Call on a student. Ideas: *There are 6 tens so I wrote 60. There are 8 single squares so I wrote 8. 60 + 8 = 68, so I wrote 68 as the answer.*)
e. Touch the picture for problem B. ✔
- How many columns of 10 does that picture show? (Signal.) 5.
- How many single squares does that picture show? (Signal.) 4.
- What does 5 tens and 4 ones equal? (Signal.) 54.
  (Repeat until firm.)
f. Listen: Why did you write 50 plus 4 equals 54 for equation B? (Call on a student. Ideas: *There are 5 tens so I wrote 50. There are 4 single squares so I wrote 4. 50 + 4 = 54, so I wrote 54 as the answer.*)

EXERCISE 5: SOLVING PROBLEMS FROM NUMBER FAMILIES
MISSING ADDEND, MINUEND, OR SUBTRAHEND [REMEDY]

a. Find part 2. ✔
  (Teacher reference:)   [R][Part E]

  a. $243 - ___ = 89$ c. $___ - 65 = 178$
  b. $___ + 89 = 154$ d. $243 - 178 = ___$

  These problems have a missing number. You're going to read these problems, and tell me if each value is a small number or the big number.
- Read problem A. (Signal.) *243 minus how many equals 89.*
- Is the missing number a small number or the big number? (Signal.) *A small number.*
- What's 243? (Signal.) *The big number.*
- What's 89? (Signal.) *A small number.*
b. Read problem B. (Signal.) *How many plus 89 equals 154.*
- Is the missing number a small number or the big number? (Signal.) *A small number.*
- What's 89? (Signal.) *A small number.*
- What's 154? (Signal.) *The big number.*

c. Read problem C. (Signal.) *How many minus 65 equals 178.*
- Is the missing number a small number or the big number? (Signal.) *The big number.*
- What's 65? (Signal.) *A small number.*
- What's 178? (Signal.) *A small number.*
d. Read problem D. (Signal.) *243 minus 178 equals how many.*
- Is the missing number a small number or the big number? (Signal.) *A small number.*
- What's 243? (Signal.) *The big number.* What's 178? (Signal.) *A small number.* (Repeat problems that were not firm.)
e. (Display:)   [113:5A]

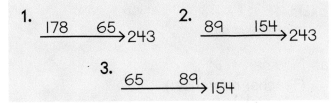

1. 178   65 →243   2. 89   154 →243

3. 65   89 →154

Here are the families for these numbers. You're going to figure out the missing number by finding the right number family.
- Go back to problem A. Raise your hand when you can tell me about 243 and 89. ✔
- Is 243 a small number or the big number? (Signal.) *The big number.*
- What's 89? (Signal.) *A small number.*
- Look at families 1, 2, and 3. Find the family with the big number of 243 and a small number of 89. ✔
- What's the number of the right family? (Signal.) *2.*
- Raise your hand when you know the missing number in problem A. ✔
- 243 minus how many equals 89? (Signal.) *154.*
f. Problem B. Raise your hand when you can tell me about 89 and 154. ✔
- Is 89 a small number or the big number? (Signal.) *A small number.*
- What's 154? (Signal.) *The big number.*
- Look at families 1, 2, and 3. Find the family with a small number of 89 and the big number of 154. ✔
- What's the number of the right family? (Signal.) *3.*
- Raise your hand when you know the missing number in problem B. ✔
- How many plus 89 equals 154? (Signal.) *65.*

g. Problem C. Look at families 1, 2, and 3 and find the family with the right numbers. ✔
- What's the number of the right family for problem C? (Signal.) *1.*
  In problem C, 65 and 178 are the small numbers. Family 1 has small numbers of 65 and 178. Raise your hand when you know the missing number in problem C. ✔
- How many minus 65 equals 178? (Signal.) *243.*
h. Problem D. Look at families 1, 2, and 3 and find the family with the right numbers. ✔
- What's the number of the right family for problem D? (Signal.) *1.*
  In problem D, 243 is the big number and 178 is a small number. Family 1 has a big number of 243 and a small number of 178.
- Raise your hand when you know the missing number in problem D. ✔
- 243 minus 178 equals how many? (Signal.) *65.* (Repeat problems that were not firm.)
i. Complete the equations in part 2. Put your pencil down when you've worked all of the problems in part 2.
  (Observe children and give feedback.)
- Check your work. You'll read each equation.
- Read equation A. (Signal.) *243 – 154 = 89.*
- (Repeat for:) B, *65 + 89 = 154;* C, *243 – 65 = 178;* D, *243 – 178 = 65.*

EXERCISE 6: TELLING TIME
ANALOG HOURS AND HALF HOURS   REMEDY

a. Think about the minute hand on clocks.
- Listen: When the minute hand is straight up, how many minutes does the clock show? (Signal.) *Zero.*
- When the minute hand is straight down, how many minutes does the clock show? (Signal.) *30.*
- For each number on a clock, how many minutes do you count? (Signal.) *5.* (Repeat until firm.)

b. (Display:) [113:6A]

The minute hand for each of these clocks shows zero minutes or 30 minutes. You're going to tell me the hour number and the minute number for each clock. Then you'll tell me the time the clock shows.

- (Point to **5:00**.) What's the hour number for this clock? (Signal.) *5*.
- What's the minute number? (Signal.) *Zero*.
- Say the time this clock shows. (Signal.) *5 o'clock*.

c. (Repeat the following tasks for the remaining clocks:)

(Point to __.) What's the hour number for this clock?	What's the minute number?	Say the time this clock shows.	
3:30	3	*Thirty*	*3:30*
10:30	10	*Thirty*	*10:30*
8:00	8	*Zero*	*8 o'clock*
7:30	7	*Thirty*	*7:30*
12:00	12	*Zero*	*12 o'clock*

(Repeat clocks that were not firm.)

d. Find part 3 on worksheet 113. ✔
(Teacher reference:) R Part 0

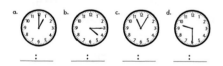

You're going to write the time for each clock. One clock shows zero minutes. Another clock shows 30 minutes. You'll count the minutes to yourself for the other two clocks.

- Write the time for each clock. Put your pencil down when you've written the hour number and the minute number for each clock in part 3.
(Observe children and give feedback.)

e. Check your work. You'll read the time for each clock.
- Clock A. (Signal.) *One o'clock*.
- (Repeat for:) B, *Four 15*; C, *Eleven oh 5*; D, *Nine 30*.
(Display:) [113:6B]

Here's what you should have for the clocks.

EXERCISE 7: FACTS
PLUS/MINUS MIX

a. Find part 4. ✔
(Teacher reference:)

a. $3 + 8$ e. $11 - 2$
b. $7 - 5$ f. $3 + 2$
c. $11 - 5$ g. $8 - 5$
d. $3 + 4$ h. $3 + 7$

These are addition and subtraction problems you'll work. Think of the number families for each problem and write the answers. Put your pencil down when you've completed the equations in part 4.
(Observe children and give feedback.)

b. Check your work. You'll read the equation for each problem.
- Problem A. (Signal.) $3 + 8 = 11$.
- (Repeat for:) B, $7 - 5 = 2$; C, $11 - 5 = 6$; D, $3 + 4 = 7$; E, $11 - 2 = 9$; F, $3 + 2 = 5$; G, $8 - 5 = 3$; H, $3 + 7 = 10$.

EXERCISE 8: 3-D OBJECTS
DECOMPOSITION (CUBE)

a. (Display:)   [113:8A]

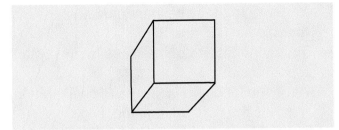

- This object is **a cube.** What's the name of this object? (Signal.) *(A) cube.*
- All of the faces of a cube have the same shape. What shape? (Signal.) *(A) square.*
- Does a cube have a top face and a bottom face? (Signal.) *Yes.*
- How many side faces does the cube have? (Signal.) *4.*
- A cube has a top face, a bottom face, and 4 side faces. How many faces does a cube have altogether? (Signal.) *6.*
- A cube has 6 faces. Say the sentence about how many faces a cube has. (Signal.) *A cube has 6 faces.*

b. (Display:)   [113:8B]

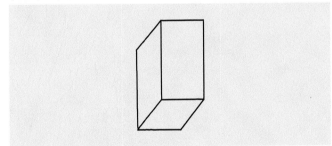

- This is not a cube. Each face is a rectangle. What shape is each face? (Signal.) *(A) rectangle.*
- How many faces are rectangles? (Signal.) *6.*
c. Find part 5. ✔
(Teacher reference:)

Part 5	Part 6
H =	H =
P =	P =
R =	R =
T =	T =

Write the number for hexagons, pentagons, rectangles, and triangles. Put your pencil down when you've completed all of the equations for part 5.
(Observe children and give feedback.)

d. Check your work.
- What number did you write for H? (Signal.) *Zero.*
- What number did you write for P? (Signal.) *Zero.*
- What number did you write for R? (Signal.) *6.*
- What number did you write for T? (Signal.) *Zero.*
e. (Display:)   [113:8C]

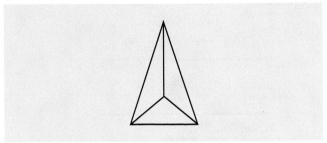

You're going to complete equations for hexagons, pentagons, rectangles, and triangles for this pyramid in part 6.
- What is the shape of the bottom face? (Signal.) *(A) triangle.*
- How many side faces does this pyramid have? (Signal.) *3.*
- What shape are the side faces? (Signal.) *Triangles.*
- Does this pyramid have a top face? (Signal.) *No.*
f. Write the number for hexagons, pentagons, rectangles, and triangles in part 6. Put your pencil down when you've completed all of the equations.
(Observe children and give feedback.)
(Answer key:)

Part 5	Part 6
H = 0	H = 0
P = 0	P = 0
R = 6	R = 0
T = 0	T = 4

g. Check your work.
- What number did you write for H? (Signal.) *Zero.*
- What number did you write for P? (Signal.) *Zero.*
- What number did you write for R? (Signal.) *Zero.*
- What number did you write for T? (Signal.) *4.*

EXERCISE 9: INDEPENDENT WORK

a. Turn to the other side of your worksheet and find part 7. ✔
(Teacher reference:)

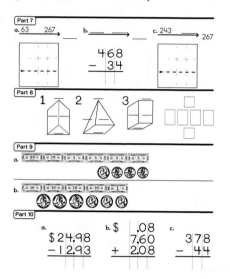

For each problem in part 7, you'll work a column problem and complete the family.

b. Find part 8. ✔
- Touch object 1. ✔
- Touch object 2. ✔
- Touch object 3. ✔
  You're going to write letters for all of the faces for objects 1, 2, and 3.
- Touch the picture of the faces for the bottom, top, and sides. ✔
  After you label the faces, you'll circle the object that has the right bottom, top, and sides.
  In part 9, you'll write the dollars and cents amount for each group of bills and coins.
  In part 10, you'll work the money problems and the column problem.

c. Complete worksheet 113.
(Observe children and mark incorrect responses on children's worksheets as you give feedback.)

Lesson 114

EXERCISE 1: FACT RELATIONSHIPS
MINUS 3

a. You'll say facts that minus 3.
- Listen: 4 minus 3. Say the fact for 4 minus 3.
  (Signal.) *4 – 3 = 1.*
- Say the fact for 5 minus 3. (Signal.) *5 – 3 = 2.*

b. (Repeat the following task for numbers
  6 through 13:)
- Say the fact for __ minus 3.

c. You'll start with 9 minus 3 and you'll say those
  facts again.
- Say the fact for 9 minus 3. (Signal.) *9 – 3 = 6.*
- Say the fact for 10 minus 3. (Signal.) *10 – 3 = 7.*
- Say the fact for 11 minus 3. (Signal.) *11 – 3 = 8.*
- Say the fact for 12 minus 3. (Signal.) *12 – 3 = 9.*
- Say the fact for 13 minus 3. (Signal.) *13 – 3 = 10.*
  (Repeat step c until firm.)

d. Now I'll mix some of them up.
- Listen: 12 minus 3. Say the fact for 12 minus 3.
  (Signal.) *12 – 3 = 9.*
- Say the fact for 10 minus 3. (Signal.) *10 – 3 = 7.*
- Say the fact for 11 minus 3. (Signal.) *11 – 3 = 8.*
- Say the fact for 9 minus 3. (Signal.) *9 – 3 = 6.*

INDIVIDUAL TURNS

(Call on individual students to perform one of
the following tasks.)

- Say the fact for 10 minus 3. (Call on a student.)
  10 – 3 = 7.
- Say the fact for 12 minus 3. (Call on a student.)
  12 – 3 = 9.
- Say the fact for 11 minus 3. (Call on a student.)
  11 – 3 = 8.

EXERCISE 2: 3-D OBJECTS
PRISM, CUBE

a. (Display:) [114:2A]

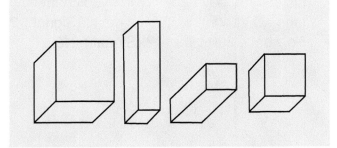

Some of these pictures are cubes.

- Does a cube have six faces? (Signal.) *Yes.*
- What shape is each of the faces of a cube?
  (Signal.) *(A) square.*
  (Repeat until firm.)

b. (Point at objects.) Do each of these objects
  have a bottom face and a top face?
  (Signal.) *Yes.*
- Raise your hand when you know how many
  side faces each of these objects has. ✔
- How many side faces does each object have?
  (Signal.) *4.*
- So each of these objects has a bottom face,
  a top face, and 4 side faces. How many total
  faces does each object have? (Signal.) *6.*

c. (Point to first cube.) This object has 6 faces. Is
  the shape of each face a square? (Signal.) *Yes.*
- So is this object a cube? (Signal.) *Yes.*

d. (Repeat the following tasks for the remaining
  objects:)

(Point to __.) Is the shape of each face a square?		So is this a cube?
First rectangular prism	*No*	*No*
Second rectangular prism	*No*	*No*
Second cube	*Yes*	*Yes*

(Repeat for objects that were not firm.)

e. (Display:) [114:2B]

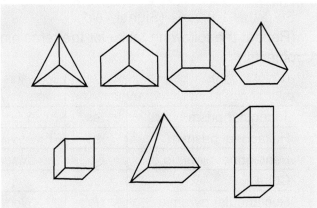

- (Point to △.) Is this a pyramid? (Touch.) *Yes.*
- How do you know? (Call on a student. Idea:
  The side faces are all triangles.)

f. (Repeat the following tasks for the remaining objects:)

(Point to __.) Is this a pyramid?		How do you know? (Call on a student. Idea:)
Triangular prism	No	The side faces are (rectangles/not triangles.)
Hexagonal prism	No	The side faces are (rectangles/not triangles.)
Pentagonal pyramid	Yes	The side faces are triangles.
Cube	No	The side faces are (rectangles/not triangles.)
Rectangular pyramid	Yes	The side faces are triangles.
Rectangular prism	No	The side faces are (rectangles/not triangles.)

(Repeat for objects that were not firm.)

g. (Point at objects.) For these objects if it's not a pyramid, it's called a prism. Say **prism.** (Signal.) *Prism.*

• What are these objects that are not pyramids called? (Signal.) *Prisms.*

h. Prisms have side faces that are rectangles. What shape are the side faces of prisms? (Signal.) *Rectangles.*

• (Point to △.) Does this object have side faces that are rectangles? (Signal.) *No.*

• So is this a prism? (Signal.) *No.*

i. (Repeat the following tasks for the remaining objects:)

(Point to __.) Does this object have side faces that are rectangles?	So is this a prism?
Triangular prism	Yes	Yes
Hexagonal prism	Yes	Yes
Pentagonal pyramid	No	No
Cube	Yes	Yes
Rectangular pyramid	No	No
Rectangular prism	Yes	Yes

(Repeat for objects that were not firm.)

j. Now you'll tell me if each object is a pyramid or a prism.

• (Point to △.) Is this object a pyramid or a prism? (Touch.) *(A) pyramid.*

k. (Repeat the following tasks for the remaining objects:)

(Point to __.) Is this object a pyramid or a prism?	
Triangular prism	(A) prism
Hexagonal prism	(A) prism
Pentagonal pyramid	(A) pyramid
Cube	(A) prism
Rectangular pyramid	(A) pyramid
Rectangular prism	(A) prism

(Repeat for objects that were not firm.)

l. Look at the objects that are prisms and get ready to tell me how many of the prisms are cubes. ✔

• How many of these prisms are cubes? (Signal.) *1.*

EXERCISE 3: MIXED COUNTING
EXPLAIN ADDING OR SUBTRACTING 10 `REMEDY`

a. Count by hundreds to 1000. Get ready. (Tap.) *100, 200, 300, 400, 500, 600, 700, 800, 900, 1000.*

b. Start with 2 and say even numbers to 20. Get ready. (Tap.) *2, 4, 6, 8, 10, 12, 14, 16, 18, 20.*

c. Start with 200 and count by tens to 250. Get 200 going. *Two huuundred.* Count. (Tap.) *210, 220, 230, 240, 250.*

d. Start with 200 and count backward by tens to 150. Get 200 going. *Two huuundred.* Count backward. (Tap.) *190, 180, 170, 160, 150.*

e. Start with 200 and count backward by ones to 190. Get 200 going. *Two huuundred.* Count backward. (Tap.) *199, 198, 197, 196, 195, 194, 193, 192, 191, 190.*

f. Start with 200 and count by fives to 250. Get 200 going. *Two huuundred.* Count. (Tap.) *205, 210, 215, 220, 225, 230, 235, 240, 245, 250.*

g. Start with 200 and count by 25s to 300. Get 200 going. *Two huuundred.* Count. (Tap.) *225, 250, 275, 300.*

h. Start with 200 and count by ones to 210. Get 200 going. *Two huuundred.* Count. (Tap.) *201, 202, 203, 204, 205, 206, 207, 208, 209, 210.* (Repeat steps a through h that were not firm.)

i. Listen: You have 200. When you count by tens, what number comes next? (Signal.) *210.*

• So what's 200 plus 10? (Signal.) *210.*

j. Listen: You have 200. When you count by 25s, what number comes next? (Signal.) *225.*
• So what's 200 plus 25? (Signal.) *225.*
k. Listen: You have 200. When you count backward by ones, what number comes next? (Signal.) *199.*
• So what's 200 minus 1? (Signal.) *199.*
l. You have 200. When you count by fives, what number comes next? (Signal.) *205.*
• So what's 200 plus 5? (Signal.) *205.*
  (Repeat steps i through l that were not firm.)
m. Listen: If you have 83, how many tens do you have? (Signal.) *8.*
• How many ones do you have? (Signal.) *3.*
• So if you subtract ten, how many tens do you end up with? (Signal.) *7*
• How many ones do you end up with? (Signal.) *3.*
• So what's 83 minus 10? (Signal.) *73.*
• How do you know that 83 minus 10 equals 73? (Call on a student. Idea: *83 is 8 tens and 3 ones. When you subtract ten, you end up with 7 tens and 3 ones, which is 73.*)
n. Listen: If you have 83, and add ten, how many tens do you end up with? (Signal.) *9.*
• How many ones do you end up with? (Signal.) *3.*
• So what's 83 plus 10? (Signal.) *93.*
• How do you know that 83 plus 10 equals 93? (Call on a student. Idea: *83 is 8 tens and 3 ones. When you add ten, you end up with 9 tens and 3 ones, which is 93.*)

EXERCISE 4: FACTS
ADDITION/SUBTRACTION

a. (Display:)   [114:4A]

5 + 5	16 − 8	7 + 3
11 − 8	10 − 6	6 + 5
4 + 3	8 − 4	14 − 7

• (Point to **5 + 5**.) Listen: 5 plus 5. Is the answer the big number or a small number? (Signal.) *The big number.*
• Say the fact. (Signal.) *5 + 5 = 10.*
b. (Point to **11 − 8**.) 11 minus 8. Is the answer the big number or a small number? (Signal.) *A small number.*
• Say the fact. (Signal.) *11 − 8 = 3.*

c. (Repeat the following tasks for the remaining problems:)

(Point to __.) __. Is the answer the big number or a small number?			Say the fact.
4 + 3	4 + 3	The big number.	4 + 3 = 7
16 − 8	16 − 8	A small number.	16 − 8 = 8
10 − 6	10 − 6	A small number.	10 − 6 = 4
8 − 4	8 − 4	A small number.	8 − 4 = 4
7 + 3	7 + 3	The big number.	7 + 3 = 10
6 + 5	6 + 5	The big number.	6 + 5 = 11
14 − 7	14 − 7	A small number.	14 − 7 = 7

(Repeat problems that were not firm.)

EXERCISE 5: EVEN AND ODD NUMBERS   [REMEDY]

a. You've learned the name for numbers you say when you count by twos. What name? (Signal.) *Even (numbers).*
• The numbers that are not even are called **odd numbers**. What do you call numbers that are not even? (Signal.) *Odd.*
• What do you call numbers that are not odd? (Signal.) *Even.*
  (Repeat until firm.)
b. I'll say numbers. You'll tell me if the number is even or odd.
• 34. What number? (Signal.) *34.*
• Is 34 a number you say when you count by twos? (Signal.) *Yes.*
• So is 34 even? (Signal.) *Yes.*
c. New number: 27. What number? (Signal.) *27.*
• Is 27 a number you say when you count by twos? (Signal.) *No.*
• So is 27 even? (Signal.) *No.*
• So what kind of number is 27? (Signal.) *Odd.*
d. New number: 11. What number? (Signal.) *11.*
• Is 11 a number you say when you count by twos? (Signal.) *No.*
• So is 11 even? (Signal.) *No.*
• So what kind of number is 11? (Signal.) *Odd.*
e. New number: 8. What number? (Signal.) *8.*
• Is 8 a number you say when you count by twos? (Signal.) *Yes.*
• So is 8 even? (Signal.) *Yes.*
f. New number: 5. What number? (Signal.) *5.*
• Is 5 a number you say when you count by twos? (Signal.) *No.*
• So is 5 even? (Signal.) *No.*
• So what kind of number is 5? (Signal.) *Odd.*
  (Repeat numbers that were not firm.)

EXERCISE 6: PLACE VALUE (TWO-DIGIT NUMBERS)
DRAWINGS AND EXPLANATIONS

a. (Distribute unopened workbooks to children.)
- Open your workbook to Lesson 114 and find part 1.
  (Observe children and give feedback.)
  (Teacher reference:)

These pictures show tens columns or rows and single squares.

- Write the place-value equation for each picture. Put your pencil down when you've completed the place-value addition for all of the pictures in part 1.
  (Observe children and give feedback.)

b. Check your work. You'll read the place-value equation for each problem.
- Equation A. (Signal.) *40 + 7 = 47.*
- (Repeat for:) B, *10 + 8 = 18;* C, *70 + 0 = 70;* D, *90 + 6 = 96.*

c. Touch the picture for problem B again. ✔
- How many columns of 10 does that picture show? (Signal.) *1.*
- How many single squares does that picture show? (Signal.) *8.*
- Why did you write 10 plus 8 equals 18 for equation B? (Call on a student. Ideas: *There is 1 ten so I wrote 10. There are 8 single squares, so I wrote 8. 10 + 8 = 18, so I wrote 18 as the answer.*)

d. Touch the picture for problem D. ✔
- How many rows of 10 does that picture show? (Signal.) *9.*
- How many single squares does that picture show? (Signal.) *6.*
- Why did you write 90 plus 6 equals 96 for equation D? (Call on a student. Ideas: *There are 9 tens so I wrote 90. There are 6 single squares, so I wrote 6. 90 + 6 = 96, so I wrote 96 as the answer.*)

EXERCISE 7: WORD PROBLEMS (COLUMNS)
CONVENTIONAL AND COMPARISON

a. Find part 2 on your worksheet. ✔
   (Teacher reference:)

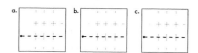

You're going to write the symbols for word problems in columns and work them. All of the problems you'll work today are money problems. You'll write the dollar signs and the dots for all problems.

- Touch where you'll write the symbols for problem A. ✔
- Listen to problem A: A box had 9 dollars and 26 cents in it. Someone put another 5 dollars and 56 cents in the box. How much money ended up in the box?
- Listen again and write the symbols for the whole problem. A box had 9 dollars and 26 cents in it. Someone put another 5 dollars and 56 cents in the box.
  (Observe children and give feedback.)
- Everybody, touch and read the problem for A. Get ready. (Signal.) *9 dollars and 26 cents plus 5 dollars and 56 cents.*

b. Work problem A. Put your pencil down when you know how much money ended up in the box.
  (Observe children and give feedback.)
  (Teacher reference:)

- Read the addition problem and the answer you wrote for A. Get ready. (Signal.) *9 dollars and 26 cents plus 5 dollars and 56 cents equals 14 dollars and 82 cents.*
- Everybody, how much money ended up in the box? (Signal.) *14 dollars and 82 cents.*

c. Touch where you'll write the symbols for problem B. ✔
You're going to tell me how much more money one person had.

- Listen to problem B: Jennifer had 18 dollars and 79 cents. Adam had 15 dollars and 33 cents. How much more money did Jennifer have than Adam?
- Listen to problem B again and write the problem for figuring out how much more money Jennifer had than Adam. Jennifer had 18 dollars and 79 cents. (Pause.) Adam had 15 dollars and 33 cents.
(Observe and give feedback.)
- Read the problem you wrote for figuring out how much more money Jennifer had. (Signal.) *18 dollars and 79 cents minus 15 dollars and 33 cents.*
- Work problem B. Put your pencil down when you know how much more money Jennifer had.
(Observe children and give feedback.)
(Teacher reference:)

b.
$$\begin{array}{r} \$18.79 \\ -15.33 \\ \hline \$\ 3.46 \end{array}$$

d. Check your work.
- Read the subtraction problem and the answer for B. Get ready. (Signal.) *18 dollars and 79 cents minus 15 dollars and 33 cents equals 3 dollars and 46 cents.*
- How much more money did Jennifer have? (Signal.) *3 dollars and 46 cents.*

e. Touch where you'll write the symbols for problem C. ✔
- Listen to problem C: Jerome had 7 dollars and 96 cents. Jerome spent 5 dollars and 91 cents of that money. How much money did Jerome end up with?
- Listen again and write the symbols for the whole problem. Jerome had 7 dollars and 96 cents. Jerome spent 5 dollars and 91 cents of that money.
(Observe children and give feedback.)
- Everybody, touch and read the problem for C. Get ready. (Signal.) *7 dollars and 96 cents minus 5 dollars and 91 cents.*
(Repeat until firm.)

f. Work problem C. Put your pencil down when you know how much money Jerome ended up with.
(Observe children and give feedback.)
(Teacher reference:)

c.
$$\begin{array}{r} \$\ 7.96 \\ -\ 5.91 \\ \hline \$\ 2.05 \end{array}$$

- Read the subtraction problem and the answer you wrote for C. Get ready. (Signal.) *7 dollars and 96 cents minus 5 dollars and 91 cents equals 2 dollars and 5 cents.*
- Everybody, how much money did Jerome end up with? (Signal.) *2 dollars and 5 cents.*

EXERCISE 8: TELLING TIME
ANALOG HOURS AND HALF HOURS

a. Think about the minute hand on clocks. Listen: When the minute hand is straight up, how many minutes does the clock show? (Signal.) *Zero.*
- When the minute hand is straight down, how many minutes does the clock show? (Signal.) *30.*
- For each number on a clock, how many minutes do you count? (Signal.) *5.*
(Repeat until firm.)

b. (Display:)   [114:7A]

The minute hand for each of these clocks shows zero minutes or 30 minutes. You're going to tell me the hour number and the minute number for each clock. Then you'll tell me the time the clock shows.

- (Point to **11:30**.) What's the hour number for this clock? (Signal.) *11.*
- What's the minute number? (Signal.) *30.*
- Say the time that this clock shows. (Signal.) *11 thirty.*

c. (Repeat the following tasks for the remaining clocks:)

(Point to __.) What's the hour number for this clock?	What's the minute number?	Say the time this clock shows.
2:00	2	Zero	2 o'clock
8:30	8	Thirty	8:30
3:30	3	Thirty	3:30
10:00	10	Zero	10 o'clock
6:30	6	Thirty	6:30

(Repeat clocks that were not firm.)

d. Find part 3 on worksheet 114. ✔
(Teacher reference:)

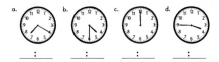

You're going to write the time for each clock. One clock shows zero minutes. Another clock shows 30 minutes. You'll count the minutes to yourself for the other two clocks.

• Write the time for each clock. Put your pencil down when you've written the hour number and the minute number for each clock in part 3.
(Observe children and give feedback.)

e. Check your work. You'll read the time for each clock.

⌈ • Clock A. (Signal.) *Seven 20.*
⌊ • (Repeat for:) B, *Four 30;* C, *12 o'clock;* D, *Three 45.*
(Display:)   [114:7B]

Here's what you should have for the clocks.

EXERCISE 9: FACTS
ADDITION

a. Find part 4 on your worksheet. ✔
(Teacher reference:)

c. 3 + 5   f. 8 + 3
a. 3 + 7   d. 3 + 2   g. 3 + 3
b. 7 + 7   e. 6 + 3   h. 2 + 8

• These are addition problems you'll work. Think of the number families for each problem and write the answers. Put your pencil down when you've completed the equations in part 4.
(Observe children and give feedback.)

b. Check your work. You'll read the equation for each problem.

⌈ • Problem A. (Signal.) *3 + 7 = 10.*
⌊ • (Repeat for:) B, *7 + 7 = 14;* C, *3 + 5 = 8;* D, *3 + 2 = 5;* E, *6 + 3 = 9;* F, *8 + 3 = 11;* G, *3 + 3 = 6;* H, *2 + 8 = 10.*

EXERCISE 10: SOLVING PROBLEMS FROM NUMBER FAMILIES
MISSING ADDEND, MINUEND, OR SUBTRAHEND   REMEDY

a. Turn to the other side of your worksheet and find part 5. ✔
(Teacher reference:)   R Part F

a. $184 - \_\_\_ = 76$ c. $\_\_\_ - 184 = 76$
b. $184 + \_\_\_ = 292$ d. $184 + 108 = \_\_\_$

These problems have a number missing. You're going to read these problems, and tell me if each value is a small number or the big number.

• Read problem A. (Signal.) *184 minus how many equals 76.*
• Is the missing number a small number or the big number? (Signal.) *A small number.*
• What's 184? (Signal.) *The big number.*
• What's 76? (Signal.) *A small number.*

b. Read problem B. (Signal.) *184 plus how many equals 292.*
• Is the missing number a small number or the big number? (Signal.) *A small number.*
• What's 184? (Signal.) *A small number.*
• What's 292? (Signal.) *The big number.*

c. Read problem C. (Signal.) *How many minus 184 equals 76.*
• Is the missing number a small number or the big number? (Signal.) *The big number.*
• What's 184? (Signal.) *A small number.*
• What's 76? (Signal.) *A small number.*

d. Read problem D. (Signal.) *184 plus 108 equals how many.*
• Is the missing number a small number or the big number? (Signal.) *The big number.*
• What's 184? (Signal.) *A small number.*
  What's 108? (Signal.) *A small number.*

e. (Display:) [114:9A]

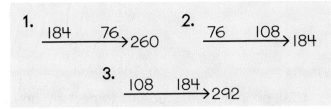

1. 184 76 →260
2. 76 108 →184
3. 108 184 →292

Here are families for these numbers. You're going to figure out the missing number by finding the right number family.

- Go back to problem A. Raise your hand when you can tell me about 184 and 76. ✔
- Is 184 a small number or the big number? (Signal.) *The big number.*
- What's 76? (Signal.) *A small number.*
- Look at families 1, 2, and 3. Find the family with the big number of 184 and a small number of 76. ✔
- What's the number of the right family? (Signal.) *2.*
- Raise your hand when you know the missing number in problem A. ✔
- 184 minus how many equals 76? (Signal.) *108.*

f. Problem B. Raise your hand when you can tell me about 184 and 292. ✔
- Is 184 a small number or the big number? (Signal.) *A small number.*
- What's 292? (Signal.) *The big number.*
- Look at families 1, 2, and 3. Find the family with a small number of 184 and the big number of 292. ✔
- What's the number of the right family? (Signal.) *3.*
- Raise your hand when you know the missing number in problem B. ✔
- 184 plus how many equals 292? (Signal.) *108.*

g. Problem C. Look at families 1, 2, and 3 and find the family with the right numbers. ✔
- What's the number of the right family for problem C? (Signal.) *1.*
  In problem C, 184 and 76 are the small numbers. Family 1 has small numbers of 184 and 76.
- Raise your hand when you know the missing number in problem C. ✔
- How many minus 184 equals 76? (Signal.) *260.*

h. Problem D. Look at families 1, 2, and 3 and find the family with the right numbers. ✔
- What's the number of the right family for problem D? (Signal.) *3.*
  In problem D, 184 and 108 are the small numbers. Family 3 has the small numbers of 184 and 108.
- Raise your hand when you know the missing number in problem D. ✔
- 184 plus 108 equals how many? (Signal.) *292.*
  (Repeat problems that were not firm.)

i. Complete the equations in part 5. Put your pencil down when you've worked all of the problems in part 5.
  (Observe children and give feedback.)
- Check your work. You'll read each equation.
- Read equation A. (Signal.) *184 – 108 = 76.*
- (Repeat for:) B, *184 + 108 = 292;* C, *260 – 184 = 76;* D, *184 + 108 = 292.*

EXERCISE 11: INDEPENDENT WORK

a. Find parts 6 and 7. ✔
  (Teacher reference:)

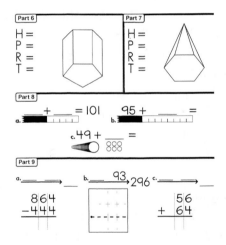

Part 6
H =
P =
R =
T =

Part 7
H =
P =
R =
T =

Part 8
a. ___ + ___ = 101   b. 95 + ___ =
c. 49 + ___ =   888

Part 9
a. ___ → ___   b. ___ 93 →296   c. ___ → ___
864   –444   56 + 64

In parts 6 and 7, you'll complete each equation to show the number of hexagons, pentagons, rectangles, and triangles.
In part 8, you'll complete the equations for the rulers and the equation for the circles.
In part 9, you'll work a column problem and complete each family.

b. Complete worksheet 114.
  (Observe children and mark incorrect responses on children's worksheets as you give feedback.)

Lesson 115

EXERCISE 1: FACT RELATIONSHIPS
ADDITION AND SUBTRACTION

a. You'll say facts for families that have a small number of 3.
   You'll start with 4 minus 3 and say the minus-3 facts to 13 minus 3.
- Say the fact for 4 minus 3. (Signal.) *4 – 3 = 1.*
- Say the minus-3 fact that starts with 5. (Signal.) *5 – 3 = 2.*
- Next fact. (Signal.) *6 – 3 = 3.*
- (Repeat for: 7 – 3 = 4, 8 – 3 = 5, 9 – 3 = 6, 10 – 3 = 7, 11 – 3 = 8, 12 – 3 = 9, 13 – 3 = 10.)
   (Repeat step a until firm.)

b. Now you'll start with 1 plus 3 and say the plus-3 facts to 10 plus 3.
- Say the fact for 1 plus 3. (Signal.) *1 + 3 = 4.*
- Say the plus-3 fact that starts with 2. (Signal.) *2 + 3 = 5.*
- Say the fact that starts with 3. (Signal.) *3 + 3 = 6.*
- Say the fact that starts with 4. (Signal.) *4 + 3 = 7.*
- Next fact. (Signal.) *5 + 3 = 8.*
- (Repeat for: 6 + 3 = 9, 7 + 3 = 10, 8 + 3 = 11, 9 + 3 = 12, 10 + 3 = 13.)
   (Repeat step b until firm.)

c. Now you'll start with 3 plus 1 and say the 3-plus facts to 3 plus 10.
- Say the fact for 3 plus 1. Get ready. (Signal.) *3 + 1 = 4.*
- Next fact. (Signal.) *3 + 2 = 5.*
- (Repeat for: 3 + 3 = 6, 3 + 4 = 7, 3 + 5 = 8, 3 + 6 = 9, 3 + 7 = 10, 3 + 8 = 11, 3 + 9 = 12, 3 + 10 = 13.)

d. Now I'll mix some of the problems up and you'll tell me the answer.
- Listen: 3 plus 8. Everybody, what's 3 plus 8? (Signal.) *11.*
- Listen: 10 minus 3. What's 10 minus 3? (Signal.) *7.*
- Listen: 8 minus 3. What's 8 minus 3? (Signal.) *5.*
- Listen: 7 plus 3. What's 7 plus 3? (Signal.) *10.*

INDIVIDUAL TURNS

(Call on individual students to perform one of the following tasks.)

- What's 10 minus 3? (Call on a student.) *7.*
- What's 3 plus 8? (Call on a student.) *11.*
- What's 8 minus 3? (Call on a student.) *5.*
- What's 7 plus 3? (Call on a student.) *10.*

EXERCISE 2: 3-D OBJECTS
PRISMS VS. PYRAMIDS

a. Last lesson you learned about prisms. Listen: Do prisms have side faces that are triangles or rectangles? (Signal.) *Rectangles.*
- What's the name for objects that have side faces that are triangles? (Signal.) *Pyramids.*
- Think about a prism with faces that are all squares. What do you call a prism that has faces that are all squares? (Signal.) *(A) cube.*
- Think about the number of faces a cube has. How many faces does a cube have? (Signal.) *6.*
   (Repeat until firm.)

b. (Display:)   [115:2A]

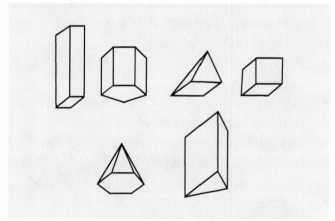

You'll tell me if each object is a pyramid or prism.
- (Point to objects.) For these objects if it's not a pyramid, it's a prism.
- These prisms have side faces that are rectangles. What shape are the side faces of these prisms? (Signal.) *Rectangles.*
- (Point to rectangular prism.) Is this object a pyramid or prism? (Signal.) *(A) prism.*

c. (Repeat the following tasks for remaining objects:)

(Point to __.) Is this object a pyramid or a prism?

Pentagonal prism	(A) prism
Rectangular pyramid	(A) pyramid
Cube	(A) prism
Hexagonal pyramid	(A) pyramid
Triangular prism	(A) prism

(Repeat for objects that were not firm.)

d. Let's do some of these again.
- (Point to rectangular prism.) Is this object a pyramid or a prism? (Signal.) (A) prism.
- How do you know it's a prism? (Call on a student. Idea: It has side faces that are rectangles.)

e. (Point to pentagonal prism.) Is this object a pyramid or a prism? (Signal.) (A) prism.
- How do you know it's a prism? (Call on a student. Idea: Side faces are rectangles.)

f. (Point to rectangular pyramid.) Is this object a pyramid or a prism? (Signal.) (A) pyramid.
- How do you know it's a pyramid? (Call on a student. Idea: Side faces are triangles.)

g. (Point to cube.) Is this object a pyramid or a prism? (Signal.) (A) prism.
- How do you know it's a prism? (Call on a student. Idea: Side faces are rectangles.)

h. (Point to hexagonal pyramid.) Is this object a pyramid or a prism? (Signal.) (A) pyramid.
- How do you know it's a pyramid? (Call on a student. Idea: Side faces are triangles.)

i. (Point to triangular prism.) Is this object a pyramid or a prism? (Signal.) (A) prism.
How do you know it's a prism? (Call on a student. Idea: Side faces are rectangles.)
(Repeat for objects that were not firm.)

j. Look at the objects that are prisms and get ready to tell me how many of the prisms are cubes. ✔
- How many of these prisms are cubes? (Signal.) 1.

EXERCISE 3: MIXED COUNTING
EXPLAIN ADDING OR SUBTRACTING 10

a. Count by hundreds to 1000. Get ready. (Tap.) 100, 200, 300, 400, 500, 600, 700, 800, 900, 1000.

b. Start with 2 and say even numbers to 20. Get ready. (Tap.) 2, 4, 6, 8, 10, 12, 14, 16, 18, 20.

c. Start with 75 and plus tens to 125. Get 75 going. Seventy-fiiive. Plus tens. (Tap.) 85, 95, 105, 115, 125.

d. Start with 75 and count by 25s to 175. Get 75 going. Seventy-fiiive. Count. (Tap.) 100, 125, 150, 175.

e. Start with 75 and count by fives to 100. Get 75 going. Seventy-fiiive. Count. (Tap.) 80, 85, 90, 95, 100.

f. Start with 75 and count by ones to 80. Get 75 going. Seventy-fiiive. Count. (Tap.) 76, 77, 78, 79, 80.
(Repeat steps a through f that were not firm.)

g. Listen: You have 75. When you count by tens, what number comes next? (Signal.) 85.
- So what's 75 plus 10? (Signal.) 85.

h. Listen: You have 75. When you count backward by ones, what number comes next? (Signal.) 74.
- So what's 75 minus 1? (Signal.) 74.

i. You have 75. When you count by fives, what number comes next? (Signal.) 80.
- So what's 75 plus 5? (Signal.) 80.

j. Listen: You have 75. When you count by 25s, what number comes next? (Signal.) 100.
- So what's 75 plus 25? (Signal.) 100.
(Repeat steps g through j that were not firm.)

k. Listen: If you have 75, how many tens do you have? (Signal.) 7.
- How many ones do you have? (Signal.) 5.
- So if you add ten to 75, how many tens do you end up with? (Signal.) 8.
- So what's 75 plus 10? (Signal.) 85.
- How do you know that 75 plus 10 equals 85? (Call on a student. Idea: 75 is 7 tens and 5 ones. When you add ten, you end up with 8 tens and 5 ones, which is 85.)

l. Listen: If you have 75, and subtract ten, how many tens do you end up with? (Signal.) 6.
- How many ones do you end up with? (Signal.) 5.
- So what's 75 minus 10? (Signal.) 65.
- How do you know that 75 minus 10 equals 65? (Call on a student. Idea: 75 is 7 tens and 5 ones. When you subtract ten, you end up with 6 tens and 5 ones, which is 65.)

EXERCISE 4: PLACE VALUE (TWO-DIGIT NUMBERS)
EXPLANATIONS

a. (Display:) [115:4A]

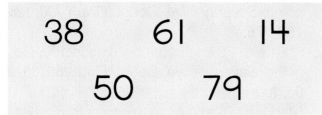

You're going to read each number. Then you'll tell me about the picture of it.

- (Point to **38.**) What number? (Touch.) *38.*
- How many groups of 10 would you draw for 38? (Signal.) *3.*
- How many single squares would you draw for 38? (Signal.) *8.*
- What would you draw for the picture of 38? (Call on a student.) *3 groups of 10. 8 single squares.*
b. (Point to **61.**) What number? (Touch.) *61.*
- How many groups of 10 would you draw for 61? (Signal.) *6.*
- How many single squares would you draw for 61? (Signal.) *1.*
- What would you draw for the picture of 61? (Call on a student.) *6 groups of 10. 1 single square.*
c. (Point to **14.**) What number? (Touch.) *14.*
- How many groups of 10 would you draw for 14? (Signal.) *1.*
- How many single squares would you draw for 14? (Signal.) *4.*
- What would you draw for the picture of 14? (Call on a student.) *1 group of 10. 4 single squares.*
d. (Point to **50.**) What number? (Touch.) *50.*
- Raise your hand when you can tell me what you'd draw for a picture of 50. ✔
- What would you draw for a picture of 50? (Call on a student.) *5 groups of 10. Zero single squares.*
e. (Point to **79.**) What number? (Touch.) *79.*
- Raise your hand when you can tell me what you'd draw for a picture of 79. ✔
- What would you draw for a picture of 79? (Call on a student.) *7 groups of 10. 9 single squares.* (Repeat steps a through e that were not firm.)

EXERCISE 5: TELLING TIME
ANALOG HOURS AND HALF HOURS

a. Think about the minute hand on clocks.
- Listen: When the minute hand is straight up, how many minutes does the clock show? (Signal.) *Zero.*

- When the minute hand is straight down, how many minutes does the clock show? (Signal.) *30.*
- For each number on a clock, how many minutes do you count? (Signal.) *5.* (Repeat until firm.)
b. (Display:) [115:5A]

The minute hand for each of these clocks shows zero minutes or 30 minutes. You're going to tell me the hour number and the minute number for each clock. Then you'll tell me the time the clock shows.

- (Point to **12:00.**) What's the hour number for this clock? (Signal.) *12.*
- What's the minute number? (Signal.) *Zero.*
- Say the time this clock shows. (Signal.) *12 o'clock.*
c. (Repeat the following tasks for the remaining clocks:)

(Point to __.) What's the hour number for this clock?	What's the minute number?	Say the time this clock shows.
4:30	4	*Thirty*	*4:30*
9:30	9	*Thirty*	*9:30*
3:00	3	*Zero*	*3 o'clock*
7:30	7	*Thirty*	*7:30*
1:00	1	*Zero*	*1 o'clock*

(Repeat clocks that were not firm.)

EXERCISE 6: EVEN AND ODD NUMBERS

a. You've learned the name for numbers you say when you count by twos. What name? (Signal.) *Even.*
- The numbers that are not even have a name, too. What do you call numbers that aren't even? (Signal.) *Odd.*
- What do you call numbers that are not odd? (Signal.) *Even.* (Repeat until firm.)

b. I'll say numbers. You'll tell me if the number is even or odd.
- Listen: 49. What number? (Signal.) *49.*
- Is 49 a number you say when you count by twos? (Signal.) *No.*
- So is 49 even? (Signal.) *No.*
- What kind of number is 49? (Signal.) *Odd.*
c. New number: 16. What number? (Signal.) *16.*
- Is 16 a number you say when you count by twos? (Signal.) *Yes.*
- So is 16 even? (Signal.) *Yes.*
d. New number: 30. What number? (Signal.) *30.*
- Is 30 a number you say when you count by twos? (Signal.) *Yes.*
- So is 30 even? (Signal.) *Yes.*
e. New number: 23. What number? (Signal.) *23.*
- Is 23 a number you say when you count by twos? (Signal.) *No.*
- So is 23 even? (Signal.) *No.*
- What kind of number is 23? (Signal.) *Odd.*
  (Repeat numbers that were not firm.)

EXERCISE 7: FACTS
PLUS/MINUS MIX

a. (Distribute unopened workbooks to children.)
- Open your workbook to Lesson 115 and find part 1.
  (Observe children and give feedback.)
  (Teacher reference:)

a. $10 - 4$	d. $4 + 4$	g. $8 + 8$
b. $14 - 7$	e. $10 - 7$	h. $11 - 8$
c. $7 - 4$	f. $10 - 5$	i. $3 + 3$

- Touch and read problem A. Get ready. (Signal.) *10 minus 4.*
- Is the answer the big number or a small number? (Signal.) *A small number.*
- What's 10 minus 4? (Signal.) *6.*
b. (Repeat the following tasks for problems B through I:)

Touch and read problem __.		Is the answer the big number or a small number?	What's __?	
B	$14 - 7$	A small number.	$14 - 7$	7
C	$7 - 4$	A small number.	$7 - 4$	3
D	$4 + 4$	The big number.	$4 + 4$	8
E	$10 - 7$	A small number.	$10 - 7$	3
F	$10 - 5$	A small number.	$10 - 5$	5
G	$8 + 8$	The big number.	$8 + 8$	16
H	$11 - 8$	A small number.	$11 - 8$	3
I	$3 + 3$	The big number.	$3 + 3$	6

(Repeat problems that were not firm.)

c. Complete each of the facts. Put your pencil down when you've written all of the equations for part 1.
  (Observe children and give feedback.)
d. Check your work.
- Touch and read the equation for problem A. (Signal.) *10 – 4 = 6.*
- (Repeat for:) B, *14 – 7 = 7;* C, *7 – 4 = 3;* D, *4 + 4 = 8;* E, *10 – 7 = 3;* F, *10 – 5 = 5;* G, *8 + 8 = 16;* H, *11 – 8 = 3;* I, *3 + 3 = 6.*

EXERCISE 8: SOLVING PROBLEMS FROM NUMBER FAMILIES
MISSING ADDEND, MINUEND, OR SUBTRAHEND

a. Find part 2 on your worksheet. ✔
  (Teacher reference:)

a. ___ – 58 = 35	c. 72 – ___ = 14
b. 14 + ___ = 93	d. 58 + 35 = ___

  These problems have a number missing. You're going to read these problems, and tell me if each value is a small number or the big number.
- Read problem A. (Signal.) *How many minus 58 equals 35.*
- Is the missing number a small number or the big number? (Signal.) *The big number.*
- What's 58? (Signal.) *A small number.*
- What's 35? (Signal.) *A small number.*
b. Read problem B. (Signal.) *14 plus how many equals 93.*
- Is the missing number a small number or the big number? (Signal.) *A small number.*
- What's 14? (Signal.) *A small number.*
- What's 93? (Signal.) *The big number.*
c. Read problem C. (Signal.) *72 minus how many equals 14.*
- Is the missing number a small number or the big number? (Signal.) *A small number.*
- What's 72? (Signal.) *The big number.*
- What's 14? (Signal.) *A small number.*
d. Read problem D. (Signal.) *58 plus 35 equals how many.*
- Is the missing number a small number or the big number? (Signal.) *The big number.*
- What's 58? (Signal.) *A small number.*
- What's 35? (Signal.) *A small number.*
  (Repeat problems that were not firm.)

e. (Display:) [115:7A]

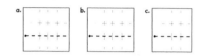

Here are families for these problems. You're going to figure out the missing number by finding the right number family.

• Go back to problem A and look at families 1, 2, and 3 and find the family with the right numbers. ✔
• What's the number of the right family for problem A? (Signal.) *2.*
  In problem A, 58 and 35 are small numbers. Family 2 has small numbers of 58 and 35.
• Raise your hand when you know the missing number for problem A. ✔
• How many minus 58 equals 35? (Signal.) *93.*
• Complete the equation for A. ✔
f. Look at families 1, 2, and 3 and find the family with the right numbers for problem B. ✔
• What's the number of the right family for problem B? (Signal.) *1.*
• Raise your hand when you know the missing number for problem B. ✔
• 14 plus how many equals 93? (Signal.) *79.*
• Complete the equation for B. Then complete the equations for the rest of the problems in part 2.
  (Observe children and give feedback.)
g. Check your work. You'll read each equation in part 2.
  ⌈• Equation A. (Signal.) *93 – 58 = 35.*
  ⌊• (Repeat for:) B, *14 + 79 = 93;* C, *72 – 58 = 14;* D, *58 + 35 = 93.*

EXERCISE 9: FACTS
ADDITION

a. Find part 3 on your worksheet. ✔
  (Teacher reference:)

c. 4 + 6   f. 6 + 2
a. 5 + 6   d. 7 + 2   g. 5 + 5
b. 7 + 7   e. 6 + 6   h. 8 + 8

These are addition problems you'll work. Think of the number families for each problem and write the answers. Put your pencil down when you've completed the equations in part 4.
(Observe children and give feedback.)

b. Check your work. You'll read the equation for each problem.
  ⌈• Problem A. (Signal.) *5 + 6 = 11.*
  ⌊• (Repeat for:) B, *7 + 7 = 14;* C, *4 + 6 = 10;* D, *7 + 2 = 9;* E, *6 + 6 = 12;* F, *6 + 2 = 8;* G, *5 + 5 = 10;* H, *8 + 8 = 16.*

EXERCISE 10: WORD PROBLEMS (COLUMNS)
CONVENTIONAL AND COMPARISON

a. Find part 4 on your worksheet. ✔
  (Teacher reference:)

You're going to write the symbols for word problems in columns and work them. Some of the problems you'll work today are money problems. You'll write the dollar signs and the dots for those problems.
• Touch where you'll write the symbols for problem A. ✔
• Listen to problem A: Darius had 7 dollars and 5 cents. Darius spent 2 dollars and 3 cents of that money. How much money did Darius have left?
• Listen again and write the symbols for the whole problem. Darius had 7 dollars and 5 cents. Darius spent 2 dollars and 3 cents of that money.
  (Observe children and give feedback.)
• Everybody, touch and read the problem for A. Get ready. (Signal.) *7 dollars and 5 cents minus 2 dollars and 3 cents.*
b. Work problem A. Put your pencil down when you know how much money Darius had left.
  (Observe children and give feedback.)
  (Teacher reference:)

a. ```
$ 7.05
+
– 2.03
$ 5.02
```

• Read the subtraction problem and the answer you wrote for A. Get ready. (Signal.) *7 dollars and 5 cents minus 2 dollars and 3 cents equals 5 dollars and 2 cents.*
• Everybody, how much money did Darius have left? (Signal.) *5 dollars and 2 cents.*
```

c. Touch where you'll write the symbols for problem B. ✔
- Listen to problem B: 5 dollars and 17 cents were in a cash register. 5 dollars and 42 cents more was put into the cash register. How much money ended up in the cash register?
- Listen again and write the symbols for the whole problem. 5 dollars and 17 cents were in a cash register. 5 dollars and 42 cents more was put into the cash register.
(Observe children and give feedback.)
- Everybody, touch and read the problem for B. Get ready. (Signal.) *5 dollars and 17 cents plus 5 dollars and 42 cents.*

d. Work problem B. Put your pencil down when you know how much money ended up in the cash register.
(Observe children and give feedback.)
(Teacher reference:)

- Read the addition problem and the answer you wrote for B. Get ready. (Signal.) *5 dollars and 17 cents plus 5 dollars and 42 cents equals 10 dollars and 59 cents.*
- Everybody, how much money ended up in the cash register? (Signal.) *10 dollars and 59 cents.*

e. Touch where you'll write the symbols for problem C. ✔
You're going to tell me how many feet taller one hill is than another hill.
- Listen to problem C: The red hill is 784 feet high. The green hill is 986 feet high. How many feet higher is the green hill?
- Listen to problem C again and write the problem for figuring out how many feet higher the green hill is than the red hill. The red hill is 784 feet high. (Pause.) The green hill is 986 feet high.
(Observe children and give feedback.)
- Read the problem you wrote for figuring out how many feet higher the green hill is.
(Signal.) *986 minus 784.*
- Work the problem. Put your pencil down when you know how many feet higher the green hill is.
(Observe children and give feedback.)
(Teacher reference:)

c.
```
  986
- 784
  202
```

f. Check your work.
- Read the subtraction problem and the answer for C. Get ready. (Signal.) *986 – 784 = 202.*
- How many feet higher is the green hill? (Signal.) *202.*

EXERCISE 11: INDEPENDENT WORK

a. Find part 5. ✔
(Teacher reference:)

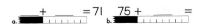

You'll complete the equations for the rulers.

b. Turn to the other side of your worksheet and find part 6. ✔
(Teacher reference:)

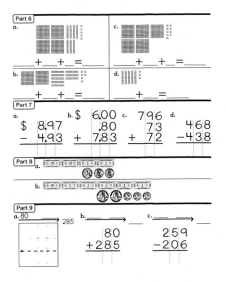

In part 6, you'll write the place-value addition equation for each picture.
In part 7, you'll work the money problems and the column problems.
In part 8, you'll write the dollars and cents amount for each group of bills and coins.
In part 9, you'll work a column problem and complete the family.

c. Complete worksheet 115. Remember to complete the equations for the rulers in part 5 on the other side of your worksheet.
(Observe children and mark incorrect responses on children's worksheets as you give feedback.)

Lessons 116–120 Planning Page

	Lesson 116	Lesson 117	Lesson 118	Lesson 119	Lesson 120
Student Learning Objectives	**Exercises** 1. Say facts for plus 3, 3 plus, and minus 3 2. Identify and explain the place values of tens and ones in 2-digit numbers 3. Identify 3-dimensional objects (prism vs. pyramid) 4. Identify even and odd numbers; mentally add 10 and subtract 10 and explain reasoning 5. Solve addition and subtraction facts 6. Write the time from an analog clock (half hours) 7. Solve problems with a missing value using multi-digit number families 8. Identify 2-dimensional faces on 3-dimensional objects 9. Solve addition facts 10. Complete work independently	**Exercises** 1. Say facts for plus 3, 3 plus, and minus 3 2. **Add and subtract tens numbers using drawings to explain reasoning** 3. Mentally add 10 and subtract 10 and explain reasoning 4. Solve addition and subtraction facts 5. **Solve problems with a missing value by writing multi-digit number families** 6. Write the symbols for word problems in columns and solve; solve comparison word problems 7. **Identify and write the number of equal parts in a shape as a preskill to fractions** 8. Solve subtraction facts 9. **Write the time from an analog clock (to the minute)** 10. Complete work independently	**Exercises** 1. Say facts for plus 3, 3 plus, and minus 3 2. Add and subtract tens numbers using drawings to explain reasoning 3. Mentally add 10 and subtract 10 and explain reasoning 4. Say addition and subtraction facts 5. Identify and write the number of equal parts in a shape as a preskill to fractions 6. **Solve problems with a missing value by writing and solving multi-digit number families** 7. Solve subtraction facts 8. Write the time from an analog clock (to the minute) 9. Complete work independently	**Exercises** 1. Say facts for plus 3, 3 plus, and minus 3 2. **Associate the number of equal parts with a name as a preskill to fractions** 3. Add and subtract tens numbers using drawings to explain reasoning 4. **Determine whether equations are true or false** 5. **Identify the missing value in a word problem** 6. Solve addition and subtraction facts 7. Solve problems with a missing value using multi-digit number families 8. Write the time from an analog clock (to the minute) 9. Complete work independently	**Exercises** 1. **Say facts for plus 3, 4, and 5** 2. Identify and write the number of equal parts in a shape as a preskill to fractions 3. **Add and subtract tens numbers using drawings to explain reasoning and write equations** 4. Determine whether equations are true or false 5. Identify the missing value in a word problem 6. Solve addition and subtraction facts 7. Solve problems with a missing value by writing and solving multi-digit number families 8. Complete work independently
Common Core State Standards for Mathematics					
1.OA 1		✔			✔
1.OA 4	✔	✔	✔	✔	✔
1.OA 5	✔	✔	✔	✔	
1.OA 6 and 8	✔	✔	✔	✔	✔
1.OA 7				✔	✔
1.NBT 1–2, 4–5	✔	✔	✔	✔	✔
1.NBT 6		✔	✔	✔	✔
1.MD 3	✔	✔	✔	✔	✔
1.G 1–2	✔	✔	✔	✔	
1.G 3		✔	✔	✔	✔
Teacher Materials	Presentation Book 3, Board Displays CD or chalkboard				
Student Materials	Workbook 2, Pencil				
Additional Practice	Student Practice Software: Block 5: Activity 1 (1.NBT.1), Activity 2 (1.NBT.1), Activity 3 (1.NBT.4 and 1.OA.6), Activity 4 (1.NBT.1), Activity 5 (1.NBT.1), Activity 6 (1.NBT.3)				
Mastery Test					Student Assessment Book (Present Mastery Test 12 following Lesson 120.)

Lesson 116

EXERCISE 1: FACT RELATIONSHIPS
ADDITION AND SUBTRACTION

a. You'll say facts for families that have a small number of 3. You'll start with 1 plus 3 and say the plus-3 facts to 10 plus 3.
- Say the fact for 1 plus 3. (Signal.) *1 + 3 = 4.*
- Say the fact that starts with 2. (Signal.) *2 + 3 = 5.*
- Next fact. (Signal.) *3 + 3 = 6.*
- (Repeat for:) *4 + 3 = 7, 5 + 3 = 8, 6 + 3 = 9, 7 + 3 = 10, 8 + 3 = 11, 9 + 3 = 12, 10 + 3 = 13.*
   (Repeat step a until firm.)

b. Now you'll start with 4 minus 3 and say the minus-3 facts to 13 minus 3.
- Say the fact for 4 minus 3. (Signal.) *4 – 3 = 1.*
- Say the minus-3 fact that starts with 5. (Signal.) *5 – 3 = 2.*
- Next fact. (Signal.) *6 – 3 = 3.*
- (Repeat for:) *7 – 3 = 4, 8 – 3 = 5, 9 – 3 = 6, 10 – 3 = 7, 11 – 3 = 8, 12 – 3 = 9, 13 – 3 = 10.*
   (Repeat step b until firm.)

c. Now you'll start with 3 plus 1 and say the 3-plus facts to 3 plus 10.
- Say the fact for 3 plus 1. Get ready. (Signal.) *3 + 1 = 4.*
- Next fact. (Signal.) *3 + 2 = 5.*
- (Repeat for:) *3 + 3 = 6, 3 + 4 = 7, 3 + 5 = 8, 3 + 6 = 9, 3 + 7 = 10, 3 + 8 = 11, 3 + 9 = 12, 3 + 10 = 13.*

d. Now I'll mix some of the problems up and you'll tell me the answer.
- Listen: 11 minus 3. Everybody, what's 11 minus 3? (Signal.) *8.*
- Listen: 3 plus 7. What's 3 plus 7? (Signal.) *10.*
- Listen: 6 minus 3. What's 6 minus 3? (Signal.) *3.*
- Listen: 8 plus 3. What's 8 plus 3? (Signal.) *11.*

=== INDIVIDUAL TURNS ===

(Call on individual students to perform one of the following tasks.)

- What's 3 plus 7? (Call on a student.) *10.*
- What's 11 minus 3? (Call on a student.) *8.*
- What's 6 minus 3? (Call on a student.) *3.*
- What's 8 plus 3? (Call on a student.) *11.*

EXERCISE 2: PLACE VALUE (TWO-DIGIT NUMBERS)
EXPLANATIONS

a. (Display:) [116:2A]

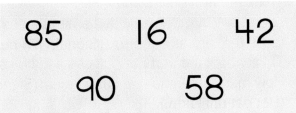

You're going to read each number. Then you'll tell me about the picture for it.
- (Point to **85**.) What number? (Signal.) *85.*
- How many groups of 10 would you draw for 85? (Signal.) *8.*
- How many single squares would you draw for 85? (Signal.) *5.*
- What would you draw for the picture of 85? (Call on a student.) *8 groups of 10. 5 single squares.*

b. (Point to **16**.) What number? (Signal.) *16.*
- Raise your hand when you can tell me what you'd draw for a picture of 16. ✔
- What would you draw for a picture of 16? (Call on a student.) *1 group of 10. 6 single squares.*

c. (Repeat the following tasks with the remaining numbers:)

(Point to __.) What number?	What would you draw for a picture of __? (Call on a student.)		
42	42	42	*4 groups of 10. 2 single squares.*
90	90	90	*9 groups of 10. Zero single squares.*
58	58	58	*5 groups of 10. 8 single squares.*

(Repeat steps a through c for numbers that were not firm.)

EXERCISE 3: 3-D OBJECTS
PRISMS VS. PYRAMIDS

a. You learned about prisms. Listen: Do prisms have side faces that are triangles or rectangles? (Signal.) *Rectangles.*
- What's the name for objects that have side faces that are triangles? (Signal.) *Pyramids.*
- Think about a prism with faces that are all squares. What do you call a prism that has faces that are all squares. (Signal.) *(A) cube.*
- Think about the number of faces a cube has. How many faces does a cube have? (Signal.) *6.*
(Repeat until firm.)

b. (Display:)   [116:3A]

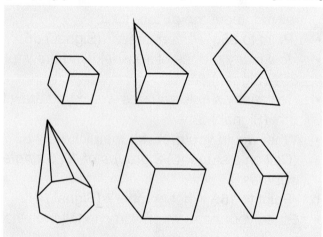

You'll tell me if each object is a pyramid or prism.
- (Point to objects.) For these objects if it's not a pyramid, it's a prism.
- These prisms have side faces that are rectangles. What shape are the side faces of these prisms? (Signal.) *Rectangles.*
- (Point to small cube.) Is this object a pyramid or prism? (Signal.) *Prism.*

c. (Repeat the following tasks for remaining objects:)

(Point to __.) Is this object a pyramid or a prism?	
Rectangular pyramid	*(A) pyramid*
Triangular prism	*(A) prism*
Octagonal pyramid	*(A) pyramid*
Larger cube	*(A) prism*
Rectangular prism	*(A) prism*

(Repeat for objects that were not firm.)

d. Let's do some of these again.
- (Point to small cube.) Is this object a pyramid or a prism? (Signal.) *Prism.*
- How do you know it's a prism? (Call on a student. Idea: *It has side faces that are rectangles.*)

e. (Repeat the following tasks for the remaining objects:)

(Point to __.) Is this object a pyramid or a prism?		How do you know it's a __? (Call on a student. Idea:
Rectangular pyramid	*(A) pyramid*	pyramid	*Side faces are triangles.)*
Triangular prism	*(A) prism*	prism	*Side faces are rectangles.)*
Octagonal pyramid	*(A) pyramid*	pyramid	*Side faces are triangles.)*
Larger cube	*(A) prism*	prism	*Side faces are rectangles.)*
Rectangular prism	*(A) prism*	prism	*Side faces are rectangles.)*

(Repeat for objects that were not firm.)

f. Look at the objects that are prisms and get ready to tell me how many of the prisms are cubes. ✔
- How many of these prisms are cubes? (Signal.) *2.*

EXERCISE 4: MIXED COUNTING
EXPLAIN ADDING OR SUBTRACTING 10   REMEDY

a. You've learned the name for numbers you say when you count by twos. What name? (Signal.) *Even.*
- The numbers that are not even have a name, too. What do you call numbers that aren't even? (Signal.) *Odd.*
- What do you call numbers that are not odd? (Signal.) *Even.*
(Repeat until firm.)

b. I'll say numbers. You'll tell me if the number is even or odd.
- Listen: 15. What number? (Signal.) *15.*
- Is 15 even or odd? (Signal.) *Odd.*
- New number: 29. What number? (Signal.) *29.*
- Is 29 even or odd? (Signal.) *Odd.*
- New number: 36. What number? (Signal.) *36.*
- Is 36 even or odd? (Signal.) *Even.*
- New number: 4. What number? (Signal.) *4.*
- Is 4 even or odd? (Signal.) *Even.*
- New number: 41. What number? (Signal.) *41.*
- Is 41 even or odd? (Signal.) *Odd.*
- New number: 10. What number? (Signal.) *10.*
- Is 10 even or odd? (Signal.) *Even.*

(Repeat numbers that were not firm.)

c. Count by hundreds to 1000. Get ready. (Tap.) *100, 200, 300, 400, 500, 600, 700, 800, 900, 1000.*

d. Start with 2 and say even numbers to 20. Get ready. (Tap.) *2, 4, 6, 8, 10, 12, 14, 16, 18, 20.*

e. Start with 350 and count by twos to 360. Get 350 going. *Three hundred fiftyyy.* Count. (Tap.) *352, 354, 356, 358, 360.*

f. Start with 350 and count backward by tens to 300. Get 350 going. *Three hundred fiftyyy.* Count backward. (Tap.) *340, 330, 320, 310, 300.*

g. Start with 350 and count by fives to 375. Get 350 going. *Three hundred fiftyyy.* Count. (Tap.) *355, 360, 365, 370, 375.*

h. Start with 350 and count by ones to 355. Get 350 going. *Three hundred fiftyyy.* Count. (Tap.) *351, 352, 353, 354, 355.*

i. Start with 350 and count by 25s to 400. Get 350 going. *Three hundred fiftyyy.* Count. (Tap.) *375, 400.*

j. Start with 350 and count backward by ones to 345. Get 350 going. *Three hundred fiftyyy.* Count backward. (Tap.) *349, 348, 347, 346, 345.* (Repeat steps c through j that were not firm.)

k. Listen: You have 350. When you count by 25s, what number comes next? (Signal.) *375.*
• So what's 350 plus 25? (Signal.) *375.*

l. Listen: You have 350. When you count backward by tens, what number comes next? (Signal.) *340.*
• So what's 350 minus 10? (Signal.) *340.*

m. Listen: You have 350. When you count by twos, what number comes next? (Signal.) *352.*
• So what's 350 plus 2? (Signal.) *352.*

n. You have 350. When you count backward by ones, what number comes next? (Signal.) *349.*
• So what's 350 minus 1? (Signal.) *349.* (Repeat steps k through n that were not firm.)

o. Listen: If you have 65, how many tens do you have? (Signal.) *6.*
• How many ones do you have? (Signal.) *5.*
• So if you add ten to 65, how many tens do you end up with? (Signal.) *7.*
• So what's 65 plus 10? (Signal.) *75.*
• How do you know that 65 plus 10 equals 75? (Call on a student. Idea: *65 is 6 tens and 5 ones. When you add ten, you end up with 7 tens and 5 ones, which is 75.*)

p. Listen: If you have 65, and subtract ten, how many tens do you end up with? (Signal.) *5.*
• How many ones do you end up with? (Signal.) *5.*
• So what's 65 minus 10? (Signal.) *55.*
• How do you know that 65 minus 10 equals 55? (Call on a student. Idea: *65 is 6 tens and 5 ones. When you subtract ten, you end up with 5 tens and 5 ones, which is 55.*)

EXERCISE 5: FACTS
PLUS/MINUS MIX

a. (Distribute unopened workbooks to children.)
• Open your workbook to Lesson 116 and find part 1.
(Observe children and give feedback.)
(Teacher reference:)

a. $8 - 4$	c. $16 - 8$	f. $10 - 3$
b. $4 + 6$	d. $11 - 8$	g. $11 - 5$
	e. $10 + 3$	h. $7 + 7$

• Touch and read problem A. Get ready. (Signal.) *8 minus 4.*
• Is the answer the big number or a small number? (Signal.) *A small number.*
• What's 8 minus 4? (Signal.) *4.*

b. (Repeat the following tasks for problems B through H:)

Touch and read problem __.	Is the answer the big number or a small number?	What's __?	
B	$4 + 6$	*The big number.*	$4 + 6$	*10*
C	$16 - 8$	*A small number.*	$16 - 8$	*8*
D	$11 - 8$	*A small number.*	$11 - 8$	*3*
E	$10 + 3$	*The big number.*	$10 + 3$	*13*
F	$10 - 3$	*A small number.*	$10 - 3$	*7*
G	$11 - 5$	*A small number.*	$11 - 5$	*6*
H	$7 + 7$	*The big number.*	$7 + 7$	*14*

(Repeat problems that were not firm.)

c. Complete each of the facts. Put your pencil down when you've written all of the equations for part 1.
(Observe children and give feedback.)

d. Check your work.
• Touch and read the equation for problem A. (Signal.) *8 – 4 = 4.*
• (Repeat for:) B, *4 + 6 = 10;* C, *16 – 8 = 8;* D, *11 – 8 = 3;* E, *10 + 3 = 13;* F, *10 – 3 = 7;* G, *11 – 5 = 6;* H, *7 + 7 = 14.*

EXERCISE 6: TELLING TIME
ANALOG HOURS AND HALF HOURS

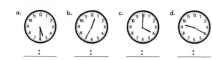

a. Find part 2 on worksheet 116. ✔
(Teacher reference:)

You're going to write the time for each clock.
One clock shows zero minutes. Another clock
shows 30 minutes. You'll count the minutes to
yourself for the other two clocks.

• Write the time for each clock. Put your pencil
down when you've written the hour number
and the minute number for each clock in part 2.
(Observe children and give feedback.)

b. Check your work. You'll read the time for
each clock.

⌐ • Clock A. (Signal.) *Five 30.*
L • (Repeat for:) B, *Twelve 35;* C, *4 o'clock;*
D, *Nine 20.*
(Display:)   [116:6A]

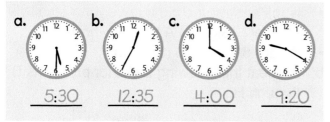

Here's what you should have for the clocks.

EXERCISE 7: SOLVING PROBLEMS FROM NUMBER FAMILIES
MISSING ADDEND, MINUEND, OR SUBTRAHEND

a. Find part 3 on your worksheet. ✔
(Teacher reference:)

a. $58 - \underline{} = 27$ c. $\underline{} - 58 = 27$
b. $\underline{} + 58 = 89$ d. $58 - 31 = \underline{}$

These problems have a missing number.
You're going to read these problems, and tell
me if each value is a small number or the big
number.

⌐ • Read problem A. (Signal.) *58 minus how
many equals 27.*
• Is the missing number a small number or the
big number? (Signal.) *A small number.*
• What's 58? (Signal.) *The big number.*
• What's 27? (Signal.) *A small number.*
L (Repeat until firm.)

b. Read problem B. (Signal.) *How many plus 58
equals 89.*
• Is the missing number a small number or the
big number? (Signal.) *A small number.*
• What's 58? (Signal.) *A small number.*
• What's 89? (Signal.) *The big number.*
c. Read problem C. (Signal.) *How many minus
58 equals 27.*
• Is the missing number a small number or the
big number? (Signal.) *The big number.*
• What's 58? (Signal.) *A small number.*
• What's 27? (Signal.) *A small number.*
d. Read problem D. (Signal.) *58 minus 31 equals
how many.*
• Is the missing number a small number or the
big number? (Signal.) *A small number.*
• What's 58? (Signal.) *The big number.*
• What's 31? (Signal.) *A small number.*
(Repeat problems that were not firm.)
e. (Display:)   [116:7A]

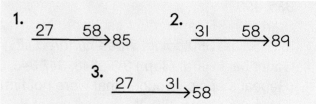

Here are families for these problems.
• Look at families 1, 2, and 3 and find the family
with the right numbers for problem A. ✔
• What's the number of the right family for
problem A? (Signal.) *3.*
• Raise your hand when you know the missing
number in problem A. ✔
• 58 minus how many equals 27? (Signal.) *31.*
• Complete the equation for A. ✔
⌐ f. Read problem B. (Signal.) *How many plus 58
equals 89.*
• Is the missing number a small number or the
big number? (Signal.) *A small number.*
• What's 58? (Signal.) *A small number.*
• What's 89? (Signal.) *The big number.*
L (Repeat step f until firm.)

g. Look at families 1, 2, and 3 and find the family with the right numbers for problem B. ✔
• Raise your hand when you know the missing number for problem B. ✔
• What's the number of the right family for problem B? (Signal.) *2.*
• Problem B: How many plus 58 equals 89. What's the missing number? (Signal.) *31.*
• Complete the equation for B. Then complete the equations for the rest of the problems in part 3.
(Observe children and give feedback.)
h. Check your work. You'll read each equation in part 3.
⌐• Equation A. (Signal.) *58 – 31 = 27.*
└• (Repeat for:) B, *31 + 58 = 89;* C, *85 – 58 = 27;* D, *58 – 31 = 27.*

EXERCISE 8: 3-D OBJECTS
DECOMPOSITION

a. Find part 4 on your worksheet. ✔
(Teacher reference:)

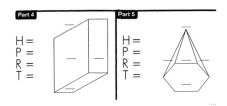

You'll tell me if the objects in parts 4 and 5 are pyramids or prisms.
• What is the object in part 4? (Signal.) *(A) prism.*
• What is the object in part 5? (Signal.) *(A) pyramid.*
b. Label the faces and complete all of the equations in part 4. Put your pencil down when you're finished.
(Observe children and give feedback.)
c. Check your work.
• What letter did you write on the bottom face of object 1? (Signal.) *R.*
• What letter did you write on side faces? (Signal.) *R.*
• Did you write a letter for the top face? (Signal.) *Yes.*
• What letter? (Signal.) *R.*
d. Touch the equation for hexagons. ✔
• What number did you write for H? (Signal.) *Zero.*
• What number did you write for P? (Signal.) *Zero.*

• What number did you write for R? (Signal.) *6.*
• What number did you write for T? (Signal.) *Zero.*
e. Label each face and complete all of the equations in part 5. Put your pencil down when you're finished.
(Observe children and give feedback.)
(Answer key:)

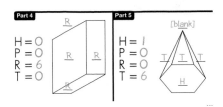

f. Check your work.
• Touch the bottom face of the object. What letter did you write on the bottom face? (Signal.) *H.*
• What letter did you write on side faces? (Signal.) *T.*
• Did you write a letter for the top face? (Signal.) *No.*
g. Touch the equation for hexagons. ✔
• What number did you write for H? (Signal.) *1.*
• What number did you write for P? (Signal.) *Zero.*
• What number did you write for R? (Signal.) *Zero.*
• What number did you write for T? (Signal.) *6.*

EXERCISE 9: FACTS
ADDITION

a. Turn to the other side of your worksheet and find part 6. ✔
(Teacher reference:)

a. 3 + 5
b. 6 + 4
c. 4 + 3
d. 3 + 7
e. 6 + 6
f. 6 + 10
g. 6 + 3
h. 2 + 3

• These are addition problems you'll work. Think of the number families for each problem and write the answers. Put your pencil down when you've completed the equations in part 6.
(Observe children and give feedback.)
b. Check your work. You'll read the equation for each problem.
⌐• Problem A. (Signal.) *3 + 5 = 8.*
└• (Repeat for:) B, *6 + 4 = 10;* C, *4 + 3 = 7;* D, *3 + 7 = 10;* E, *6 + 6 = 12;* F, *6 + 10 = 16;* G, *6 + 3 = 9;* H, *2 + 3 = 5.*

EXERCISE 10: INDEPENDENT WORK

a. Find part 7. ✔
   (Teacher reference:)

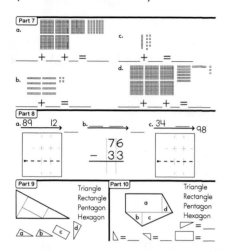

b. Complete worksheet 116.

   (Observe children and mark incorrect responses on children's worksheets as you give feedback.)

In part 7, you'll write the place-value addition equation for each picture.

In part 8, you'll work a column problem and complete the family.

In part 9, you'll circle the name for the whole shape. The names are triangle, rectangle, pentagon, and hexagon. Then you'll write the letter for each part.

In part 10, you'll circle the name for the whole shape. Then you'll complete each equation by writing the letter to show which part it is.

Lesson 117

EXERCISE 1: FACT RELATIONSHIPS
ADDITION AND SUBTRACTION   [REMEDY]

a. You'll say facts for families that have a small number of 3. First you'll start with 3 plus 1 and say the 3-plus facts to 3 plus 10.
  - • Say the fact for 3 plus 1. Get ready. (Signal.) *3 + 1 = 4.*
  - • Next fact. Get ready. (Signal.) *3 + 2 = 5.*
  - • (Repeat for:) *3 + 3 = 6, 3 + 4 = 7, 3 + 5 = 8, 3 + 6 = 9, 3 + 7 = 10, 3 + 8 = 11, 3 + 9 = 12, 3 + 10 = 13.*
    (Repeat until firm.)

b. Now you'll start with 4 minus 3 and say the minus-3 facts to 13 minus 3.
  - • Say the fact for 4 minus 3. (Signal.) *4 – 3 = 1.*
  - • Say the minus-3 fact that starts with 5. (Signal.) *5 – 3 = 2.*
  - • Next fact. (Signal.) *6 – 3 = 3.*
  - • (Repeat for:) *7 – 3 = 4, 8 – 3 = 5, 9 – 3 = 6, 10 – 3 = 7, 11 – 3 = 8, 12 – 3 = 9, 13 – 3 = 10.*
    (Repeat until firm.)

c. Now you'll start with 1 plus 3 and say the plus-3 facts to 10 plus 3.
  - • Say the fact for 1 plus 3. Get ready. (Signal.) *1 + 3 = 4.*
  - • Next fact. (Signal.) *2 + 3 = 5.*
  - • (Repeat for:) *3 + 3 = 6, 4 + 3 = 7, 5 + 3 = 8, 6 + 3 = 9, 7 + 3 = 10, 8 + 3 = 11, 9 + 3 = 12, 10 + 3 = 13.*
    (Repeat until firm.)

d. Now I'll mix some of the problems up and you'll tell me the answer.
  - • Listen: 3 plus 8. What's 3 plus 8? (Signal.) *11.*
  - • Listen: 10 minus 3. What's 10 minus 3? (Signal.) *7.*
  - • Listen: 4 plus 3. What's 4 plus 3? (Signal.) *7.*
  - • Listen: 11 minus 3. What's 11 minus 3? (Signal.) *8.*

=========== INDIVIDUAL TURNS ===========

(Call on individual students to perform one of the following tasks.)

  - • What's 11 minus 3? (Call on a student.) *8.*
  - • What's 4 plus 3? (Call on a student.) *7.*
  - • What's 10 minus 3? (Call on a student.) *7.*
  - • What's 3 plus 8? (Call on a student.) *11.*

EXERCISE 2: ADDING AND SUBTRACTING TENS NUMBERS
DRAWINGS AND REASONING

a. You're going to read problems. For each problem, you'll tell me about the picture for it. Then I'll show you the picture and you'll explain it.
    (Display:)   [117:2A]

```
  80
- 30
```

  - • (Point to **80**.) Read the problem. (Signal.) *80 minus 30.*
  - • How many groups of ten does it start with? (Signal.) *8.*
  - • How many groups of ten does it subtract? (Signal.) *3.*
  - • So how many groups of ten does it equal? (Signal.) *5.*
  - • What number does it equal? (Signal.) *50.*
    (Add to show:)   [117:2B]

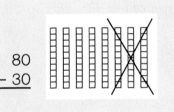

Here's the picture. The problem starts with 8 groups of ten (circle all columns), it subtracts 3 groups of ten (touch crossed out columns), and equals 5 groups of ten (touch columns not crossed out) which equals 50.

  - • What does this picture of tens groups show? (Call on a student.) *Starts with 8 groups of 10. Subtracts 3 groups of ten. Equals 5 groups of ten, which is 50.*

b. (Display:) [117:2C]

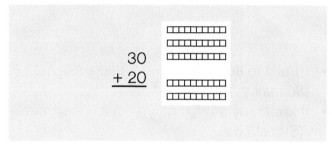

$$30 + 20$$

- (Point to **30**.) Read the problem. (Signal.) *30 plus 20.*
- How many groups of ten does it start with? (Signal.) *3.*
- How many groups of ten does it add? (Signal.) *2.*
- So how many groups of ten does it equal? (Signal.) *5.*
- What number does it equal? (Signal.) *50.* (Add to show:) [117:2D]

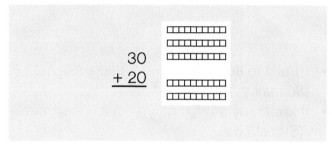

Here's the picture. The problem starts with 3 groups of ten (circle first 3 rows), it adds 2 groups of ten (touch 2 rows), and equals 5 groups of ten (circle all rows) which equals 50.
- What does this picture show? (Call on a student.) *Starts with 3 groups of 10. Adds 2 groups of ten. Equals 5 groups of ten, which is 50.*

c. (Display:) [117:2E]

$$70 - 50$$

- (Point to **70**.) Read the problem. (Signal.) *70 minus 50.*
- How many groups of ten does it start with? (Signal.) *7.*
- How many groups of ten does it subtract? (Signal.) *5.*
- So how many groups of ten does it equal? (Signal.) *2.*

- What number does it equal? (Signal.) *20.* (Add to show:) [117:2F]

$$70 - 50$$

Here's the picture. The problem starts with 7 groups of ten (circle all rows), it subtracts 5 groups of ten (touch crossed out rows), and equals 2 groups of ten (touch rows not crossed out) which equals 20.
- What does this picture of tens groups show? (Call on a student.) *Starts with 7 groups of 10. Subtracts 5 groups of ten. Equals 2 groups of ten, which is 20.*

d. (Display:) [117:2G]

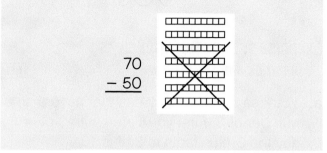

$$10 + 40$$

- (Point to **10**.) Read the problem. (Signal.) *10 plus 40.*
- How many groups of ten does it start with? (Signal.) *1.*
- How many groups of ten does it add? (Signal.) *4.*
- So how many groups of ten does it equal? (Signal.) *5.*
- What number does it equal? (Signal.) *50.* (Add to show:) [117:2H]

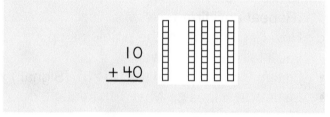

$$10 + 40$$

Here's the picture. The problem starts with 1 group of ten (circle 1 column), it adds 4 groups of ten (touch 4 columns), and equals 5 groups of ten (circle all columns) which equals 50.
- What does this picture of tens groups show? (Call on a student.) *Starts with 1 group of 10. Adds 4 groups of ten. Equals 5 groups of ten, which is 50.*

Connecting Math Concepts

EXERCISE 3: MIXED COUNTING
EXPLAIN ADDING OR SUBTRACTING 10

a. Start with 200 and count by twos to 210. Get 200 going. *Two huuundred.* Count. (Tap.) *202, 204, 206, 208, 210.*

b. Start with 200 and count backward by tens to 100. Get 200 going. *Two huuundred.* Count backward. (Tap.) *190, 180, 170, 160, 150, 140, 130, 120, 110, 100.*

c. Start with 200 and count by fives to 225. Get 200 going. *Two huuundred.* Count. (Tap.) *205, 210, 215, 220, 225.*

d. Start with 200 and count by 25s to 300. Get 200 going. *Two huuundred.* Count. (Tap.) *225, 250, 275, 300.*

e. Start with 200 and count backward by ones to 195. Get 200 going. *Two huuundred.* Count backward. (Tap.) *199, 198, 197, 196, 195.*

f. Start with 200 and count by ones to 205. Get 200 going. *Two huuundred.* Count. (Tap.) *201, 202, 203, 204, 205.*
   (Repeat steps a through f that were not firm.)

g. Listen: You have 200. When you count backward by tens, what number comes next? (Signal.) *190.*
   • So what's 200 minus 10? (Signal.) *190.*

h. Listen: You have 200. When you count by twos, what number comes next? (Signal.) *202.*
   • So what's 200 plus 2? (Signal.) *202.*

i. Listen: You have 200. When you count by 25s, what number comes next? (Signal.) *225.*
   • So what's 200 plus 25? (Signal.) *225.*

j. You have 200. When you count backward by ones, what number comes next? (Signal.) *199.*
   • So what's 200 minus 1? (Signal.) *199.*

k. Listen: You have 200. When you count by fives, what number comes next? (Signal.) *205.*
   • So what's 200 plus 5? (Signal.) *205.*
   (Repeat steps g through k that were not firm.)

l. Listen: If you have 18, how many tens do you have? (Signal.) *1.*
   • How many ones do you have? (Signal.) *8.*
   • So if you add ten to 18, tell me about the tens and ones you end up with. (Call on a student) *2 tens and 8 ones.*
   • Everybody, what's 18 plus 10? (Signal.) *28.*
   • How do you know that 18 plus 10 equals 28? (Call on a student. Idea: *18 is 1 ten and 8 ones. When you add ten, you end up with 2 tens and 8 ones, which is 28.*)

m. Listen: You have 18, and subtract ten. What's the answer? (Signal.) *8.*
   • So what's 18 minus 10? (Signal.) *8.*
   • How do you know that 18 minus 10 equals 8? (Call on a student. Idea: *18 is 1 ten and 8 ones. When you subtract ten, you end up with zero tens and 8 ones, which is 8.*)

EXERCISE 4: FACTS
PLUS/MINUS MIX

a. (Distribute unopened workbooks to children.)
   • Open your workbook to Lesson 117 and find part 1.
     (Observe children and give feedback.)
     (Teacher reference:)

   a. $9 - 6$ c. $8 + 8$ f. $4 + 4$
   b. $12 - 6$ d. $10 - 7$ g. $10 - 6$
                e. $11 - 6$ h. $5 + 5$

   • Touch and read problem A. Get ready. (Signal.) *9 minus 6.*
   • Is the answer the big number or a small number? (Signal.) *A small number.*
   • What's 9 minus 6? (Signal.) *3.*

b. (Repeat the following tasks for problems B through H:)

Touch and read problem __.		Is the answer the big number or a small number?	What's __?	
B	12 – 6	A small number.	12 – 6	6
C	8 + 8	The big number.	8 + 8	16
D	10 – 7	A small number.	10 – 7	3
E	11 – 6	A small number.	11 – 6	5
F	4 + 4	The big number.	4 + 4	8
G	10 – 6	A small number.	10 – 6	4
H	5 + 5	The big number.	5 + 5	10

(Repeat problems that were not firm.)

c. Complete each of the facts. Put your pencil down when you've written all of the equations for part 1.
   (Observe children and give feedback.)

d. Check your work.
   • Touch and read the equation for problem A. (Signal.) *9 – 6 = 3.*
   • (Repeat for:) B, *12 – 6 = 6;* C, *8 + 8 = 16;* D, *10 – 7 = 3;* E, *11 – 6 = 5;* F, *4 + 4 = 8;* G, *10 – 6 = 4;* H, *5 + 5 = 10.*

EXERCISE 5: SOLVING PROBLEMS FROM NUMBER FAMILIES

MISSING ADDEND, MINUEND, OR SUBTRAHEND   REMEDY

a. Find part 2 on your worksheet. ✔
(Teacher reference:)   R Parts C and J

These problems have a missing number.
You're going to make a family for each
problem, and say the problem you'll work to
find the missing number.

• Read problem A. (Signal.) *79 minus how
  many equals 18.*
• Is the missing number a small number or the
  big number? (Signal.) *A small number.*
• What's 79? (Signal.) *The big number.*
• What's 18? (Signal.) *A small number.*
  (Repeat until firm.)

b. Put 79 and 18 in the family for problem A. Put
your pencil down when you've done that much.
(Observe children and give feedback.)
(Display:)   [117:5A]

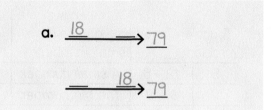

• You should have one of these families for
  problem A. Make sure your family has a small
  number of 18 and the big number of 79. ✔

c. Read problem B. (Signal.) *How many plus
31 equals 81.*
• Is the missing number a small number or the
  big number? (Signal.) *A small number.*
• What's 31? (Signal.) *A small number.*
• What's 81? (Signal.) *The big number.*
  (Repeat until firm.)

d. Put 31 and 81 in the family for problem B. Put
your pencil down when you've done that much.
(Observe children and give feedback.)
(Display:)   [117:5B]

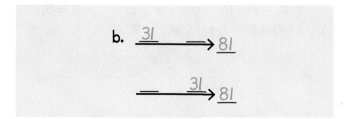

• You should have one of these families for
  problem B. Make sure your family has a small
  number of 31 and the big number of 81. ✔

e. Read problem C. (Signal.) *How many minus
  63 equals 18.*
• Is the missing number a small number or the
  big number? (Signal.) *The big number.*
• What's 63? (Signal.) *A small number.*
• What's 18? (Signal.) *A small number.*
  (Repeat until firm.)

f. Put 63 and 18 in the family for problem C.
Then put the numbers that are shown in the
families for the rest of the problems in part 2.
Put your pencil down when you've written two
numbers in all of the families.
(Observe children and give feedback.)

g. Check your work.
• Family C: Is the big number or a small number
  missing in family C? (Signal.) *The big number.*
• What are the small numbers? (Signal.)
  63 and 18.
  (Display:)   [117:5C]

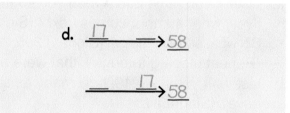

• You should have one of these families for
  problem C. Make sure your family has the
  small numbers of 63 and 18. ✔

h. Family D: Is the big number or a small number
missing in family D? (Signal.) *A small number.*
• What's the other small number? (Signal.) *17.*
• What's the big number? (Signal.) *58.*
  (Display:)   [117:5D]

d. $\underline{17} \Longrightarrow \underline{58}$

$\underline{}\,\underline{17} \Longrightarrow \underline{58}$

• You should have one of these families for
  problem D. Make sure your family has a small
  number of 17 and the big number of 58. ✔

i. Family E: Is the big number or a small number missing in family E? (Signal.) *The big number.*

• What are the small numbers? (Signal.) *63 and 32.*

(Display:) [117:5E]

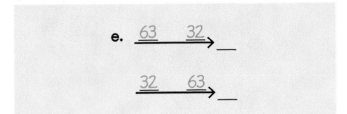

e. 63   32 ⟶ __

    32   63 ⟶ __

• You should have one of these families for problem E. Make sure your family has the small numbers of 63 and 32. ✔

j. Family F: Is the big number or a small number missing in family F? (Signal.) *A small number.*

• What's the other small number? (Signal.) *18.*

• What's the big number? *89.*

(Display:) [117:5F]

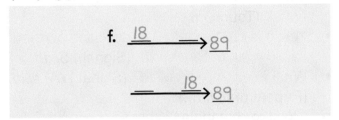

f. 18 ⟶ 89

    18 ⟶ 89

• You should have one of these families for problem F. Make sure your family has a small number of 18 and the big number of 89. ✔

EXERCISE 6: WORD PROBLEMS (COLUMNS)
CONVENTIONAL AND COMPARISON

a. Find part 3 on your worksheet. ✔
(Teacher reference:)

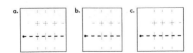

You're going to write the symbols for word problems in columns and work them. Some of the problems you'll work today are money problems. You'll write the dollar signs and the dots for those problems.

• Touch where you'll write the symbols for problem A. ✔

• Listen to problem A: Gerald had 68 cents. Gerald earned another 9 dollars and 34 cents. How much money did Gerald end up with?

• Listen again and write the symbols for the whole problem: Gerald had 68 cents. Gerald earned another 9 dollars and 34 cents. (Observe children and give feedback.)

• Everybody, touch and read the problem for A. Get ready. (Signal.) *68 cents plus 9 dollars and 34 cents.*

b. Work problem A. Put your pencil down when you know how much money Gerald ended up with.
(Observe children and give feedback.)
(Teacher reference:)

a. $ .68
+ 9.34
$10.02

• Read the addition problem and the answer you wrote for A. Get ready. (Signal.) *68 cents plus 9 dollars and 34 cents equals 10 dollars and 2 cents.*

• Everybody, how much money did Gerald end up with? (Signal.) *10 dollars and 2 cents.*

c. Touch where you'll write the symbols for problem B. ✔
You're going to tell me how many pounds lighter one dog was.

• Listen to problem B: Linda's dog weighed 23 pounds. Ben's dog weighed 149 pounds. How much lighter was Linda's dog?

• Listen to problem B again and write the problem for figuring out how much lighter Linda's dog was than Ben's dog. Linda's dog weighed 23 pounds. (Pause.) Ben's dog weighed 149 pounds.
(Observe children and give feedback.)

• Read the problem you wrote for figuring out how much lighter Linda's dog was. (Signal.) *149 minus 23.*

• Work the problem. Put your pencil down when you know how much lighter Linda's dog was.
(Observe children and give feedback.)
(Teacher reference:)

b. 149
− 23
1.26

d. Check your work.
- Read the subtraction problem and the answer for B. Get ready. (Signal.) *149 – 23 = 126.*
- How much lighter was Linda's dog? (Signal.) *126.*
e. Touch where you'll write the symbols for problem C. ✔
- Listen to problem C: There was 23 dollars and 85 cents in a jar. 3 dollars and 81 cents of that money was taken out of the jar. How much money was left in the jar?
- Listen again and write the symbols for the whole problem: There was 23 dollars and 85 cents in a jar. 3 dollars and 81 cents of that money was taken out of the jar.
  (Observe children and give feedback.)
- Everybody, touch and read the problem for C. Get ready. (Signal.) *23 dollars and 85 cents minus 3 dollars and 81 cents.*
f. Work problem C. Put your pencil down when you know how much money was left in the jar.
  (Observe children and give feedback.)
  (Teacher reference:)

$$\begin{array}{r} c. \quad \$23.85 \\ -\quad 3.81 \\ \hline \$20.04 \end{array}$$

- Read the subtraction problem and the answer you wrote for C. Get ready. (Signal.) *23 dollars and 85 cents minus 3 dollars and 81 cents equals 20 dollars and 4 cents.*
- Everybody, how much money was left in the jar? (Signal.) *20 dollars and 4 cents.*

EXERCISE 7: FRACTIONS   REMEDY

a. (Display:)   [117:7A]

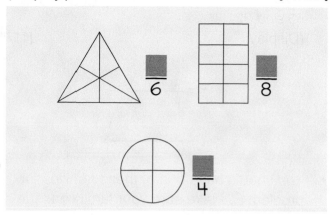

- (Point to triangle.) What shape is this? (Touch.) *(A) triangle.*
  This triangle is divided into parts that are the same size.
- (Point to **6**.) This number shows how many parts the triangle is divided into. How many parts? (Touch.) *6.*
- Yes, it is divided into 6 parts. Each part is called a sixth. Say **sixth**. (Signal.) *Sixth.*
- What is each part called? (Signal.) *(A) sixth.*
  (Repeat until firm.)
b. (Point to rectangle.) What shape is this? (Touch.) *(A) rectangle.*
- (Point to **8**.) How many equal parts is this rectangle divided into? (Touch.) *8.*
- So each part is called **an eighth.** What is each part called? (Signal.) *(An) eighth.*
c. (Point to circle.) What shape is this? (Touch.) *(A) circle.*
- (Point to **4**.) How many equal parts is this circle divided into? (Touch.) *4.*
- What is each part called? (Signal.) *(A) fourth.*
  (Repeat steps b and c until firm.)
d. Look at the triangle again and get ready to tell me what each part is called. ✔
- What is a part of the triangle called? (Signal.) *(A) sixth.*
e. Look at the rectangle and get ready to tell me what each part is called. ✔
- What is a part of the rectangle called? (Signal.) *(An) eighth.*
f. Look at the circle and get ready to tell me what each part is called. ✔
- What is a part of the circle called? (Signal.) *(A) fourth.*
  (Repeat steps d through f until firm.)

g. I'll tell you how many equal parts a shape is divided into. You'll tell me what a part of that shape is called.
- Listen: A shape is divided into 10 equal parts. How many parts? (Signal.) *10.*
- What is a part of that shape called? (Signal.) *(A) tenth.*
- Listen: A shape is divided into 14 equal parts. How many parts? (Signal.) *14.*
- What is a part of that shape called? (Signal.) *(A) fourteenth.*
- Listen: A shape is divided into 100 equal parts. How many parts? (Signal.) *100.*
- What is a part of that shape called? (Signal.) *(A) one hundredth.*
- Listen: A shape is divided into 9 equal parts. How many parts? (Signal.) *9.*
- What is a part of that shape called? (Signal.) *(A) ninth.*
(Repeat shapes that were not firm.)

h. (Display:) [117:7B]

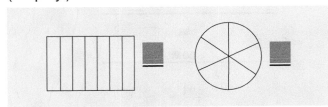

For each of these shapes, you're going to tell me what to write for the number of parts, and what each part is called.
- (Point to rectangle.) What shape is this? (Signal.) *(A) rectangle.*
- I'll touch each part. Count the parts to yourself. (Touch each part.)
- How many equal parts are in this rectangle? (Signal.) *7.*
(Add to show:) [117:7C]

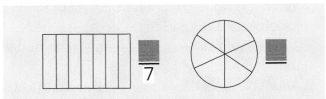

- What is a part of the rectangle called? (Signal.) *(A) seventh.*
i. (Point to circle.) What shape is this? (Signal.) *(A) circle.*
- I'll touch each part. Count the parts to yourself. (Touch each part.)

- How many equal parts are in this circle? (Signal.) *6.*
(Add to show:) [117:7D]

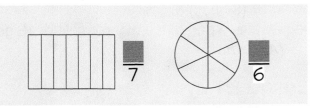

- What is a part of the circle called? (Signal.) *(A) sixth.*
j. Let's do those again.
- What is a part of the rectangle called? (Signal.) *(A) seventh.*
- What is a part of the circle called? (Signal.) *(A) sixth.*
k. Find part 4 on your worksheet. ✔
(Teacher reference:) R Parts A and Q

You're going to write the number of equal parts for each shape. Then you'll tell me what the parts for each shape are called.
- Touch shape A. It's a square. ✔
- Count the parts for the square and write the number.
(Observe children and give feedback.)
- Everybody, how many equal parts is shape A divided into? (Signal.) *9.*
(Display:) [117:7E]

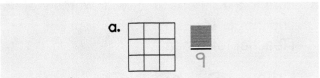

Here's what you should have.
- What is a part of that square called? (Signal.) *(A) ninth.*
l. Count the parts for shapes B and C and write the numbers. Put your pencil down when you've written the number of parts for each shape.
(Observe children and give feedback.)
m. Check your work.
- Shape B. What shape is it? (Signal.) *(A) circle.*
- How many equal parts is the circle divided into? (Signal.) *7.*
- What is a part of the circle called? (Signal.) *(A) seventh.*

n. Shape C. What shape is it? (Signal.)
(A) triangle.
• How many equal parts is the triangle divided into? (Signal.) *6.*
• What is a part of the triangle called? (Signal.)
(A) sixth.

EXERCISE 8: FACTS
SUBTRACTION

a. Find part 5. ✔
(Teacher reference:)

a. $8 - 5$	c. $11 - 5$	f. $9 - 7$
b. $11 - 8$	d. $14 - 7$	g. $10 - 5$
	e. $10 - 4$	h. $10 - 7$

• These are subtraction problems you'll work. Think of the number families for each problem and write the answers. Put your pencil down when you've completed the equations in part 5.
(Observe children and give feedback.)

b. Check your work. You'll read the equation for each problem.
• Problem A. (Signal.) $8 - 5 = 3$.
• (Repeat for:) B, $11 - 8 = 3$; C, $11 - 5 = 6$; D, $14 - 7 = 7$; E, $10 - 4 = 6$; F, $9 - 7 = 2$; G, $10 - 5 = 5$; H, $10 - 7 = 3$.

EXERCISE 9: TELLING TIME
ANALOG HOURS AND HALF HOURS

a. Turn to the other side of your worksheet and find part 6. ✔
(Teacher reference:)

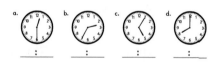

You're going to write the time for each clock. One clock shows zero minutes. Another clock shows 30 minutes. You'll count the minutes to yourself for the other two clocks.
• Write the time for each clock. Put your pencil down when you've written the hour number and the minute number for each clock in part 6.
(Observe children and give feedback.)

b. Check your work. You'll read the time for each clock.
• Clock A. (Signal.) *Twelve 30.*
• (Repeat for:) B, *Two 35;* C, *Five oh 5;* D, *8 o'clock.*
(Display:)   [117:9A]

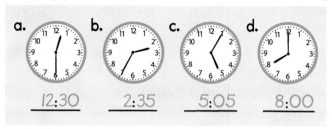

Here's what you should have for the clocks.

EXERCISE 10: INDEPENDENT WORK

a. Find part 7. ✔
(Teacher reference:)

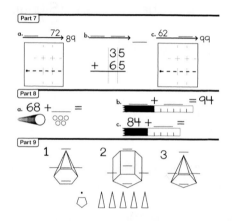

In part 7, you'll work a column problem and complete the family.
In part 8, you'll complete the equation for the circles and the equations for the rulers.

b. Find part 9. ✔
• Touch object 1. ✔
• Touch object 2. ✔
• Touch object 3. ✔
You're going to write letters for all of the faces for objects 1, 2, and 3.
• Touch the picture of the faces for the bottom, top, and sides. ✔
After you label the faces, you'll circle the object that has the right bottom, top, and sides.

c. Complete worksheet 117.
(Observe children and mark incorrect responses on children's worksheets as you give feedback.)

Lesson  118

EXERCISE 1: FACT RELATIONSHIPS
ADDITION AND SUBTRACTION

a. You'll say facts for families that have a small number of 3. First you'll start with 1 plus 3 and say the plus-3 facts to 10 plus 3.
- Say the fact for 1 plus 3. Get ready. (Signal.) *1 + 3 = 4.*
- Next fact. (Signal.) *2 + 3 = 5.*
- (Repeat for:) *3 + 3 = 6, 4 + 3 = 7, 5 + 3 = 8, 6 + 3 = 9, 7 + 3 = 10, 8 + 3 = 11, 9 + 3 = 12, 10 + 3 = 13.*
  (Repeat until firm.)

b. Now you'll start with 4 minus 3 and say the minus-3 facts to 13 minus 3.
- Say the fact for 4 minus 3. (Signal.) *4 – 3 = 1.*
- Say the minus-3 fact that starts with 5. (Signal.) *5 – 3 = 2.*
- Next fact. (Signal.) *6 – 3 = 3.*
- (Repeat for:) *7 – 3 = 4, 8 – 3 = 5, 9 – 3 = 6, 10 – 3 = 7, 11 – 3 = 8, 12 – 3 = 9, 13 – 3 = 10.*
  (Repeat until firm.)

c. Now you'll start with 3 plus 1 and say the facts that start with 3.
- Say the fact for 3 plus 1. Get ready. (Signal.) *3 + 1 = 4.*
- Next fact. Get ready. (Signal.) *3 + 2 = 5.*
- (Repeat for:) *3 + 3 = 6, 3 + 4 = 7, 3 + 5 = 8, 3 + 6 = 9, 3 + 7 = 10, 3 + 8 = 11, 3 + 9 = 12, 3 + 10 = 13.*
  (Repeat until firm.)

d. Now I'll mix some of the problems up and you'll tell me the answer.
- Listen: 8 plus 3. What's 8 plus 3? (Signal.) *11.*
- Listen: 8 minus 3. What's 8 minus 3? (Signal.) *5.*
- Listen: 3 plus 7. What's 3 plus 7? (Signal.) *10.*
- Listen: 11 minus 3. What's 11 minus 3? (Signal.) *8.*
  (Repeat until firm.)

■■■■■■■■ INDIVIDUAL TURNS ■■■■■■■■

(Call on individual students to perform one of the following tasks.)

- What's 11 minus 3? (Call on a student.) *8.*
- What's 3 plus 7? (Call on a student.) *10.*
- What's 8 minus 3? (Call on a student.) *5.*
- What's 8 plus 3? (Call on a student.) *11.*

EXERCISE 2: ADDING AND SUBTRACTING TENS NUMBERS
DRAWINGS AND REASONING

a. You're going to read problems. For each problem, you'll tell me about the picture for it. Then I'll show you the picture and you'll explain it.
   (Display:)   [118:2A]

$$\begin{array}{r} 10 \\ +\,40 \\ \hline \end{array}$$

- (Point to **10.**) Read the problem. (Signal.) *10 plus 40.*
- How many groups of ten does it start with? (Signal.) *1.*
- How many groups of ten does it add? (Signal.) *4.*
- So how many groups of ten does it equal? (Signal.) *5.*
- What number does it equal? (Signal.) *50.*
  (Add to show:)   [118:2B]

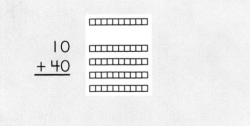

Here's the picture. The problem starts with 1 group of ten (touch 1st row), it adds 4 groups of ten (circle group of 4 rows), and equals 5 groups of ten (circle all rows) which equals 50.
- What does this picture of tens groups show? (Call on a student.) *Starts with 1 group of ten. Adds 4 groups of ten. Equals 5 groups of ten, which is 50.*

b. (Display:) [118:2C]

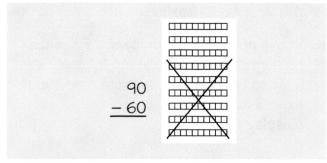

90
− 60

- (Point to **90**.) Read the problem. (Signal.) *90 minus 60.*
- How many groups of ten does it start with? (Signal.) *9.*
- How many groups of ten does it subtract? (Signal.) *6.*
- So how many groups of ten does it equal? (Signal.) *3.*
- What number does it equal? (Signal.) *30.* (Add to show:) [118:2D]

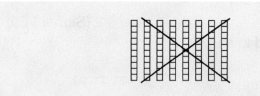

90
− 60

Here's the picture. The problem starts with 9 groups of ten (circle all rows), it subtracts 6 groups of ten (touch crossed out rows), and equals 3 groups of ten (touch rows not crossed out) which equals 30.

- What does this picture of tens groups show? (Call on a student.) *Starts with 9 groups of ten. Subtracts 6 groups of ten. Equals 3 groups of ten, which is 30.*

c. (Display:) [118:2E]

Here's a picture for a new problem. You'll tell me what the picture shows and I'll write the equation.

- What does this picture show? (Call on a student.) *Starts with 8 groups of ten. Subtracts 7 groups of ten. Equals 1 group of ten, which is 10.*

- Everybody, say the equation for this picture. (Signal.) *80 − 70 = 10.* (Add to show:) [118:2F]

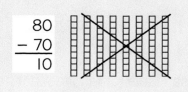

80
− 70
10

Here's the equation for the picture.

d. (Display:) [118:2G]

Here's a picture for a new problem. You'll tell me what the picture shows and I'll write the equation.

- What does this picture show? (Call on a student.) *Starts with 5 groups of ten. Adds 3 groups of ten. Equals 8 groups of ten, which is 80.*
- Everybody, say the equation for this picture. (Signal.) *50 + 30 = 80.* (Add to show:) [118:2H]

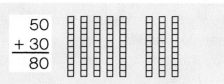

50
+ 30
80

Here's the equation for the picture.

EXERCISE 3: MIXED COUNTING
EXPLAIN ADDING OR SUBTRACTING 10

a. Start with 80 and count by twos to 90. Get 80 going. *Eightyyy.* Count. (Tap.) *82, 84, 86, 88, 90.*
b. Start with 80 and count backward by tens. Get 80 going. *Eightyyy.* Count backward. (Tap.) *70, 60, 50, 40, 30, 20, 10.*
c. Start with 80 and count by fives to 100. Get 80 going. *Eightyyy.* Count. (Tap.) *85, 90, 95, 100.*
d. Start with 80 and count by tens to 100. Get 80 going. *Eightyyy.* Count. (Tap.) *90, 100.*
e. Start with 80 and count backward by ones to 75. Get 80 going. *Eightyyy.* Count backward. (Tap.) *79, 78, 77, 76, 75.*
f. Start with 80 and count by ones to 85. Get 80 going. *Eightyyy.* Count. (Tap.) *81, 82, 83, 84, 85.* (Repeat steps a through f that were not firm.)

g. Listen: You have 80. When you count by twos, what number comes next? (Signal.) *82.*
• So what's 80 plus 2? (Signal.) *82.*
h. Listen: You have 80. When you count by tens, what number comes next? (Signal.) *90.*
• So what's 80 plus 10? (Signal.) *90.*
i. Listen: You have 80. When you count backward by ones, what number comes next? (Signal.) *79.*
• So what's 80 minus 1? (Signal.) *79.*
j. You have 80. When you count backward by tens, what number comes next? (Signal.) *70.*
• So what's 80 minus 10? (Signal.) *70.*
k. Listen: You have 80. When you count by fives, what number comes next? (Signal.) *85.*
• So what's 80 plus 5? (Signal.) *85.*
(Repeat steps g through k that were not firm.)
l. Listen: If you have 46, how many tens do you have? (Signal.) *4.*
• How many ones do you have? (Signal.) *6.*
• So if you add ten, tell me about the tens and ones you end up with. (Call on a student.) *5 tens and 6 ones.*
• Everybody, what's 46 plus 10? (Signal.) *56.*
• How do you know that 46 plus 10 equals 56? (Call on a student. Idea: *46 is 4 tens and 6 ones. When you add ten, you end up with 5 tens and 6 ones, which is 56.*)
m. Listen: If you have 46, and subtract ten, what's the answer? (Signal.) *36.*
• So what's 46 minus 10? (Signal.) *36.*
• How do you know that 46 minus 10 equals 36? (Call on a student. Idea: *46 is 4 tens and 6 ones. When you subtract ten, you end up with 3 tens and 6 ones, which is 36.*)

EXERCISE 4: FACTS
ADDITION/SUBTRACTION

a. (Display:) [118:4A]

7 – 3	11 – 8	16 – 8
7 + 3	11 – 5	5 – 5
7 + 7	6 + 6	5 + 5

• (Point to **7 – 3**.) Listen: 7 – 3. Is the answer the big number or a small number? (Signal.) *A small number.*
• Say the fact. (Signal.) *7 – 3 = 4.*

b. (Point to **7 + 3**.) 7 + 3. Is the answer the big number or a small number? (Signal.) *The big number.*
• Say the fact. (Signal.) *7 + 3 = 10.*
c. (Repeat the following tasks for the remaining problems:)

(Point to __.) __. Is the answer the big number or a small number?		Say the fact.
7 + 7	The big number.	7 + 7 = 14
11 – 8	A small number.	11 – 8 = 3
11 – 5	A small number.	11 – 5 = 6
6 + 6	The big number.	6 + 6 = 12
16 – 8	A small number.	16 – 8 = 8
5 – 5	A small number.	5 – 5 = 0
5 + 5	The big number.	5 + 5 = 10

(Repeat problems that were not firm.)

EXERCISE 5: FRACTIONS [REMEDY]

a. (Display:) [118:5A]

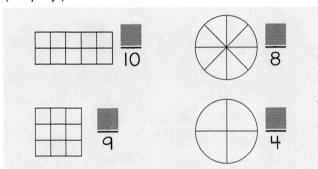

The number next to each shape shows how many equal parts the shape is divided into. For each shape, you're going to tell me the number of parts and what a part is called.
b. (Point to rectangle.) How many equal parts is this shape divided into? (Signal.) *10.*
• What is each part called? (Signal.) *(A) tenth.*
c. (Repeat the following tasks for the remaining shapes:)

(Point to __.) How many equal parts is this shape divided into?		What is each part called?
First circle	8	(An) eighth
Square	9	(A) ninth
Second circle	4	(A) fourth

(Repeat shapes that were not firm.)

d. (Distribute unopened workbooks to children.)
- Open your workbook to Lesson 118 and find part 1.
  (Observe children and give feedback.)
  (Teacher reference:) **R Part R**

You're going to write the number of equal parts for each shape. Then you'll tell me what the parts for each shape are called.
- Count the parts for shape A and write the number.
  (Observe children and give feedback.)
- Everybody, how many equal parts is shape A divided into? (Signal.) *6.*
- What is a part of shape A called? (Signal.) *(A) sixth.*
  (Display:) [118:5B]

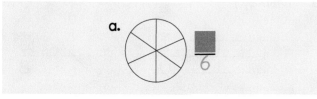

Here's what you should have.
e. Count the parts for shapes B and C and write the numbers. Put your pencil down when you've written the number of parts for each shape.
  (Observe children and give feedback.)
f. Check your work.
- Shape B. How many equal parts is shape B divided into? (Signal.) *4.*
- What is a part of shape B called? (Signal.) *(A) fourth.*
g. Shape C. How many equal parts is shape C divided into? (Signal.) *7.*
- What is a part of shape C called? (Signal.) *(A) seventh.*
- h. Let's do those again. You'll look at the number of parts for each shape and tell me what a part is called.
- Shape A. What is a part called? (Signal.) *(A) sixth.*
- Shape B. What is a part called? (Signal.) *(A) fourth.*
- Shape C. What is a part called? (Signal.) *(A) seventh.*
  (Repeat step h until firm.)

EXERCISE 6: SOLVING PROBLEMS FROM NUMBER FAMILIES
MISSING ADDEND, MINUEND, OR SUBTRAHEND **REMEDY**

a. Find part 2 on your worksheet. ✔
  (Teacher reference:) **R Part K**

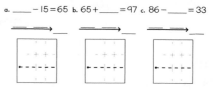

These problems have a missing number. You're going to make a family for each problem, write the problem you'll work for the missing number, and find it.
- Read problem A. (Signal.) *How many minus 15 equals 65.*
- Is the missing number a small number or the big number? (Signal.) *The big number.*
- So what are 15 and 65? (Signal.) *Small numbers.*
b. Read problem B. (Signal.) *65 plus how many equals 97.*
- Is the missing number a small number or the big number? (Signal.) *A small number.*
- What's 65? (Signal.) *A small number.*
- What's 97? (Signal.) *The big number.*
c. Read problem C. (Signal.) *86 minus how many equals 33.*
- Is the missing number a small number or the big number? (Signal.) *A small number.*
- What's 86? (Signal.) *The big number.*
- What's 33? (Signal.) *A small number.*
  (Repeat until firm.)

d. Make the family for each problem with two numbers. Put your pencil down when you've written the family with two numbers for each problem in part 2.
(Observe children and give feedback.)
(Display:) [118:6A]

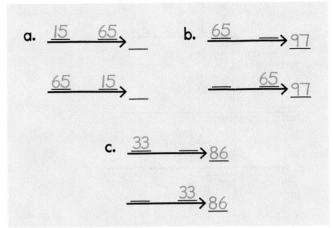

a. 15 —→ 65, __   b. 65 —→ 97

   65 —→ 15, __   __ —→ 65, 97

c. 33 —→ 86

   __ —→ 33, 86

- You should have one of these families for each problem. Make sure your families have the numbers in the correct places. ✔

e. Now you're going to say the problem for finding the missing number for each family.
- Family A. Say the problem for the missing number. (Signal.) *15 plus 65.*
- Family B. (Signal.) *97 minus 65.*
- Family C. (Signal.) *86 minus 33.*
(Repeat until firm.)

f. In the box below family A, write the column problem and work it. Then write the missing number in the family, and complete the equation above family A. Put your pencil down when you've completed problem A.
(Observe children and give feedback.)

g. Check your work.
- Read the column problem and the answer for problem A. (Signal.) *15 + 65 = 80.*
- Look at the equation above family A. How many minus 15 equals 65? (Signal.) *80.*
(Display:) [118:6B]

a. 80 − 15 = 65

   15 —→ 65, 80

   ```
     1 5
   + 6 5
   ─────
     8 0
   ```

- Here's what you should have for problem A. Check your work. ✔

h. Write the column problems for the rest of the families and work them. Then write the missing number for each family and complete the equation above it. Put your pencil down when you've completed part 2.
(Observe children and give feedback.)

i. Check your work.
- Problem B. Read the column problem and the answer. (Signal.) *97 − 65 = 32.*
- Look at the equation above family B. 65 plus how many equals 97? (Signal.) *32.*
- Problem C. Read the column problem and the answer. (Signal.) *86 − 33 = 53.*
- Look at the equation above family C. 86 minus how many equals 33? (Signal.) *53.*
(Display:) [118:6C]

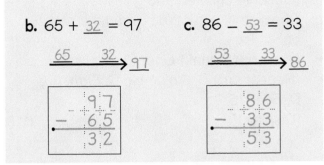

b. 65 + 32 = 97   c. 86 − 53 = 33

   65 —→ 32, 97   53 —→ 33, 86

   ```
     9 7
   − 6 5
   ─────
     3 2
   ```
   ```
     8 6
   − 3 3
   ─────
     5 3
   ```

Here's what you should have for problems B and C.

EXERCISE 7: FACTS
SUBTRACTION

a. Find part 3. ✔
(Teacher reference:)

a. 11 − 8   c. 11 − 6   f. 10 − 7
b. 9 − 6   d. 14 − 7   g. 10 − 4
               e. 7 − 4   h. 8 − 4

- These are subtraction problems you'll work. Think of the number families for each problem and write the answers. Put your pencil down when you've completed the equations in part 3.
(Observe children and give feedback.)

b. Check your work. You'll read the equation for each problem.
- Problem A. (Signal.) *11 − 8 = 3.*
- (Repeat for:) B, *9 − 6 = 3;* C, *11 − 6 = 5;* D, *14 − 7 = 7;* E, *7 − 4 = 3;* F, *10 − 7 = 3;* G, *10 − 4 = 6;* H, *8 − 4 = 4.*

EXERCISE 8: TELLING TIME
ANALOG HOURS AND HALF HOURS

a. Find part 4. ✔
    (Teacher reference:)

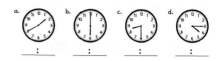

You're going to write the time for each clock. One clock shows zero minutes. Another clock shows 30 minutes. You'll count the minutes to yourself for the other two clocks.

*   Write the time for each clock. Put your pencil down when you've written the hour number and the minute number for each clock in part 4.
    (Observe children and give feedback.)

b. Check your work. You'll read the time for each clock.

*   Clock A. (Signal.) *One 40.*
*   (Repeat for:) B, *6 o'clock;* C, *Eight 30;* D, *Four 15.*
    (Display:)   [118:8A]

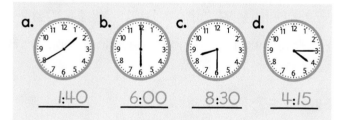

Here's what you should have for the clocks.

EXERCISE 9: INDEPENDENT WORK

a. Turn to the other side of your worksheet and find part 5. ✔
    (Teacher reference:)

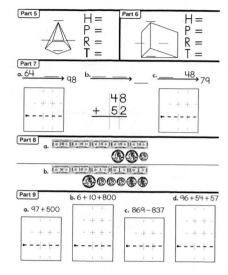

In parts 5 and 6, you'll label the faces and complete each equation to show the number of hexagons, pentagons, rectangles, and triangles.

In part 7, you'll work column problems and complete the families.

In part 8, you'll write the dollars and cents amount for each group of bills and coins.

In part 9, you'll write each problem in a column and work it.

b. Complete worksheet 118.
    (Observe children and mark incorrect responses on children's worksheets as you give feedback.)

Lesson

EXERCISE 1: FACT RELATIONSHIPS
ADDITION AND SUBTRACTION

a. You'll say facts for families that have a small number of 3. First you'll start with 3 plus 1 and say the 3-plus facts to 3 plus 10.
- Say the fact for 3 plus 1. Get ready. (Signal.) $3 + 1 = 4$.
- Next fact. Get ready. (Signal.) $3 + 2 = 5$.
- (Repeat for:) $3 + 3 = 6, 3 + 4 = 7, 3 + 5 = 8, 3 + 6 = 9, 3 + 7 = 10, 3 + 8 = 11, 3 + 9 = 12, 3 + 10 = 13$.
 (Repeat until firm.)

b. Now you'll start with 4 minus 3 and say the minus-3 facts to 13 minus 3.
- Say the fact for 4 minus 3. (Signal.) $4 - 3 = 1$.
- Say the minus-3 fact that starts with 5. (Signal.) $5 - 3 = 2$.
- Next fact. (Signal.) $6 - 3 = 3$.
- (Repeat for:) $7 - 3 = 4, 8 - 3 = 5, 9 - 3 = 6, 10 - 3 = 7, 11 - 3 = 8, 12 - 3 = 9, 13 - 3 = 10$.
 (Repeat until firm.)

c. Now you'll start with 1 plus 3 and say the plus-3 facts to 10 plus 3.
- Say the fact for 1 plus 3. Get ready. (Signal.) $1 + 3 = 4$.
- Next fact. (Signal.) $2 + 3 = 5$.
- (Repeat for:) $3 + 3 = 6, 4 + 3 = 7, 5 + 3 = 8, 6 + 3 = 9, 7 + 3 = 10, 8 + 3 = 11, 9 + 3 = 12, 10 + 3 = 13$.
 (Repeat until firm.)

d. Now I'll mix some of the problems up and you'll tell me the answer.
- Listen: 3 plus 5. What's 3 plus 5? (Signal.) 8.
- Listen: 10 minus 3. What's 10 minus 3? (Signal.) 7.
- Listen: 3 plus 8. What's 3 plus 8? (Signal.) 11.
- Listen: 11 minus 3. What's 11 minus 3?
 (Signal.) 8.
 (Repeat until firm.)

INDIVIDUAL TURNS

(Call on individual students to perform one of the following tasks.)

- What's 11 minus 3? (Call on a student.) 8.
- What's 3 plus 8? (Call on a student.) 11.
- What's 10 minus 3? (Call on a student.) 7.
- What's 3 plus 5? (Call on a student.) 8.

EXERCISE 2: FRACTIONS   REMEDY

a. You've learned the name for parts of shapes that are divided up equally. I'll tell you a name for the parts. You'll tell me how many equal parts the shape is divided into.
- Listen: Fourths. What name? (Signal.) *Fourths.*
- A shape with fourths is divided into how many equal parts? (Signal.) 4.

b. (Repeat the following tasks for the remaining names:)

Listen: __. What name?		A shape with __ is divided into how many equal parts?	
Elevenths	*Elevenths*	Elevenths	11
Eighths	*Eighths*	Eighths	8
Sixths	*Sixths*	Sixths	6
Ninths	*Ninths*	Ninths	9

(Repeat shapes that were not firm.)

c. The names for some parts are tricky because they don't tell how many parts the shape is divided into.
- Listen: Fifths. What name? (Signal.) *Fifths.*
- Who thinks they know how many equal parts a shape with fifths is divided into? (Call on a student.) 5.
 Yes, if the parts of a shape are fifths, the shape is divided into 5 parts.

d. Again: Fifths. What name? (Signal.) *Fifths.*
- A shape with fifths is divided into how many parts? (Signal.) 5.

e. Here's another tricky name: Halves. What name? (Signal.) *Halves.*
- A shape with halves is divided into 2 equal parts. A shape with halves is divided into how many parts? (Signal.) 2.

f. Let's do those again.
- Listen: Fifths. What name? (Signal.) *Fifths.*
- A shape with fifths is divided into how many equal parts? (Signal.) 5.

g. Listen: Halves. What name? (Signal.) *Halves.*
- A shape with halves is divided into how many equal parts? (Signal.) 2.
 (Repeat steps f and g until firm.)

h. Listen: Sevenths. What name? (Signal.) *Sevenths.*
- A shape with sevenths is divided into how many equal parts? (Signal.) 7.

i. (Repeat the following tasks for the following names:)

Listen: __. What name?		A shape with __ is divided into how many equal parts?	
Halves	*Halves*	Halves	2
Tenths	*Tenths*	Tenths	10
Fifths	*Fifths*	Fifths	5

(Repeat names that were not firm.)

EXERCISE 3: ADDING AND SUBTRACTING TENS NUMBERS

DRAWINGS AND REASONING

a. You're going to look at pictures of groups of 10. You'll explain each picture. Then you'll tell me the equation for it.
(Display:)   [119:3A]

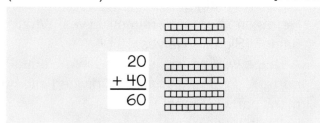

- (Point to first 2 rows.) How many tens does it start with? (Signal.) *2.*
- (Point to next 4 rows.) How many tens does it add? (Signal.) *4.*
- (Circle all the rows.) How many tens does it equal? (Signal.) *6.*
- What number does it equal? (Signal.) *60.*
- Everybody, say the equation for this picture. (Signal.) *20 + 40 = 60.*
(Add to show:)   [119:3B]

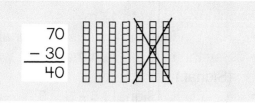

Here's the equation for the picture.

b. (Display:)   [119:3C]

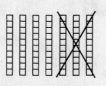

- (Circle all the columns.) How many groups of ten does it start with? (Signal.) *7.*
- (Point to crossed out columns.) How many groups of ten does it subtract? (Signal.) *3.*
- (Point to columns that are not crossed out.) How many groups of ten does it equal? (Signal.) *4.*
- What number does it equal? (Signal.) *40.*
- Everybody, say the equation for this picture. (Signal.) *70 − 30 = 40.*
(Add to show:)   [119:3D]

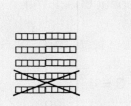

Here's the equation for the picture.

c. (Display:)   [119:3E]

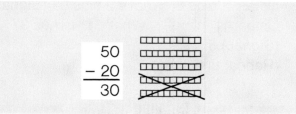

Here's a picture for a new problem.
- What does this picture show? (Call on a student.) *Starts with 5 groups of ten. Subtracts 2 groups of ten. Equals 3 groups of ten, which is 30.*
- Everybody, say the equation for this picture. (Signal.) *50 − 20 = 30.*
(Add to show:)   [119:3F]

Here's the equation for the picture.

d. (Display:)   [119:3G]

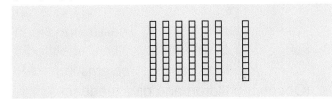

Here's a picture for a new problem.
- What does this picture show? (Call on a student.) *Starts with 6 groups of ten. Adds 1 group of ten. Equals 7 groups of ten, which is 70.*
- Everybody, say the equation for this picture. (Signal.) *60 + 10 = 70.*
(Add to show:)   [119:3H]

Here's the equation for the picture.

EXERCISE 4: EQUATIONS
EQUALITY/INEQUALITY

a. (Display:)   [119:4A]

$$8 = 9 - 1$$
$$8 = 9 + 1$$
$$5 + 2 = 8 - 2$$

Here are equations with a number you add or subtract on the other side.
- (Point to **8 = 9 – 1**.) Read the equation. (Signal.) *8 = 9 – 1.*
- What does 9 minus 1 equal? (Signal.) *8.*
- So are the sides equal? (Signal.) *Yes.*
- So this equation is true. Is this equation true? (Signal.) *Yes.*
b. (Point to **8 = 9 + 1**.) Read the equation. (Signal.) *8 = 9 + 1.*
- What does 9 plus 1 equal? (Signal.) *10.*
- Does 8 equal 10? (Signal.) *No.*
- So are the sides equal? (Signal.) *No.*
- So is this equation true? (Signal.) *No.*
c. (Point to **5**.) Read the equation. (Signal.) *5 + 2 = 8 – 2.*
- What does 5 plus 2 equal? (Signal.) *7.*
- What does 8 minus 2 equal? (Signal.) *6.*
- Does 7 equal 6? (Signal.) *No.*
- So are the sides equal? (Signal.) *No.*
- So is this equation true? (Signal.) *No.*

d. Let's do those again and I'll fix the sign.
- (Point to **9 – 1.**) What does this side equal? (Signal.) *8.*
(Add to show:)   [119:4B]

$$8 = 9 \overset{8}{-} 1$$

- Are the sides of this equation equal? (Signal.) *Yes.*
- So is this equation true? (Signal.) *Yes.*
e. (Point to **9 + 1.**) What does this side equal? (Signal.) *10.*
(Add to show:)   [119:4C]

$$8 = 9 \overset{10}{+} 1$$

- Are the sides of this equation equal? (Signal.) *No.*
- So is this equation true? (Signal.) *No.*
(Add to show:)   [119:4D]

$$8 \overset{10}{\neq} 9 + 1$$

A single line through the equals shows that the sides are not equal. I'll read the statement for the sides the problem started with. (Touch symbols.) 8 is not equal to 9 plus 1.
- (Point to **8 ≠ 9 + 1.**) Read the statement. (Touch.) *8 is not equal to 9 plus 1.*
f. (Point to **5 + 2.**) What does this side equal? (Signal.) *7.*
- (Point to **8 – 2.**) What does this side equal? (Signal.) *6.*
(Add to show:)   [119:4E]

$$5 \overset{7}{+} 2 = 8 \overset{6}{-} 2$$

- Are the sides of this equation equal? (Signal.) *No.*
- So is this equation true? (Signal.) *No.*
(Add to show:)   [119:4F]

$$5 \overset{7}{+} 2 \neq 8 \overset{6}{-} 2$$

- (Point to **5 + 2 ≠ 8 – 2.**) Read the statement. (Touch.) *5 plus 2 is not equal to 8 minus 2.*

g. (Distribute unopened workbooks to children.)
• Open your workbook to Lesson 119 and find part 1.
    (Observe children and give feedback.)
    (Teacher reference:)

a. 10 = 6 + 5   c. 9 – 1 = 7 + 1   e. 5 = 8 – 5
b. 19 = 10 – 9   d. 7 – 2 = 5 + 2   f. 13 + 1 = 15 – 1

Some of these equations are true and some of them are false. You're going to fix the signs for the equations that are false so that all of the statements in part 1 are true.
• Read problem A. (Signal.) *10 = 6 + 5.*
• What does 6 plus 5 equal? (Signal.) *11.*
• Write 11 above 6 plus 5 to show what the side equals. ✔
• Are the sides of equation A equal? (Signal.) *No.*
• So is equation A true? (Signal.) *No.*
• Make a single line down through the equals, to show the sides are not equal. ✔
    (Display:)   [119:4G]

$$a. \quad 10 \neq 6 + 5$$

Here's what you should have for problem A.
• Touch and read the statement for A. (Signal.) *10 is not equal to 6 plus 5.*
h. Read problem B. (Signal.) *19 = 10 – 9.*
• What does 10 minus 9 equal? (Signal.) *1.*
• Write 1 above 10 minus 9 to show what the side equals. ✔
• Are the sides of equation B equal? (Signal.) *No.*
• So is equation B true? (Signal.) *No.*
• What do you make through the equals to show the sides are not equal? (Call on a student.) *A single line.*
• Fix the equals for problem B. ✔
• Make a single line down through the equals, to show the sides are not equal. ✔
    (Display:)   [119:4H]

$$b. \quad 19 \neq 10 - 9$$

Here's what you should have for problem B.
• Touch and read the statement for B. (Signal.) *19 is not equal to 10 minus 9.*

i. Read the rest of the problems to yourself. If a side adds or subtracts, write the number that the side equals above it. Then fix the sign. Put your pencil down when you've completed the statements for all of the problems in part 1.
    (Observe children and give feedback.)
j. Check your work.
• Problem C: 9 minus 1 equals 7 plus 1. Is that equation true or false? (Signal.) *True.*
• Read the statement for C. (Signal.) *9 – 1 = 7 + 1.*
k. (Repeat the following tasks for the remaining problems:)

Problem __. Is that equation true or false?		Read the statement for __.	
D: 7 – 2 = 5 + 2	*False*	D	*7 – 2 is not equal to 5 + 2.*
E: 5 = 8 – 5	*False*	E	*5 is not equal to 8 – 5.*
F: 13 + 1 = 15 – 1	*True*	F	*13 + 1 = 15 – 1.*

EXERCISE 5: WORD PROBLEMS
MISSING ADDEND, MINUEND, OR SUBTRAHEND

a. You've worked word problems that add or subtract. Today we'll work word problems that tell you the number after the equals sign but ask you about the number you start with or the number you add or subtract.
• Listen to the problem: There were 12 apples in a basket. Jamie put some more apples in the basket. 19 apples ended up in the basket. How many apples did Jamie put in the basket?
• Listen again and tell me the symbols: There were 12 apples in a basket. What do I write for that part? (Signal.) *12.*
    (Display:)   [119:5A]

    12 _____

• Next part: Jamie put some more apples in the basket. What sign do I write for that part? (Signal.) *Plus.*
• I don't know how many apples Jamie added to the basket, so I make a space for how many. What do I make after the plus sign? (Signal.) *A space.*
    (Add to show:)   [119:5B]

    12 + __ _____

b. Listen to the next part: 19 apples ended up in the basket. It tells what the basket equals, so I write an equals. Then I write a number. What symbol do I write first? (Signal.) *An equals.* (Add to show:) [119:5C]

$$12 + \underline{\quad} = \underline{\qquad}$$

- Then what number do I write? (Signal.) *19.* (Add to show:) [119:5D]

$$12 + \underline{\quad} = 19$$

c. We can make a number family to figure out the missing number.
- Is the big number or a small number missing? (Signal.) *A small number.*
- What's 12? (Signal.) *A small number.*
- What's 19? (Signal.) *The big number.*

d. (Add to show:) [119:5E]

$$12 + \underline{\quad} = 19$$

$$\underset{\longrightarrow}{12} \, 19$$

- (Point to $\underset{\longrightarrow}{12}$19.) Say the problem for the missing number. (Signal.) *19 minus 12.*
- What's 19 minus 12? (Signal.) *7.*
- So how many apples did Jamie put in the basket? (Signal.) *7.*

e. New problem. There was water in a tank. 4 gallons of water was poured out of the tank. The tank ended up with 12 gallons of water. How many gallons did the tank start with?
- Listen to the first part again and tell me the symbols. There was water in a tank. We don't know how much water was in the tank, so what do I make for that part? (Signal.) *A space.* (Display:) [119:5F]

$$\underline{\qquad\qquad}$$

- Next part: 4 gallons of water was poured out of the tank. Tell me the symbols I write for that part. (Signal.) *Minus 4.* (Add to show:) [119:5G]

$$\underline{\quad} - 4 \underline{\qquad}$$

- Next part: The tank ended up with 12 gallons of water in it. Tell me the symbols I write for that part. (Signal.) *Equals 12.* (Add to show:) [119:5H]

$$\underline{\quad} - 4 = 12$$

- (Point to ____ – **4 = 12.**) Read this problem. (Signal.) *How many minus 4 equals 12.*

f. We can make a number family to figure out the missing number.
- Is the big number or a small number missing? (Signal.) *The big number.*
- So what are 4 and 12? (Signal.) *The small numbers.* (Add to show:) [119:5I]

$$\underline{\quad} - 4 = 12$$

$$\underset{\longrightarrow}{4 \quad 12} \, \underline{\quad}$$

- (Point to $\underset{\longrightarrow}{4 \, 12}$_.) Say the problem for the missing number. (Signal.) *4 plus 12.*
- What's 4 plus 12? (Signal.) *16.*
- So how many gallons of water did the tank start with? (Signal.) *16 (gallons).*

g. Find part 2 on your worksheet. ✔ (Teacher reference:)

a. _____ b. _____

I'll tell you word problems. You'll write the symbols for each problem.
- Problem A. There were 14 people on a bus. Some people got off of the bus. 9 people were still on the bus. How many people got off of the bus?
- Listen to the first part again: There were 14 people on a bus. What number do you write for that part? (Signal.) *14.*
- Write it. ✔
- Next part: Some people got off of the bus. It doesn't tell how many people got off the bus. What sign do you write for that part? (Signal.) *Minus.*
- What do you make after the minus sign? (Signal.) *A space.*
- Write the minus sign and make the space. ✔ (Display:) [119:5J]

a. $14 - \underline{\qquad}$

So far, here's what you should have for problem A.

h. Next part: 9 people were still on the bus. Tell me the symbols for that part. (Signal.) *Equals 9.*
- Write equals 9. ✔
  (Add to show:) [119:5K]

a. $14 - \underline{} = 9$

Here's what you should have for problem A. Read the problem you wrote for A. Get ready. (Signal.) *14 minus how many equals 9.*

i. Problem B. A bush had flowers on it. 11 more flowers grew on the bush. The bush ended up with 17 flowers. How many flowers did the bush have to start with?
- Listen to the first part again: A bush had flowers on it. It doesn't tell how many flowers were on the bush. So what do you make for how many? (Signal.) *A space.*
- Make it. ✔
- Next part: 11 more flowers grew on the bush. What symbols do you write for that part? (Signal) *Plus 11.*
- Write plus 11. ✔
  (Display:) [119:5L]

b. $\underline{} + 11$

So far, here's what you should have for problem B.

j. Next part: The bush ended up with 17 flowers. Tell me the symbols for that part? (Signal.) *Equals 17.*
- Write equals 17. ✔
  (Add to show:) [119:5M]

b. $\underline{} + 11 = 17$

Here's what you should have for problem B.
- Read the problem you wrote for B. Get ready. (Signal.) *How many plus 11 equals 17.*

EXERCISE 6: FACTS
PLUS/MINUS MIX

a. Find part 3. ✔
(Teacher reference:)

a. 3 + 8	c. 4 + 3	f. 8 − 4
b. 6 − 3	d. 10 − 7	g. 12 − 6
	e. 8 + 8	h. 9 − 6

- These are addition and subtraction problems you'll work. Think of the number families for each problem and write the answers. Put your pencil down when you've completed the equations in part 3.
  (Observe children and give feedback.)

b. Check your work. You'll read the equation for each problem.
- Problem A. (Signal.) *3 + 8 = 11.*
- (Repeat for:) B, *6 − 3 = 3;* C, *4 + 3 = 7;* D, *10 − 7 = 3;* E, *8 + 8 = 16;* F, *8 − 4 = 4;* G, *12 − 6 = 6;* H, *9 − 6 = 3.*

EXERCISE 7: SOLVING PROBLEMS FROM NUMBER FAMILIES
MISSING ADDEND, MINUEND, OR SUBTRAHEND REMEDY

a. Find part 4 on your worksheet. ✔
(Teacher reference:)   R Parts D and L

a. $\underline{} + 27 = 68$ c. $54 + \underline{} = 88$ e. $88 - \underline{} = 23$

b. $\underline{} - 23 = 67$ d. $68 - 23 = \underline{}$ f. $\underline{} - 23 = 27$

These problems have a missing number. You're going to make a family for each problem, write the problem you'll work to find the missing number, and find it.

- Read problem A. (Signal.) *How many plus 27 equals 68.*
- Is the missing number a small number or the big number? (Signal.) *A small number.*
- What is 27? (Signal.) *A small number.*
- What is 68? (Signal.) *The big number.*

b. Read problem B. (Signal.) *How many minus 23 equals 67.*
- Is the missing number a small number or the big number? (Signal.) *The big number.*
- What's 23? (Signal.) *A small number.*
- What's 67? (Signal.) *A small number.*

c. Read problem C. (Signal.) *54 plus how many equals 88.*
- Is the missing number a small number or the big number? (Signal.) *A small number.*
- What's 54? (Signal.) *A small number.*
- What's 88? (Signal.) *The big number.*

d. Read problem D. (Signal.) *68 minus 23 equals how many.*
- Is the missing number a small number or the big number? (Signal.) *A small number.*
- What's 68? (Signal.) *The big number.*
- What's 23? (Signal.) *A small number.*

e. Read problem E. (Signal.) *88 minus how many equals 23.*
- Is the missing number a small number or the big number? (Signal.) *A small number.*
- What's 88? (Signal.) *The big number.*
- What's 23? (Signal.) *A small number.*

f. Read problem F. (Signal.) *How many minus 23 equals 27.*
- Is the missing number a small number or the big number? (Signal.) *The big number.*
- What's 23? (Signal.) *A small number.*
- What's 27? (Signal.) *A small number.*
(Repeat for problems that were not firm.)

g. Make the family for each problem with two numbers. Put your pencil down when you've written the family with two numbers for each problem in part 4.
(Observe children and give feedback.)
(Display:)   [119:7A]

h. Now you're going to say the problem for finding the missing number for each family.
- Family A. Say the problem for the missing number. (Signal.) *68 minus 27.*
- (Repeat for:) B, *23 plus 67;* C, *88 minus 54;* D, *68 minus 23;* E, *88 minus 23;* F, *23 plus 27.* (Repeat problems that were not firm.)

i. In the box below family A, write the column problem and work it. Then write the missing number in the family, and complete the equation above family A. Put your pencil down when you've completed problem A.
(Observe children and give feedback.)

j. Check your work.
- Read the column problem and the answer for problem A. (Signal.) *68 − 27 = 41.*
- Look at the equation above family A. How many plus 27 equals 68? (Signal.) *41.*
(Display:)   [119:7B]

- Here's what you should have for problem A. Check your work. ✔

k. Write the column problems for the rest of the families and work them. Then write the missing number for each family and complete the equation above it. Put your pencil down when you've completed part 4.
(Observe children and give feedback.)

l. Check your work.
- Problem B. Read the column problem and the answer. (Signal.) *23 + 67 = 90.*
- Look at the equation above family B. How many minus 23 equals 67? (Signal.) *90.*

You should have one of these families for each problem. Make sure your families have the numbers in the correct places. ✔

m. (Repeat the following tasks for problems C through F:)

Problem __. Read the column problem and the answer.		Look at the equation above family __. __?		
C	88 – 54 = 34	C	54 + how many = 88	34
D	68 – 23 = 45	D	68 – 23 = how many	45
E	88 – 23 = 65	E	88 – how many = 23	65
F	23 + 27 = 50	F	How many – 23 = 27	50

Good using number families to complete addition and subtraction equations that have a number missing.

EXERCISE 8: TELLING TIME
ANALOG HOURS AND HALF HOURS

a. Turn to the other side of worksheet 119 and find part 5. ✔
(Teacher reference:)

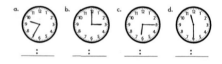

You're going to write the time for each clock. One clock shows zero minutes. Another clock shows 30 minutes. You'll count the minutes to yourself for the other two clocks.

• Write the time for each clock. Put your pencil down when you've written the hour number and the minute number for each clock in part 5. (Observe children and give feedback.)

b. Check your work. You'll read the time for each clock.

• Clock A. (Signal.) *Nine 35.*
• (Repeat for:) B, *3 o'clock;* C, *Six 15;* D, *Eleven 30.*
(Display:) [119:8A]

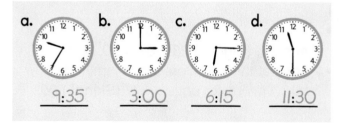

Here's what you should have for the clocks.

EXERCISE 9: INDEPENDENT WORK

a. Find part 6. ✔
(Teacher reference:)

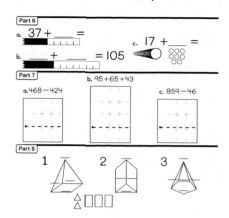

In part 6, you'll complete the equations for the rulers and the equation for the circles.
In part 7, you'll write each problem in a column and work it.

b. Find part 8. ✔
• Touch object 1. ✔
• Touch object 2. ✔
• Touch object 3. ✔
You're going to write letters for all of the faces for objects 1, 2, and 3.
• Touch the picture of the faces for the bottom, top, and sides. ✔
After you label the faces, you'll circle the object that has the right bottom, top, and sides.

c. Complete worksheet 119.
(Observe children and mark incorrect responses on children's worksheets as you give feedback.)

Lesson 120

EXERCISE 1: FACT RELATIONSHIPS
PLUS 3, PLUS 4, AND PLUS 5

a. You'll say facts for families that have a small number of 3. Start with 1 plus 3 and say the plus-3 facts to 10 plus 3.
- Say the fact for 1 plus 3. Get ready. (Signal.) *1 + 3 = 4.*
- Next fact. (Signal.) *2 + 3 = 5.*
- (Repeat for:) *3 + 3 = 6, 4 + 3 = 7, 5 + 3 = 8, 6 + 3 = 9, 7 + 3 = 10, 8 + 3 = 11, 9 + 3 = 12, 10 + 3 = 13.*
  (Repeat until firm.)

b. Now you'll say facts for families that have a small number of 4. Start with 1 plus 4 and say the plus-4 facts to 10 plus 4.
- Say the fact for 1 plus 4. (Signal.) *1 + 4 = 5.*
- Next fact. (Signal.) *2 + 4 = 6.*
- (Repeat for:) *3 + 4 = 7, 4 + 4 = 8, 5 + 4 = 9, 6 + 4 = 10, 7 + 4 = 11, 8 + 4 = 12, 9 + 4 = 13, 10 + 4 = 14.*
  (Repeat until firm.)

c. Now you'll say facts for families that have a small number of 5. Start with 1 plus 5 and say the plus-5 facts to 10 plus 5.
- Say the fact for 1 plus 5. (Signal.) *1 + 5 = 6.*
- Next fact. (Signal.) *2 + 5 = 7.*
- (Repeat for:) *3 + 5 = 8, 4 + 5 = 9, 5 + 5 = 10, 6 + 5 = 11, 7 + 5 = 12, 8 + 5 = 13, 9 + 5 = 14, 10 + 5 = 15.*
  (Repeat until firm.)

d. Now I'll mix some of the problems up and you'll tell me the answer.
- Listen: 7 plus 3. What's 7 plus 3? (Signal.) *10.*
- Listen: 7 plus 4. What's 7 plus 4? (Signal.) *11.*
- Listen: 7 plus 5. What's 7 plus 5? (Signal.) *12.*
- Listen: 8 plus 5. What's 8 plus 5? (Signal.) *13.*
- Listen: 9 plus 5. What's 9 plus 5? (Signal.) *14.*
  (Repeat until firm.)

EXERCISE 2: FRACTIONS   [REMEDY]

a. Last time you learned the name for parts of shapes that are divided up into 5 equal parts and 2 equal parts.
- Listen: Fifths. What name? (Signal.) *Fifths.*
- A shape with fifths is divided into how many equal parts? (Signal.) *5.*

b. Listen: Halves. What name? (Signal.) *Halves.*
- A shape with halves is divided into how many equal parts? (Signal.) *2.*
c. Listen: Eighths. What name? (Signal.) *Eighths.*
- A shape with eighths is divided into how many equal parts? (Signal.) *8.*
d. (Repeat the following tasks with the following names:)

Listen: __. What name?		A shape with __ is divided into how many equal parts?	
Halves	*Halves*	Halves	2
Fourths	*Fourths*	Fourths	4
Fifths	*Fifths*	Fifths	5

(Repeat names that were not firm.)
e. (Distribute unopened workbooks to children.)
- Open your workbook to Lesson 120 and find part 1.
  (Observe children and give feedback.)
  (Teacher reference:)   [R] Parts B and S

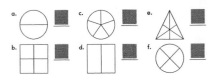

You're going to write the number of equal parts for each shape. Then you'll tell me what the parts for each shape are called.
- Count the parts for shape A and write it.
  (Observe children and give feedback.)
- Everybody, how many equal parts is shape A divided into? (Signal.) *2.*
- What are the parts for shape A called? (Signal.) *Halves.*
  (Display:)   [120:2A]

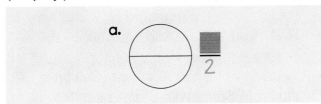

Here's what you should have.

f. The parts of shape A are called halves. One part is called a half. What is one part called? (Signal.) *A half.*

• What are the parts of shape A called? (Signal.) *Halves.*
(Repeat step f until firm.)

g. Count the parts for the rest of the shapes and write them. Put your pencil down when you've written the number of parts for each shape in part 1.
(Observe children and give feedback.)

h. Check your work.

• Shape B. How many equal parts is shape B divided into? (Signal.) *4.*

• What are the parts of shape B called? (Signal.) *Fourths.*

i. (Repeat the following tasks for problems C through F:)

How many equal parts is shape __ divided into?		What are the parts of shape __ called?	
C	5	C	Fifths
D	2	D	Halves
E	6	E	Sixths
F	4	F	Fourths

(Repeat names that were not firm.)

EXERCISE 3: ADDING AND SUBTRACTING TENS NUMBERS
DRAWINGS, REASONING, AND WRITING EQUATIONS

a. Find part 2 on your worksheet. ✔
(Teacher reference:)

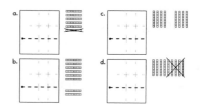

You're going to explain each picture. Then you'll tell me the equation for it and write it.

• Raise your hand when you can tell me what picture A shows. ✔

• What does picture A show? (Call on a student.) *Starts with 6 groups of ten. Subtracts 1 group of ten. Equals 5 groups of ten, which is 50.*

• Everybody, say the equation for this picture. (Signal.) *60 – 10 = 50.*

• Write the equation for picture A. ✔
(Display:)   [120:3A]

a. $$\begin{array}{r} 6\,0 \\ -\,1\,0 \\ \hline 5\,0 \end{array}$$

Here's what you should have for picture A.

b. Raise your hand when you can tell me what picture B shows.

• What does picture B show? (Call on a student.) *Starts with 6 groups of ten. Adds 2 groups of ten. Equals 8 groups of ten, which is 80.*

• Everybody, say the equation for picture B. (Signal.) *60 + 20 = 80.*

• Write the equation for picture B. ✔

c. Write the equations for pictures C and D. Put your pencil down when you've written equations for all of the pictures in part 2.
(Observe children and give feedback.)

d. Check your work. You'll read the equation for pictures C and D.

• Picture C. Read the equation. (Signal.) *40 + 40 = 80.*

• Picture D. Read the equation. (Signal.) *90 – 50 = 40.*

EXERCISE 4: EQUATIONS
EQUALITY/INEQUALITY

a. (Display:)   [120:4A]

$$11 - 2 = 9 + 2$$

Here is an equation with a number you add or subtract on both sides.

• (Point to **11**.) Read the equation. (Signal.) *11 – 2 = 9 + 2.*

• What does 11 minus 2 equal? (Signal.) *9.*

• What does 9 plus 2 equal? (Signal.) *11.*
(Add to show:)   [120:4B]

$$\overset{9}{11} - 2 = \overset{11}{9} + 2$$

• Does 9 equal 11? (Signal.) *No.*

• So are the sides equal? (Signal.) *No.*

• So is this equation true? (Signal.) *No.*

b. Let's do those again and I'll fix the sign.
- (Point to **11 – 2.**) What does this side equal? (Signal.) 9.
- (Point to **9 + 2.**) What does this side equal? (Signal.) 11.
- Are the sides of this equation equal? (Signal.) No.
- So is this equation true or false? (Signal.) False.
- What do I do to the equals to make this statement true? (Call on a student. Idea: *Cross it out with a single line.*)
  (Add to show:)   [120:4C]

$$11 \overset{9}{-} 2 \neq 9 \overset{11}{+} 2$$

- (Point to **11 – 2 ≠ 9 + 2.**) Read this statement. (Signal.) *11 minus 2 is not equal to 9 plus 2.*

c. Find part 3 on your worksheet. ✔
   (Teacher reference:)

   a. $14 = 7 + 7$ c. $30 + 8 = 28 + 10$ e. $9 = 6 - 3$
   b. $10 - 1 = 9 + 1$ d. $14 - 4 = 24 - 10$ f. $15 + 1 = 18 - 1$

Some of these equations are true and some of them are false. You're going to fix the equals in problems that are false so that all of the statements in part 3 are true.

- Read problem A. (Signal.) *14 = 7 + 7.*
- What does 7 plus 7 equal? (Signal.) *14.*
- Write 14 above 7 plus 7 to show what the side equals. ✔
- Are the sides of equation A equal? (Signal.) Yes.
- So is equation A true or false? (Signal.) *True.*
- Touch and read the statement for A. (Signal.) *14 = 7 + 7.*

d. Read problem B. (Signal.) *10 – 1 = 9 + 1.*
- What does 10 minus 1 equal? (Signal.) *9.*
- Write 9 above 10 minus 1 to show what the side equals. ✔
- What does 9 plus 1 equal? (Signal.) *10.*
- Write 10 above 9 plus 1. ✔
- Are the sides of equation B equal? (Signal.) No.

- So is equation B true or false? (Signal.) *False.*
- What do you make through the equals to show the sides are not equal? (Call on a student.) *A single line.*
- Fix the equals for problem B. ✔
  (Display:)   [120:4D]

$$10 \overset{9}{-} 1 \neq 9 \overset{10}{+} 1$$
b.

Here's what you should have for problem B.
- Touch and read the statement for B. (Signal.) *10 minus 1 is not equal to 9 plus 1.*

e. Read the rest of the problems to yourself. If a side adds or subtracts, write the number that the side equals above it. Then fix the sign. Put your pencil down when you've completed the statements for all of the problems in part 2. (Observe children and give feedback.)

f. Check your work.
- Problem C: 30 plus 8 equals 28 plus 10. Is that equation true or false? (Signal.) *True.*
- Read the statement for C. (Signal.) *30 + 8 = 28 + 10.*

g. (Repeat the following tasks for problems D through F:)

Problem __. Is that equation true or false?		Read the statement for __.	
D: $14 - 4 = 24 - 10$	*False*	D	*14 – 4 is not = to 24 – 10.*
E: $9 = 6 - 3$	*False*	E	*9 is not = to 6 – 3.*
F: $15 + 1 = 18 - 1$	*False*	F	*15 + 1 is not = to 18 – 1.*

EXERCISE 5: WORD PROBLEMS
MISSING ADDEND, MINUEND, OR SUBTRAHEND

a. I'll tell you word problems. We'll write the symbols for each problem.
Problem A: A string started out a certain length. 8 feet of the string was cut off. The string ended up 11 feet long. How long did the string start out?

• Listen to the first part again: A string started out a certain length. We don't know how long the string was to start out, so what do I make for that part? (Signal.) *A space.*
(Display:)   [120:5A]

$$\underline{\underline{\qquad\qquad}}$$

• Next part: 8 feet of the string was cut off. Tell me the symbols I write for that part. (Signal.) *Minus 8.*
(Add to show:)   [120:5B]

$$\underline{\underline{\qquad}} - 8 \underline{\qquad}$$

• Next part: The string ended up 11 feet long. Tell me the symbols I write for that part. (Signal.) *Equals 11.*
(Add to show:)   [120:5C]

$$\underline{\underline{\qquad}} - 8 = 11$$

• (Point to __ **– 8 = 11.**) Read this problem. (Signal.) *How many minus 8 equals 11.*

b. We can make a number family to figure out the missing number.
• Is the big number or a small number missing? (Signal.) *The big number.*
• So what are 8 and 11? (Signal.) *Small numbers.*
(Add to show:)   [120:5D]

$$\underline{\underline{\qquad}} - 8 = 11$$
$$\overset{8 \qquad 11}{\xrightarrow{\hspace{1cm}}} \underline{\quad}$$

• (Point to $\overset{8 \quad 11}{\longrightarrow}$_.) Say the problem for the missing number. (Signal.) *8 plus 11.*
• What's 8 plus 11? (Signal.) *19.*
• So how long did the string start out? (Signal.) *19 feet.*

c. Find part 4 on your worksheet. ✔
(Teacher reference:)

a. _____   c. _____
b. _____

I'll tell you word problems. You'll write the symbols for each problem.

• Problem A: There were 9 people in a room. Some more people went into the room. 11 people ended up in the room. How many people went into the room?
• Listen to the first part again: There were 9 people in a room. What number do you write for that part? (Signal.) *9.*
• Write it. ✔
• Next part: Some more people went into the room. It doesn't tell how many people went into the room. What sign do you write for that part? (Signal.) *Plus.*
• What do you make after the plus sign? (Signal.) *A space.*
• Write the plus sign and the space. ✔
(Display:)   [120:5E]

a. $9 + \underline{\qquad}$

So far, here's what you should have for problem A.

d. Next part: 11 people ended up in the room. Tell me the symbols for that part. (Signal.) *Equals 11.*
• Write equals 11. ✔
(Add to show:)   [120:5F]

a. $9 + \underline{\qquad} = 11$

Here's what you should have for problem A.
• Read the problem you wrote for A. Get ready. (Signal.) *9 plus how many equals 11.*

e. Problem B: A bush had flowers on it. 5 of those flowers were picked off of the bush. The bush ended up with 15 flowers on it. How many flowers did the bush have to start with?
• Listen to the first part again: A bush had flowers on it. It doesn't tell how many flowers were on the bush. So what do you make for that part? (Signal.) *A space.*
• Make it. ✔
• Next part: 5 of those flowers were picked off of the bush. What symbols do you write for that part? (Signal) *Minus 5.*
• Write minus 5. ✔

f. Next part: The bush ended up with 15 flowers on it. Tell me the symbols for that part. (Signal.) *Equals 15.*
- Write equals 15. ✔
  (Display:) [120:5G]

> b. ___ − 5 = 15

Here's what you should have for problem B.
- Read the problem you wrote for B. Get ready. (Signal.) *How many minus 5 equals 15.*

g. Problem C: A car had boxes in it. 6 more boxes were put in the car. The car ended up with 10 boxes in it. How many boxes were in the car to start with?
- Listen to the first part again: A car had boxes in it. What do you make for that part? (Signal.) *A space.*
- Make it. ✔
- Next part: 6 more boxes were put in the car. What symbols do you write for that part? (Signal) *Plus 6.*
- Write plus 6. ✔

h. Next part: The car ended up with 10 boxes in it. Tell me the symbols for that part. (Signal.) *Equals 10.*
- Write equals 10. ✔
  (Display:) [120:5H]

> c. ___ + 6 = 10

Here's what you should have for problem C.
- Read the problem you wrote for C. Get ready. (Signal.) *How many plus 6 equals 10.*

EXERCISE 6: FACTS
PLUS/MINUS MIX

a. Turn to the other side of your worksheet and find part 5.
  (Observe children and give feedback.)
  (Teacher reference:)

a. 11 − 3
b. 6 + 3
c. 6 − 3
d. 8 − 3
e. 8 + 3
f. 7 + 7
g. 10 − 7
h. 10 − 5

- Touch and read problem A. Get ready. (Signal.) *11 minus 3.*
- Is the answer the big number or a small number? (Signal.) *A small number.*
- What's 11 minus 3? (Signal.) *8.*

b. (Repeat the following tasks for problems B through H:)

Touch and read problem __.		Is the answer the big number or a small number?	What's __?	
B	6 + 3	*The big number.*	6 + 3	9
C	6 − 3	*A small number.*	6 − 3	3
D	8 − 3	*A small number.*	8 − 3	5
E	8 + 3	*The big number.*	8 + 3	11
F	7 + 7	*The big number.*	7 + 7	14
G	10 − 7	*A small number.*	10 − 7	3
H	10 − 5	*A small number.*	10 − 5	5

(Repeat problems that were not firm.)

c. Complete each of the facts. Put your pencil down when you've written all of the equations for part 5.
  (Observe children and give feedback.)

d. Check your work.
- Touch and read the equation for problem A. (Signal.) *11 − 3 = 8.*
- (Repeat for:) B, *6 + 3 = 9;* C, *6 − 3 = 3;* D, *8 − 3 = 5;* E, *8 + 3 = 11;* F, *7 + 7 = 14;* G, *10 − 7 = 3;* H, *10 − 5 = 5.*

EXERCISE 7: SOLVING PROBLEMS FROM NUMBER FAMILIES
MISSING ADDEND, MINUEND, OR SUBTRAHEND

a. Find part 6 on your worksheet. ✔
  (Teacher reference:)

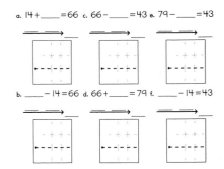

a. 14 + ___ = 66   c. 66 − ___ = 43   e. 79 − ___ = 43
b. ___ − 14 = 66   d. 66 + ___ = 79   f. ___ − 14 = 43

These problems have a missing number. You're going to make a family for each problem, and find the missing number.
- Read problem A. (Signal.) *14 plus how many equals 66.*
- Is the missing number a small number or the big number? (Signal.) *A small number.*
- What is 14? (Signal.) *A small number.*
- What is 66? (Signal.) *The big number.*

b. Make the family for problem A. Put your pencil
    down when you've written the family with two
    numbers from problem A.
    (Observe children and give feedback.)
    (Display:)   [120:7A]

a. 14 + __ = 66

$$\underline{\underset{14}{}}\Longrightarrow \underline{66}$$

•   Here's what you should have for family A. Say
    the problem for finding the missing number in
    family A. Get ready. (Signal.) *66 minus 14.*
c. In the box below family A, write the column
    problem and work it. Then write the missing
    number in the family and complete the
    equation above family A. Put your pencil
    down when you've completed problem A.
    (Observe children and give feedback.)
d. Check your work.
•   Read the column problem and the answer for
    problem A. (Signal.) *66 − 14 = 52.*
•   Look at the equation above family A. 14 plus
    how many equals 66? (Signal.) *52.*
    (Add to show:)   [120:7B]

a. 14 + <u>52</u> = 66

$$\underline{\underset{14}{}\quad\underset{52}{}}\Longrightarrow \underline{66}$$

$$\begin{array}{r} 6\,6 \\ -\ 1\,4 \\ \hline 5\,2 \end{array}$$

•   Here's what you should have for problem A.
    Check your work. ✔
e. Read problem B. (Signal.) *How many minus
    14 equals 66.*
•   Is the missing number a small number or the
    big number? (Signal.) *The big number.*
•   Make the family for problem B. Then write
    the column problem for finding the missing
    number and work it. Write the missing number
    in the family and complete the equation above
    family B. Put your pencil down when you've
    completed problem B.
    (Observe children and give feedback.)

f. Check your work.
•   Read the column problem and the answer for
    problem B. (Signal.) *14 + 66 = 80.*
•   Look at the equation above family B. How
    many minus 14 equals 66? (Signal.) *80.*
    (Display:)   [120:7C]

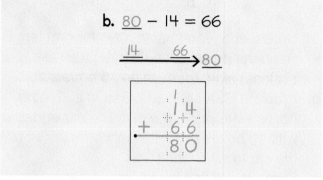

b. <u>80</u> − 14 = 66

•   Here's what you should have for problem B.
    Check your work. ✔
g. Read problem C. (Signal.) *66 minus how
    many equals 43.*
•   Make the family for problem C. Write the
    column problem and work it. Write the missing
    number in the family and complete the
    equation above. Put your pencil down when
    you've completed problem C.
    (Observe children and give feedback.)
h. Check your work.
•   Read the column problem and the answer for
    problem C. (Signal.) *66 − 43 = 23.*
•   Look at the equation above family C. 66 minus
    how many equals 43? (Signal.) *23.*
    (Display:)   [120:7D]

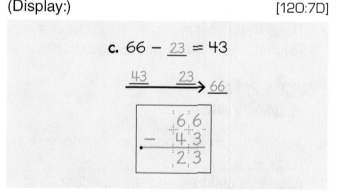

c. 66 − <u>23</u> = 43

•   Here's what you should have for problem C.
    Check your work. ✔
i. Read problem D. (Signal.) *66 plus how many
    equals 79.*
•   Make the family for problem D. Write the
    column problem and work it. Write the missing
    number in the family and complete the
    equation above. Put your pencil down when
    you've completed problem D.
    (Observe children and give feedback.)

j. Check your work.
*   Read the column problem and the answer for problem D. (Signal.) *79 – 66 = 13.*
*   Look at the equation above family D. 66 plus how many equals 79? (Signal.) *13.*
k. Work problems E and F. Put your pencil down when you've completed the families and the equations for part 6.
    (Observe children and give feedback.)
l. Check your work.
*   Read the column problem and the answer for problem E. (Signal.) *79 – 43 = 36.*
*   Look at the equation above family E. 79 minus how many equals 43? (Signal.) *36.*
*   Read the column problem and the answer for problem F. (Signal.) *14 + 43 = 57.*
*   Look at the equation above family F. How many minus 14 equals 43? (Signal.) *57.*

EXERCISE 8: INDEPENDENT WORK

a. Find part 7. ✔
    (Teacher reference:)

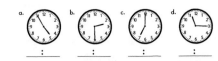

*   Write the time for each clock in part 7.
    (Observe children and mark incorrect responses on children's worksheets as you give feedback.)

Mastery Test 12

> *Note:* Mastery Tests are administered to all students in the group. Each student will need a pencil and a *Student Assessment Book.* Try to arrange students so they cannot look at other students' responses.

Teacher Presentation

a. Find Test 12 in your test booklet. ✔
•   Touch part 1. ✔
    (Teacher reference:)

    a. _____   b. _____   c. _____   d. _____   e. _____

    I'm going to tell you about shapes that are divided up into equal parts. You're going to write the number of parts that shape is divided into.
•   Touch the space for A. ✔
•   Listen: Fourths. What name? (Signal.) *Fourths.*
•   For A, write the number of equal parts a shape with fourths is divided into.
    (Observe children.)
b. Touch the space for B. ✔
•   Listen: Halves. What name? (Signal.) *Halves.*
•   For B, write the number of equal parts a shape with halves is divided into.
    (Observe children.)
c. Touch the space for C. ✔
•   Listen: Fifths. What name? (Signal.) *Fifths.*
•   For C, write the number of equal parts a shape with fifths is divided into.
    (Observe children.)
d. Touch the space for D. ✔
•   Listen: Eighths. What name? (Signal.) *Eighths.*
•   For D, write the number of equal parts a shape with eighths is divided into.
    (Observe children.)
e. Touch the space for E. ✔
•   Listen: Elevenths. What name? (Signal.) *Elevenths.*
•   For E, write the number of equal parts a shape with elevenths is divided into.
    (Observe children.)

f. Touch part 2 on test 12. ✔
    (Teacher reference:)

    a. ___ − 18 = 63   c. 68 − 23 = ___   e. 54 + ___ = 88
    b. ___ + 27 = 68   d. 79 − ___ = 18   f. ___ − 23 = 27

    For most of the equations in part 2, the number it starts with or the number that's added or subtracted is missing. You're going to write the numbers that are shown in each equation where they go in a number family.
•   Write the numbers that are shown in each equation in the family below it. Put your pencil down when you've written the two numbers for each family in part 2.
    (Observe children.)
g. Touch part 3 on test 12 and touch space A. ✔
    (Teacher reference:)

    a. _____   b. _____

    In space A, you'll write all of the even numbers from 10 to 20.
•   Write the even numbers from 10 to 20 in space A.
    (Observe children.)
h. In space B, you'll write all of the odd numbers from 10 to 20.
•   Write the odd numbers from 10 to 20 in space B.
    (Observe children.)
i. Touch part 4 on test 12. ✔
    (Teacher reference:)

    a. 184 + 108 = ___   c. 184 + ___ = 292   e. ___ − 76 = 108
    b. ___ − 184 = 76   d. 184 − ___ = 76   f. ___ + 184 = 260

    The equations in part 4 have a missing number. (Display:)   [Test 12]

    $$\underrightarrow{184 \quad 76} 260 \qquad \underrightarrow{76 \quad 108} 184$$

    $$\underrightarrow{108 \quad 184} 292$$

    These are the families for the equations in part 4.
•   Find the family for each equation and write the missing number. Put your pencil down when you've completed all of the equations in part 4.
    (Observe children.)

j. Touch part 5 on test 12. ✔
(Teacher reference:)

You're going to complete the equations for hexagons, pentagons, rectangles, and triangles for the objects in part 5.
- Complete the equations for H, P, R, and T in part 5. Put your pencil down when you've completed the equations for both objects.
(Observe children.)

k. Turn to the other side of test sheet 12 and touch part 6. ✔
(Teacher reference:)

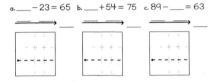

a. ___ − 23 = 65   b. ___ + 54 = 75   c. 89 − ___ = 63

You're going to complete all of the equations in part 6.
- Write the numbers that are shown in each equation where they go in the number family below it. Put your pencil down when you've written the two numbers for each family in part 6.
(Observe children.)
- Below each family, write the column problem to figure out the missing number. Put your pencil down when you've figured out the missing number for each family in part 6.
(Observe children.)
- Write the missing number to complete the equation you started with for each problem.
(Observe children.)

l. Touch part 7. ✔
(Teacher reference:)

	g. 3 + 7 =	n. 6 − 3 =
a. 3 + 1 =	h. 3 + 8 =	o. 7 − 3 =
b. 3 + 2 =	i. 3 + 9 =	p. 8 − 3 =
c. 3 + 3 =	j. 3 + 10 =	q. 9 − 3 =
d. 3 + 4 =	k. 3 − 3 =	r. 10 − 3 =
e. 3 + 5 =	l. 4 − 3 =	s. 11 − 3 =
f. 3 + 6 =	m. 5 − 3 =	t. 12 − 3 =

You'll complete the equations in part 7.

m. Turn to the next page of test 12 and touch part 8. ✔
(Teacher reference:)

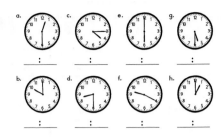

You're going to write the time for each clock. Some of the clocks show zero minutes. Some of the clocks show 30 minutes. You'll count the minutes to yourself for the other clocks.
- Write the time for each clock. Put your pencil down when you've written the hour number and the minute number for each clock in part 8.
(Observe children.)

n. Touch part 9. ✔
(Teacher reference:)

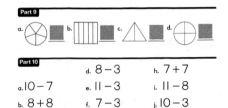

Part 10		
	d. 8 − 3	h. 7 + 7
a. 10 − 7	e. 11 − 3	i. 11 − 8
b. 8 + 8	f. 7 − 3	j. 10 − 3
c. 8 + 3	g. 7 + 3	

In part 9, you'll count the equal parts for each shape and write the number where it goes.
- All of the problems in part 10 are from families you know. You'll complete equations in part 10.
- Your turn: Work the rest of the problems on Test 12.

(Direct students where to put their assessment books when they are finished.)

Scoring Notes

a. Collect test booklets. Use the Answer Key and Passing Criteria Table to score the tests.

Passing Criteria Table — Mastery Test 12			
Part	Score	Possible Score	Passing Score
1	2 for each number	10	8
2	1 for each problem	6	5
3	2 for each correct odd/even number	11	10
4	1 for each problem	6	5
5	1 for each equation	8	7
6	3 for each problem	9	6
7	1 for each equation	20	19
8	2 for each problem	16	12
9	1 for each problem	4	3
10	1 for each equation	10	9
	Total	100	

b. Complete the Mastery Test 12 Remedy Summary Sheet to determine whether group remedies are needed. Reproducible Remedy Summary Sheets are at the back of the Answer Key and at the back of the *Teacher's Guide.*

• If ¼ or more of the students did not pass a test part, present the remedy for that part before beginning Lesson 121. The Remedy Table follows and is also at the end of the Mastery Test 12 Answer Key. Remedies worksheets follow Mastery Test 12 in the *Student Assessment Book.*

Remedy Table — Mastery Test 12				
Part	Test Items	Remedy		Student Material Remedies Worksheet
		Lesson	Exercise	
1	Fractions	117	7	Part A
		119	2	—
		120	2	Part B
2	Writing Number Families from Problems	117	5	Part C
		119	7	Part D
3	Odd and Even Numbers	111	3	—
		114	3	—
		114	5	—
		116	4	—
4	Number Families– Completing Equations with Missing Numbers	113	5	Part E
		114	10	Part F
5	3-D Objects Decompostion	110	4	Part G
		111	8	Part H
		112	6	Part I
6	Number Families from Problems	117	5	Part J
		118	6	Part K
		119	7	Part L
7	3 Plus, Minus-3 Fact Relationships	111	1	—
		113	1	—
		117	1	—
8	Writing Time (Analog Clocks)	108	5	Part M
		109	7	Part N
		113	6	Part O
		116	6	Part P
9	Fractions (Denominator)	117	7	Part Q
		118	5	Part R
		120	2	Part S
10	Facts (Mix)	109	2	—
		110	3	—
		112	4	—

Retest

Retest individual students on any part failed.

	Lesson 121	Lesson 122	Lesson 123	Lesson 124	Lesson 125
Student Learning Objectives	**Exercises** 1. **Figure out unknown facts for plus 4, 5, 6** 2. **Learn about the associative property** 3. **Say fractions** 4. Solve addition and subtraction facts 5. Identify the missing value in a word problem 6. Determine whether equations are true or false 7. Add and subtract tens numbers using drawings to explain reasoning and write equations 8. Solve problems with a missing value by writing and solving multi-digit number families 9. Complete work independently	**Exercises** 1. Figure out unknown facts for plus 4, 5, 6 2. **Use the associative property to determine whether equations are true or false** 3. **Organize, represent, and interpret data** 4. **Write <, >, or = between two statements to make the equation true using the associative property** 5. **Write fractions** 6. Solve addition and subtraction facts 7. Identify the missing value in a word problem 8. Solve problems with a missing value by writing and solving multi-digit number families 9. Complete work independently	**Exercises** 1. Figure out unknown facts for plus 4, 5, 6 2. Use the associative property to determine whether equations are true or false 3. Write <, >, or = between two statements to make the equation true using the associative property 4. Organize, represent, and interpret data 5. Identify the missing value in a word problem 6. **Write fractions; learn about quarters** 7. Solve problems with a missing value by writing and solving multi-digit number families 8. **Add 2-digit to 1-digit or tens numbers using drawings to explain reasoning and write equations** 9. Complete work independently	**Exercises** 1. **Figure out unknown facts for plus 7, 8** 2. Use the associative property to determine whether equations are true or false 3. Write fractions 4. Solve problems with a missing value by writing and solving multi-digit number families 5. Add 2-digit to 1-digit or tens numbers using drawings to explain reasoning and write equations 6. Write <, >, or = between two statements to make the equation true using the associative property 7. **Write the equations for word problems and solve** 8. Organize, represent, and interpret data	**Exercises** 1. **Figure out unknown facts for plus 7, 8, 9** 2. Use the associative property to determine whether equations are true or false 3. Write fractions 4. Solve problems with a missing value by writing and solving multi-digit number families 5. Organize, represent, and interpret data 6. Write <, >, or = between two statements to make the equation true using the associative property 7. Write the equations for word problems and solve 8. Complete work independently

Common Core State Standards for Mathematics					
1.OA 1		✔	✔	✔	✔
1.OA 3–4, 6, 8	✔	✔	✔	✔	✔
1.OA 5		✔			
1.OA 7	✔				
1.NBT 1 and 4	✔	✔	✔	✔	✔
1.NBT 2	✔	✔	✔	✔	
1.NBT 5–6	✔	✔	✔		
1.MD 3	✔	✔			✔
1.MD 4		✔	✔	✔	✔
1.G 3	✔	✔	✔	✔	✔
Teacher Materials	Presentation Book 3, Workbook 2, Board Displays CD or chalkboard				
Student Materials	Workbook 2, Pencil				
Additional Practice	• Student Practice Software: Block 6: Activity 1 (1.G.2), Activity 2 (1.G.3), Activity 3 (1.G.3), Activity 4 (1.OA.2), Activity 5 (1.MD.2), Activity 6 (1.OA.2) • Math Fact Worksheets 73–74 (After Lesson 121), 75 (After Lesson 123), 76–80 (After Lesson 124)				
Mastery Test					

Lesson 121

EXERCISE 1: FACT RELATIONSHIPS
FIGURING OUT UNKNOWN FACTS

a. You'll say facts for families that have a small number of 4. Start with 1 plus 4 and say the plus-4 facts to 10 plus 4.
- Say the fact for 1 plus 4. Get ready. (Signal.) *1 + 4 = 5.*
- Next fact. (Signal.) *2 + 4 = 6.*
- (Repeat for:) *3 + 4 = 7, 4 + 4 = 8, 5 + 4 = 9, 6 + 4 = 10, 7 + 4 = 11, 8 + 4 = 12, 9 + 4 = 13, 10 + 4 = 14.*
  (Repeat until firm.)

b. Now you'll say facts for families that have a small number of 5. Start with 1 plus 5 and say the plus-5 facts to 10 plus 5.
- Say the fact for 1 plus 5. Get ready. (Signal.) *1 + 5 = 6.*
- Next fact. (Signal.) *2 + 5 = 7.*
- (Repeat for:) *3 + 5 = 8, 4 + 5 = 9, 5 + 5 = 10, 6 + 5 = 11, 7 + 5 = 12, 8 + 5 = 13, 9 + 5 = 14, 10 + 5 = 15.*
  (Repeat until firm.)

c. Now you'll say facts for families that have a small number of 6. Start with 1 plus 6 and say the plus-6 facts to 10 plus 6.
- Say the fact for 1 plus 6. Get ready. (Signal.) *1 + 6 = 7.*
- Next fact. (Signal.) *2 + 6 = 8.*
- (Repeat for:) *3 + 6 = 9, 4 + 6 = 10, 5 + 6 = 11, 6 + 6 = 12, 7 + 6 = 13, 8 + 6 = 14, 9 + 6 = 15, 10 + 6 = 16.*
  (Repeat until firm.)

d. (Display:)   [121:1A]

$$7 + 4 \qquad 9 + 6$$
$$8 + 5$$

Here are problems we've never worked before. I'll show you how to figure out the fact for each problem.

- (Point to **7 +.**) Read this problem. (Signal.) *7 plus 4.*
- Start with 4 plus 4 and say the plus-4 facts until you've said the fact for 7 plus 4. Get ready. (Signal.) *4 + 4 = 8.*
- Next fact. (Signal.) *5 + 4 = 9.*
- (Repeat for:) *6 + 4 = 10, 7 + 4 = 11.*
- Everybody, what's 7 plus 4? (Signal.) *11.*
  (Repeat until firm.)

e. (Point to **9 +.**) Read this problem. (Signal.) *9 plus 6.*
- Start with 6 plus 6 and say the plus-6 facts until you've said the fact for 9 plus 6. Get ready. (Signal.) *6 + 6 = 12.*
- Next fact. (Signal.) *7 + 6 = 13.*
- (Repeat for:) *8 + 6 = 14, 9 + 6 = 15.*
- Everybody, what's 9 plus 6? (Signal.) *15.*
  (Repeat step e until firm.)

f. (Point to **8 +.**) Read this problem. (Signal.) *8 plus 5.*
- Start with 5 plus 5 and say the plus-5 facts until you've said the fact for 8 plus 5. Get ready. (Signal.) *5 + 5 = 10.*
- Next fact. (Signal.) *6 + 5 = 11.*
- Next fact. (Signal.) *7 + 5 = 12.*
- Next fact. (Signal.) *8 + 5 = 13.*
- Everybody, what's 8 plus 5? (Signal.) *13.*
  (Repeat step f until firm.)

=== INDIVIDUAL TURNS ===

(Call on individual students to perform one of the following tasks.)

- (Point to **7 + 4.**) Start with 4 plus 4 and say the plus-4 facts until you've said the fact for 7 plus 4. (Call on a student.) *4 + 4 = 8, 5 + 4 = 9, 6 + 4 = 10, 7 + 4 = 11.*
  What's 7 plus 4? *11.*
- (Point to **9 + 6.**) Start with 6 plus 6 and say the plus-6 facts until you've said the fact for 9 plus 6. (Call on a student.) *6 + 6 = 12, 7 + 6 = 13, 8 + 6 = 14, 9 + 6 = 15.*
  What's 9 plus 6? *15.*
- (Point to **8 + 5.**) Start with 5 plus 5 and say the plus-5 facts until you've said the fact for 8 plus 5. (Call on a student.) *5 + 5 = 10, 6 + 5 = 11, 7 + 5 = 12, 8 + 5 = 13.*
  What's 8 plus 5? *13.*

EXERCISE 2: ASSOCIATIVE PROPERTY
EQUALITY

a. (Display:) [121:2A]

$$10 + 6 = 10 + 3 + 3 \qquad 47 + 28 = 40 + 7 + 28$$
$$50 + 10 = 50 + 9 + 1 \qquad 54 + 10 = 44 + 10 + 10$$

All of these equations have numbers that are added on both sides. All of them are equal and I'll show you how to tell.

- (Point to **10 + 6.**) Read this equation. (Signal.) *10 + 6 = 10 + 3 + 3.*
- (Point to **10 + 6.**) What is 10 added to on this side of the equation? (Signal.) *6.*
- (Point to **10 + 3.**) 10 is added to 3 plus 3 on this side of the equation. What is 10 added to on this side of the equation? (Signal.) *3 plus 3.*
- What is 3 plus 3? (Signal.) *6.*
  (Add to show:) [121:2B]

$$10 + 6 = 10 + \overset{6}{(3 + 3)} \qquad 47 + 28 = 40 + 7 + 28$$
$$50 + 10 = 50 + 9 + 1 \qquad 54 + 10 = 44 + 10 + 10$$

- Are both sides equal? (Signal.) *Yes.* Both sides show 10 plus 6.

b. (Point to **50 + 10.**) Read this equation. (Signal.) *50 + 10 = 50 + 9 + 1.*
- (Point to **50 + 10.**) What is 50 added to on this side of the equation? (Signal.) *10.*
- (Point to **50 + 9.**) 50 is added to 9 plus 1 on this side of the equation. What is 50 added to on this side of the equation? (Signal.) *9 plus 1.*
- What is 9 plus 1? (Signal.) *10.*
  (Add to show:) [121:2C]

$$10 + 6 = 10 + \overset{6}{(3 + 3)} \qquad 47 + 28 = 40 + 7 + 28$$
$$50 + 10 = 50 + \overset{10}{(9 + 1)} \qquad 54 + 10 = 44 + 10 + 10$$

- Are both sides equal? (Signal.) *Yes.* Both sides show 50 plus 10.

c. (Point to **47 + 28.**) Read this equation. (Signal.) *47 + 28 = 40 + 7 + 28.*
- (Point to **47 + 28.**) What is 28 added to? (Signal.) *47.*
- (Point to **40 + 7 + 28.**) 28 is added to 40 plus 7 on this side of the equation. What is 28 added to on this side of the equation? (Signal.) *40 plus 7.*
- What is 40 plus 7? (Signal.) *47.*
  (Add to show:) [121:2D]

$$10 + 6 = 10 + \overset{6}{(3 + 3)} \qquad 47 + 28 = \overset{47}{(40 + 7)} + 28$$
$$50 + 10 = 50 + \overset{10}{(9 + 1)} \qquad 54 + 10 = 44 + 10 + 10$$

- Are both sides equal? (Signal.) *Yes.*
- What do both sides show? (Signal.) *47 plus 28.*

d. (Point to **54 + 10.**) Read this equation. (Signal.) *54 + 10 = 44 + 10 + 10.*
- (Point to **54 + 10.**) What is 10 added to? (Signal.) *54.*
- (Point to **44 + 10 + 10.**) 10 is added to 44 plus 10 on this side of the equation. What is 10 added to on this side of the equation? (Signal.) *44 plus 10.*
- What is 44 plus 10? (Signal.) *54.*
  (Add to show:) [121:2E]

$$10 + 6 = 10 + \overset{6}{(3 + 3)} \qquad 47 + 28 = \overset{47}{(40 + 7)} + 28$$
$$50 + 10 = 50 + \overset{10}{(9 + 1)} \qquad 54 + 10 = \overset{54}{(44 + 10)} + 10$$

- Are both sides equal? (Signal.) *Yes.*
- What do both sides show? (Signal.) *54 plus 10.*

e. (Point to **10 + 6.**) This equation is true because both sides show 10 plus 6. How do you know that the equation is true? (Signal.) *Both sides show 10 plus 6.*
- (Point to **50 + 10.**) How do you know this equation is true? (Signal.) *Both sides show 50 plus 10.*
- (Point to **47 + 28.**) How do you know this equation is true? (Signal.) *Both sides show 47 plus 28.*
- (Point to **54 + 10.**) How do you know this equation is true? (Signal.) *Both sides show 54 plus 10.*
  (Repeat equations that were not firm.)

EXERCISE 3: FRACTIONS

a. You've learned the names for parts of shapes that are divided up equally. I'll tell you a name for the parts. You tell me how many equal parts the shape is divided into.
- Listen: Halves. What name? (Signal.) *Halves.*
- A shape with halves is divided into how many equal parts? (Signal.) *2.*
- The name for the parts is halves. What do you call one of them? (Signal.) *A half.*

b. (Repeat the following tasks with the following names:)

Listen: __. What name?		A shape with __ is divided into how many equal parts?	
Fourths	*Fourths*	fourths	4
Fifths	*Fifths*	fifths	5
Sixths	*Sixths*	sixths	6

(Repeat names that were not firm.)

c. (Display:) [121:3A]

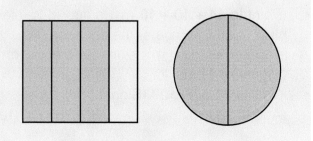

Each of these shapes shows a fraction. First you're going to tell me what the parts for each shape are called and how many equal parts are shaded. Then you'll say the fraction for the shape.
- (Point to rectangle.) How many equal parts is this shape divided into? (Signal.) *4.*
- What are the parts called? (Signal.) *Fourths.*
- How many fourths are shaded? (Signal.) *3.*
- So the fraction for this shape is **3 fourths.** What's the fraction for this shape? (Signal.) *3 fourths.*

d. (Point to circle.) How many equal parts is this shape divided into? (Signal.) *2.*
- What are the parts called? (Signal.) *Halves.*
- How many halves are shaded? (Signal.) *2.*
- So what's the fraction for this shape? (Signal.) *2 halves.*

e. (Display:) [121:3B]

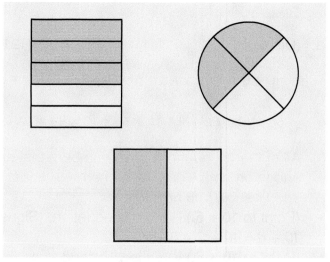

Here are some more shapes that show fractions.
- (Point to 1st rectangle.) Count the parts for this shape to yourself. ✔
- How many equal parts is this shape divided into? (Signal.) *5.*
- What are the parts called? (Signal.) *Fifths.*
- How many fifths are shaded? (Signal.) *3.*
- So the fraction for this shape is 3 fifths. What's the fraction for this shape? (Signal.) *3 fifths.*

f. (Point to circle.) How many equal parts is this shape divided into? (Signal.) *4.*
- What are the parts called? (Signal.) *Fourths.*
- How many fourths are shaded? (Signal.) *2.*
- So what's the fraction for this shape? (Signal.) *2 fourths.*

g. (Point to 2nd rectangle.) How many equal parts is this shape divided into? (Signal.) *2.*
- What are the parts called? (Signal.) *Halves.*
- How many halves are shaded? (Signal.) *1.*
- So what's the fraction for this shape? (Signal.) *1 half.*
  Yes, this shape is 1 half.

h. (Point to 1st rectangle.) How many parts in this picture are shaded? (Signal.) *3.*
- How many parts are there in the rectangle? (Signal.) *5.*
- This picture shows 3 of 5 parts shaded. What does this picture show? (Signal.) *3 of 5 parts shaded.*

i. (Point to circle.) How many parts are shaded? (Signal.) *2.*
- How many parts are there in the circle? (Signal.) *4.*
- What does this picture show? (Signal.) *2 of 4 parts shaded.*

j. (Point to 2nd rectangle.) How many parts in this picture are shaded? (Signal.) *1.*
• How many parts are there in the rectangle? (Signal.) *2.*
• What does this picture show? (Signal.) *1 of 2 parts shaded.*

EXERCISE 4: FACTS
PLUS/MINUS MIX

a. (Distribute unopened workbooks to children.)
• Open your workbook to Lesson 121 and find part 1.
  (Observe children and give feedback.)
  (Teacher reference:)

	c. 8 − 3	f. 7 − 4
a. 11 − 3	d. 10 − 3	g. 7 + 7
b. 8 + 8	e. 6 + 6	h. 10 − 6

You'll tell me if the missing number in each problem is the big number or a small number. Then you'll say the fact.
• Problem A. 11 minus 3. Is the big number or a small number missing? (Signal.) *A small number.*
• Say the fact. (Signal.) *11 − 3 = 8.*
b. (Repeat the following tasks for problems B through H:)

Problem __. __. Is the big number or a small number missing?			Say the fact.
B	8 + 8	The big number.	8 + 8 = 16
C	8 − 3	A small number.	8 − 3 = 5
D	10 − 3	A small number.	10 − 3 = 7
E	6 + 6	The big number.	6 + 6 = 12
F	7 − 4	A small number.	7 − 4 = 3
G	7 + 7	The big number.	7 + 7 = 14
H	10 − 6	A small number.	10 − 6 = 4

(Repeat problems that were not firm.)
c. Complete each of the facts. Put your pencil down when you've written all of the equations for part 1.
  (Observe children and give feedback.)
d. Check your work. You'll read the fact for each problem.
⌈• Problem A. (Signal.) *11 − 3 = 8.*
⌊• (Repeat for:) B, *8 + 8 = 16;* C, *8 − 3 = 5;* D, *10 − 3 = 7;* E, *6 + 6 = 12;* F, *7 − 4 = 3;* G, *7 + 7 = 14;* H, *10 − 6 = 4.*

EXERCISE 5: WORD PROBLEMS
MISSING ADDEND, MINUEND, OR SUBTRAHEND

a. Find part 2 on your worksheet. ✔
  (Teacher reference:)

  a. _____   c. _____
  b. _____

I'll tell you word problems. You'll write the symbols for each problem.
• Problem A. Julie's hair was 14 inches long. Julie's hair grew. Julie's hair ended up 19 inches long. How much did Julie's hair grow?
• Listen to the first part again: Julie's hair was 14 inches long. What number do you write for that part? (Signal.) *14.*
• Write it. ✔
• Next part: Julie's hair grew. It doesn't tell how many inches it grew. What sign do you write for that part? (Signal.) *Plus.*
• What do you make after the plus sign? (Signal.) *A space.*
• Write the plus sign and the space. ✔
  (Display:)   [121:5A]

  a. $14 + \underline{\hspace{1cm}}$

So far, here's what you should have for problem A.
b. Next part: Julie's hair ended up 19 inches long. Tell me the symbols for that part. (Signal.) *Equals 19.*
• Write equals 19. ✔
  (Add to show:)   [121:5B]

  a. $14 + \underline{\hspace{0.5cm}} = 19$

Here's what you should have for problem A.
• Read the problem you wrote for A. Get ready. (Signal.) *14 plus how many equals 19.*
c. Problem B. There were some horses in a barn. 6 of those horses were taken out of the barn. The barn ended up with 11 horses in it. How many horses were in the barn to start with?
• Listen to the first part again: There were some horses in a barn. It doesn't tell how many horses were in the barn. So what do you make for that part? (Signal.) *A space.*
• Make it. ✔
• Next part: 6 of those horses were taken out of the barn. What symbols do you write for that part? (Signal) *Minus 6.*
• Write minus 6. ✔

d. Next part: The barn ended up with 11 horses in it. Tell me the symbols for that part. **(Signal.)** *Equals 11.*
- Write equals 11. ✔
    (Display:)   [121:5C]

b. ___ $- 6 = 11$ ___

Here's what you should have for problem B.
- Read the problem you wrote for B. Get ready. **(Signal.)** *How many minus 6 equals 11.*
e. Problem C. Darnell had some money. He found 8 more dollars. Darnell ended up with 16 dollars. How much money did Darnell have to start with?
- Listen to the first part again: Darnell had some money. What do you make for that part? **(Signal.)** *A space.*
- Make it. ✔
- Next part: He found 8 more dollars. What symbols do you write for that part? **(Signal)** *Plus 8.*
- Write plus 8. ✔
f. Next part: Darnell ended up with 16 dollars. Tell me the symbols for that part. **(Signal.)** *Equals 16.*
- Write equals 16. ✔
    (Display:)   [121:5D]

c. ___ $+ 8 = 16$ ___

Here's what you should have for problem C.
- Read the problem you wrote for C. Get ready. **(Signal.)** *How many plus 8 equals 16.*

EXERCISE 6: EQUATIONS
EQUALITY/INEQUALITY

a. Find part 3 on your worksheet. ✔
    (Teacher reference:)

a. $10 + 6 = 8 - 8$ c. $25 + 10 = 40 + 5$ e. $9 + 10 = 20 - 1$
b. $9 - 1 = 7 + 1$ d. $26 = 36 - 6$ f. $15 + 2 = 15 + 1$

Some of these equations are true and some of them are false. You're going to fix the equals in problems that are false so that all of the statements in part 3 are true.
- Read problem A. **(Signal.)** $10 + 6 = 8 - 8.$
- What does 10 plus 6 equal? **(Signal.)** *16.*
- What does 8 minus 8 equal? **(Signal.)** *Zero.*
- Write 16 above 10 plus 6 and zero above 8 minus 8 to show what the sides equal. ✔
- Are the sides of equation A equal? **(Signal.)** *No.*
- So is equation A true or false? **(Signal.)** *False.*
- What do you make through the equals to show the sides are not equal? **(Call on a student.)** *A single line.*
- Fix the equals for problem A. ✔
- Touch and read the statement for A. **(Signal.)** *10 plus 6 is not equal to 8 minus 8.*
b. Read problem B. **(Signal.)** $9 - 1 = 7 + 1.$
- What does 9 minus 1 equal? **(Signal.)** *8.*
- What does 7 plus 1 equal? **(Signal.)** *8.*
- Write 8 above both sides to show what they equal. ✔
- Are the sides of equation B equal? **(Signal.)** *Yes.*
- So is equation B true or false? **(Signal.)** *True.*
- Touch and read the equation for B. **(Signal.)** $9 - 1 = 7 + 1.$
c. Read the rest of the problems to yourself. If a side adds or subtracts, write the number that the side equals above it. Then fix the sign. Put your pencil down when you've completed the statements for all of the problems in part 3. **(Observe children and give feedback.)**
d. Check your work.
- Problem C: 25 plus 10 equals 40 plus 5. Is that equation true or false? **(Signal.)** *False.*
- Read the statement for C. **(Signal.)** *25 plus 10 is not equal to 40 plus 5.*
e. (Repeat the following tasks for the remaining problems:)

Problem __. Is that equation true or false?		Read the statement for __.	
D: $26 = 36 - 6$	*False*	D	*26 is not = to 36 – 6*
E: $9 + 10 = 20 - 1$	*True*	E	*9 + 10 equals 20 – 1*
F: $15 + 2 = 15 + 1$	*False*	F	*15 + 2 is not = to 15 + 1*

EXERCISE 7: ADDING AND SUBTRACTING TENS NUMBERS

DRAWINGS, REASONING, AND WRITING EQUATIONS

a. Find part 4 on your worksheet. ✔
(Teacher reference:)

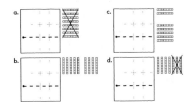

You're going to explain each picture. Then you'll tell me the equation for it and write it.

- Raise your hand when you can tell me what picture A shows. ✔
- What does picture A show? (Call on a student.) *Starts with 8 groups of ten. Subtracts 8 groups of ten. Equals zero groups of ten, which is zero.*
- Everybody, say the equation for this picture. (Signal.) *80 – 80 = 0.*
- Write the equation for picture A. ✔
(Display:)   [121:7A]

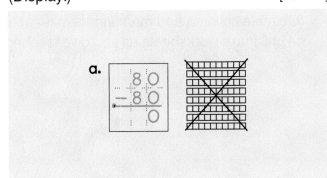

Here's the equation for picture A.

b. Raise your hand when you can tell me what picture B shows. ✔

- What does picture B show? (Call on a student.) *Starts with 5 groups of ten. Adds 4 groups of ten. Equals 9 groups of ten, which is 90.*
- Everybody, say the equation for picture B. (Signal.) *50 + 40 = 90.*
- Write the equation for picture B. ✔

c. Write the equations for pictures C and D. Put your pencil down when you've written equations for all of the pictures in part 4.
(Observe children and give feedback.)

d. Check your work. You'll read the equation for pictures C and D.

- Picture C. Read the equation. (Signal.) *20 + 50 = 70.*
- Picture D. Read the equation. (Signal.) *70 – 30 = 40.*

EXERCISE 8: SOLVING PROBLEMS FROM NUMBER FAMILIES

MISSING ADDEND, MINUEND, OR SUBTRAHEND

a. Turn to the other side of your worksheet and find part 5. ✔
(Teacher reference:)

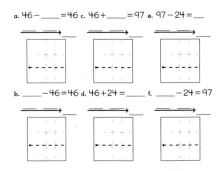

These problems have a missing number. You're going to make a family for each problem, and find the missing number.

- Read problem A. (Signal.) *46 minus how many equals 46.*
- Is the missing number a small number or the big number? (Signal.) *A small number.*
- What is the first 46? (Signal.) *The big number.*
- What is the other 46? (Signal.) *A small number.*
(Repeat until firm.)

b. Make the family for problem A. Put your pencil down when you've written the family with two numbers for problem A.
(Observe children and give feedback.)
(Display:)   [121:8A]

$$a. \quad 46 - \underline{\quad} = 46$$

$$\underline{46} \longrightarrow \underline{46}$$

- Here's what you should have for family A. Say the problem for finding the missing number in family A. Get ready. (Signal.) *46 minus 46.*

c. In the box below family A, write the column problem and work it. Then write the missing number in the family and complete the equation above family A. Put your pencil down when you've completed problem A.
(Observe children and give feedback.)

d. Check your work.

- Read the column problem and the answer for problem A. (Signal.) *46 – 46 = 0.*
- 46 minus how many equals 46? (Signal.) *Zero.*

e. Read problem B. (Signal.) *How many minus 46 equals 46.*

- Is the missing number a small number or the big number? (Signal.) *The big number.*
- Make the family for problem B. Then write the column problem for finding the missing number and work it. Write the missing number in the family, and complete the equation above family B. Put your pencil down when you've completed problem B.
  (Observe children and give feedback.)

f. Check your work.

- Read the column problem and the answer for problem B. (Signal.) *46 + 46 = 92.*
- How many minus 46 equals 46? (Signal.) *92.*

g. Work the rest of the problems. Put your pencil down when you've completed the families and the equations for part 5.
  (Observe children and give feedback.)

h. Check your work.

- Problem C. Read the column problem and the answer. (Signal.) *97 – 46 = 51.*
- 46 plus how many equals 97? (Signal.) *51.*

i. (Repeat the following tasks for the remaining problems:)

Problem __. Read the column problem and the answer.		__?	
D	46 + 24 = 70	46 + 24 = how many	70
E	97 – 24 = 73	97 – 24 = how many	73
F	24 + 97 = 121	How many – 24 = 97	121

EXERCISE 9: INDEPENDENT WORK

a. Find part 6. ✔
(Teacher reference:)

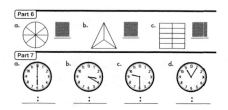

In part 6, you'll count the equal parts for each picture and write the number.
In part 7, you'll write the time for each clock.

b. Complete worksheet 121.
(Observe children and mark incorrect responses on children's worksheets as you give feedback.)

Lesson 122

EXERCISE 1: FACT RELATIONSHIPS
FIGURING OUT UNKNOWN FACTS

a. You'll say facts for families that have a small number of 4. Start with 1 plus 4 and say the plus-4 facts to 10 plus 4.
- Say the fact for 1 plus 4. Get ready. (Signal.) *1 + 4 = 5.*
- Next fact. (Signal.) *2 + 4 = 6.*
- (Repeat for:) *3 + 4 = 7, 4 + 4 = 8, 5 + 4 = 9, 6 + 4 = 10, 7 + 4 = 11, 8 + 4 = 12, 9 + 4 = 13, 10 + 4 = 14.*
- (Repeat until firm.)

b. Now you'll say facts for families that have a small number of 5. Start with 1 plus 5 and say the plus-5 facts to 10 plus 5.
- Say the fact for 1 plus 5. Get ready. (Signal.) *1 + 5 = 6.*
- Next fact. (Signal.) *2 + 5 = 7.*
- (Repeat for:) *3 + 5 = 8, 4 + 5 = 9, 5 + 5 = 10, 6 + 5 = 11, 7 + 5 = 12, 8 + 5 = 13, 9 + 5 = 14, 10 + 5 = 15.*
- (Repeat until firm.)

c. Now you'll say facts for families that have a small number of 6. Start with 1 plus 6 and say the plus-6 facts to 10 plus 6.
- Say the fact for 1 plus 6. Get ready. (Signal.) *1 + 6 = 7.*
- Next fact. (Signal.) *2 + 6 = 8.*
- (Repeat for:) *3 + 6 = 9, 4 + 6 = 10, 5 + 6 = 11, 6 + 6 = 12, 7 + 6 = 13, 8 + 6 = 14, 9 + 6 = 15, 10 + 6 = 16.*
- (Repeat until firm.)

d. (Display:)   [122:1A]

$$9 + 4 \qquad 8 + 6$$
$$7 + 5$$

Here are some new problems we've never worked before.

- (Point to **9 +.**) Read this problem. (Signal.) *9 plus 4.*
- Start with 4 plus 4 and say the plus-4 facts until you've said the fact for 9 plus 4. Get ready. (Signal.) *4 + 4 = 8.*
- Next fact. (Signal.) *5 + 4 = 9.*
- (Repeat for:) *6 + 4 = 10, 7 + 4 = 11, 8 + 4 = 12, 9 + 4 = 13.*
- Everybody, what's 9 plus 4? (Signal.) *13.*
- (Repeat until firm.)

e. (Point to **8 +.**) Read this problem. *(Signal.)* *8 plus 6.*
- Start with 6 plus 6 and say the plus-6 facts until you've said the fact for 8 plus 6. Get ready. (Signal.) *6 + 6 = 12.*
- Next fact. (Signal.) *7 + 6 = 13.*
- Next fact. (Signal.) *8 + 6 = 14.*
- Everybody, what's 8 plus 6? (Signal.) *14.*
- (Repeat until firm.)

f. (Point to **7 +.**) Read this problem. (Signal.) *7 plus 5.*
- Start with 5 plus 5 and say the plus-5 facts until you've said the fact for 7 plus 5. Get ready. (Signal.) *5 + 5 = 10.*
- Next fact. (Signal.) *6 + 5 = 11.*
- Next fact. (Signal.) *7 + 5 = 12.*
- Everybody, what's 7 plus 5? (Signal.) *12.*
- (Repeat until firm.)

INDIVIDUAL TURNS

(Call on individual students to perform one of the following tasks.)

- (Point to **9 + 4.**) Start with 4 plus 4 and say the plus-4 facts until you've said the fact for 9 plus 4. (Call on a student.) *4 + 4 = 8, 5 + 4 = 9, 6 + 4 = 10, 7 + 4 = 11, 8 + 4 = 12, 9 + 4 = 13.* What's 9 plus 4? *13.*
- (Point to **8 + 6.**) Start with 6 plus 6 and say the plus-6 facts until you've said the fact for 8 plus 6. (Call on a student.) *6 + 6 = 12, 7 + 6 = 13, 8 + 6 = 14.* What's 8 plus 6? *14.*
- (Point to **7 + 5.**) Start with 5 plus 5 and say the plus-5 facts until you've said the fact for 7 plus 5. (Call on a student.) *5 + 5 = 10, 6 + 5 = 11, 7 + 5 = 12.* What's 7 plus 5? *12.*

EXERCISE 2: ASSOCIATIVE PROPERTY
INEQUALITY

a. (Display:) [122:2A]

$$18 + 7 = 10 + 8 + 7$$

$$70 + 10 = 70 + 9 + 1$$

These equations have numbers that are added on both sides.
- (Point to **18 + 7.**) Read this equation. (Signal.) *18 + 7 = 10 + 8 + 7.*
- (Point to **18 + 7.**) What is 7 added to on this side of the equation? (Signal.) *18.*
- (Point to **10 + 8.**) What is 7 added to on this side of the equation? (Signal.) *10 plus 8.*
- So I make parentheses around that part. What do I make around that part? (Signal.) *Parentheses.*
(Add to show:) [122:2B]

$$18 + 7 = (10 + 8) + 7$$

$$70 + 10 = 70 + 9 + 1$$

- What is 10 plus 8? (Signal.) *18.*
(Add to show:) [122:2C]

$$18 + 7 = \overset{18}{(10 + 8)} + 7$$

$$70 + 10 = 70 + 9 + 1$$

- Are both sides equal? (Signal.) *Yes.*
- What do both sides show? (Signal.) *18 plus 7.* Yes, both sides show 18 minus 7.

b. (Point to **70 + 10.**) Read this equation. (Signal.) *70 + 10 = 70 + 9 + 1.*
- (Point to **70 +10.**) What is 70 added to on this side of the equation? (Signal.) *10.*
- (Point to **70 + 9.**) What is 70 added to on this side of the equation? (Signal.) *9 plus 1.*
- What is 9 plus 1? (Signal.) *10.*
(Add to show:) [122:2D]

$$18 + 7 = \overset{18}{(10 + 8)} + 7$$

$$70 + 10 = 70 + \overset{10}{(9 + 1)}$$

- Are both sides equal? (Signal.) *Yes.*
- What do both sides show? (Signal.) *70 plus 10.* Yes, both sides show 70 plus 10.

c. (Point to **18 + 7.**) This equation is true because both sides show 18 plus 7. How do you know the equation is true? (Signal.) *Both sides show 18 plus 7.*
- (Point to **70 + 10.**) How do you know this equation is true? (Signal.) *Both sides show 70 plus 10.*

d. (Display:) [122:2E]

$18 + 7 = 10 + 9 + 7$	$70 + 10 = 70 + 8 + 1$
$5 + 5 + 28 = 10 + 28$	$15 + 40 + 8 = 15 + 49$

Some of these equations are not true. We'll figure out if each equation is true. Then we'll fix the sign for the equations that are false.
- (Point to **18 + 7.**) Read this equation. (Signal.) *18 + 7 = 10 + 9 + 7.*
- (Point to **18 + 7.**) What is 7 added to on this side of the equation? (Signal.) *18.*
- (Point to **10 + 9.**) What is 7 added to on this side of the equation? (Signal.) *10 plus 9.*
- What do I make around 10 plus 9? (Signal.) *Parentheses.*
- What is 10 plus 9? (Signal.) *19.*
(Add to show:) [122:2F]

$18 + 7 = \overset{19}{(10 + 9)} + 7$	$70 + 10 = 70 + 8 + 1$
$5 + 5 + 28 = 10 + 28$	$15 + 40 + 8 = 15 + 49$

- Are both sides equal? (Signal.) *No.*
- So is this equation true or false? (Signal.) *False.*
(Add to show:) [122:2G]

$18 + 7 \neq \overset{19}{(10 + 9)} + 7$	$70 + 10 = 70 + 8 + 1$
$5 + 5 + 28 = 10 + 28$	$15 + 40 + 8 = 15 + 49$

- Read the statement. (Signal.) *18 plus 7 is not equal to 10 plus 9 plus 7.*

e. (Point to **5 + 5 + 28.**) Read this equation. (Signal.) *5 + 5 + 28 = 10 + 28.*
- (Point to **5 + 5 + 28.**) What is 28 added to on this side of the equation? (Signal.) *5 plus 5.*
- What do I make around 5 plus 5? (Signal.) *Parentheses.*
- What's 5 plus 5? (Signal.) *10.*
  (Add to show:) [122:2H]

$$\overset{19}{18 + 7} \neq (10 + 9) + 7 \qquad 70 + 10 = 70 + 8 + 1$$

$$\overset{10}{(5 + 5)} + 28 = 10 + 28 \qquad 15 + 40 + 8 = 15 + 49$$

- (Point to **10 + 28.**) What is 28 added to on this side of the equation? (Signal.) *10.*
- Are both sides equal? (Signal.) *Yes.*
- So is this equation true or false? (Signal.) *True.*
- What do both sides show? (Signal.) *10 plus 28.*

f. (Point to **70 + 10.**) Read this equation. (Signal.) *70 + 10 = 70 + 8 + 1.*
- (Point to **70 + 10.**) What is 70 added to on this side of the equation? (Signal.) *10.*
- (Point to **70 + 8.**) What is 70 added to on this side of the equation? (Signal.) *8 plus 1.*
- What do I make around 8 plus 1? (Signal.) *Parentheses.*
- What is 8 plus 1? (Signal.) *9.*
  (Add to show:) [122:2I]

$$\overset{19}{18 + 7} \neq (10 + 9) + 7 \qquad \overset{9}{70 + 10} = 70 + (8 + 1)$$

$$\overset{10}{(5 + 5)} + 28 = 10 + 28 \qquad 15 + 40 + 8 = 15 + 49$$

- Are both sides equal? (Signal.) *No.*
- So is this equation true or false? (Signal.) *False.*
  (Add to show:) [122:2J]

$$\overset{19}{18 + 7} \neq (10 + 9) + 7 \qquad \overset{9}{70 + 10} \neq 70 + (8 + 1)$$

$$\overset{10}{(5 + 5)} + 28 = 10 + 28 \qquad 15 + 40 + 8 = 15 + 49$$

- Read the statement. (Signal.) *70 plus 10 is not equal to 70 plus 9 plus 1.*

g. (Point to **15 + 40 + 8.**) Read this equation. (Signal.) *15 + 40 + 8 = 15 + 49.*
- (Point to **15 + 40 + 8.**) What is 15 added to on this side of the equation? (Signal.) *40 plus 8.*
- What is 40 plus 8? (Signal.) *48.*
- What do I make around 40 plus 8? (Signal.) *Parentheses.*
  (Add to show:) [122:2K]

$$\overset{19}{18 + 7} \neq (10 + 9) + 7 \qquad \overset{9}{70 + 10} \neq 70 + (8 + 1)$$

$$\overset{10}{(5 + 5)} + 28 = 10 + 28 \qquad 15 + \overset{48}{(40 + 8)} = 15 + 49$$

- (Point to **15 + 49.**) What is 15 added to on this side of the equation? (Signal.) *49.*
- Are both sides equal? (Signal.) *No.*
- So is this equation true or false? (Signal.) *False.*
  (Add to show:) [122:2L]

$$\overset{19}{18 + 7} \neq (10 + 9) + 7 \qquad \overset{9}{70 + 10} \neq 70 + (8 + 1)$$

$$\overset{10}{(5 + 5)} + 28 = 10 + 28 \qquad 15 + \overset{48}{(40 + 8)} \neq 15 + 49$$

- Read the statement. (Signal.) *15 plus 40 plus 8 is not equal to 15 plus 49.*

h. Let's read those statements again.
- (Point to **18 + 7.**) Read this equation. (Signal.) *18 plus 7 is not equal to 10 plus 9 plus 7.*
- (Point to **5 + 5 + 28.**) Read this equation. (Signal.) *5 + 5 + 28 = 10 + 28.*
- (Point to **70 + 10.**) Read this equation. (Signal.) *70 plus 10 is not equal to 70 plus 8 plus 1.*
- (Point to **15 + 40 + 8.**) Read this equation. (Signal.) *15 plus 40 plus 8 is not equal to 15 plus 49.*
  (Repeat step h until firm.)

EXERCISE 3: DATA
ORGANIZE, REPRESENT, INTERPRET

a. (Display:) [122:3A]

> b a c b a c a c c
>
> a =   b =   c =

(Point to **b a c b.**) Here's a row of letters.
- Read each letter when I touch under it. Get ready. (Touch.) *B, a, c, b, a, c, a, c, c.*

b. We're going to complete the equations that show how many As there are, how many Bs there are, and how many Cs there are.
- (Point to **a =.**) To complete this equation, we write the number of As here. (Touch.)
- (Point to **b a c b.**) Count the As in this row. Raise your hand when you know how many As there are. ✔
- How many As are there? (Signal.) *3.* (Add to show:) [122:3B]

> b a c b a c a c c
>
> a = 3   b =   c =

c. (Point to **b a c b.**) Count the Bs in this row. Raise your hand when you know how many Bs there are. ✔
- How many Bs are there? (Signal.) *2.* (Add to show:) [122:3C]

> b a c b a c a c c
>
> a = 3   b = 2   c =

d. (Point to **b a c b.**) Count the Cs in this row. Raise your hand when you know how many Cs there are. ✔
- How many Cs are there? (Signal.) *4.* (Add to show:) [122:3D]

> b a c b a c a c c
>
> a = 3   b = 2   c = 4

e. (Point to **a = 3.**) I'll say the equation for figuring out the total number of letters. (Touch.) 3 plus 2 plus 4. Say the problem. (Touch.) *3 plus 2 plus 4.*
- 3 plus 2 plus 4 equals 9. How many total letters are there? (Signal.) *9.*

f. (Add to show:) [122:3E]

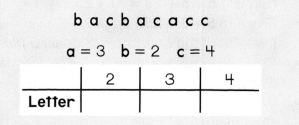

	2	3	4
Letter			

- (Point to **2** in table.) This row shows numbers.
- (Point to spaces in the letter row.) In this row, you're going to write the letter each number tells about.

g. (Point to **2** in table.) What number is this? (Touch.) 2.
- (Point to **a =.**) One of these equations equals 2. Look at that equation. ✔
- What letter equals 2? (Signal.) *B.* (Point to:)

	2	3	4
Letter	☐		

So the letter B goes here. (Touch.)

h. (Point to **3** in table.) What number is this? (Touch.) 3.
- (Point to **a =.**) One of these equations equals 3. What letter equals 3? (Signal.) *A.* (Point to:)

	2	3	4
Letter		☐	

So the letter A goes here. (Touch.)

i. (Point to **4** in table.) What number is this? (Touch.) 4.
- (Point to **a =.**) One of these equations equals 4. What letter equals 4? (Signal.) *C.* (Point to:)

	2	3	4
Letter			☐

So the letter C goes here. (Touch.)

j. Tell me the letters that go in the bottom row again, and I'll write them.
- (Point to **2** in table.) Look at the equations. Raise your hand when you know what letter goes below 2. ✔
- What letter goes below 2? (Touch.) *B.*
  (Add to show:) [122:3F]

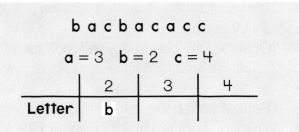

k. (Point to **3** in table.) Look at the equations. Raise your hand when you know what letter goes below 3. ✔
- What letter goes below 3? (Touch.) *A.*
  (Add to show:) [122:3G]

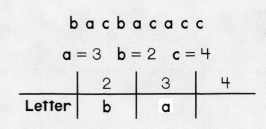

l. (Point to **4** in table.) Look at the equations. Raise your hand when you know what letter goes below 4. ✔
- What letter goes below 4? (Touch.) *C.*
  (Add to show:) [122:3H]

b a c b a c a c c
a = 3 b = 2 c = 4

	2	3	4
Letter	b	a	c

(Point to **2** in table.) This table says there are (touch 2) 2 (touch b) Bs, there are (touch 3) 3 (touch a) As, and there are (touch 4) 4 (touch c) Cs.

m. Now I'll ask questions about more than and less than.
- Listen to the first question. How many more Cs are there than Bs?
- How many Cs are there? (Signal.) *4.*
- How many Bs are there? (Signal.) *2.*

- Say the problem and the answer for figuring out how many more Cs there are. (Signal.) *4 – 2 = 2.*
- How many more Cs than Bs are there? (Signal.) *2.*

EXERCISE 4: ASSOCIATIVE PROPERTY
<, >, =

a. (Distribute unopened workbooks to children.)
- Open your workbook to Lesson 122 and find part 1.
  (Observe children and give feedback.)
  (Teacher reference:)

  a. 18 + 7   10 + 9 + 7   c. 70 + 10   70 + 8 + 1
  b. 5 + 5 + 28   10 + 28   d. 15 + 40 + 8   15 + 49

These are the statements you worked with today, but there are no signs between the sides.
(Display:) [122:4A]

> = <

You're going to write one of these signs between the sides of each problem to make the statement true.
- Read the sides of problem A to yourself. Raise your hand when you know what part you'll make parentheses around.
  (Observe children and give feedback.)
- What part of problem A will you make parentheses around? (Signal.) *10 plus 9.*
- What does 10 plus 9 equal? (Signal.) *19.*
- Make parentheses around 10 plus 9 and write what it equals above.
  (Observe children and give feedback.)
- b. Touch the side that shows 18 plus 7. ✔
- Touch the side that shows 19 plus 7. ✔
  (Repeat until firm.)
- Which is more, 18 or 19? (Signal.) *19.*
- So which is more, 18 plus 7 or 19 plus 7? (Signal.) *19 plus 7.*
- Make the sign with the bigger end next to 19 to show 19 plus 7 is more.
  (Observe children and give feedback.)
  (Display:) [122:4B]

$$a. \quad 18 + 7 < \overset{19}{(10 + 9)} + 7$$

Here's what you should have for problem A.

c. Read the sides of problem B to yourself. Raise your hand when you know what part you'll make parentheses around.
(Observe children and give feedback.)
- What part of problem B will you make parentheses around? (Signal.) *5 plus 5.*
- What does 5 plus 5 equal? (Signal.) *10.*
- Make parentheses around 5 plus 5 and write what it equals above.
(Observe children and give feedback.)
d. What do both sides show? (Signal.) *10 plus 28.*
- So are both sides equal? (Signal.) *Yes.*
- Make the sign to complete the equation.
(Observe children and give feedback.)
(Display:) [122:4C]

$$\overset{10}{\textbf{b.} \quad (5 + 5) + 28 = 10 + 28}$$

Here's what you should have for problem B.
e. Read the sides of problem C to yourself. Raise your hand when you know what part you'll make parentheses around.
(Observe children and give feedback.)
- What part of problem C will you make parentheses around? (Signal.) *8 plus 1.*
- What does 8 plus 1 equal? (Signal.) *9.*
- Make parentheses around 8 plus 1 and write what it equals above.
(Observe children and give feedback.)
f. Touch the side that shows 70 plus 10. ✔
- Touch the side that shows 70 plus 9. ✔
(Repeat until firm.)
- Which is more, 10 or 9? (Signal.) *10.*
- So which is more, 70 plus 10 or 70 plus 9? (Signal.) *70 plus 10.*
- Make the sign with the bigger end next to 10 to show 70 plus 10 is more.
(Observe children and give feedback.)
(Display:) [122:4D]

$$\overset{9}{\textbf{c.} \quad 70 + 10 > 70 + (8 + 1)}$$

Here's what you should have for problem C.

g. Read the sides of problem D to yourself. Raise your hand when you know what part you'll make parentheses around.
(Observe children and give feedback.)
- What part of problem D will you make parentheses around? (Signal.) *40 plus 8.*
- What does 40 plus 8 equal? (Signal.) *48.*
- Make parentheses around 40 plus 8 and write what it equals above.
(Observe children and give feedback.)
h. Touch the side that shows 15 plus 48. ✔
- Touch the side that shows 15 plus 49. ✔
(Repeat until firm.)
- Which is more, 48 or 49? (Signal.) *49.*
- So which is more, 15 plus 48 or 15 plus 49? (Signal.) *15 plus 49.*
- Make the sign to show which side is more.
(Observe children and give feedback.)
(Display:) [122:4E]

$$\overset{48}{\textbf{d.} \quad 15 + (40 + 8) < 15 + 49}$$

Here's what you should have for problem D.

EXERCISE 5: FRACTIONS

a. You've learned the names for parts of shapes that are divided up equally. I'll tell you a name for the parts. You tell me how many equal parts the shape is divided into.
- Listen: Fourths. What name? (Signal.) *Fourths.*
- A shape with fourths is divided into how many equal parts? (Signal.) *4.*
b. (Repeat the following tasks with the following names:)

Listen: __. What name?		A shape with __ is divided into how many equal parts?	
Sevenths	*Sevenths*	sevenths	7
Fifths	*Fifths*	fifths	5
Halves	*Halves*	halves	2

(Repeat names that were not firm.)

c. (Display:) [122:5A]

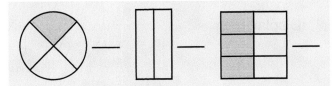

Each of these shapes shows a fraction. You'll say the fraction for each shape and I'll write the fraction.

- (Point to circle.) How many equal parts is this shape divided into? (Signal.) *4.*
(Add to show:) [122:5B]

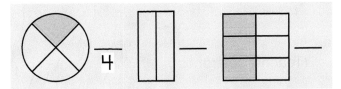

- What are the parts called? (Signal.) *Fourths.*
- How many fourths are shaded? (Signal.) *1.*
(Add to show:) [122:5C]

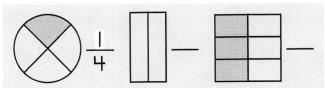

- So what's the fraction for this shape? (Signal.) *1 fourth.*
I wrote the number of equal parts for the shape below the bar.
d. (Point to rectangle.) How many equal parts is this shape divided into? (Signal.) *2.*
(Add to show:) [122:5D]

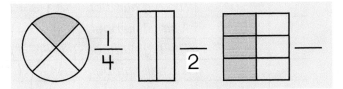

- What are the parts called? (Signal.) *Halves.*
- How many halves are shaded? (Signal.) *Zero.*
(Add to show:) [122:5E]

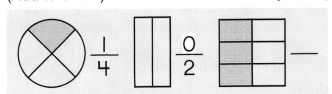

- So what's the fraction for this shape? (Signal.) *Zero halves.*
- Did I write the number of equal parts for the shape above or below the bar? (Signal.) *Below (the bar).*

e. (Point to square.) Count the parts for this shape to yourself. ✔
- How many equal parts is this shape divided into? (Signal.) *6.*
- Do I write 6 above or below the bar? (Signal.) *Below (the bar).*
(Add to show:) [122:5F]

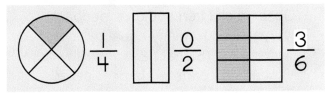

- What are the parts called? (Signal.) *Sixths.*
- How many sixths are shaded? (Signal.) *3.*
(Add to show:) [122:5G]

- So what's the fraction for this shape? (Signal.) *3 sixths.*
f. Find part 2 on your worksheet. ✔
(Teacher reference:)

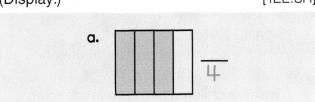

You're going to write the number of equal parts for each shape and how many of those parts are shaded. Then you'll tell me the fraction.
- Count the equal parts for shape A and write it in the space below the bar.
(Observe children and give feedback.)
- Everybody, how many equal parts is shape A divided into? (Signal.) *4.*
- What are the parts of shape A called? (Signal.) *Fourths.*
(Display:) [122:5H]

a.

Here's what you should have.

g. Count the parts that are shaded for shape A
    and write that number above the bar.
    (Observe children and give feedback.)
*   Shape A: How many fourths are shaded?
    (Signal.) *3.*
*   So what's the fraction for shape A? (Signal.)
    3 fourths.
    (Add to show:)   [122:5I]

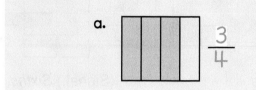

    Here's the fraction you should have for shape A.
h. Write the equal parts for shape B below the bar.
    (Observe children and give feedback.)
*   Everybody, how many equal parts is shape B
    divided into? (Signal.) *2.*
*   What are the parts of shape B called?
    (Signal.) *Halves.*
i. Write the number of parts that are shaded for
    shape B above the bar.
    (Observe children and give feedback.)
*   Shape B: How many halves are shaded?
    (Signal.) *2.*
*   So what's the fraction for shape B? (Signal.)
    2 halves.
    (Display:)   [122:5J]

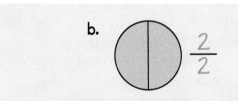

    Here's the fraction you should have for shape B.
j. Write the equal parts for shape C below the bar.
    (Observe children and give feedback.)
*   Everybody, how many equal parts is shape C
    divided into? (Signal.) *6.*
*   What are the parts of shape C called?
    (Signal.) *Sixths.*
k. Write the number of parts that are shaded for
    shape C above the bar.
    (Observe children and give feedback.)
*   Shape C: How many sixths are shaded?
    (Signal.) *1.*

*   So what's the fraction for shape C? (Signal.)
    1 sixth.
    (Display:)   [122:5K]

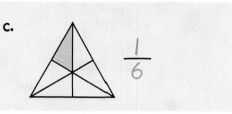

    Here's the fraction you should have for shape C.

EXERCISE 6: FACTS
PLUS/MINUS MIX

a. Find part 3 on your worksheet. ✔
    (Teacher reference:)

	c. $8 + 3$	f. $14 - 7$
a. $6 + 3$	d. $8 - 3$	g. $12 - 6$
b. $6 - 3$	e. $5 + 5$	h. $10 - 7$

    You'll tell me if the missing number in each
    problem is the big number or a small number.
    Then you'll say the fact.
*   Problem A. 6 plus 3. Is the big number or
    a small number missing? (Signal.) *The big
    number.*
*   Say the fact. (Signal.) *6 + 3 = 9.*
b. (Repeat the following tasks for problems B
    through H:)

Problem __. __. Is the big number or a small number missing?			Say the fact.
B	$6 - 3$	*A small number.*	$6 - 3 = 3$
C	$8 + 3$	*The big number.*	$8 + 3 = 11$
D	$8 - 3$	*A small number.*	$8 - 3 = 5$
E	$5 + 5$	*The big number.*	$5 + 5 = 10$
F	$14 - 7$	*A small number.*	$14 - 7 = 7$
G	$12 - 6$	*A small number.*	$12 - 6 = 6$
H	$10 - 7$	*A small number.*	$10 - 7 = 3$

    (Repeat problems that were not firm.)
c. Complete each of the facts. Put your pencil
    down when you've written all of the equations
    for part 3.
    (Observe children and give feedback.)
d. Check your work.
*   Touch and read the equation for problem A.
    (Signal.) *6 + 3 = 9.*
*   (Repeat for:) B, *6 − 3 = 3*; C, *8 + 3 = 11*;
    D, *8 − 3 = 5*; E, *5 + 5 = 10*; F, *14 − 7 = 7*;
    G, *12 − 6 = 6*; H, *10 − 7 = 3.*

EXERCISE 7: WORD PROBLEMS
MISSING ADDEND, MINUEND, OR SUBTRAHEND

a. Find part 4 on your worksheet. ✔
(Teacher reference:)

a. _____   b. _____
═══⟶__   ═══⟶__

I'll tell you word problems. You'll write the symbols for each problem. Then you'll make the number family for the problem and figure out the answer.
Problem A: A board was 11 feet long. A carpenter cut part of the board off. The board ended up 6 feet long. How much of the board was cut off?

• Listen to the first part again: A board was 11 feet long. What number do you write for that part? (Signal.) *11.*
• Write it. ✔
• Next part: A carpenter cut part of the board off. Tell me the symbols you write for that part. (Signal.) *Minus space.*
• Write the minus sign and the space. ✔

b. Next part: The board ended up 6 feet long. Tell me the symbols for that part. (Signal.) *Equals 6.*
• Write equals 6. ✔
(Display:)   [122:7A]

a. $11 - \underline{\quad} = 6$

Here's what you should have for problem A.
• Read the problem you wrote for A. Get ready. (Signal.) *11 minus how many equals 6.*

c. Look at the problem you wrote for A and get ready to tell me about the numbers.
• Is the missing number the big number or a small number? (Signal.) *A small number.*
• What is 11? (Signal.) *The big number.*
• What is 6? (Signal.) *A small number.*
• Make the family for A.
(Observe children and give feedback.)
(Add to show:)   [122:7B]

a. $11 - \underline{\quad} = 6$

$\dfrac{6}{\quad\quad} \longrightarrow 11$

Here's what you should have.
• Say the problem for the missing number in family A. (Signal.) *11 minus 6.*
• What's the answer? (Signal.) *5.*
5 tells how many feet of the board was cut off.

• Listen to problem A again: A board was 11 feet long. A carpenter cut part of the board off. The board ended up 6 feet long. How much of the board was cut off?
• Everybody, what's the answer? (Signal.) *5 (feet).*

d. Problem B. There were some people in a room. 8 of those people left the room. 11 people were still in the room. How many people were in the room to start with?
• Listen to the first part again: There were some people in a room. What do you make for that part? (Signal.) *A space.*
• Write it. ✔
• Next part: 8 of those people left the room. Tell me the symbols you write for that part. (Signal.) *Minus 8.*
• Write minus 8. ✔

e. Next part: 11 people were still in the room. Tell me the symbols for that part. (Signal.) *Equals 11.*
• Write equals 11. ✔
(Add to show:)   [122:7C]

b. $\underline{\quad} - 8 = 11$

Here's what you should have for problem B.
• Read the problem you wrote for B. Get ready. (Signal.) *How many minus 8 equals 11.*

f. Get ready to tell me about the numbers in problem B.
• Is the missing number the big number or a small number? (Signal.) *The big number.*
• What are 8 and 11? (Signal.) *Small numbers.*
• Make the family for B.
(Observe children and give feedback.)
(Display:)   [122:7D]

b. $\underline{\quad} - 8 = 11$

$\dfrac{8 \quad\quad 11}{\quad\quad} \longrightarrow \underline{\quad}$

Here's what you should have.
• Say the problem for the missing number in family B. (Signal.) *8 plus 11.*
• What's the answer? (Signal.) *19.*
19 tells how many people were in the room to start with.
• Listen to problem B again: There were some people in a room. 8 of those people left the room. 11 people were still in the room. How many people were in the room to start with?
• Everybody, what's the answer? (Signal.) *19 (people).*

EXERCISE 8: SOLVING PROBLEMS FROM NUMBER FAMILIES

MISSING ADDEND, MINUEND, OR SUBTRAHEND

a. Find part 5. ✔
(Teacher reference:)

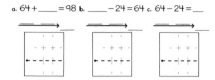

These problems have a missing number. You're going to make a family for each problem, and find the missing number.

• Read problem A. (Signal.) *64 plus how many equals 98.*
• Is the missing number a small number or the big number? (Signal.) *A small number.*
• What is 64? (Signal.) *A small number.*
• What is 98? (Signal.) *The big number.*
(Repeat until firm.)

b. Make the family for problem A. Put your pencil down when you've written the family with two numbers for problem A.
(Observe children and give feedback.)
(Display:)   [122:8A]

$$a. \quad 64 + \underline{\quad} = 98$$

$$\underset{\longrightarrow}{\overset{64}{}} 98$$

• Here's what you should have for family A. Say the problem for finding the missing number in family A. Get ready. (Signal.) *98 minus 64.*

c. In the box below family A, write the column problem and work it. Then write the missing number in the family and complete the equation above family A. Put your pencil down when you've completed problem A.
(Observe children and give feedback.)

d. Check your work.
• Read the column problem and the answer for problem A. (Signal.) *98 – 64 = 34.*
• 64 plus how many equals 98? (Signal.) *34.*

e. Read problem B. (Signal.) *How many minus 24 equals 64.*
• Is the missing number a small number or the big number? (Signal.) *The big number.*
• Make the family for problem B. Then write the column problem for finding the missing number and work it. Write the missing number in the family and complete the equation above family B. Put your pencil down when you've completed problem B.
(Observe children and give feedback.)

f. Check your work.
• Read the column problem and the answer for problem B. (Signal.) *24 + 64 = 88.*
• How many minus 24 equals 64? (Signal.) *88.*

g. Work the rest of the problems in part 5. Problems D, E, and F of part 5 are on the top of the other side of worksheet 122.
(Teacher reference:)

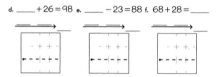

Put your pencil down when you've completed the families and the equations for part 5.
(Observe children and give feedback.)

h. Check your work.
• Touch problem C. ✔
• Read the column problem for C and the answer. (Signal.) *64 – 24 = 40.*
• 64 minus 24 equals how many? (Signal.) *40.*

i. Turn to the other side of your worksheet and find problems D through F.
• (Repeat the following tasks for problems D through F:)

Problem __. Read the column problem and the answer.	__ ?	
D	98 – 26 = 72	How many + 26 = 98	72
E	23 + 88 = 111	How many – 23 = 88	111
F	68 + 28 = 96	68 + 28 = how many	96

EXERCISE 9: INDEPENDENT WORK

a. Find part 6. ✔
   (Teacher reference:)

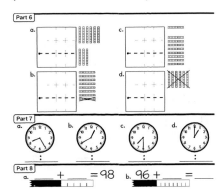

For each picture you're going to write the equation in a column as part of your independent work.
In part 7, write the time for each clock.
In part 8, you'll complete the equations for the rulers.

b. Complete worksheet 122.
   (Observe children and mark incorrect responses on children's worksheets as you give feedback.)

Lesson 123

EXERCISE 1: FACT RELATIONSHIPS
FIGURING OUT UNKNOWN FACTS

a. You'll say facts for families that have a small number of 4. Start with 1 plus 4 and say the plus-4 facts to 10 plus 4.
- Say the fact for 1 plus 4. Get ready. (Signal.) $1 + 4 = 5$.
- Next fact. (Signal.) $2 + 4 = 6$.
- (Repeat for:) $3 + 4 = 7, 4 + 4 = 8, 5 + 4 = 9$, $6 + 4 = 10, 7 + 4 = 11, 8 + 4 = 12, 9 + 4 = 13$, $10 + 4 = 14$.
- (Repeat until firm.)

b. Now you'll say facts for families that have a small number of 5. Start with 1 plus 5 and say the plus-5 facts to 10 plus 5.
- Say the fact for 1 plus 5. Get ready. (Signal.) $1 + 5 = 6$.
- Next fact. (Signal.) $2 + 5 = 7$.
- (Repeat for:) $3 + 5 = 8, 4 + 5 = 9, 5 + 5 = 10$, $6 + 5 = 11, 7 + 5 = 12, 8 + 5 = 13, 9 + 5 = 14$, $10 + 5 = 15$.
- (Repeat until firm.)

c. Now you'll say facts for families that have a small number of 6. Start with 1 plus 6 and say the plus-6 facts to 10 plus 6.
- Say the fact for 1 plus 6. Get ready. (Signal.) $1 + 6 = 7$.
- Next fact. (Signal.) $2 + 6 = 8$.
- (Repeat for:) $3 + 6 = 9, 4 + 6 = 10, 5 + 6 = 11$, $6 + 6 = 12, 7 + 6 = 13, 8 + 6 = 14, 9 + 6 = 15$, $10 + 6 = 16$.
- (Repeat until firm.)

d. (Display:)   [123:1A]

$$8 + 4 \qquad 7 + 6$$
$$9 + 5$$

Here are some new problems we've never worked before.
- (Point to **8 +.**) Read this problem. (Signal.) *8 plus 4.*

- You're going to say the plus-4 facts to yourself to figure out the answer. What kind of plus facts will you say? (Signal.) *Plus 4.*
- What plus-4 fact will you start with? (Call on a student. Idea: *3 plus 4 or 4 plus 4.*)
- Start with 4 plus 4 and say the plus-4 facts to yourself until you've said the fact for 8 plus 4. ✔
- Everybody, what's 8 plus 4? (Signal.) *12.*
- Say the fact. (Signal.) *8 + 4 = 12.*

e. (Point to **7 +.**) Read this problem. (Signal.) *7 plus 6.*
- What kind of plus facts will you say to figure out the answer? (Signal.) *Plus 6.*
- What plus-6 fact will you start with? (Call on a student. Idea: *5 plus 6 or 6 plus 6.*)
- Start with 6 plus 6 and say the plus-6 facts to yourself until you've said the fact for 7 plus 6. ✔
- Everybody, what's 7 plus 6? (Signal.) *13.*
- Say the fact. (Signal.) *7 + 6 = 13.*

f. (Point to **9 +.**) Read this problem. (Signal.) *9 plus 5.*
- What kind of plus facts will you say to figure out the answer? (Signal.) *Plus 5.*
- What plus-5 fact will you start with? (Call on a student. Idea: *5 plus 5 or 6 plus 5.*)
- Start with 6 plus 5 and say the plus-5 facts to yourself until you've said the fact for 9 plus 5. ✔
- Everybody, what's 9 plus 5? (Signal.) *14.*
- Say the fact. (Signal.) *9 + 5 = 14.*
(Repeat steps that were not firm.)

INDIVIDUAL TURNS

(Call on individual students to perform one of the following tasks.)

- (Point to **8 + 4.**) Say plus-4 facts to yourself until you've figured out 8 plus 4.
(Pause.)
What's 8 plus 4? (Call on a student.) *12.*
- (Point to **7 + 6.**) Say plus-6 facts to yourself until you've figured out 7 plus 6.
(Pause.)
What's 7 plus 6? (Call on a student.) *13.*
- (Point to **9 + 5.**) Say plus-5 facts to yourself until you've figured out 9 plus 5.
(Pause.)
What's 9 plus 5? (Call on a student.) *14.*

EXERCISE 2: ASSOCIATIVE PROPERTY
INEQUALITY

a. (Display:)   [123:2A]

> $10 + 8 + 9 = 17 + 9$ $56 + 10 = 50 + 6 + 10$
>
> $25 + 4 = 25 + 3 + 2$ $22 + 1 + 7 = 21 + 7$

Some of these equations are not true. We'll figure out if each equation is true. Then we'll fix the sign for the equations that are false.

- (Point to **10 + 8.**) Read this equation. (Signal.) *10 + 8 + 9 = 17 + 9.*
- (Point to **10 + 8 + 9.**) What is 9 added to on this side of the equation? (Signal.) *10 plus 8.*
- (Point to **17.**) What is 9 added to on this side of the equation? (Signal.) *17.*
- Say the part I make parentheses around. Get ready. (Signal.) *10 + 8.*
- What's 10 plus 8? (Signal.) *18.*
(Add to show:)   [123:2B]

>   18
> $(10 + 8) + 9 = 17 + 9$ $56 + 10 = 50 + 6 + 10$
>
> $25 + 4 = 25 + 3 + 2$ $22 + 1 + 7 = 21 + 7$

- Are both sides equal? (Signal.) *No.*
- So is this equation true or false? (Signal.) *False.*
(Add to show:)   [123:2C]

>   18
> $(10 + 8) + 9 \neq 17 + 9$ $56 + 10 = 50 + 6 + 10$
>
> $25 + 4 = 25 + 3 + 2$ $22 + 1 + 7 = 21 + 7$

- Read the statement. (Signal.) *10 plus 8 plus 9 is not equal to 17 plus 9.*
b. (Point to **25 + 4.**) Read this equation. (Signal.) *25 + 4 = 25 + 3 + 2.*
- (Point to **25 + 4.**) What is 25 added to on this side of the equation? (Signal.) *4.*
- (Point to **25 + 3.**) What is 25 added to on this side of the equation? (Signal.) *3 plus 2.*
- Say the part I make parentheses around. Get ready. (Signal.) *3 plus 2.*

- What is 3 plus 2? (Signal.) *5.*
(Add to show:)   [123:2D]

>   18
> $(10 + 8) + 9 \neq 17 + 9$ $56 + 10 = 50 + 6 + 10$
>
>   5
> $25 + 4 = 25 + (3 + 2)$ $22 + 1 + 7 = 21 + 7$

- Are both sides equal? (Signal.) *No.*
- So is this equation true or false? (Signal.) *False.*
(Add to show:)   [123:2E]

>   18
> $(10 + 8) + 9 \neq 17 + 9$ $56 + 10 = 50 + 6 + 10$
>
>   5
> $25 + 4 \neq 25 + (3 + 2)$ $22 + 1 + 7 = 21 + 7$

- Read the statement. (Signal.) *25 plus 4 is not equal to 25 plus 3 plus 2.*
c. (Point to **56 + 10.**) Read this equation. (Signal.) *56 + 10 = 50 + 6 + 10.*
- (Point to **56 + 10.**) What is 10 added to on this side of the equation? (Signal.) *56.*
- (Point to **50 + 6.**) What is 10 added to on this side of the equation? (Signal.) *50 plus 6.*
- Say the part I make parentheses around. Get ready. (Signal.) *50 plus 6.*
- What is 50 plus 6? (Signal.) *56.*
(Add to show:)   [123:2F]

>   18   56
> $(10 + 8) + 9 \neq 17 + 9$ $56 + 10 = (50 + 6) + 10$
>
>   5
> $25 + 4 \neq 25 + (3 + 2)$ $22 + 1 + 7 = 21 + 7$

- Are both sides equal? (Signal.) *Yes.*
- So is this equation true or false? (Signal.) *True.*
- What do both sides show? (Signal.) *56 plus 10.*

d. (Point to **22 + 1 + 7.**) Read this equation. (Signal.) *22 + 1 + 7 = 21 + 7.*

• (Point to **22 + 1 + 7.**) What is 7 added to on this side of the equation? (Signal.) *22 plus 1.*

• (Point to **21 + 7.**) What is 7 added to on this side of the equation? (Signal.) *21.*

• Say the part I make parentheses around. Get ready. (Signal.) *22 plus 1.*

• What is 22 plus 1? (Signal.) *23.*
   (Add to show:)   [123:2G]

   18   56
   $(10 + 8) + 9 \neq 17 + 9$ $56 + 10 = (50 + 6) + 10$

                      5   23
   $25 + 4 \neq 25 + (3 + 2)$ $(22 + 1) + 7 = 21 + 7$

• Are both sides equal? (Signal.) *No.*

• So is this equation true or false? (Signal.) *False.*
   (Add to show:)   [123:2H]

   18   56
   $(10 + 8) + 9 \neq 17 + 9$ $56 + 10 = (50 + 6) + 10$

                      5   23
   $25 + 4 \neq 25 + (3 + 2)$ $(22 + 1) + 7 \neq 21 + 7$

• Read the statement. (Signal.) *22 plus 1 plus 7 is not equal to 21 plus 7.*

e. Let's read those statements again.

• (Point to **10 + 8 + 9.**) Read this statement. (Signal.) *10 plus 8 plus 9 is not equal to 17 plus 9.*

• (Point to **25 + 4.**) Read this statement. (Signal.) *25 plus 4 is not equal to 25 plus 3 plus 2.*

• (Point to **56 + 10.**) Read this statement. (Signal.) *56 + 10 = 50 + 6 + 10.*

• (Point to **22 + 1 + 7.**) Read this statement. (Signal.) *22 plus 1 plus 7 is not equal to 21 plus 7.*
   (Repeat step e until firm.)

EXERCISE 3: ASSOCIATIVE PROPERTY
<, >, =

a. (Distribute unopened workbooks to children.)

• Open your workbook to Lesson 123 and find part 1.
   (Observe children and give feedback.)
   (Teacher reference:)

   a. $10 + 8 + 9$ $17 + 9$ c. $56 + 10$ $50 + 6 + 10$
   b. $25 + 4$ $25 + 3 + 2$ d. $22 + 1 + 7$ $21 + 7$

These are the statements you just worked with, but there are no signs between the sides.
(Display:)   [123:3A]

                      > = <

You're going to write one of these signs between the sides of each problem to make the statement true.

• Read the sides of problem A to yourself. Raise your hand when you know what part you'll make parentheses around. ✔

• What part of problem A will you make parentheses around? (Signal.) *10 plus 8.*

b. (Repeat the following tasks for problems B through D:)

Read the sides of problem __ to yourself. Raise your hand when you know what part you'll make parentheses around. ✔	What part of problem __ will you make parentheses around?	
B	B	3 + 2
C	C	50 + 6
D	D	22 + 1

(Repeat problems that were not firm.)

c. Go back to problem A and get ready to tell me the part you'll make parentheses around.

• What part will you make parentheses around for problem A? (Signal.) *10 plus 8.*

• Make parentheses around 10 plus 8 and write what it equals above.
   (Observe children and give feedback.)

d. The first side shows 18 plus 9. What does the first side show? (Signal.) *18 plus 9.*
- What does the other side show? (Signal.) *17 plus 9.*
(Repeat until firm.)
- Is 18 more than, less than, or equal to 17? (Signal.) *More than.*
- So is 18 plus 9 more than, less than, or equal to 17 plus 9? (Signal.) *More than.*
- Make the sign to show 18 plus 9 is more. (Observe children and give feedback.) (Display:) [123:3B]

a. $\overset{18}{(10 + 8)} + 9 > 17 + 9$

Here's what you should have for problem A.

e. Make parentheses for problem B and write what it equals above. ✔
- What part of problem B did you make parentheses around? (Signal.) *3 plus 2.*
- What does 3 plus 2 equal? (Signal.) *5.*

f. What does the first side show? (Signal.) *25 plus 4.*
- What does the other side show? (Signal.) *25 plus 5.*
(Repeat until firm.)
- Is 4 more than, less than, or equal to 5? (Signal.) *Less than.*
- So is 25 plus 4 more than, less than, or equal to 25 plus 5? (Signal.) *Less than.*
- Make the sign to show 25 plus 4 is less. (Observe children and give feedback.) (Display:) [123:3C]

b. $25 + 4 < 25 + \overset{5}{(3 + 2)}$

Here's what you should have for problem B.

g. Make parentheses for problem C and write what it equals above. ✔
- What part of problem C did you make parentheses around? (Signal.) *50 plus 6.*
- What does 50 plus 6 equal? (Signal.) *56.*

h. What does the first side show? (Signal.) *56 plus 10.*
- What does the other side show? (Signal.) *56 plus 10.*
(Repeat until firm.)
- Is 56 more than, less than, or equal to 56? (Signal.) *Equal to.*
- So is 56 plus 10 more than, less than, or equal to 56 plus 10? (Signal.) *Equal to.*
- Make the sign to show the sides are equal. (Observe children and give feedback.) (Display:) [123:3D]

c. $56 + 10 = \overset{56}{(50 + 6)} + 10$

Here's what you should have for problem C.

i. Make parentheses for problem D and write what it equals above. ✔
- What part of problem D did you make parentheses around? (Signal.) *22 plus 1.*
- What does 22 plus 1 equal? (Signal.) *23.*

j. What does the first side show? (Signal.) *23 plus 7.*
- What does the other side show? (Signal.) *21 plus 7.*
(Repeat until firm.)
- Is 23 more than, less than, or equal to 21? (Signal.) *More than.*
- So is 23 plus 7 more than, less than, or equal to 21 plus 7? (Signal.) *More than.*
- Make the sign to show 23 plus 7 is more. (Observe children and give feedback.) (Display:) [123:3E]

d. $\overset{23}{(22 + 1)} + 7 > 21 + 7$

Here's what you should have for problem D.

EXERCISE 4: DATA
ORGANIZE, REPRESENT, INTERPRET

a. Find part 2 on your worksheet. ✔
(Teacher reference:)

There's a row of letters in part 2. I'll read the letters. Touch each letter as I read it. (Children touch letters.) P, R, R, T, P, S, P, P, R, T, P.
- Touch the equation you'll complete to show how many Ps. ✔
- Count the Ps in the row of letters.
- Then complete the equation to show how many Ps there are. Put your pencil down when you've completed the equation for P.
(Observe children and give feedback.)
- How many Ps are there? (Signal.) 5.
- Complete the rest of the equations to show the number of Rs, Ss, and Ts. Put your pencil down when you've completed the equations for all the letters.
(Observe children and give feedback.)
(Teacher reference:)

b. Touch the equation for R. ✔
- How many Rs are there? (Signal.) 3.
- How many Ss are there? (Signal.) 1.
- How many Ts are there? (Signal.) 2.
c. Say the equation for figuring the total number of letters. (Signal.) 5 plus 3 plus 1 plus 2.
- Raise your hand when you know what 5 plus 3 plus 1 plus 2 equals. ✔
- Everybody, how many total letters are there? (Signal.) 11.
d. Touch the row in the table that has the numbers 1, 2, 3, 4, 5, and 6. ✔
- You're going to write the letters that go in the bottom row.
- Touch 1 in the top row of the table. ✔
- Look at the equations. Raise your hand when you know what letter equals 1. ✔
- What letter equals 1? (Signal.) S.
Later, you'll write S below 1 in the table.

e. Touch 2 in the top row of the table. ✔
- Raise your hand when you know what letter equals 2. ✔
- What letter equals 2? (Signal.) T.
- Touch where you'll write T. ✔
f. Touch 3 in the top row. ✔
- Raise your hand when you know what letter equals 3. ✔
- What letter equals 3? (Signal.) R.
- Touch where you'll write R. ✔
g. Touch 4 in the top row. ✔
- Look at the equations. Raise your hand when you know if any letter equals 4. ✔
- Do any letters equal 4? (Signal.) No.
- So you won't write any letter below 4.
h. Touch 5 in the top row. ✔
- Raise your hand when you know if any letter equals 5. ✔
- What letter equals 5? (Signal.) P.
- Touch where you'll write P. ✔
i. Touch 6 in the top row. ✔
- Raise your hand when you know if any letter equals 6. ✔
- Does any letter equal 6? (Signal.) No.
- So do you write anything below 6? (Signal.) No.
j. Write the letters in the bottom row of the table. Remember, don't write anything below numbers that are not in the equations. Put your pencil down when you've completed the table.
(Observe children and give feedback.)
k. Check your work.
- Touch number 1 in the table. ✔
- What did you write below 1? (Signal.) S.
- What did you write below 2? (Signal.) T.
- What did you write below 3? (Signal.) R.
- What did you write below 4? (Signal.) Nothing.
- What did you write below 5? (Signal.) P.
- What did you write below 6? (Signal.) Nothing.
l. Now you're going to answer questions that compare.
Listen to the first question: How many more Ps are there than Ss?
- How many Ps are there? (Signal.) 5.
- How many Ss are there? (Signal.) 1.
- Say the problem and the answer for figuring out how many more Ps there are than Ss. (Signal.) 5 − 1 = 4.
- How many more Ps than Ss are there? (Signal.) 4.

m. Listen to the next question: How many fewer
Ts are there than Rs?
• Look at the number for Ts and the number for
Rs and raise your hand when you can say the
problem for figuring out the answer. ✔
• How many Ts are there? (Signal.) 2.
• How many Rs are there? (Signal.) 3.
• Say the problem and the answer for how many
fewer Ts there are than Rs. (Signal.) 3 – 2 = 1.
• How many fewer Ts are there than Rs?
(Signal.) 1.
n. Listen to the next question: How many more
Ps are there than Rs?
• Look at the number for Ps and the number for
Rs and raise your hand when you can say the
problem for figuring out the answer. ✔
• How many Ps are there? (Signal.) 5.
• How many Rs are there? (Signal.) 3.
• Say the problem for figuring out how many
more Ps there are than Rs. (Signal.) 5 – 3 = 2.
• How many more Ps are there than Rs?
(Signal.) 2.

EXERCISE 5: WORD PROBLEMS
MISSING ADDEND, MINUEND, OR SUBTRAHEND

a. Find part 3 on your worksheet. ✔
(Teacher reference:)

a. _____   b. _____
═ ═⟶   ═ ═⟶

I'll tell you word problems. You'll write the
symbols for each problem. Then you'll make
the number family for the problem and figure
out the answer.
• Problem A: There were some cookies in a jar.
Paul ate 4 of the cookies. 14 cookies were still
in the jar. How many cookies were in the jar to
start with?
• Listen to the first part again: There were some
cookies in a jar. What do you make for that
part? (Signal.) (A) space.
• Write it. ✔
• Next part: Paul ate 4 of the cookies. Tell me
the symbols you write for that part. (Signal.)
Minus 4.
• Write minus 4. ✔

b. Next part: 14 cookies were still in the jar.
Tell me the symbols for that part. (Signal.)
Equals 14.
• Write equals 14. ✔
(Display:)   [123:5A]

a. ___ – 4 = 14

Here's what you should have for problem A.
• Read the problem you wrote for A. Get ready.
(Signal.) How many minus 4 equals 14.
c. Look at the problem you wrote for A and get
ready to tell me about the numbers.
• Is the missing number the big number or a
small number? (Signal.) The big number.
• What are 4 and 14? (Signal.) Small numbers.
• Make the family for A.
(Observe children and give feedback.)
(Add to show:)   [123:5B]

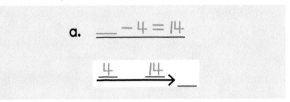

a. ___ – 4 = 14

4 ___ 14 ⟶ ___

Here's what you should have.
d. Say the problem for the missing number in
family A. (Signal.) 4 plus 14.
• What's the answer? (Signal.) 18.
18 tells how many cookies were in the jar to
start with.
• Listen to problem A again: There were some
cookies in a jar. Paul ate 4 of the cookies.
14 cookies were still in the jar. How many
cookies were in the jar to start with?
• Everybody, what's the answer? (Signal.)
18 (cookies).
e. Problem B: A tank had 17 gallons of water in
it. Michael poured some water out of the tank.
The tank ended up with 5 gallons of water in
it. How much water did Michael pour out of
the tank?
• Listen to the first part again: A tank had
17 gallons of water in it. What number do
you write for that part? (Signal.) 17.
• Write it. ✔
• Next part: Michael poured some water out of
the tank. What symbols do you write for that
part? (Signal.) Minus space.
• Write minus space. ✔

f. Next part: The tank ended up with 5 gallons of water in it. Tell me the symbols for that part. (Signal.) *Equals 5.*
- Write equals 5. ✔
  (Add to show:)   [123:5C]

b. $17 - \underline{\quad} = 5$

Here's what you should have for problem B.
- Read the problem you wrote for B. Get ready. (Signal.) *17 minus how many equals 5.*

g. Get ready to tell me about the numbers in problem B.
- Is the missing number the big number or a small number? (Signal.) *A small number.*
- What is 5? (Signal.) *A small number.*
- What is 17? (Signal.) *The big number.*
- Make the family for B.
  (Observe children and give feedback.)
  (Display:)   [123:5D]

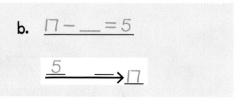

Here's what you should have.
h. Say the problem for the missing number in family B. (Signal.) *17 minus 5.*
- What's the answer? (Signal.) *12.*
  12 tells how much water Michael poured out of the tank.
- Listen to problem B again: A tank had 17 gallons of water in it. Michael poured some of the water out of the tank. The tank ended up with 5 gallons of water in it. How much water did Michael pour out of the tank?
- Everybody, what's the answer? (Signal.) *12 (gallons).*

EXERCISE 6: FRACTIONS

a. You've learned the names for parts of shapes that are divided up equally. I'll tell you a name for the parts. You tell me how many equal parts the shape is divided into.
- Listen: Halves. What name? (Signal.) *Halves.*
- A shape with halves is divided into how many equal parts? (Signal.) *2.*

b. (Repeat the following tasks with the following names:)

Listen: __. What name?		A shape with __ is divided into how many equal parts?	
Eighths	*Eighths*	eighths	8
Fourths	*Fourths*	fourths	4
Sixths	*Sixths*	sixths	6

(Repeat names that were not firm.)

c. Now I'm going to teach you another name for fourths.
- Another way to say fourths is **quarters.** What's another way to say fourths? (Signal.) *Quarters.*
- What's another way to say quarters? (Signal.) *Fourths.*
  (Repeat until firm.)

d. Another way to say 3 fourths is 3 quarters.
- What's another way to say 3 fourths? (Signal.) *3 quarters.*
- What's another way to say 1 fourth? (Signal.) *1 quarter.*
- What's another way to say 2 quarters? (Signal.) *2 fourths.*
- What's another way to say 4 quarters? (Signal.) *4 fourths.*
  (Repeat until firm.)
- You can remember the new name for fourths, because 1 dollar equals 4 quarters. How many quarters equal a dollar? (Signal.) *4.*

e. Find part 4 on your worksheet. ✔
(Teacher reference:)

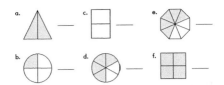

You're going to write the number of equal parts
for each shape and how many of those parts
are shaded. Then you'll tell me the fraction.

• Count the equal parts for shape A and write it
in the space below the bar.
(Observe children and give feedback.)
• Everybody, how many equal parts is shape A
divided into? (Signal.) *2.*
• What are the parts of shape A called?
(Signal.) *Halves.*
(Display:)   [123:6A]

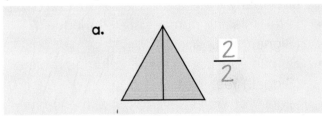

Here's what you should have.

f. Count the parts that are shaded for shape A
and write that number above the bar.
(Observe children and give feedback.)
• Shape A: How many halves are shaded?
(Signal.) *2.*
• So what's the fraction for shape A? (Signal.)
2 halves.
(Add to show:)   [123:6B]

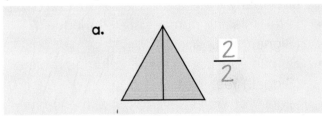

Here's the fraction you should have for shape A.

g. Write the equal parts for shape B below the
bar. Then write the number of parts that are
shaded above the bar.
(Observe children and give feedback.)
• Everybody, how many equal parts is shape B
divided into? (Signal.) *4.*
• You learned a new name for the parts. What's
the new name? (Signal.) *Quarters.*
• How many quarters are shaded? (Signal.) *1.*

• So what's the fraction for shape B? (Signal.)
1 quarter.
• What's another way to say 1 quarter? (Signal.)
1 fourth.
(Display:)   [123:6C]

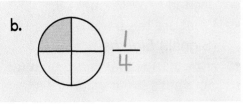

Here's the fraction you should have for shape B.

h. Write the equal parts for shape C below the
bar. Then write the number of parts that are
shaded above the bar.
(Observe children and give feedback.)
• Everybody, how many equal parts is shape C
divided into? (Signal.) *2.*
• What are the parts of shape C called?
(Signal.) *Halves.*
• How many halves are shaded? (Signal.) *Zero.*
• So what's the fraction for shape C? (Signal.)
Zero halves.
(Display:)   [123:6D]

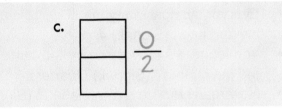

Here's the fraction you should have for
shape C.

i. Write the equal parts for shape D below the
bar. Then write the number of parts that are
shaded above the bar.
(Observe children and give feedback.)
• Everybody, how many equal parts is shape D
divided into? (Signal.) *6.*
• What are the parts of shape D called?
(Signal.) *Sixths.*
• How many sixths are shaded? (Signal.) *4.*
• So what's the fraction for shape D? (Signal.)
4 sixths.
(Display:)   [123:6E]

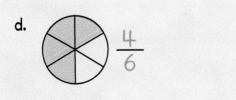

Here's the fraction you should have for
shape D.

j. Write the equal parts for shape E below the bar. Then write the number of parts that are shaded above the bar.
(Observe children and give feedback.)
• Everybody, how many equal parts is shape E divided into? (Signal.) *8.*
• What are the parts of shape E called? (Signal.) *Eighths.*
• How many eighths are shaded? (Signal.) *5.*
• So what's the fraction for shape D? (Signal.) *5 eighths.*
(Display:)   [123:6F]

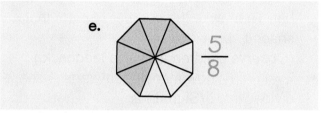

Here's the fraction you should have for shape E.

k. Write the equal parts for shape F below the bar. Then write the number of parts that are shaded above the bar.
(Observe children and give feedback.)
• Everybody, how many equal parts is shape F divided into? (Signal.) *4.*
• You learned a new name for the parts. What's the new name? (Signal.) *Quarters.*
• How many quarters are shaded? (Signal.) *4.*
• So what's the fraction for shape F? (Signal.) *4 quarters.*
• What's another way to say 4 quarters? (Signal.) *4 fourths.*
(Display:)   [123:6G]

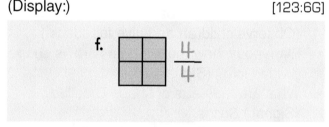

Here's the fraction you should have for shape F.

EXERCISE 7: SOLVING PROBLEMS FROM NUMBER FAMILIES
MISSING ADDEND, MINUEND, OR SUBTRAHEND

a. Turn to the other side of your worksheet and find part 5. ✔
(Teacher reference:)

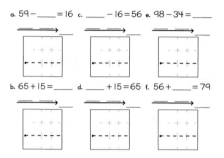

These problems have a missing number. You're going to write a column problem to figure out the missing number. For some of these problems you don't have to make a number family to write the column problem. Here's the rule. If the missing number in a problem is after the equals, you don't need to make the family. You just write the column problem and work it.
• Read problem A. (Signal.) *59 minus how many equals 16.*
• Is the missing number after the equals? (Signal.) *No.*
• So can you just write the column problem? (Signal.) *No.*
Right, you have to make a family to figure out the column problem.
b. Read problem B. (Signal.) *65 plus 15 equals how many.*
• Is the missing number after the equals? (Signal.) *Yes.*
• So can you just write the column problem? (Signal.) *Yes.*
Right, you don't have to make a number family. The column problem you write is the problem that's shown.
c. Problem C is how many minus 16 equals 56. Is the missing number after the equals? (Signal.) *No.*
• So can you just write the column problem? (Signal.) *No.*
• What do you make to figure out the column problem? (Signal.) *A number family.*

d. Read problem D. (Signal.) *How many plus 15 equals 65.*
- Is the missing number after the equals? (Signal.) *No.*
- So can you just write the column problem? (Signal.) *No.*
- What do you make to figure out the column problem? (Signal.) *A number family.*

e. Read problem E. (Signal.) *98 minus 34 equals how many.*
- Is the missing number after the equals? (Signal.) *Yes.*
- So can you just write the column problem? (Signal.) *Yes.*

f. Read problem F. (Signal.) *56 plus how many equals 79.*
- Is the missing number after the equals? (Signal.) *No.*
- So can you just write the column problem? (Signal.) *No.*
- What do you make to figure out the column problem? (Signal.) *A number family.*
  (Repeat problems that were not firm.)

g. Work the problems. Put your pencil down when you've written a column problem and an answer and completed the equation for each problem in part 5.
  (Observe children and give feedback.)

h. Check your work. For each problem, you'll read the column problem and the answer, and answer the question you started with.
- Problem A. Read the column problem and the answer. (Signal.) *59 – 16 = 43.*
- 59 minus how many equals 16? (Signal.) *43.*

i. (Repeat the following tasks for problems B through F:)

	Problem __. Read the column problem and the answer.	__?	
B	65 + 15 = 80	65 + 15 = how many	80
C	16 + 56 = 72	How many – 16 = 56	72
D	65 – 15 = 50	How many + 15 = 65	50
E	98 – 34 = 64	98 – 34 = how many	64
F	79 – 56 = 23	56 + how many = 79	23

EXERCISE 8: ADDING 2-DIGIT WITH 1-DIGIT OR TENS NUMBERS
DRAWINGS, REASONING, AND WRITING EQUATIONS

a. (Display:) [123:8A]

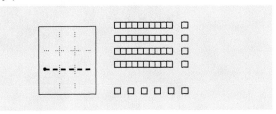

This picture shows a 1-digit number added to a 2-digit number.
- (Point to 4 tens.) How many groups of 10 does this problem start with? (Signal.) *4.*
- (Point to 4 squares.) How many single squares does this problem start with? (Signal.) *4.*
- So how many squares does this problem start with? (Signal.) *44.*
  (Add to show:) [123:8B]

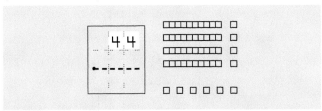

- (Point to 6 squares.) How many single squares does this problem add? (Signal.) *6.*
  (Add to show:) [123:8C]

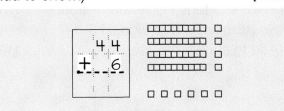

- Start with 44 and count the squares that are added to yourself. Raise your hand when you know what 44 plus 6 equals. ✔
- Everybody, what does 44 plus 6 equal? (Signal.) *50.*
  (Add to show:) [123:8D]

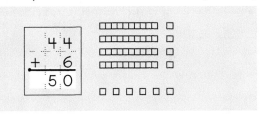

- (Point to squares.) Say the equation for this picture. (Signal.) *44 + 6 = 50.*
- What does this picture show? (Call on a student.) *Starts with 44 squares. Adds 6 squares. Equals 50 squares.*

b. (Display:) [123:8E]

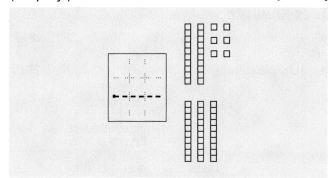

This picture shows a tens number added to a 2-digit number.

- (Point to 2 tens.) How many groups of 10 does this problem start with? (Signal.) *2.*
- (Point to 6 squares.) How many single squares does this problem start with? (Signal.) *6.*
- So how many squares does this problem start with? (Signal.) *26.*
  (Add to show:) [123:8F]

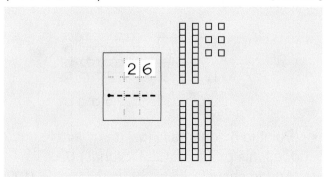

- (Point to 3 tens.) How many groups of 10 does this problem add? (Signal.) *3.*
  (Add to show:) [123:8G]

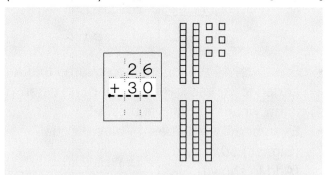

- Raise your hand when you know what 26 plus 30 equals. ✔

- Everybody, what does 26 plus 30 equal? (Signal.) *56.*
  (Add to show:) [123:8H]

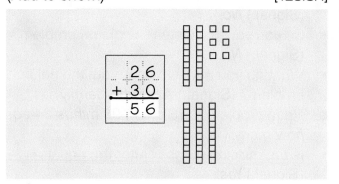

- (Point to squares.) Say the equation for this picture. (Signal.) *26 + 30 = 56.*
- What does this picture show? (Call on a student.) *Starts with 26 squares. Adds 30 squares. Equals 56 squares.*

c. Find part 6 on your worksheet. ✔
  (Teacher reference:)

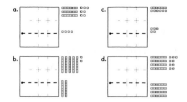

You'll write an equation for each picture.
- Raise your hand when you can explain the equation picture A shows.
  (Observe children and give feedback.)
- What number does the equation for picture A start with? (Signal.) *37.*
- What number does it add? (Signal.) *4.*
- How many squares does this picture show in all? (Signal.) *41.*
- Say the equation for picture A. (Signal.) *37 + 4 = 41.*
- Write the equation for picture A. ✔
- What does picture A show? (Call on a student.) *Starts with 37 squares. Adds 4 squares. Equals 41 squares.*

d. Raise your hand when you can explain the equation picture B shows.
(Observe children and give feedback.)
- What number does the equation for picture B start with? (Signal.) *52.*
- What number does it add? (Signal.) *20.*
- How many squares does this picture show in all? (Signal.) *72.*
- Say the equation for picture B. (Signal.) *52 + 20 = 72.*
- Write the equation for picture B. ✔
- What does picture B show? (Call on a student.) *Starts with 52 squares. Adds 20 squares. Equals 72 squares.*

e. Write equations for the rest of the pictures. Put your pencil down when you've written equations for all the pictures in part 6.
(Observe children and give feedback.)

f. Check your work. Read the equations for pictures C and D.
- Picture C. (Signal.) *16 + 5 = 21.*
- Picture D. (Signal.) *49 + 40 = 89.*

Lesson

EXERCISE 1: FACT RELATIONSHIPS
FIGURING OUT UNKNOWN FACTS

a. You'll say facts for families that have a small number of 5. Start with 1 plus 5 and say the plus-5 facts to 10 plus 5.
- Say the fact for 1 plus 5. Get ready. **(Signal.)** *1 + 5 = 6.*
- Next fact. **(Signal.)** *2 + 5 = 7.*
- (Repeat for:) *3 + 5 = 8, 4 + 5 = 9, 5 + 5 = 10, 6 + 5 = 11, 7 + 5 = 12, 8 + 5 = 13, 9 + 5 = 14, 10 + 5 = 15.*
- (Repeat until firm.)

b. Now you'll say facts for families that have a small number of 6. Start with 1 plus 6 and say the plus-6 facts to 10 plus 6.
- Say the fact for 1 plus 6. Get ready. **(Signal.)** *1 + 6 = 7.*
- Next fact. **(Signal.)** *2 + 6 = 8.*
- (Repeat for:) *3 + 6 = 9, 4 + 6 = 10, 5 + 6 = 11, 6 + 6 = 12, 7 + 6 = 13, 8 + 6 = 14, 9 + 6 = 15, 10 + 6 = 16.*
- (Repeat until firm.)

c. Now you'll say facts for families that have a small number of 7. Start with 1 plus 7 and say the plus-7 facts to 10 plus 7.
- Say the fact for 1 plus 7. Get ready. **(Signal.)** *1 + 7 = 8.*
- Next fact. **(Signal.)** *2 + 7 = 9.*
- (Repeat for:) *3 + 7 = 10, 4 + 7 = 11, 5 + 7 = 12, 6 + 7 = 13, 7 + 7 = 14, 8 + 7 = 15, 9 + 7 = 16, 10 + 7 = 17.*
- (Repeat until firm.)

d. Now you'll say facts for families that have a small number of 8. Start with 1 plus 8 and say the plus-8 facts to 10 plus 8.
- Say the fact for 1 plus 8. Get ready. **(Signal.)** *1 + 8 = 9.*
- Next fact. **(Signal.)** *2 + 8 = 10.*
- (Repeat for:) *3 + 8 = 11, 4 + 8 = 12, 5 + 8 = 13, 6 + 8 = 14, 7 + 8 = 15, 8 + 8 = 16, 9 + 8 = 17, 10 + 8 = 18.*
- (Repeat until firm.)

e. (Display:) [124:1A]

$$9 + 7 \qquad 7 + 8$$
$$9 + 8$$

Here are some new problems we've never worked before.
- (Point to **9 + 7**.) Read this problem. **(Signal.)** *9 plus 7.*
- You're going to say the plus facts to yourself to figure out the answer. What kind of plus facts will you say? **(Signal.)** *Plus 7.*
- What plus-7 fact will you start with? **(Call on a student. Idea:** *7 plus 7.***)**
- Start with 7 plus 7 and say the plus-7 facts to yourself until you've said the fact for 9 plus 7. ✔
- Everybody, what's 9 plus 7? **(Signal.)** *16.*
- Say the fact. **(Signal.)** *9 + 7 = 16.*

f. (Point to **7 + 8**.) Read this problem. **(Signal.)** *7 plus 8.*
- What kind of plus facts will you say to figure out the answer? **(Signal.)** *Plus 8.*
- What plus-8 fact will you start with? **(Call on a student. Idea:** *2 plus 8 or 3 plus 8.***)**
- Start with 3 plus 8 and say the plus-8 facts to yourself until you've said the fact for 7 plus 8. ✔
- Everybody, what's 7 plus 8? **(Signal.)** *15.*
- Say the fact. **(Signal.)** *7 + 8 = 15.*

g. (Point to **9 + 8**.) Read this problem. **(Signal.)** *9 plus 8.*
- What kind of plus facts will you say? **(Signal.)** *Plus 8.*
- What plus-8 fact will you start with? **(Call on a student. Idea:** *3 plus 8 or 6 plus 8.***)**
- Start with 8 plus 8 and say the plus-8 facts to yourself until you've said the fact for 9 plus 8. ✔
- Everybody, what's 9 plus 8? **(Signal.)** *17.*
- Say the fact. **(Signal.)** *9 + 8 = 17.*
(Repeat steps that were not firm.)

(Call on individual students to perform one of the following tasks.)

- (Point to **9 + 7.**) Say the plus-7 facts to yourself until you've figured out 9 plus 7. (Pause.)
  What's 9 plus 7? (Call on a student.) *16.*
- (Point to **7 + 8.**) Say the plus-8 facts to yourself until you've figured out 7 plus 8. (Pause.)
  What's 7 plus 8? (Call on a student.) *15.*
- (Point to **9 + 8.**) Say the plus-8 facts to yourself until you've figured out 9 plus 8. (Pause.)
  What's 9 plus 8? (Call on a student.) *17.*

EXERCISE 2: ASSOCIATIVE PROPERTY
INEQUALITY

a. (Distribute unopened workbooks to children.)
- Open your workbook to Lesson 124 and find part 1.
  (Observe children and give feedback.)
  (Teacher reference:)

  a. $10 + 2 + 8 = 13 + 8$ c. $19 + 10 = 10 + 9 + 10$
  b. $25 + 40 + 7 = 25 + 46$ d. $20 + 1 + 7 = 18 + 7$

  Some of these equations are not true. You'll figure out if each equation is true. Then you'll fix the sign for the equations that are false.
- Read equation A. (Signal.) *10 + 2 + 8 = 13 + 8.*
- Raise your hand when you know the part you'll put parentheses around. ✔
- Say the part you'll make parentheses around. Get ready. (Signal.) *10 plus 2.*
- What is 10 plus 2? (Signal.) *12.*
- Make parentheses around 10 plus 2 and write what it equals above.
  (Observe children and give feedback.)
- Are both sides equal? (Signal.) *No.*
- So is this equation true or false? (Signal.) *False.*
- Fix the sign. ✔
- Read the statement. (Signal.) *10 plus 2 plus 8 is not equal to 13 plus 8.*

b. Read equation B. (Signal.) *25 + 40 + 7 = 25 + 46.*
- Raise your hand when you know the part you'll put parentheses around. ✔
- Say the part you'll make parentheses around. Get ready. (Signal.) *40 plus 7.*
- What is 40 plus 7? (Signal.) *47.*
- Make parentheses around 40 plus 7 and write what it equals above.
  (Observe children and give feedback.)
- Are both sides equal? (Signal.) *No.*
- So is this equation true or false? (Signal.) *False.*
- Fix the sign. ✔
- Read the statement. (Signal.) *25 plus 40 plus 7 is not equal to 25 plus 46.*
c. Read equation C. (Signal.) *19 + 10 = 10 + 9 + 10.*
- Raise your hand when you know the part you'll put parentheses around. ✔
- Say the part you'll make parentheses around. Get ready. (Signal.) *10 plus 9.*
- What is 10 plus 9? (Signal.) *19.*
- Make parentheses around 10 plus 9 and write what it equals above.
  (Observe children and give feedback.)
- Are both sides equal? (Signal.) *Yes.*
- So is this equation true or false? (Signal.) *True.*
- Read the statement. (Signal.) *19 + 10 = 10 + 9 + 10.*
d. Read equation D. (Signal.) *20 + 1 + 7 = 18 + 7.*
- Raise your hand when you know the part you'll put parentheses around. ✔
- Say the part you'll make parentheses around. Get ready. (Signal.) *20 plus 1.*
- What is 20 plus 1? (Signal.) *21.*
- Make parentheses around 20 plus 1 and write what it equals above.
  (Observe children and give feedback.)
- Are both sides equal? (Signal.) *No.*
- So is this equation true or false? (Signal.) *False.*
- Fix the sign. ✔
- Read the statement. (Signal.) *20 plus 1 plus 7 is not equal to 18 plus 7.*

EXERCISE 3: FRACTIONS

a. You learned about quarters last time.
- Listen: Quarters. What name? (Signal.) *Quarters.*
- A shape with quarters is divided into how many equal parts? (Signal.) *4.*

b. (Repeat the following tasks with the following names:)

Listen: __. What name?		A shape with __ is divided into how many equal parts?	
Halves	*Halves*	halves	2
Tenths	*Tenths*	tenths	10
Fourths	*Fourths*	fourths	4

(Repeat names that were not firm.)

c. What's another way to say zero fourths? (Signal.) *Zero quarters.*
- What's another way to say 1 quarter? (Signal.) *1 fourth.*
- What's another way to say 3 fourths? (Signal.) *3 quarters.*
- What's another way to say 4 quarters? (Signal.) *4 fourths.*

d. Find part 2 on your worksheet. ✔
(Teacher reference:)

You're going to write the number of equal parts for each shape and how many of those parts are shaded. Then you'll tell me the fraction.
- Count the equal parts for shape A and write it in the space below the bar.
(Observe children and give feedback.)
- Everybody, how many equal parts is shape A divided into? (Signal.) *5.*
- What are the parts of shape A called? (Signal.) *Fifths.*
(Display:) [124:3A]

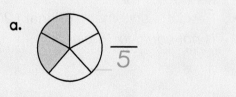

Here's what you should have.

e. Count the parts that are shaded for shape A and write that number above the bar.
(Observe children and give feedback.)

- Shape A: How many fifths are shaded? (Signal.) *2.*
- So what's the fraction for shape A? (Signal.) *2 fifths.*
(Add to show:) [124:3B]

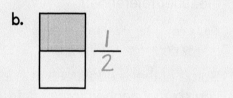

Here's the fraction you should have for shape A.

f. Write the equal parts for shape B below the bar. Then write the number of parts that are shaded above the bar.
(Observe children and give feedback.)
- Everybody, how many equal parts is shape B divided into? (Signal.) *2.*
- What are the parts of shape B called? (Signal.) *Halves.*
- How many halves are shaded? (Signal.) *1.*
- So what's the fraction for shape B? (Signal.) *1 half.*
(Display:) [124:3C]

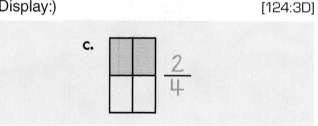

Here's the fraction you should have for shape B.

g. Write the equal parts for shape C below the bar. Then write the number of parts that are shaded above the bar.
(Observe children and give feedback.)
- Everybody, how many equal parts is shape C divided into? (Signal.) *4.*
- You learned a new name for the parts. What's the new name? (Signal.) *Quarters.*
- How many quarters are shaded? (Signal.) *2.*
- So what's the fraction for shape C? (Signal.) *2 quarters.*
- What's another way to say 2 quarters? (Signal.) *2 fourths.*
(Display:) [124:3D]

c.

Here's the fraction you should have for shape C.

h. Now I'll tell you about shapes, and you'll tell me the fractions of those shapes.
- Listen: A shape has 4 equal parts. How many parts does the shape have? (Signal.) *4.*
- What's the new name for the parts? (Signal.) *Quarters.*
- 2 of the 4 quarters are shaded. How many of the quarters are shaded? (Signal.) *2.*
- What's the fraction? (Signal.) *2 quarters.*
- What's another way of saying 2 quarters? (Signal.) *2 fourths.*
  (Repeat until firm.)
i. New shape: It has 2 equal parts. How many parts does the shape have? (Signal.) *2.*
- What's the name for the parts? (Signal.) *Halves.*
- 1 of the halves is shaded. How many of the halves are shaded? (Signal.) *1.*
- What's the fraction? (Signal.) *1 half.*
j. New shape: It has 7 equal parts. How many parts does the shape have? (Signal.) *7.*
- What's the name for the parts? (Signal.) *Sevenths.*
- 5 of the sevenths are shaded. How many of the sevenths are shaded? (Signal.) *5.*
- What's the fraction? (Signal.) *5 sevenths.*
k. New shape: It has 4 equal parts. How many parts does the shape have? (Signal.) *4.*
- What's the new name for the parts? (Signal.) *Quarters.*
- 4 of the quarters are shaded. How many of the quarters are shaded? (Signal.) *4.*
- What's the fraction? (Signal.) *4 quarters.*
- What's another way of saying 4 quarters? (Signal.) *4 fourths.*

EXERCISE 4: SOLVING PROBLEMS FROM NUMBER FAMILIES
MISSING ADDEND, MINUEND, OR SUBTRAHEND

a. Find part 3. ✔
(Teacher reference:)

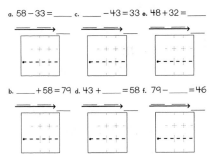

a. $58 - 33 =$ ____   c. ____ $- 43 = 33$ e. $48 + 32 =$ ____

b. ____ $+ 58 = 79$ d. $43 +$ ____ $= 58$ f. $79 -$ ____ $= 46$

These problems have a missing number. You're going to write a column problem to figure out the missing number. For some of these problems you don't have to make a number family to write the column problem.
- Read problem A. (Signal.) *58 minus 33 equals how many.*
- Is the missing number after the equals? (Signal.) *Yes.*
- So can you just write the column problem? (Signal.) *Yes.*
  Right, you don't have to make a family. The column problem you write is the problem that's shown.
b. Read problem B. (Signal.) *How many plus 58 equals 79.*
- Is the missing number after the equals? (Signal.) *No.*
- So can you just write the column problem? (Signal.) *No.*
- What do you make to figure out the column problem? (Signal.) *A number family.*
c. For the rest of the problems, you'll tell me if you'll make a number family or just write the column problem.

d. Problem C. Will you make a number family or just write the column problem? (Signal.) *Make a number family.*
• Problem D. Will you make a number family or just write the column problem? (Signal.) *Make a number family.*
• Problem E. Will you make a number family or just write the column problem? (Signal.) *(Just) write the column problem.*
• Problem F. Will you make a number family or just write the column problem? (Signal.) *Make a number family.*
    (Repeat problems that were not firm.)
e. Work the problems. Put your pencil down when you've written a column problem and an answer and completed the equation for each problem in part 3.
    (Observe children and give feedback.)
f. Check your work. For each problem, you'll read the column problem and the answer, and answer the question you started with.
• Problem A. Read the column problem and the answer. (Signal.) *58 – 33 = 25.*
• 58 minus 33 equals how many? (Signal.) *25.*
g. (Repeat the following tasks for problems B through F:)

Problem __. Read the column problem and the answer.		__?	
B	79 – 58 = 21	How many + 58 = 79	21
C	43 + 33 = 76	How many – 43 = 33	76
D	58 – 43 = 15	43 + how many = 58	15
E	48 + 32 = 80	48 + 32 = how many	80
F	79 – 46 = 33	79 – how many = 46	33

EXERCISE 5: ADDING 2-DIGIT WITH 1-DIGIT OR TENS NUMBERS
DRAWINGS, REASONING, AND WRITING EQUATIONS

a. Turn to the other side of your worksheet and find part 4. ✔
    (Teacher reference:)

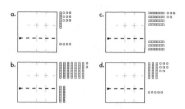

Some of these pictures show a one-digit number added to a two-digit number. The other pictures show a tens number added to a two-digit number. You'll write an equation for each picture.
• Raise your hand when you can explain the equation picture A shows. ✔
• What number does the equation for picture A start with? (Signal.) *19.*
• What number does it add? (Signal.) *4.*
• How many squares does this picture show in all? (Signal.) *23.*
• Say the equation for picture A. (Signal.) *19 + 4 = 23.*
• Write the equation for picture A. ✔
• What does picture A show? (Call on a student.) *Starts with 19 squares. Adds 4 squares. Equals 23 squares.*
b. Raise your hand when you can explain the equation picture B shows. ✔
• What number does the equation for picture B start with? (Signal.) *72.*
• What number does it add? (Signal.) *20.*
• How many squares does this picture show in all? (Signal.) *92.*
• Say the equation for picture B. (Signal.) *72 + 20 = 92.*
• Write the equation for picture B. ✔
• What does picture B show? (Call on a student.) *Starts with 72 squares. Adds 20 squares. Equals 92 squares.*
c. Write equations for the rest of the pictures. Put your pencil down when you've written equations for all of the pictures in part 4.
    (Observe children and give feedback.)
d. Check your work. Read the equations for pictures C and D.
• Picture C. (Signal.) *45 + 30 = 75.*
• Picture D. (Signal.) *38 + 5 = 43.*

EXERCISE 6: ASSOCIATIVE PROPERTY

<, >, =

a. Find part 5.
(Teacher reference:)

a. 10 + 2 + 8   13 + 8   c. 19 + 10   10 + 9 + 10
b. 25 + 40 + 7   25 + 46   d. 20 + 1 + 7   18 + 7

These are the statements you worked with
earlier, but there are no signs between the
sides.
(Display:)   [124:6A]

> = <

You're going to write one of these signs
between the sides of each problem to make
the statement true.
• Read the sides of problem A to yourself. Raise
  your hand when you know what part you'll
  make parentheses around. ✔
• What part of problem A will you make
  parentheses around? (Signal.) *10 plus 2.*

b. (Repeat the following tasks for problems
B through D:)

Read the sides of problem __ to yourself. Raise your hand when you know what part you'll make parentheses around. ✔	What part of problem __ will you make parentheses around?	
B	B	40 + 7
C	C	10 + 9
D	D	20 + 1

(Repeat problems that were not firm.)

c. Make parentheses for each problem and
write what it equals above. Put your pencil
down when you've written parentheses and a
number for all of the problems in part 5.
(Observe children and give feedback.)

d. Check your work.
• Problem A: What does the first side show?
  (Signal.) *12 plus 8.*
• What does the other side show? (Signal.)
  13 plus 8.

e. (Repeat the following tasks for problems
B through D:)

Problem __: What does the first side show?	What does the other side show?
B	25 + 47	25 + 46
C	19 + 10	19 + 10
D	21 + 7	18 + 7

f. Go back to problem A. ✔
• The sides show 12 plus 8 and 13 plus 8. Is 12
  more than, less than, or equal to 13? (Signal.)
  Less than.
• So is 12 plus 8 more than, less than, or equal
  to 13 plus 8? (Signal.) *Less than.*
• Make the sign to show 12 plus 8 is less. ✔
g. Problem B. The sides show 25 plus 47 and 25
plus 46. Is 47 more than, less than, or equal to
46? (Signal.) *More than.*
• So is 25 plus 47 more than, less than, or equal
  to 25 plus 46? (Signal.) *More than.*
• Make the sign to show 25 plus 47 is more. ✔
h. Problem C. The sides show 19 plus 10 and 19
plus 10. Is 19 more than, less than, or equal to
19? (Signal.) *Equal to.*
• So is 19 plus 10 more than, less than, or equal
  to 19 plus 10? (Signal.) *Equal to.*
• Make the sign to show the sides are equal. ✔
i. Problem D. The sides show 21 plus 7 and 18
plus 7. Is 21 more than, less than, or equal to
18? (Signal.) *More than.*
• So is 21 plus 7 more than, less than, or equal
  to 18 plus 7? (Signal.) *More than.*
• Make the sign to show 21 plus 7 is more. ✔
(Display:)   [124:6B]

$$a. \overset{12}{(10 + 2)} + 8 < 13 + 8 \quad c. 19 + 10 = \overset{19}{(10 + 9)} + 10$$

$$b. 25 + \overset{47}{(40 + 7)} > 25 + 46 \quad d. \overset{21}{(20 + 1)} + 7 > 18 + 7$$

Here's what you should have for the problems
in part 5.

EXERCISE 7: WORD PROBLEMS
UNKNOWNS

a. Find part 6 on your worksheet. ✔
(Teacher reference:)

a._____ b._____ c._____
═══⟶__ ═══⟶__ ═══⟶__

I'll tell you word problems. You'll write the symbols for each problem. Then you'll figure out the answer. For some of the problems you'll make a number family. For other problems, you'll just work the number problem. Problem A. 7 ducks were on a lake. Some more ducks flew onto the lake. 19 ducks ended up on the lake. How many ducks flew onto the lake?
I'll say each part of problem A again. Write the symbols for the parts as I say them.
- Listen to the first part again: 7 ducks were on a lake. ✔
- Next part: Some more ducks flew onto the lake. ✔
- Next part: 19 ducks ended up on the lake. ✔
(Repeat if necessary.)

b. Read the problem you wrote for A. Get ready. (Signal.) *7 plus how many equals 19.*
- Is the missing number after the equals? (Signal.) *No.*
So you make a number family.
- Look at the problem you wrote for A. Is the missing number a big number or a small number? (Signal.) *A small number.*
- Which number is the big number? (Signal.) *19.*
- Make the family for A.
(Observe children and give feedback.)
(Display:)   [124:7A]

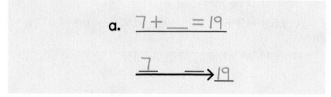

a. $7 + \underline{} = 19$

$\underline{}7\Longrightarrow 19$

Here's what you should have.

c. Say the problem for the missing number in family A. (Signal.) *19 minus 7.*
- What's the answer? (Signal.) *12.*
- Complete the equation. ✔
- Listen to problem A again and get ready to tell me the answer. 7 ducks were on a lake. Some more ducks flew onto the lake. 19 ducks ended up on the lake. How many ducks flew onto the lake?
- Everybody, what's the answer? (Signal.) *12 (ducks).*

d. Problem B. There were 13 bushes in a garden. 7 more bushes were planted in the garden. How many bushes ended up in the garden? I'll say each part of problem B again. Write the symbols for the parts as I say them.
- Listen to the first part: There were 13 bushes in a garden. ✔
- Next part: 7 more bushes were planted in the garden. ✔
- Next part: How many bushes ended up in the garden? ✔
(Repeat if necessary.)

e. Read the problem you wrote for B. Get ready. (Signal.) *13 plus 7 (equals how many).*
- Is the missing number after the equals? (Signal.) *Yes.*
So you just work the problem.
- What's 13 plus 7? (Signal.) *20.*
- Complete the equation. ✔
- Listen to problem B again and get ready to tell me the answer. There were 13 bushes in a garden. 7 more bushes were planted in the garden. How many bushes ended up in the garden?
- Everybody, what's the answer? (Signal.) *20 (bushes).*

f. Problem C. Serena had some money. She spent 12 dollars of her money. Serena had 6 dollars left. How much money did Serena have to start with? I'll say each part of problem C again. Write the symbols for the parts as I say them.
- Listen to the first part: Serena had some money. ✔
- Next part: She spent 12 dollars of her money. ✔
- Next part: Serena had 6 dollars left? ✔
(Repeat if necessary.)

g. Read the problem you wrote for C. Get ready. (Signal.) *How many minus 12 equals 6.*
- Is the missing number after the equals? (Signal.) *No.*
- So do you work the problem or make a number family? (Signal.) *Make a number family.*
- Look at the problem you wrote for C. Is the missing number a big number or a small number? (Signal.) *The big number.*
- Make the family for C.
(Observe children and give feedback.)
(Display:) [124:7B]

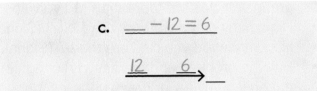

Here's what you should have.

h. Say the problem for the missing number in family C. (Signal.) *12 plus 6.*
- What's the answer? (Signal.) *18.*
- Complete the equation. ✔
- Listen to problem C again and get ready to tell me the answer. Serena had some money. She spent 12 dollars of her money. Serena had 6 dollars left. How much money did Serena have to start with?
- Everybody, what's the answer? (Signal.) *18 (dollars).*

EXERCISE 8: DATA
ORGANIZE, REPRESENT, INTERPRET

a. (Display:) [124:8A]

These are pictures of vehicles that are different colors. You're going to tell me the color for each vehicle. Then you'll complete equations and a table to show the data.

- (Point to blue car.) Tell me the color for each vehicle when I touch it. Get ready. (Touch.) *Blue.*
- (Repeat for remaining vehicles:) *Red, yellow, green, yellow, red, yellow, red, yellow, blue, yellow, red.*
(Repeat until firm.)

b. Find part 7 on your worksheet. ✔
(Teacher reference:)

- First, you'll complete the equations for letters B, R, Y, and G. B tells about vehicles that are blue. What color will the equation for B tell about? (Signal.) *Blue.*
- What color will the equation for R tell about? (Signal.) *Red.*
- What color will the equation for Y tell about? (Signal.) *Yellow.*
- What color will the equation for G tell about? (Signal.) *Green.*
(Repeat until firm.)

c. (Point to blue car.) Your turn to complete the equations. Count the vehicles that are blue and complete the equation for B. Put your pencil down when you've written what B equals.
(Observe children and give feedback.)
- Everybody, how many blue vehicles are there? (Signal.) *2.*

d. (Point to red truck.) Count the vehicles that are red and complete the equation for R. Put your pencil down when you've written what R equals.
(Observe children and give feedback.)
- Everybody, how many red vehicles are there? (Signal.) *4.*

e. (Point to yellow car.) Count the vehicles that are yellow and complete the equation for Y. Put your pencil down when you've written what Y equals.
(Observe children and give feedback.)
- Everybody, how many yellow vehicles are there? (Signal.) *5.*

f. (Point to green motorcycle.) Count the vehicles that are green and complete the equation for G. Put your pencil down when you've written what G equals.
    (Observe children and give feedback.)
- Everybody, how many green vehicles are there? (Signal.) *1.*
    (Display:)   [124:8B]

$$B = 2 \quad R = 4 \quad Y = 5 \quad G = 1$$

Here are the equations you should have for part 7.
g. Say the equation for figuring out the total number of vehicles. Get ready. (Signal.) *2 plus 4 plus 5 plus 1.*
- Raise your hand when you know what 2 plus 4 plus 5 plus 1 equals. ✔
- Everybody, how many total vehicles are there? (Signal.) *12.*
h. Touch the row in the table that has the numbers 1 through 6. ✔
    You're going to write the letter for each color where they go in the bottom row.
- Touch 1 in the top row of the table. ✔
- Are you going to write a letter below the number 1 in the table? (Signal.) *Yes.*
- What letter will you write below 1? (Signal.) *G.*
- So how many vehicles are green? (Signal.) *1.*
i. (Repeat the following tasks for the remaining numbers in the top row:)

Touch __ in the top row of the table.	Are you going to write a letter below the number __ in the table?	What letter will you write below __?		So how many vehicles are __?	
2	2	Yes	2	B	Blue	2
3	3	No				
4	4	Yes	4	R	Red	4
5	5	Yes	5	Y	Yellow	5
6	6	No				

(Repeat numbers that were not firm.)
j. Write the letters in the bottom row of the table. Remember, don't write anything below numbers that are not in the equations. Put your pencil down when you've completed the table.
    (Observe children and give feedback.)

k. Check your work.
    (Add to show:)   [124:8C]

$$B = 2 \quad R = 4 \quad Y = 5 \quad G = 1$$

	1	2	3	4	5	6
Color	G	B		R	Y	

Here's what you should have for part 7.
l. Now you're going to say the problem and the answer for questions about the vehicles. Listen to the first question. How many fewer green vehicles are there than red vehicles?
- How many green vehicles are there? (Signal.) *1.*
- How many red vehicles are there? (Signal.) *4.*
- Say the problem and the answer for figuring out how many fewer green vehicles there are. (Signal.) *4 – 1 = 3.*
- How many fewer green vehicles are there than red vehicles? (Signal.) *3.*
m. Listen to the next question: How many more yellow vehicles are there than green vehicles?
- Look at the number for yellow vehicles and the number for green vehicles and raise your hand when you can say the problem for figuring out the answer. ✔
- How many yellow vehicles are there? (Signal.) *5.*
- How many green vehicles are there? (Signal.) *1.*
- Say the problem and the answer for how many more yellow vehicles there are than green vehicles. (Signal.) *5 – 1 = 4.*
- How many more yellow vehicles are there than green vehicles? (Signal.) *4.*
n. Listen to the next question: How many more red vehicles are there than blue vehicles?
- Look at the number for red vehicles and the number for blue vehicles and raise your hand when you can say the problem for figuring out the answer. ✔
- How many red vehicles are there? (Signal.) *4.*
- How many blue vehicles are there? (Signal.) *2.*
- Say the problem for figuring out how many more red vehicles there are than blue vehicles. (Signal.) *4 – 2 = 2.*
- How many more red vehicles are there than blue vehicles? (Signal.) *2.*

Lesson

EXERCISE 1: FACT RELATIONSHIPS
FIGURING OUT UNKNOWN FACTS

a. You'll say facts for families that have a small number of 6. Start with 1 plus 6 and say the plus-6 facts to 10 plus 6.
- Say the fact for 1 plus 6. Get ready. (Signal.) *1 + 6 = 7.*
- Next fact. (Signal.) *2 + 6 = 8.*
- (Repeat for:) *3 + 6 = 9, 4 + 6 = 10, 5 + 6 = 11, 6 + 6 = 12, 7 + 6 = 13, 8 + 6 = 14, 9 + 6 = 15, 10 + 6 = 16.*
- (Repeat until firm.)

b. Now you'll say facts for families that have a small number of 7. Start with 1 plus 7 and say the plus-7 facts to 10 plus 7.
- Say the fact for 1 plus 7. Get ready. (Signal.) *1 + 7 = 8.*
- Next fact. (Signal.) *2 + 7 = 9.*
- (Repeat for:) *3 + 7 = 10, 4 + 7 = 11, 5 + 7 = 12, 6 + 7 = 13, 7 + 7 = 14, 8 + 7 = 15, 9 + 7 = 16, 10 + 7 = 17.*
- (Repeat until firm.)

c. Now you'll say facts for families that have a small number of 8. Start with 1 plus 8 and say the plus-8 facts to 10 plus 8.
- Say the fact for 1 plus 8. Get ready. (Signal.) *1 + 8 = 9.*
- Next fact. (Signal.) *2 + 8 = 10.*
- (Repeat for:) *3 + 8 = 11, 4 + 8 = 12, 5 + 8 = 13, 6 + 8 = 14, 7 + 8 = 15, 8 + 8 = 16, 9 + 8 = 17, 10 + 8 = 18.*
- (Repeat until firm.)

d. Now you'll say facts for families that have a small number of 9. Start with 1 plus 9 and say the plus-9 facts to 10 plus 9.
- Say the fact for 1 plus 9. Get ready. (Signal.) *1 + 9 = 10.*
- Next fact. (Signal.) *2 + 9 = 11.*
- (Repeat for:) *3 + 9 = 12, 4 + 9 = 13, 5 + 9 = 14, 6 + 9 = 15, 7 + 9 = 16, 8 + 9 = 17, 9 + 9 = 18, 10 + 9 = 19.*
- (Repeat until firm.)

e. (Display:) [125:1A]

$$4 + 7 \qquad 5 + 8$$
$$6 + 9$$

Here are some new problems we've never worked before.
- (Point to **4 +**.) Read this problem. (Signal.) *4 plus 7.*
- You're going to say the plus facts to yourself to figure out the answer. What kind of plus facts will you say? (Signal.) *Plus 7.*
- What plus-7 fact will you start with? (Call on a student. Idea: *2 plus 7 or 3 plus 7.*)
- Start with 3 plus 7 and say the plus-7 facts to yourself until you've said the fact for 4 plus 7. ✔
- Everybody, what's 4 plus 7? (Signal.) *11.*
- Say the fact. (Signal.) *4 + 7 = 11.*

f. (Point to **5 +**.) Read this problem. (Signal.) *5 plus 8.*
- What kind of plus facts will you say to figure out the answer? (Signal.) *Plus 8.*
- What plus-8 fact will you start with? (Call on a student. Idea: *2 plus 8 or 3 plus 8.*)
- Start with 3 plus 8 and say the plus-8 facts to yourself until you've said the fact for 5 plus 8. ✔
- Everybody, what's 5 plus 8? (Signal.) *13.*
- Say the fact. (Signal.) *5 + 8 = 13.*

g. (Point to **6 +**.) Read this problem. (Signal.) *6 plus 9.*
- What kind of plus fact will you say? (Signal.) *Plus 9.*
- What plus-9 fact will you start with? (Call on a student. Idea: *2 plus 9 or 3 plus 9.*)
- Start with 3 plus 9 and say the plus-9 facts to yourself until you've said the fact for 6 plus 9. ✔
- Everybody, what's 6 plus 9? (Signal.) *15.*
- Say the fact. (Signal.) *6 + 9 = 15.*
(Repeat steps that were not firm.)

(Call on individual students to perform one of the following tasks.)

- (Point to **4 + 7.**) Say the plus-7 facts to yourself until you've figured out 4 plus 7. (Pause.)
  What's 4 plus 7? (Call on a student.) *11.*
- (Point to **5 + 8.**) Say the plus-8 facts to yourself until you've figured out 5 plus 8. (Pause.)
  What's 5 plus 8? (Call on a student.) *13.*
- (Point to **6 + 9.**) Say the plus-9 facts to yourself until you've figured out 6 plus 9. (Pause.)
  What's 6 plus 9? (Call on a student.) *15.*

EXERCISE 2: ASSOCIATIVE PROPERTY
INEQUALITY

a. (Distribute unopened workbooks to children.)
- Open your workbook to Lesson 125 and find part 1.
  (Observe children and give feedback.)
  (Teacher reference:)

  a. $80 + 3 = 70 + 10 + 3$ c. $8 + 30 + 5 = 8 + 34$

  b. $16 + 1 + 5 = 18 + 5$ d. $13 + 3 + 6 = 16 + 6$

  Some of these equations are not true. You'll figure out if each equation is true. Then you'll fix the sign for the equations that are false.
- Read equation A. (Signal.) *80 + 3 = 70 + 10 + 3.*
- Raise your hand when you know the part you'll put parentheses around. ✔
- Say the part you'll make parentheses around. Get ready. (Signal.) *70 plus 10.*
- What is 70 plus 10? (Signal.) *80.*
- Make parentheses around 70 plus 10 and write what it equals above.
  (Observe children and give feedback.)
- Are both sides equal? (Signal.) *Yes.*
- So is this equation true or false? (Signal.) *True.*
- Read the statement. (Signal.) *80 + 3 = 70 + 10 + 3.*

b. Read equation B to yourself. Put parentheses around a part, and write what it equals above. Then fix the sign if the equation is false. Put your pencil down when you've worked problem B.
  (Observe children and give feedback.)

- Read the part of problem B you made parentheses around. Get ready. (Signal.) *16 plus 1.*
- What did you write above 16 plus 1? (Signal.) *17.*
- Is equation B true or false? (Signal.) *False.*
- Read the statement. (Signal.) *16 plus 1 plus 5 is not equal to 18 plus 5.*

c. Read equation C to yourself. Put parentheses around a part, and write what it equals above. Then fix the sign if the equation is false. Put your pencil down when you've worked problem C.
  (Observe children and give feedback.)

- Read the part of problem C you made parentheses around. Get ready. (Signal.) *30 plus 5.*
- What did you write above 30 plus 5? (Signal.) *35.*
- Is equation C true or false? (Signal.) *False.*
- Read the statement. (Signal.) *8 plus 30 plus 5 is not equal to 8 plus 34.*

d. Read equation D to yourself. Put parentheses around a part, and write what it equals above. Then fix the sign if the equation is false. Put your pencil down when you've worked problem D.
  (Observe children and give feedback.)

- Read the part of problem D you made parentheses around. Get ready. (Signal.) *13 plus 3.*
- What did you write above 13 plus 3? (Signal.) *16.*
- Is equation D true or false? (Signal.) *True.*
- Read the statement. (Signal.) *13 + 3 + 6 = 16 + 6.*

EXERCISE 3: FRACTIONS

a. Listen: Halves. What name? (Signal.) *Halves.*
- A shape with halves is divided into how many equal parts? (Signal.) *2.*

b. (Repeat the following tasks with the following names:)

Listen: __. What name?		A shape with __ is divided into how many equal parts?	
Quarters	*Quarters*	quarters	4
Fifths	*Fifths*	fifths	5
Ninths	*Ninths*	ninths	9
Fourths	*Fourths*	fourths	4

(Repeat names that were not firm.)

c. Find part 2 on your worksheet. ✔
(Teacher reference:)

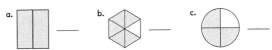

You're going to write the number of equal parts
for each shape and how many of those parts
are shaded. Then you'll tell me the fraction.
- Count the equal parts for shape A and write it
  in the space below the bar.
  (Observe children and give feedback.)
- Everybody, how many equal parts is shape
  A divided into? (Signal.) *2.*
- What are the parts of shape A called?
  (Signal.) *Halves.*
  (Display:)   [125:3A]

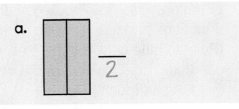

Here's what you should have.
d. Count the parts that are shaded for shape A
  and write that number above the bar.
  (Observe children and give feedback.)
- Shape A: How many halves are shaded?
  (Signal.) *2.*
- So what's the fraction for shape A? (Signal.)
  2 halves.
  (Add to show:)   [125:3B]

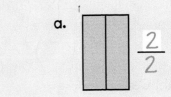

Here's the fraction you should have for shape A.
e. Write the equal parts for shape B below the
  bar. Then write the number of parts that are
  shaded above the bar.
  (Observe children and give feedback.)
- Everybody, how many equal parts is shape B
  divided into? (Signal.) *6.*
- How many sixths are shaded? (Signal.) *5.*

- So what's the fraction for shape B? (Signal.)
  5 sixths.
  (Display:)   [125:3C]

Here's the fraction you should have for shape B.
f. Write the equal parts for shape C below the
  bar. Then write the number of parts that are
  shaded above the bar.
  (Observe children and give feedback.)
- Everybody, how many equal parts is shape C
  divided into? (Signal.) *4.*
- You learned a new name for the parts. What's
  the new name? (Signal.) *Quarters.*
- How many quarters are shaded? (Signal.) *3.*
- So what's the fraction for shape C? (Signal.)
  3 quarters.
- What's another way to say 3 quarters?
  (Signal.) *3 fourths.*
  (Display:)   [125:3D]

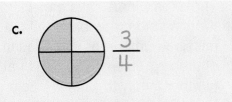

Here's the fraction you should have for shape C.
g. Now I'll tell you about shapes, and you'll tell
  me the fractions of those shapes.
- Listen: A shape has 2 equal parts. How many
  parts does the shape have? (Signal.) *2.*
- What's the name for the parts? (Signal.) *Halves.*
- Zero of the halves are shaded. How many of
  the halves are shaded? (Signal.) *Zero.*
- What's the fraction? (Signal.) *Zero halves.*
  (Repeat until firm.)
h. New shape: It has 4 equal parts. How many
  parts does the shape have? (Signal.) *4.*
- What's the new name for the parts? (Signal.)
  Quarters.
- 1 of the quarters is shaded. How many of the
  quarters are shaded? (Signal.) *1.*
- What's the fraction? (Signal.) *1 quarter.*
- What's another way of saying 1 quarter?
  (Signal.) *1 fourth.*

i. New shape: It has 9 equal parts. How many parts does the shape have? (Signal.) *9.*
•   What's the name for the parts? (Signal.) *Ninths.*
•   7 of the ninths are shaded. How many of the ninths are shaded? (Signal.) *7.*
•   What's the fraction? (Signal.) *7 ninths.*

EXERCISE 4: SOLVING PROBLEMS FROM NUMBER FAMILIES
MISSING ADDEND, MINUEND, OR SUBTRAHEND

a. Find part 3. ✔
    (Teacher reference:)

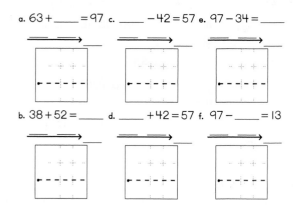

a. $63 + \underline{\quad} = 97$ c. $\underline{\quad} - 42 = 57$ e. $97 - 34 = \underline{\quad}$

b. $38 + 52 = \underline{\quad}$ d. $\underline{\quad} + 42 = 57$ f. $97 - \underline{\quad} = 13$

These problems have a missing number.
•   Write a column problem to figure out the missing number for each equation. For some of these problems you don't have to make a number family to write the column problem. Put your pencil down when you've completed the equation for each problem.
    (Observe children and give feedback.)
b. Check your work. For each problem, you'll read the column problem and the answer, and answer the question for the equation you started with.
•   Problem A. Read the column problem and the answer. (Signal.) *97 – 63 = 34.*
•   63 plus how many equals 97? (Signal.) *34.*
c. (Repeat the following tasks for problems B through F:)

Problem __. Read the column problem and the answer.		__?
B	38 + 52 = 90	38 + 52 = how many	90
C	42 + 57 = 99	How many – 42 = 57	99
D	57 – 42 = 15	How many + 42 = 57	15
E	97 – 34 = 63	97 – 34 = how many	63
F	97 – 13 = 84	97 – how many = 13	84

EXERCISE 5: DATA
ORGANIZE, REPRESENT, INTERPRET

a. (Display:) [125:5A]

These are pictures of vehicles. You're going to tell me if each vehicle is a car, truck, boat, or motorcycle. Then you'll complete the equations and a table to show the data.
•   (Point to blue car.) Tell me if each vehicle is a car, truck, boat, or motorcycle when I touch it. Get ready. (Touch.) *Car.*
•   (Repeat for remaining vehicles:) *Truck, car, motorcycle, truck, motorcycle, car, boat, motorcycle, car, motorcycle, car.*
    (Repeat until firm.)
b. Turn to the other side of your worksheet and find part 4. ✔
    (Teacher reference:)

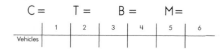

C =   T =   B =   M =

Vehicles	1	2	3	4	5	6

•   First, you'll complete the equations for letters C, T, B, and M. C tells about vehicles that are cars. What kind of vehicle will the equation for C tell about? (Signal.) *Cars.*
•   What kind of vehicle will the equation for T tell about? (Signal.) *Trucks.*
•   What kind of vehicle will the equation for B tell about? (Signal.) *Boats.*
•   What kind of vehicle will the equation for M tell about? (Signal.) *Motorcycles.*
    (Repeat until firm.)
c. (Point to blue car.) Your turn to complete the equations. Count the vehicles that are cars and complete the equation for C. Put your pencil down when you've written what C equals.
    (Observe children and give feedback.)
•   Everybody, how many vehicles are cars? (Signal.) *5.*

d. Count the vehicles that are trucks, boats, and motorcycles, and complete equations for T, B, and M. Put your pencil down when you've completed all the equations.
e. Check your work.
- Look at the equation for T. How many trucks are there? (Signal.) *2.*
- Look at the equation for B. How many boats are there? (Signal.) *1.*
- Look at the equation for M. How many motorcycles are there? (Signal.) *4.*
(Display:)   [125:5B]

$$C = 5 \quad T = 2 \quad B = 1 \quad M = 4$$

Here are the equations you should have for part 4.
f. Say the equation for figuring out the total number of vehicles. Get ready. (Signal.) *5 plus 2 plus 1 plus 4.*
- Raise your hand when you know what 5 plus 2 plus 1 plus 4 equals. ✔
- How many total vehicles are there? (**Call on a student.**) *12.*
g. Touch the row in the table that has the numbers 1 through 6. ✔
  You're going to write the letter for each kind of vehicle where it goes in the bottom row.
- Touch 1 in the top row of the table. ✔
- Are you going to write a letter below the number 1 in the table? (Signal.) *Yes.*
- What letter? (Signal.) *B.*
- So how many vehicles are boats? (Signal.) *1.*
h. Touch 2 in the top row of the table. ✔
- Are you going to write a letter below the number 2? (Signal.) *Yes.*
- What letter? (Signal.) *T.*
- So how many vehicles are trucks? (Signal.) *2.*
i. (Repeat the following tasks for the remaining numbers in the top row:)

Touch __ in the top row of the table.	Are you going to write a letter below the number __?	What letter?	So how many vehicles are __?	
3	3	*No*		
4	4	*Yes*	*M*	Motorcycles	*4*
5	5	*Yes*	*C*	Cars	*5*
6	6	*No*		

(Repeat numbers that were not firm.)

j. Write the letters in the bottom row of the table. Remember, don't write anything below numbers that are not in the equations. Put your pencil down when you've completed the table.
(Observe children and give feedback.)
(Add to show:)   [125:5C]

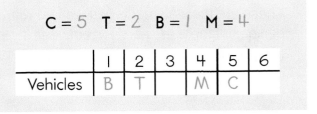

Here's what you should have for part 4.
k. Now you're going to say the problem and the answer for equations about the vehicles.
- Listen to the first question: How many fewer boats were there than cars?
- Look at the number for boats and the number for cars and raise your hand when you can say the problem for figuring out the answer. ✔
- Say the problem and the answer for how many fewer boats there were than cars. (Signal.) *5 – 1 = 4.*
- How many fewer boats were there than cars? (Signal.) *4.*
l. Listen to the next question: How many more motorcycles were there than trucks?
- Look at the number for motorcycles and the number for trucks and raise your hand when you can say the problem for figuring out the answer. ✔
- Say the problem and the answer for how many more motorcycles there were than trucks. (Signal.) *4 – 2 = 2.*
- How many more motorcycles were there than trucks? (Signal.) *2.*
m. Listen to the next question: How many more cars were there than motorcycles?
- Look at the number for cars and the number for motorcycles and raise your hand when you can say the problem for figuring out the answer. ✔
- Say the problem and the answer for how many more cars there were than motorcycles. (Signal.) *5 – 4 = 1.*
- How many more cars were there than motorcycles? (Signal.) *1.*

EXERCISE 6: ASSOCIATIVE PROPERTY

<, >, =

a. Find part 5. ✔
(Teacher reference:)

a. 80 + 3   70 + 10 + 3   c. 8 + 30 + 5   8 + 34
b. 16 + 1 + 5   18 + 5   d. 13 + 3 + 6   16 + 6

These are the statements you worked with earlier, but there are no signs between the sides.
(Display:)   [125:6A]

$$> \quad = \quad <$$

You're going to write one of these signs between the sides of each problem to make the statement true.
- Read the sides of problem A to yourself. Raise your hand when you know what part you'll make parentheses around. ✔
- What part of problem A will you make parentheses around? (Signal.) *70 plus 10.*

b. (Repeat the following tasks for problems B through D:)

Read the sides of problem __ to yourself. Raise your hand when you know what part you'll make parentheses around. ✔	What part of problem __ will you make parentheses around?	
B	B	*16 + 1*
C	C	*30 + 5*
D	D	*13 + 3*

(Repeat problems that were not firm.)

c. Make parentheses for each problem and write what it equals above. Then write the sign to show which side is more or if the sides are equal. Put your pencil down when you've worked all of the problems in part 5.
(Observe children and give feedback.)

d. Check your work.
- Problem A: What does the first side show? (Signal.) *80 plus 3.*
- What does the other side show? (Signal.) *80 plus 3.*
- Is 80 plus 3 more than, less than, or equal to 80 plus 3? (Signal.) *Equal to.*

e. (Repeat the following tasks for problems B through D:)

Problem __: What does the first side show?	What does the other side show?	Is __ more than, less than, or equal to __?		
B	*17 + 5*	*18 + 5*	17 + 5	18 + 5	*Less than*
C	*8 + 35*	*8 + 34*	8 + 35	8 + 34	*More than*
D	*16 + 6*	*16 + 6*	16 + 6	16 + 6	*Equal to*

(Display:)   [125:6B]

a. $80 + 3 \overset{80}{=} (70 + 10) + 3$ c. $8 + \overset{35}{(30 + 5)} > 8 + 34$

b. $\overset{17}{(16 + 1)} + 5 < 18 + 5$ d. $\overset{16}{(13 + 3)} + 6 = 16 + 6$

Here's what you should have for the problems in part 5.

EXERCISE 7: WORD PROBLEMS

UNKNOWNS

a. Find part 6 on your worksheet. ✔
(Teacher reference:)

a. _____
  ==⟶ _

b. _____
  ==⟶ _

c. _____
  ==⟶ _

I'll tell you word problems. You'll write the symbols for each problem. Then you'll figure out the answer. For some of the problems you'll make a number family. For other problems, you'll just work the number problem. Problem A. Eric worked 8 math problems. Then Eric worked some science problems. Eric worked 16 problems in all. How many science problems did Eric work?
I'll say each part of problem A again. Write the symbols for the parts as I say them.
- Listen to the first part: Eric worked 8 math problems. ✔
- Next part: Then Eric worked some science problems. ✔
- Next part: Eric worked 16 problems in all. ✔
(Repeat if necessary.)

b. Read the problem you wrote for A. Get ready.
(Signal.) *8 plus how many equals 16.*
- Is the missing number after the equals?
(Signal.) *No.*
So do you work the problem or make a number family? (Signal.) *Make a number family.*
- Make the family for A. Then complete the equation. Put your pencil down when you can answer the question.
(Observe children and give feedback.)
c. Say the problem for the missing number in family A. (Signal.) *16 minus 8.*
- What's the answer? (Signal.) *8.*
- Listen to problem A again and get ready to tell me the answer. Eric worked 8 math problems. Then Eric worked some science problems. Eric worked 16 problems in all. How many science problems did Eric work?
- Everybody, what's the answer? (Signal.)
8 (science problems).
d. Problem B. There were some apples in a box. 3 of those apples were eaten. 14 apples were left in the box. How many apples were in the box to start with?
I'll say each part of problem B again. Write the symbols for the parts as I say them.
- Listen to the first part: There were some apples in a box. ✔
- Next part: 3 of those apples were eaten. ✔
- Next part: 14 apples were left in the box. ✔
(Repeat if necessary.)
e. Read the problem you wrote for B. Get ready.
(Signal.) *How many minus 3 equals 14.*
- Is the missing number after the equals?
(Signal.) *No.*
So do you work the problem or make a number family? (Signal.) *Make a number family.*
- Make the family for B. Then complete the equation. Put your pencil down when you can answer the question.
f. Say the problem for the missing number in family B. (Signal.) *3 plus 14.*
- What's the answer? (Signal.) *17.*
- Listen to problem B again and get ready to tell me the answer. There were some apples in a box. 3 of those apples were eaten. 14 apples were left in the box. How many apples were in the box to start with?
- Everybody, what's the answer? (Signal.)
17 (apples).

g. Problem C. Jasmine had 19 dollars. She spent 10 of those dollars. How many dollars did Jasmine have left?
I'll say each part of problem C again. Write the symbols for the parts as I say them.
- Listen to the first part: Jasmine had 19 dollars. ✔
- Next part: She spent 10 of those dollars. ✔
- Next part: How many dollars did Jasmine have left? ✔
(Repeat if necessary.)
h. Read the problem you wrote for C. Get ready.
(Signal.) *19 minus 10 (equals how many).*
- Is the missing number after the equals?
(Signal.) *Yes.*
- So do you work the problem or make a number family? (Signal.) *Work the problem.*
- Complete the equation. Put your pencil down when you can answer the question.
(Observe children and give feedback.)
i. Read the problem and the answer for C.
(Signal.) *19 – 10 = 9.*
- Listen to problem C again and get ready to tell me the answer. Jasmine had 19 dollars. She spent 10 of those dollars. How many dollars did Jasmine have left?
- Everybody, what's the answer? (Signal.)
9 (dollars).

EXERCISE 8: INDEPENDENT WORK

a. Find part 7 on your worksheet. ✔
(Teacher reference:)

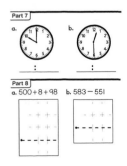

You'll write the time for each clock in part 7. In part 8, you'll write each problem in a column and work it.
b. Complete worksheet 125.
(Observe children and mark incorrect responses on children's worksheets as you give feedback.)

Cumulative Test 2

Note: Cumulative Tests are group-administered. Each student will need a pencil and the *Student Assessment Book.* Try to arrange students so they cannot look at other students' responses.

Do not try to present this test all in one day. Present related portions of it over the course of a few days.

Note: Children who have performed acceptably on Cumulative Test 1 do not have to be tested on Section A of Cumulative Test 2. Begin testing with Section B. After presenting Section B, grade children's papers and follow the recommendations in Scoring Notes for Section B.

Children who have not taken Cumulative Test 1 should begin testing with Section A. After presenting Section A, grade children's papers and follow the recommendations in Scoring Notes for Section A.

Note: Do not administer gray-screened parts to children who have not received instruction in CMC Level B.

Teacher Presentation

Section A (Parts 1–5)
Part 1:
a. Find Cumulative Test 2 in your test booklet. ✔
 • Touch part 1. ✔
 (Teacher reference:)

> a. ___
> b. ___
> c. ___
> d. ___
> e. ___
> f. ___
> g. ___
> h. ___

 You're going to write numbers in a column.
 • Look at the spaces where you'll write the numbers for part 1. ✔
 These columns don't have letters for hundreds, tens, and ones.

 • Touch the space for the hundreds digit in row A. ✔
 • Touch the space for the tens digit in row A. ✔
 • Touch the space for the ones digit in row A. ✔
 (Repeat until firm.)
b. Listen: You're going to write 346 in row A. What number? (Signal.) *346.*
 • Write 346 in row A.
 (Observe children.)
c. You're going to write 179 in row B. What number? (Signal.) *179.*
 • Write 179 in row B.
 (Observe children.)
d. (Repeat step c for: C, 805; D, 3; E, 514; F, 72; G, 835; H, 18.)

Part 2:
a. Touch part 2. ✔
 (Teacher reference:)

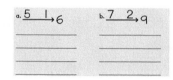

 In the spaces below each number family you'll write the two plus facts and the two minus facts the family tells about.
 • Write the four facts for each family.
 (Observe children.)

Part 3:
a. Touch part 3. ✔
 (Teacher reference:)

> a. 2+7=9   b. 30+8=38   c. 15+1=16

 For each problem, I'll read the equation you'll turn-around. The turn-around is the other plus equation that has the same three numbers.
 • Equation A: 2 plus 7 equals 9. ·
 • Equation B: 30 plus 8 equals 38.
 • Equation C: 15 plus 1 equals 16.
 For each problem in part 3, write the turn-around equation below. That's the equation with the same three numbers.
 (Observe children.)

Part 4:

a. Touch part 4. ✔
(Teacher reference:)

$$\underline{90} + \underline{6} = 96$$

a. ___ + __ = 37
b. ___ + __ = 81
c. ___ + __ = 90
d. $300 + 40 + 8 =$ ___
e. ___ + __ + __ = 813
f. ___ + __ + __ = 209

g.
```
  100
   70
+   6
```

h.
```
  ___
+ ___
 640
```

In part 4, you'll complete the place-value equations for 2-digit and 3-digit numbers.

• I'll read the place-value equation for the 2-digit number that's shown. 90 plus 6 equals 96.
• Complete the place-value equations in part 4. (Observe children.)

Part 5:

a. Touch part 5. ✔
(Teacher reference:)

1 + 7	70 + 10	8 + 10
7 − 0	6 − 2	9 + 1 + 6
4 + 1	72 + 10	17 − 10
7 − 1	2 + 6	10 + 5
1 + 8	9 − 7	8 − 3
10 − 10	25 + 10	10 − 6
12 + 1	9 + 2	9 − 3
1 + 5	8 − 2	3 + 5
9 − 8	1 + 7 + 2	6 + 4
18 + 1	13 − 3	9 − 6

• Complete the equations in part 5. Put your pencil down when you've completed the equations in part 5.
(Observe children.)

Scoring Notes: Section A

a. Collect test booklets. Use the Answer Key to score the tests. Children can earn 1 point for each problem in all parts.
Record children's performance on the Summary Sheet. Reproducible Summary Sheets are at the back of the *Teacher's Guide.*

b. If children fail a part in Section A, present the Cumulative Test 1 parts specified in the online Remedy Table for Cumulative Test 2. After children perform acceptably on Cumulative Test 1 parts, present Section B of Cumulative Test 2. If children fail more than one part of Section A, presenting Section B is not appropriate.

Section B (Parts 6–29)
Part 6:

a. Touch part 6. ✔
(Teacher reference:)

a.
```
  857
−  37
```
b.
```
  386
− 381
```
c.
```
  459
− 402
```
d.
```
  738
− 532
```

You'll work the column problems that minus in part 6.

Part 7:

a. Touch part 7. ✔
(Teacher reference:)

a.
```
   34
   21
+  23
```
d.
```
   72
+  18
```
g.
```
  297
  312
+  70
```

b.
```
   32
    6
+  21
```
e.
```
   56
+  14
```
h.
```
  412
    8
+ 153
```

c.
```
   18
  201
+  40
```
f.
```
    8
   52
+  37
```
i.
```
   81
  124
+ 562
```

You'll work the column problems that plus in part 7.

• Work the problems in parts 6 and 7. Put your pencil down when you've completed the column problems in both parts.
(Observe children.)

Part 8:

a. Touch part 8. ✔
(Teacher reference:)

a. ___ b. ___ c. ___ d. ___
e. ___ f. ___ g. ___ h. ___

You're going to write symbols for dollars and cents. For each amount, you'll write a dollar sign and a dot.

• What two symbols will you write for each amount? (Signal.) *(A) dollar sign and (a) dot.* (Repeat until firm.)
• Touch the space for A. ✔
• You'll write 9 dollars and 13 cents in space A. What will you write in space A? (Signal.) *9 dollars and 13 cents.*
• Write 9 dollars and 13 cents for A. (Observe children.)

b. Touch the space for B. ✔
- You'll write 10 dollars and 52 cents in space B. What will you write in space B? **(Signal.)** *10 dollars and 52 cents.*
- Write 10 dollars and 52 cents for B. (Observe children.)
c. (Repeat the following tasks for amounts C through H:)

Touch the space for __.	You'll write __ in space __.		What will you write in space __?		Write __ for __.	
C	3 dollars and 56 cents	C	C	*3 dollars and 56 cents*	3 dollars and 56 cents	C
D	4 dollars and 19 cents	D	D	*4 dollars and 19 cents*	4 dollars and 19 cents	D
E	4 dollars and 2 cents	E	E	*4 dollars and 2 cents*	4 dollars and 2 cents	E
F	15 dollars and 60 cents	F	F	*15 dollars and 60 cents*	15 dollars and 60 cents	F
G	18 dollars and 6 cents	G	G	*18 dollars and 6 cents*	18 dollars and 6 cents	G
H	18 dollars and 9 cents	H	H	*18 dollars and 9 cents*	18 dollars and 9 cents	H

Part 9:

a. Touch space A of part 9. ✔
(Teacher reference:)

a. _____   b. _____

In space A, you'll write all of the even numbers from 10 to 20.
- Write the even numbers from 10 to 20 in space A. (Observe children.)
b. In space B, you'll write all of the odd numbers from 10 to 20.
- Write the odd numbers from 10 to 20 in space B. (Observe children.)

Part 10:

a. Touch part 10. ✔
(Teacher reference:)

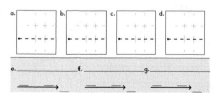

You're going to write the symbols for word problems and work them.
Problem A tells about two lines. You're going to write the symbols and figure out how much longer one line is than the other line.

- Touch where you'll write the column problem for A. ✔
- Listen to problem A: The red line is 89 inches long. The blue line is 38 inches long. How much longer is the red line than the blue line?
- Listen to problem A again and write the symbols for the problem as I say it: The red line is 89 inches long. The blue line is 38 inches long. How much longer is the red line than the blue line?
- Work problem A. Remember to make the equals bar. Put your pencil down when you've written the number for how many inches longer the red line is than the blue line. (Observe children.)
b. Touch where you'll write the symbols for problem B. ✔
- Listen to problem B: 685 students were in a school. 613 of those students left the school. How many students ended up in the school?
- Listen again and write the symbols for the whole problem: 685 students were in a school. 613 of those students left the school. (Observe children.)
- Work problem B. Put your pencil down when you know how many students ended up in the school. (Observe children.)
c. For problem C you're going to tell me how much shorter one building was.
- Touch where you'll write problem C. ✔
- Listen to problem C: The white building was 173 feet tall. The blue building was 50 feet tall. How many feet shorter was the blue building?
- Listen to problem C again and write the symbols for the problem as I say it: The white building was 173 feet tall. The blue building was 50 feet tall. How many feet shorter was the blue building? (Observe children.)
- Work problem C. Put your pencil down when you know how many feet shorter the blue building was. (Observe children.)

d. Touch where you'll write the symbols for problem D. ✔
- Listen to problem D: There were 218 children in the school. 72 more children came to the school. How many children ended up in the school?
- Listen again and write the symbols for both parts: There were 218 children in the school. 72 more children came to the school. **(Observe children.)**
- Work problem D. Put your pencil down when you know how many children ended up in the school. **(Observe children.)**
e. Now, I'll tell you word problems E, F, and G. You'll write the symbols for each problem. Then you'll figure out the answer. For these problems, making a number family helps to solve the problem.
- Problem E: Eric worked 8 math problems. Then Eric worked some science problems. Eric worked 16 problems in all. How many science problems did Eric work?
- Touch the space where you'll write the symbols for problem E. ✔
  I'll say each part of problem E again. Write the symbols for the parts as I say them.
- Listen to the first part: Eric worked 8 math problems. ✔
- Next part: Then Eric worked some science problems. ✔
- Next part: Eric worked 16 problems in all. ✔
  (Repeat if necessary.)
- Make the family for E if you need to. Then complete the equation. Put your pencil down when you've written the answer to the question: How many science problems did Eric work?
  (Observe children.)
f. Problem F. There were some apples in a box. 3 of those apples were eaten. 14 apples were left in the box. How many apples were in the box to start with?
- Touch the space where you'll write the symbols for problem F. ✔
  I'll say each part of problem F again. Write the symbols for the parts as I say them.
- Listen to the first part: There were some apples in a box. ✔
- Next part: 3 of those apples were eaten. ✔
- Next part: 14 apples were left in the box. ✔
  (Repeat if necessary.)

- Make the family for F if you need to. Then complete the equation. Put your pencil down when you've written the answer to the question: How many apples were in the box to start with?
g. Problem G. There were 9 flowers on a bush. Some of those flowers were picked off of the bush. The bush ended up with 3 flowers. How many flowers were picked off of the bush?
- Touch the space where you'll write the symbols for problem G. ✔
  I'll say each part of problem G again. Write the symbols for the parts as I say them.
- Listen to the first part: There were 9 flowers on a bush. ✔
- Next part: Some of those flowers were picked off of the bush. ✔
- Next part: The bush ended up with 3 flowers. ✔
  (Repeat if necessary.)
- Make the family for G if you need to. Then complete the equation. Put your pencil down when you've written the answer to the question: How many flowers were picked off of the bush?

Part 11:
a. Touch part 11. ✔
  (Teacher reference:)

$$= \qquad \neq$$

a. $20 + 35 = 20 + 30 + 4$ d. $70 + 11 + 3 = 80 + 3$

b. $16 + 1 + 5 = 18 + 5$ e. $43 + 9 = 43 + 8 + 1$

c. $19 + 10 = 10 + 9 + 10$ f. $25 + 40 + 7 = 25 + 49$

Some of these equations are not true. You'll figure out if each equation is true. Then you'll fix the sign for the equations that are false.
- Touch the equals above the problems. ✔
- Touch the sign for not equals. ✔
  That's the sign you'll change the equals into if the equation is false.
b. Read equation A to yourself. Put parentheses around a part, and write what it equals above. Then fix the sign if the equation is false. Leave the equals alone if the equation is true. Put your pencil down when you've worked problem A.
  (Observe children.)

c. For the rest of the problems, read the equation to yourself. Put parentheses around a part and write what it equals above. Then fix the equal sign if the equation is false. Leave the equation alone if it's true. Put your pencil down when you've completed the problems in part 11.
(Observe children.)

Part 12:
a. Find part 12. ✔
(Teacher reference:)

Blue =	Red =		Yellow =		Green =	
	1	2	3	4	5	6
Color						

b. (Display:) [CT2:12A]

These are pictures of vehicles that are different colors. You're going to complete the equation for each color to show the number of vehicles that are blue, red, yellow, and green. Then you'll complete the table to show the data.

c. Touch the word blue. ✔
• (Point to blue car.) Your turn to complete the equation for blue. Count the vehicles that are blue and complete the equation for blue. Put your pencil down when you've completed the equation.
(Observe children.)

d. Touch the word red. ✔
• (Point to red truck.) Count the vehicles that are red and complete the equation. Put your pencil down when you've written what red equals.
(Observe children.)

e. Touch the word yellow. ✔
• (Point to yellow car.) Count the vehicles that are yellow and complete the equation. Put your pencil down when you've written what yellow equals.
(Observe children.)

f. Touch the word green. ✔
• (Point to green motorcycle.) Count the vehicles that are green and complete the equation. Put your pencil down when you've written what green equals.
(Observe children.)

g. Touch the row in the table that has the numbers 1 through 6. ✔
• You're going to write the color where they go in the bottom row.
• Touch 1 in the top row of the table. ✔ If blue, red, yellow, or green equals 1, write that color below 1 in the bottom row. If none of the colors equals 1, don't write anything below 1.

h. Touch 2 in the top row of the table. ✔ If a color equals 2, write that color in the bottom row below 2. If none of the colors equal 2, don't write anything below 2.

i. Write the colors in the bottom row of the table. Remember, don't write anything below the numbers that are not in the equations. Put your pencil down when you've completed the table for part 12.
(Observe children.)

Part 13:
a. Touch part 13. ✔
(Teacher reference:)

• Touch shape A. ✔
b. Write the number of equal parts for shape A below the bar. Then write the number of parts that are shaded above the bar.
(Observe children.)
c. Write the number of equal parts for shape B below the bar. Then write the number of parts that are shaded above the bar.
(Observe children.)
d. For shapes C and D, write the number of equal parts below the bar. Write the number of parts that are shaded above the bar. Put your pencil down when you've completed the fractions for all of the shapes in part 13.
(Observe children.)

Part 14:

a. Touch part 14. ✔
(Teacher reference:)

> < =

a.	7 5	f.	16 + 1 + 5	18 + 5
b.	8 9	g.	19 + 10	10 + 9 + 10
c.	6 3	h.	70 + 11 + 3	80 + 3
d.	4 4	i.	43 + 9	43 + 8 + 1
e.	20 + 35 20 + 30 + 4	j.	25 + 40 + 7	25 + 49

• Touch the signs at the top of part 14. ✔
You're going to write one of those signs between the sides of each problem to make the statement true.
For problems A through D, there's only one number on each side. You'll just make the sign between the numbers for each problem.
Problems E through J are the statements you worked with earlier, but there are no signs between the sides.
For those problems, you'll make parentheses for each problem and write what it equals above. Then write the sign to show which side is more or if the sides are equal.

• Make the sign for each problem. Put your pencil down when you've worked all of the problems in part 14.
(Observe children.)

Part 15:

a. Touch part 15. ✔
(Teacher reference:)

a. ___:___ b. ___:___ c. ___:___ d. ___:___

I'll say a time for each space. You'll write the time. The dots for each time are already shown.

• Time A is 12 thirty. What's time A? (Signal.)
12 thirty.

• Write 12 thirty in space A.
(Observe children.)

b. Time B is 2 fifteen. What's time B? (Signal.)
2 fifteen.

• Write 2 fifteen in space B.
(Observe children.)

c. Time C is 6 o'clock. What's time C? (Signal.)
6 o'clock.

• Write 6 o'clock in space C.
(Observe children.)

d. Time D is 11 oh 3. What's time D? (Signal.)
11 oh 3.

• Write 11 oh 3 in space D.
(Observe children.)

Part 16:

a. Touch part 16. ✔
(Teacher reference:)

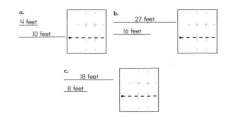

For each problem in part 16, you'll write the column problem to figure out how much longer one line is than the other line.

• Write the column problems and work them.
(Observe children.)

Part 17:

a. Touch part 17. ✔
(Teacher reference:)

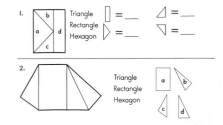

Next to each figure are the names: triangle, rectangle, and hexagon. You'll circle the word for the shape of each whole figure. Then, for figure 1, you'll write the letter to complete the equation for each part. For figure 2, you'll write the letter in each part for the whole figure.

• Circle the shape for each figure in part 17. Then write the letters for each figure.
(Observe children.)

Part 18:

a. Touch part 18. ✔
(Teacher reference:)

a. ▭ ▭ ▭ ▭
b. ▭ ▭ ▭ ▭ ▭
c. $61 + 🪙🪙🪙🪙
d. $33 + 🪙🪙🪙🪙🪙
e. ▭ ▭ 🪙🪙🪙🪙
f. ▭ ▭ ▭ ▭ 🪙🪙

Problems A and B show bills. Problems C and D show a dollar amount and coins. Problems E and F show bills and coins.

• Figure out how much money there is for each problem. Write an equals and the amount. Remember the dollar sign.
(Observe children.)

Part 19:

a. Touch part 19. ✔
(Teacher reference:)

Each ruler in part 19 is supposed to show centimeters. You're going to complete the equation for each ruler in part 19. For ruler A, you'll write the number for the shaded part and the unshaded part. For the rest of the rulers, you'll write the number for the unshaded parts. For rulers A, B, and C, you'll count on to complete the equations.
For rulers D, E, and F, you'll count backward to complete the equations.

• Complete the equations for the rulers in part 19. Put your pencil down when you've completed the equation for each ruler.
(Observe children.)

Part 20:

a. Touch part 20. ✔
(Teacher reference:)

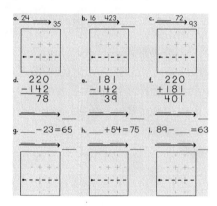

Problems A, B, and C have a missing number in the family. You'll write and work the problems to find the missing numbers and write them in the family.
For problems D, E, and F you're going to make number families for the equations.
For problems G, H, and I, you're going to write the numbers that are shown in each problem where they go in the number family below it. Then you'll write the column problem to figure out the missing number in each family and complete the family and the equation you started with.

• Work each problem. Put your pencil down when you've completed the problems in part 20.
(Observe children.)

Part 21:

a. Touch part 21. ✔
(Teacher reference:)

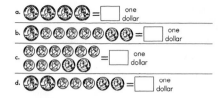

• Touch the words after the box in group A. ✔
• The words are one dollar. What are they?
(Signal.) *One dollar.*
• If a group of coins is worth one dollar, circle the words one dollar. What words will you circle if a group is worth one dollar? (Signal.)
One dollar.
• If a group of coins is not worth one dollar, write the number of cents in the box. Where will you write the number of cents if a group is not worth one dollar? (Signal.) *In the box.*

b. Count the cents for each group of coins. If the group is worth one dollar, circle the words one dollar. If the group isn't worth one dollar, write the number of cents in the box.
(Observe children.)

Part 22:

a. Touch part 22. ✔
(Teacher reference:)

a. $184 + 108 =$ ___   c. $184 +$ ___ $= 292$ e. ___ $- 76 = 108$
b. ___ $- 184 = 76$ d. $184 -$ ___ $= 76$ f. ___ $+ 184 = 260$

The equations in part 22 have a missing number.
(Display:)   [CT2:22A]

$$1. \ \underrightarrow{184 \quad 76} 260 \qquad 2. \ \underrightarrow{76 \quad 108} 184$$

$$3. \ \underrightarrow{108 \quad 184} 292$$

These are the families for the equations in part 22.

• Find the family for each equation and write the missing number. Put your pencil down when you've completed all of the equations in part 22.
(Observe children.)

Part 23:

a. Touch part 23. ✔
   (Teacher reference:)

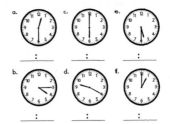

You're going to write the time for each clock. Some of the clocks show zero minutes. Some of the clocks show 30 minutes. You'll count the minutes to yourself for the other clocks.

• Write the time for each clock. Put your pencil down when you've written the hour number and the minute number for each clock in part 23.
   (Observe children.)

Part 24:

a. Touch part 24. ✔
   (Teacher reference:)

You're going to write letters for the faces of objects 1, 2, 3, and 4.

• For faces that are shaped like triangles, what letter will you write? (Signal.) *T.*
• For faces that are shaped like rectangles, what letter will you write? (Signal.) *R.*
• For faces that are shaped like pentagons, what letter will you write? (Signal.) *P.*
• For faces that are shaped like hexagons, what letter will you write? (Signal.) *H.*
   (Repeat until firm)

b. Write the letters for all of the faces for objects 1, 2, 3, and 4. Remember, if an object doesn't have a top face, leave the space for the top face blank. Put your pencil down when you've labeled all of the faces with the letter for its shape.
   (Observe children.)

c. (Display:)   [CT2:24A]

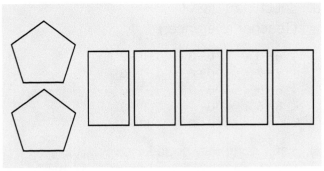

Here are the bottom, top, and side faces of one of the objects. Circle the object that has the correct bottom, top, and sides.
(Observe children.)

Part 25:

a. Touch part 25. ✔
   (Teacher reference:)

a. $9-3$	f. $10-2$	k. $3+5$
b. $9+2$	g. $3+6$	l. $6+4$
c. $8-5$	h. $6+6$	m. $12-10$
d. $11-5$	i. $11-6$	n. $12-6$
e. $10+2$	j. $11-9$	o. $5+6$

You'll complete equations in part 25.

Part 26:

a. Touch part 26. ✔
   (Teacher reference:)

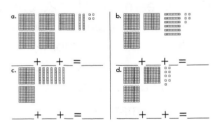

___ + ___ + ___ = ___

In part 26, you'll write place-value equations for each picture.

Part 27:

a. Touch part 27. ✔
   (Teacher reference:)

a. $4+3$	f. $7-3$	k. $5-5$
b. $4-3$	g. $6-3$	l. $5+5$
c. $4+6$	h. $6+3$	m. $8-4$
d. $4+4$	i. $10+5$	n. $3+3$
e. $10-4$	j. $10-5$	o. $8-3$

You'll complete equations in part 27.

Part 28:

a. Touch part 28. ✔
   (Teacher reference:)

a. $16-8$	f. $6-3$	k. $11-3$
b. $7+7$	g. $6+6$	l. $7-3$
c. $8-4$	h. $14-7$	m. $7+3$
d. $8+8$	i. $10-7$	n. $11-8$
e. $10-5$	j. $8+3$	o. $10-3$

You'll complete equations in part 28.

• Your turn: Work the rest of the problems on Cumulative Test 2.
   (Direct students where to put their assessment books when they are finished.)

Scoring Notes: Section B

a. Collect test booklets. Use the Answer Key to score the parts in Section B. Children can earn 1 point for each response in all parts except Part 12 and Part 24.

Part 12: Children can earn a total of 10 points for part 12. Children earn 1 point for completing each equation correctly. Children earn 1 point for writing the correct color for each number in the table. Award 1 point for leaving the cell below 3 blank and 1 point for leaving the cell below 6 blank.

Part 24: Children can earn a total of 20 points for part 24. Children earn 1 point for labeling each face correctly. Award 1 point for leaving top space for each pyramid blank. Children earn 3 points for circling the correct object.

Record children's performance on the Summary Sheet. Reproducible Summary Sheets are at the back of the *Teacher's Guide*.

b. Use the Remedy Table for Cumulative Test 2, available online, to determine whether remedies are needed. Use the online Remedies Worksheet when presenting remedies for Cumulative Test 2 Section B.

Retest

Retest individual students on any part failed.

Answer Key for Cumulative Test 2

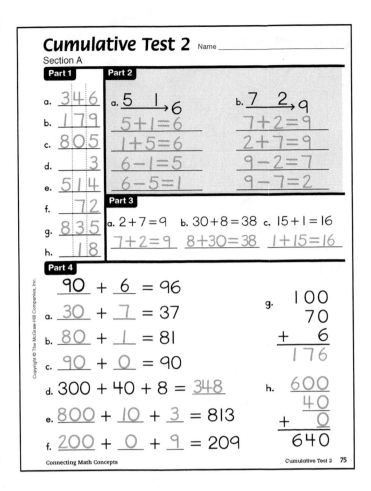

Cumulative Test 2   Name _____

Section A

Part 1

a. 346
b. 179
c. 805
d. 3
e. 514
f. 72
g. 835
h. 18

Part 2

a. $5 \quad 1 \to 6$
$5+1=6$
$1+5=6$
$6-1=5$
$6-5=1$

b. $7 \quad 2 \to 9$
$7+2=9$
$2+7=9$
$9-2=7$
$9-7=2$

Part 3

a. $2+7=9$ b. $30+8=38$ c. $15+1=16$
$7+2=9$ $8+30=38$ $1+15=16$

Part 4

a. $90 + 6 = 96$
$30 + 7 = 37$
b. $80 + 1 = 81$
c. $90 + 0 = 90$
d. $300 + 40 + 8 = 348$
e. $800 + 10 + 3 = 813$
f. $200 + 0 + 9 = 209$

g. $\begin{array}{r} 100 \\ 70 \\ +\ 6 \\ \hline 176 \end{array}$

h. $\begin{array}{r} 600 \\ 40 \\ +\ 0 \\ \hline 640 \end{array}$

Connecting Math Concepts   Cumulative Test 2   75

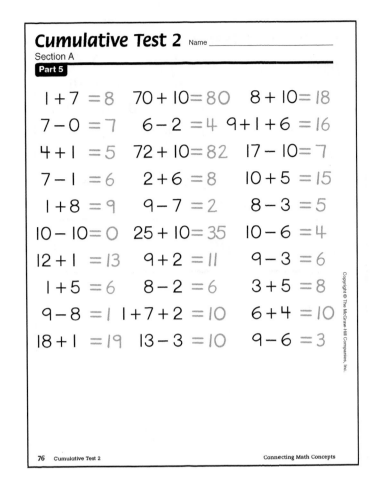

Cumulative Test 2   Name _____

Section A

Part 5

$1+7=8$	$70+10=80$	$8+10=18$
$7-0=7$	$6-2=4$	$9+1+6=16$
$4+1=5$	$72+10=82$	$17-10=7$
$7-1=6$	$2+6=8$	$10+5=15$
$1+8=9$	$9-7=2$	$8-3=5$
$10-10=0$	$25+10=35$	$10-6=4$
$12+1=13$	$9+2=11$	$9-3=6$
$1+5=6$	$8-2=6$	$3+5=8$
$9-8=1$	$1+7+2=10$	$6+4=10$
$18+1=19$	$13-3=10$	$9-6=3$

76   Cumulative Test 2   Connecting Math Concepts

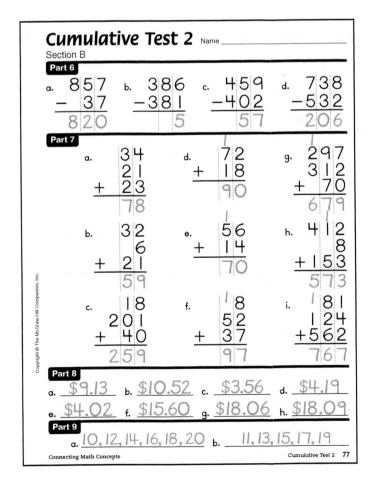

Cumulative Test 2   Name _____

Section B

Part 6

a. $\begin{array}{r} 857 \\ -\ 37 \\ \hline 820 \end{array}$
b. $\begin{array}{r} 386 \\ -381 \\ \hline 5 \end{array}$
c. $\begin{array}{r} 459 \\ -402 \\ \hline 57 \end{array}$
d. $\begin{array}{r} 738 \\ -532 \\ \hline 206 \end{array}$

Part 7

a. $\begin{array}{r} 34 \\ 21 \\ +\ 23 \\ \hline 78 \end{array}$
b. $\begin{array}{r} 32 \\ 6 \\ +\ 21 \\ \hline 59 \end{array}$
c. $\begin{array}{r} 18 \\ 201 \\ +\ 40 \\ \hline 259 \end{array}$
d. $\begin{array}{r} 72 \\ +\ 18 \\ \hline 90 \end{array}$
e. $\begin{array}{r} 56 \\ +\ 14 \\ \hline 70 \end{array}$
f. $\begin{array}{r} 8 \\ 52 \\ +\ 37 \\ \hline 97 \end{array}$
g. $\begin{array}{r} 297 \\ 312 \\ +\ 70 \\ \hline 679 \end{array}$
h. $\begin{array}{r} 412 \\ 8 \\ +153 \\ \hline 573 \end{array}$
i. $\begin{array}{r} 81 \\ 124 \\ +562 \\ \hline 767 \end{array}$

Part 8

a. $\$9.13$ b. $\$10.52$ c. $\$3.56$ d. $\$4.19$
e. $\$4.02$ f. $\$15.60$ g. $\$18.06$ h. $\$18.09$

Part 9

a. $10, 12, 14, 16, 18, 20$ b. $11, 13, 15, 17, 19$

Connecting Math Concepts   Cumulative Test 2   77

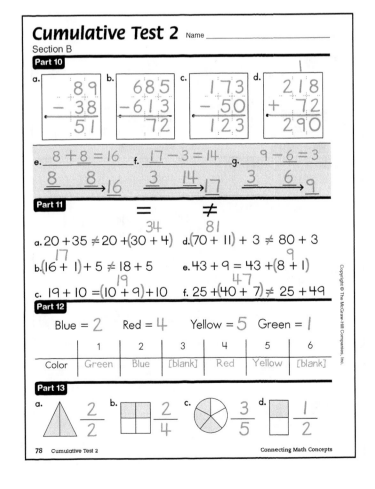

Cumulative Test 2   Name _____

Section B

Part 10

a. $\begin{array}{r} 89 \\ -\ 38 \\ \hline 51 \end{array}$
b. $\begin{array}{r} 685 \\ -613 \\ \hline 72 \end{array}$
c. $\begin{array}{r} 173 \\ -\ 50 \\ \hline 123 \end{array}$
d. $\begin{array}{r} 218 \\ +\ 72 \\ \hline 290 \end{array}$

e. $8+8=16$
$8 \quad 8 \to 16$

f. $17-3=14$
$3 \quad 14 \to 17$

g. $9-6=3$
$3 \quad 6 \to 9$

Part 11

$=$ $\neq$
34 81

a. $20+35 \neq 20+(30+4)$ d. $(70+11)+3 \neq 80+3$
17 9
b. $(16+1)+5 \neq 18+5$ e. $43+9 = 43+(8+1)$
19 47
c. $19+10 = (10+9)+10$ f. $25+(40+7) \neq 25+49$

Part 12

Blue = 2 Red = 4 Yellow = 5 Green = 1

	1	2	3	4	5	6
Color	Green	Blue	[blank]	Red	Yellow	[blank]

Part 13

a. $\frac{2}{2}$ b. $\frac{2}{4}$ c. $\frac{3}{5}$ d. $\frac{1}{2}$

78   Cumulative Test 2   Connecting Math Concepts

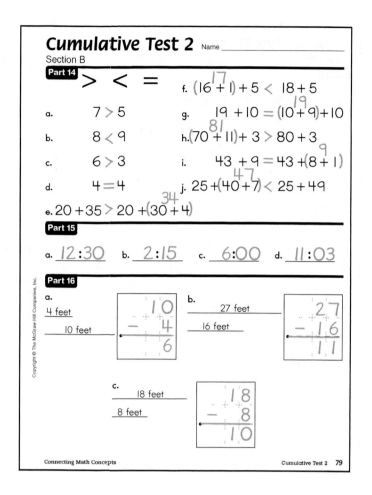

Cumulative Test 2
Section B

Part 14

> < =

       f. $(16 + 1) + 5 < 18 + 5$ ⁽¹⁷⁾

a. $7 > 5$ g. $19 + 10 = (10 + 9) + 10$ ⁽¹⁹⁾

b. $8 < 9$ h. $(70 + 11) + 3 > 80 + 3$ ⁽⁸¹⁾

c. $6 > 3$ i. $43 + 9 = 43 + (8 + 1)$ ⁽⁹⁾

d. $4 = 4$ j. $25 + (40 + 7) < 25 + 49$ ⁽⁴⁷⁾

e. $20 + 35 > 20 + (30 + 4)$ ⁽³⁴⁾

Part 15

a. $12{:}30$ b. $2{:}15$ c. $6{:}00$ d. $11{:}03$

Part 16

a.
4 feet
10 feet
$$\begin{array}{r} 10 \\ -\ 4 \\ \hline 6 \end{array}$$

b.
27 feet
16 feet
$$\begin{array}{r} 27 \\ -16 \\ \hline 11 \end{array}$$

c.
18 feet
8 feet
$$\begin{array}{r} 18 \\ -\ 8 \\ \hline 10 \end{array}$$

Connecting Math Concepts   Cumulative Test 2   79

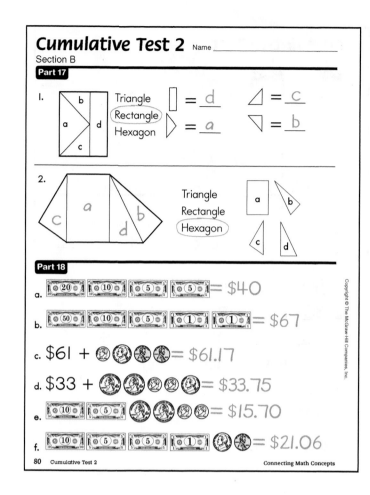

Cumulative Test 2
Section B

Part 17

1.
Triangle ▯ = d ◁ = c
Rectangle ▷ = a ◺ = b
Hexagon

2.
Triangle
Rectangle
Hexagon

Part 18

a. $= \$40$

b. $= \$67$

c. $\$61 + $ ◯◯◯◯ $= \$61.17$

d. $\$33 + $ ◯◯◯◯◯ $= \$33.75$

e. $= \$15.70$

f. $= \$21.06$

80   Cumulative Test 2   Connecting Math Concepts

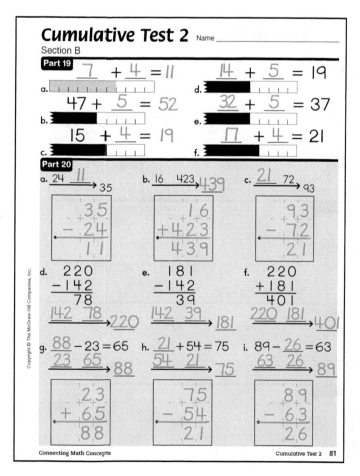

Cumulative Test 2
Section B

Part 19

a. $7 + 4 = 11$ d. $14 + 5 = 19$

b. $47 + 5 = 52$ e. $32 + 5 = 37$

c. $15 + 4 = 19$ f. $17 + 4 = 21$

Part 20

a. $24 \xrightarrow{11} 35$ b. $16 \xrightarrow{423} 439$ c. $21 \xrightarrow{72} 93$

$$\begin{array}{r} 3.5 \\ -\ 2.4 \\ \hline 1.1 \end{array}$$
$$\begin{array}{r} 1.6 \\ +4.2.3 \\ \hline 4.3.9 \end{array}$$
$$\begin{array}{r} 9.3 \\ -\ 7.2 \\ \hline 2.1 \end{array}$$

d.
$$\begin{array}{r} 220 \\ -142 \\ \hline 78 \end{array}$$
e.
$$\begin{array}{r} 181 \\ -142 \\ \hline 39 \end{array}$$
f.
$$\begin{array}{r} 220 \\ +181 \\ \hline 401 \end{array}$$

$142 \xrightarrow{78} 220$ $142 \xrightarrow{39} 181$ $220 \xrightarrow{181} 401$

g. $88 - 23 = 65$ h. $21 + 54 = 75$ i. $89 - 26 = 63$

$23 \xrightarrow{65} 88$ $54 \xrightarrow{21} 75$ $63 \xrightarrow{26} 89$

$$\begin{array}{r} 2.3 \\ +\ 6.5 \\ \hline 88 \end{array}$$
$$\begin{array}{r} 7.5 \\ -\ 5.4 \\ \hline 2.1 \end{array}$$
$$\begin{array}{r} 8.9 \\ -\ 6.3 \\ \hline 2.6 \end{array}$$

Connecting Math Concepts   Cumulative Test 2   81

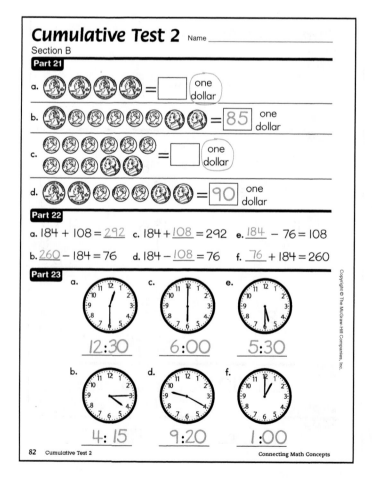

Cumulative Test 2
Section B

Part 21

a. ◯◯◯◯ $=$ ▢ one dollar

b. ◯◯◯◯◯◯◯ $= 85$ one dollar

c. ◯◯◯◯◯◯◯ $=$ ▢ one dollar

d. ◯◯◯◯◯ $= 90$ one dollar

Part 22

a. $184 + 108 = 292$ c. $184 + 108 = 292$ e. $184 - 76 = 108$

b. $260 - 184 = 76$ d. $184 - 108 = 76$ f. $76 + 184 = 260$

Part 23

a. $12{:}30$ c. $6{:}00$ e. $5{:}30$

b. $4{:}15$ d. $9{:}20$ f. $1{:}00$

82   Cumulative Test 2   Connecting Math Concepts

Part 24

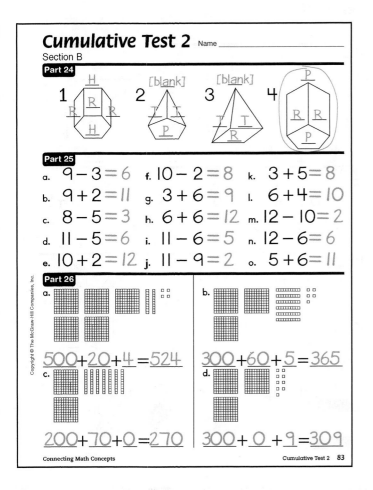

Part 25

a. $9 - 3 = 6$ f. $10 - 2 = 8$ k. $3 + 5 = 8$

b. $9 + 2 = 11$ g. $3 + 6 = 9$ l. $6 + 4 = 10$

c. $8 - 5 = 3$ h. $6 + 6 = 12$ m. $12 - 10 = 2$

d. $11 - 5 = 6$ i. $11 - 6 = 5$ n. $12 - 6 = 6$

e. $10 + 2 = 12$ j. $11 - 9 = 2$ o. $5 + 6 = 11$

Part 26

a. $\underline{500} + \underline{20} + \underline{4} = \underline{524}$

b. $\underline{300} + \underline{60} + \underline{5} = \underline{365}$

c. $\underline{200} + \underline{70} + \underline{0} = \underline{270}$

d. $\underline{300} + \underline{0} + \underline{9} = \underline{309}$

Part 27

a. $4 + 3 = 7$ f. $7 - 3 = 4$ k. $5 - 5 = 0$

b. $4 - 3 = 1$ g. $6 - 3 = 3$ l. $5 + 5 = 10$

c. $4 + 6 = 10$ h. $6 + 3 = 9$ m. $8 - 4 = 4$

d. $4 + 4 = 8$ i. $10 + 5 = 15$ n. $3 + 3 = 6$

e. $10 - 4 = 6$ j. $10 - 5 = 5$ o. $8 - 3 = 5$

Part 28

a. $16 - 8 = 8$ f. $6 - 3 = 3$ k. $11 - 3 = 8$

b. $7 + 7 = 14$ g. $6 + 6 = 12$ l. $7 - 3 = 4$

c. $8 - 4 = 4$ h. $14 - 7 = 7$ m. $7 + 3 = 10$

d. $8 + 8 = 16$ i. $10 - 7 = 3$ n. $11 - 8 = 3$

e. $10 - 5 = 5$ j. $8 + 3 = 11$ o. $10 - 3 = 7$

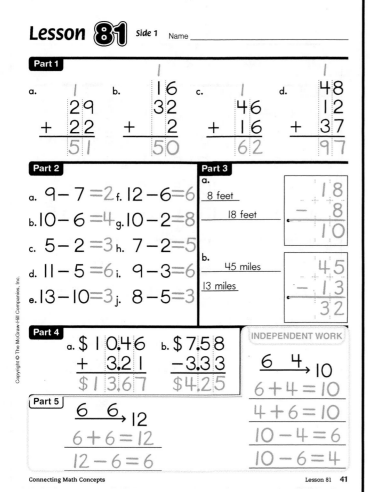

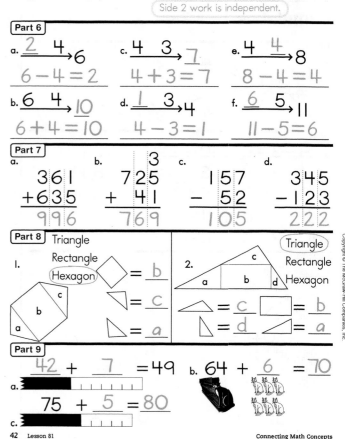

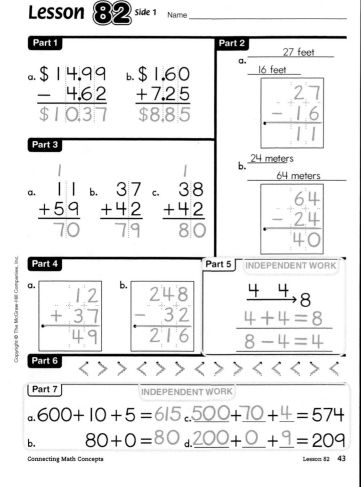

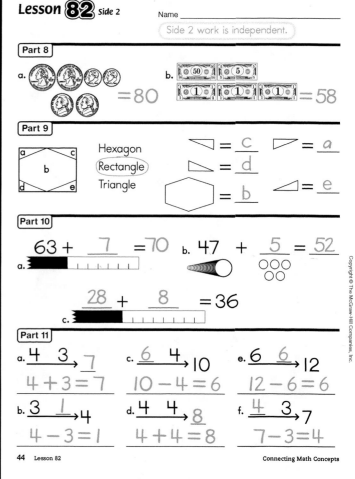

Lesson 81 Side 1 Name _____

Part 1

a.
```
  1
 29
+22
 51
```

b.
```
 16
 32
+ 2
 50
```

c.
```
  1
 46
+16
 62
```

d.
```
  1
 48
 12
+37
 97
```

Part 2

a. $9-7=2$ f. $12-6=6$

b. $10-6=4$ g. $10-2=8$

c. $5-2=3$ h. $7-2=5$

d. $11-5=6$ i. $9-3=6$

e. $13-10=3$ j. $8-5=3$

Part 3

a.
8 feet
18 feet
```
  18
-  8
  10
```

b.
45 miles
13 miles
```
  45
- 13
  32
```

Part 4

a.
```
$10.46
+  3.21
$13.67
```

b.
```
$7.58
-3.33
$4.25
```

INDEPENDENT WORK
```
 6  4
      →10
6+4=10
4+6=10
10-4=6
10-6=4
```

Part 5

```
 6  6
      →12
6+6=12
12-6=6
```

Connecting Math Concepts   Lesson 81 41

Lesson 81 Side 2 Name _____

Part 6

a. $\underline{2}\ \ 4 →6$ $6-4=2$

c. $4\ \ 3 →\underline{7}$ $4+3=7$

e. $4\ \ \underline{4} →8$ $8-4=4$

b. $6\ \ 4 →\underline{10}$ $6+4=10$

d. $\underline{1}\ \ 3 →4$ $4-3=1$

f. $\underline{6}\ \ 5 →11$ $11-5=6$

Part 7

a.
```
 361
+635
 996
```

b.
```
  3
 725
+ 41
 769
```

c.
```
 157
- 52
 105
```

d.
```
 345
-123
 222
```

Part 8

1.
Triangle
Rectangle
(Hexagon)
= b
= c
= a

2.
(Triangle)
Rectangle
Hexagon
= c = b
= d = a

Part 9

a. $\underline{42} + \underline{7} = 49$

$75 + \underline{5} = 80$

b. $64 + \underline{6} = \underline{70}$

42 Lesson 81   Connecting Math Concepts

Lesson 82 Side 1 Name _____

Part 1

a.
```
$14.99
-  4.62
$10.37
```

b.
```
$1.60
+7.25
$8.85
```

Part 2

a.
27 feet
16 feet
```
  27
- 16
  11
```

b.
24 meters
64 meters
```
  64
- 24
  40
```

Part 3

a.
```
  1
 11
+59
 70
```

b.
```
 37
+42
 79
```

c.
```
  1
 38
+42
 80
```

Part 4

a.
```
  12
+ 37
  49
```

b.
```
 248
- 32
 216
```

Part 5 INDEPENDENT WORK

```
 4  4
      →8
4+4=8
8-4=4
```

Part 6

< > < > < > > < < < > <

Part 7 INDEPENDENT WORK

a. $600+10+5=615$

c. $\underline{500}+\underline{70}+\underline{4}=574$

b. $80+0=80$

d. $\underline{200}+\underline{0}+\underline{9}=209$

Connecting Math Concepts   Lesson 82 43

Lesson 82 Side 2 Name _____

Part 8

a. $=80$

b. $=58$

Part 9

Hexagon
(Rectangle)
Triangle

= c = a
= d
= b = e

Part 10

a. $63+\underline{7}=70$

b. $47+\underline{5}=\underline{52}$

c. $\underline{28}+\underline{8}=36$

Part 11

a. $4\ \ 3 →\underline{7}$ $4+3=7$

c. $\underline{6}\ \ 4 →10$ $10-4=6$

e. $6\ \ 6 →12$ $12-6=6$

b. $3\ \ \underline{1} →4$ $4-3=1$

d. $4\ \ 4 →\underline{8}$ $4+4=8$

f. $\underline{4}\ \ 3 →7$ $7-3=4$

44 Lesson 82   Connecting Math Concepts

Lesson 83 Side 1   Name _____

Part 1
a. 19 < 20
b. 33 > 22
c. 53 > 48

Part 2

a.
```
  26
+ 53
  79
```
b.
```
  1
  58
+ 12
  70
```
c.
```
   14
   63
 + 12
   89
```
```
 1
   43
    6
 + 22
   71
```

Part 3
a. $5.04 b. $5.40 c. $5.60 d. $5.06

Part 4

a.
```
   89
 - 38
   51
```
b.
```
   66
 - 24
   42
```

Part 5 — INDEPENDENT WORK

a.
```
  $6.45
 +3.12
  $9.57
```
b.
```
  $25.49
 -14.17
  $11.32
```

Part 6

88 + 6 = 94   67 + 6 = 73

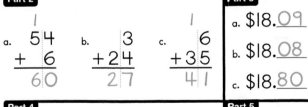

Part 7

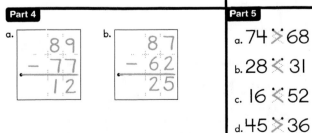

a. = 46
b. = 57
c. = 70   d. = 26

Connecting Math Concepts   Lesson 83   45

Lesson 83 Side 2   Name _____

Side 2 work is independent.

Part 8

a.
```
4  3
    → 7
4 + 3 = 7
3 + 4 = 7
7 - 3 = 4
7 - 4 = 3
```
b.
```
6  5
    → 11
6 + 5 = 11
5 + 6 = 11
11 - 5 = 6
11 - 6 = 5
```
c.
```
6  3
    → 9
6 + 3 = 9
3 + 6 = 9
9 - 3 = 6
9 - 6 = 3
```

Part 9

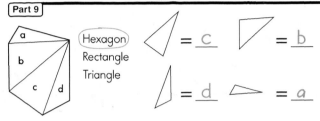

Hexagon
Rectangle
Triangle

= c   = b

= d   = a

Part 10

a.
```
4  4
   → 8
8 - 4 = 4
```
c.
```
7  2
   → 9
7 + 2 = 9
```
e.
```
10  4
    → 14
10 + 4 = 14
```
b.
```
5  2
   → 7
7 - 2 = 5
```
d.
```
6  4
   → 10
10 - 4 = 6
```
f.
```
4  3
   → 7
7 - 4 = 3
```

46   Lesson 83   Connecting Math Concepts

Lesson 84 Side 1   Name _____

Part 1
a. 7 - 3 = 4   d. 14 - 4 = 10   g. 11 - 5 = 6
b. 6 - 4 = 2   e. 10 - 4 = 6   h. 8 - 4 = 4
c. 9 - 3 = 6   f. 11 - 9 = 2   i. 12 - 6 = 6

Part 2

a.
```
 1
   54
 +  6
   60
```
b.
```
   3
 + 24
   27
```
c.
```
 1
    6
 + 35
   41
```

Part 3
a. $18.09
b. $18.08
c. $18.80

Part 4

a.
```
   89
 - 77
   12
```
b.
```
   87
 - 62
   25
```

Part 5
a. 74 > 68
b. 28 < 31
c. 16 < 52
d. 45 > 36

Part 6
a. 4 + 3 = 7   d. 9 + 2 = 11   g. 4 + 10 = 14
b. 2 + 8 = 10   e. 3 + 6 = 9   h. 5 + 6 = 11
c. 4 + 4 = 8   f. 6 + 6 = 12   i. 3 + 5 = 8

Connecting Math Concepts   Lesson 84   47

Lesson 84 Side 2   Name _____

Side 2 work is independent.

Part 7

32 + 7 = 39   18 + 5 = 23

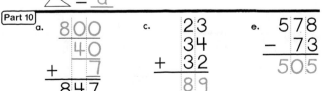

Part 8

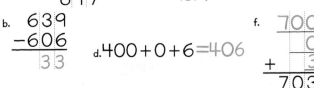

a. = 66
b. = 31
c. = [ ] one dollar
d. = 33 one dollar

Part 9

= a
= c
= b
= d

Hexagon
Rectangle
Triangle

Part 10

a.
```
  800
   40
 +  7
  847
```
c.
```
   23
   34
 + 32
   89
```
e.
```
  578
 - 73
  505
```
b.
```
  639
 -606
   33
```
d. 400 + 0 + 6 = 406
f.
```
  700
    0
 +  3
  703
```

48   Lesson 84   Connecting Math Concepts

Connecting Math Concepts   Answer Key   355

Lesson 85 Side 1   Name _____

Part 1

a.
$$\begin{array}{r} \$27.89 \\ -\ 3.83 \\ \hline \$24.06 \end{array}$$

b.
$$\begin{array}{r} \$\ \ 2.45 \\ +12.32 \\ \hline \$14.77 \end{array}$$

Part 2

a. $6-4=2$ d. $8-4=4$ g. $9-6=3$
b. $19-9=10$ e. $11-5=6$ h. $10-4=6$
c. $7-2=5$ f. $7-3=4$ i. $11-9=2$

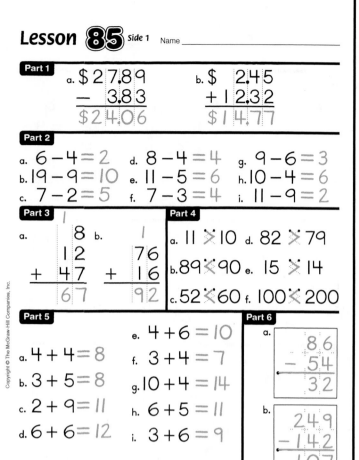

Part 3

a.
$$\begin{array}{r} 1\ \ \\ 8 \\ 12 \\ +\ 47 \\ \hline 67 \end{array}$$

b.
$$\begin{array}{r} 1\ \ \\ 76 \\ +\ 16 \\ \hline 92 \end{array}$$

Part 4

a. $11 > 10$ d. $82 > 79$
b. $89 < 90$ e. $15 > 14$
c. $52 < 60$ f. $100 < 200$

Part 5

a. $4+4=8$ e. $4+6=10$
b. $3+5=8$ f. $3+4=7$
c. $2+9=11$ g. $10+4=14$
d. $6+6=12$ h. $6+5=11$
            i. $3+6=9$

Part 6

a.
$$\begin{array}{r} 86 \\ -\ 54 \\ \hline 32 \end{array}$$

b.
$$\begin{array}{r} 249 \\ -142 \\ \hline 107 \end{array}$$

Connecting Math Concepts   Lesson 85 **49**

Lesson 85 Side 2   Name _____

Side 2 work is independent.

Part 7

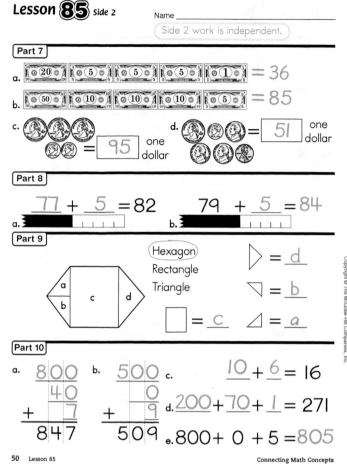

a. [20][5][5][5][1] = 36
b. [50][10][10][10][5] = 85
c. (coins) = 95 one dollar
d. (coins) = 51 one dollar

Part 8

a. $\underline{77} + \underline{5} = 82$ b. $79 + \underline{5} = 84$

Part 9

Hexagon
Rectangle
Triangle

▷ = d
◁ = b
□ = c
◿ = a

Part 10

a.
$$\begin{array}{r} 800 \\ 40 \\ +\ \ 7 \\ \hline 847 \end{array}$$

b.
$$\begin{array}{r} 500 \\ 0 \\ +\ \ 9 \\ \hline 509 \end{array}$$

c. $\underline{10} + \underline{6} = 16$
d. $\underline{200} + \underline{70} + \underline{1} = 271$
e. $800 + 0 + 5 = 805$

50   Lesson 85   Connecting Math Concepts

Lesson 86 Side 1   Name _____

Part 1

a. $7-4=3$ e. $8-5=3$
b. $9-3=6$ f. $4-3=1$
c. $8-4=4$ g. $11-5=6$
d. $7-2=5$ h. $10-4=6$

Part 2

a. $\$\ .88$
b. $\$\ .05$
c. $\$\ .17$
d. $\$\ 3.42$

Part 3

a. $3+4=7$ g. $2+4=6$
b. $5+6=11$ h. $6+6=12$
c. $10+7=17$ i. $8+2=10$
d. $4+4=8$ j. $2+9=11$
e. $6+4=10$ k. $3+6=9$
f. $5+3=8$

Part 4

a. $47 > 39$
b. $82 < 91$
c. $57 > 48$
d. $22 > 17$
e. $107 < 111$
f. $65 > 56$

Part 5

a.
$$\begin{array}{r} \$4.66 \\ -1.02 \\ \hline \$3.64 \end{array}$$

b.
$$\begin{array}{r} \$\ .74 \\ +2.02 \\ \hline \$2.76 \end{array}$$

c.
$$\begin{array}{r} \$9.81 \\ -9.01 \\ \hline \$\ .80 \end{array}$$

INDEPENDENT WORK

Part 6

a. $37 + \underline{8} = \underline{45}$ b. $\underline{23} + \underline{9} = 32$

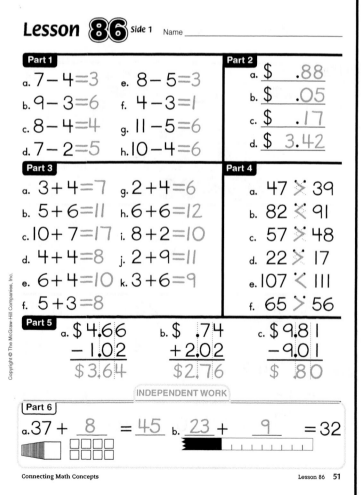

Connecting Math Concepts   Lesson 86 **51**

Lesson 86 Side 2   Name _____

Side 2 work is independent.

Part 7

a. $743 - 603$
$$\begin{array}{r} 743 \\ -603 \\ \hline 140 \end{array}$$

b. $68 + 2 + 316$
$$\begin{array}{r} 1\ \ \\ 68 \\ 2 \\ +316 \\ \hline 386 \end{array}$$

Part 8

$$\begin{array}{c} 4\ \ \ 3 \to 7 \end{array}$$
$4+3=7$
$3+4=7$
$7-3=4$
$7-4=3$

Part 9

a. $4\ \ 4 \to 8$ $4+4=8$
b. $3\ \ \underline{1} \to 4$ $4-3=1$
c. $6\ \ 4 \to 10$ $10-4=6$
d. $10\ \ 4 \to \underline{14}$ $10+4=14$
e. $6\ \ 6 \to 12$ $12-6=6$
f. $6\ \ 3 \to 9$ $9-3=6$

Part 10

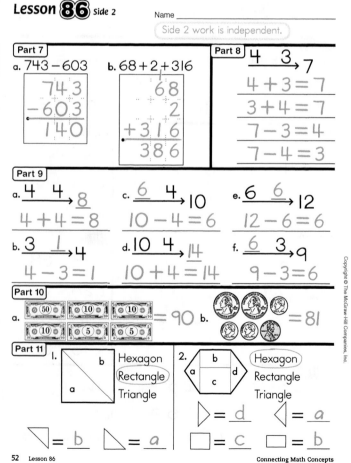

a. [50][10][10][10] = 90
b. (coins) = 81

Part 11

1. (figure) Hexagon / Rectangle / Triangle
   ◿ = b   ◺ = a   □ = c

2. (figure) Hexagon / Rectangle / Triangle
   ▷ = d   ◁ = a   □ = c   ▭ = b

52   Lesson 86   Connecting Math Concepts

---

356   **Answer Key**   **Connecting Math Concepts**

Lesson 87 Side 1   Name _____

Part 1
a. $7 - 3 = 4$ d. $11 - 6 = 5$ g. $14 - 10 = 4$
b. $10 - 4 = 6$ e. $5 - 3 = 2$ h. $7 - 5 = 2$
c. $9 - 3 = 6$ f. $8 - 3 = 5$ i. $8 - 4 = 4$

Part 2
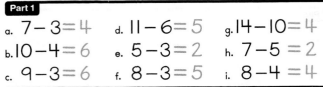

Part 3
a.
$$\begin{array}{r} 1 \\ 122 \\ +486 \\ \hline 608 \end{array}$$

Part 4
a. $64 < 71$
b. $37 > 28$
c. $36 > 19$
d. $78 < 90$
e. $27 > 25$
f. $58 < 61$

Part 5
e. $6 + 3 = 9$
a. $4 + 3 = 7$ f. $4 + 4 = 8$
b. $10 + 5 = 15$ g. $3 + 2 = 5$
c. $2 + 5 = 7$ h. $3 + 5 = 8$
d. $5 + 6 = 11$ i. $6 + 6 = 12$

INDEPENDENT WORK

Part 6
a. $ .09
b. $ 15.60
c. $ .06

Part 7
a. $\underline{78} + \underline{6} = 84$ b. $24 + \underline{7} = 31$
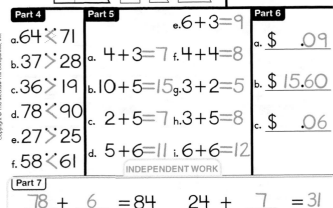

Connecting Math Concepts   Lesson 87 53

Lesson 87 Side 2   Name _____

Side 2 work is independent.

Part 8
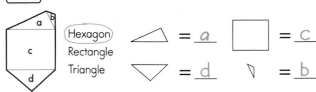
Hexagon
Rectangle
Triangle

$\triangle = \underline{a}$ $\square = \underline{c}$
$\bigtriangledown = \underline{d}$ $\diagdown = \underline{b}$

Part 9

a. $= 70$ b. $= 36$
c. $= 66$
d. $= 55$

Part 10
a.
$$\begin{array}{r} 36.57 \\ -32.55 \\ \hline \$ 4.02 \end{array}$$
b.
$$\begin{array}{r} 1 \\ \$ 6.24 \\ +13.46 \\ \hline \$19.70 \end{array}$$
c.
$$\begin{array}{r} \$29.88 \\ -13.36 \\ \hline \$16.52 \end{array}$$

Part 11
a. $\underline{6} \quad \underline{6} \rightarrow 12$ b. $\underline{4} \quad \underline{4} \rightarrow 8$ c. $\underline{6} \quad \underline{4} \rightarrow 10$
$12 - 6 = 6$ $4 + 4 = 8$ $10 - 4 = 6$

54   Lesson 87   Connecting Math Concepts

Lesson 88 Side 1   Name _____

Part 1
a.
$$\begin{array}{r} 1 \\ 394 \\ +112 \\ \hline 506 \end{array}$$
b.
$$\begin{array}{r} 1 \\ 106 \\ +856 \\ \hline 962 \end{array}$$

Part 2
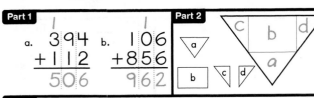

Part 3
a. $8 - 4 = 4$ e. $8 - 6 = 2$ i. $11 - 9 = 2$
b. $3 - 2 = 1$ f. $11 - 6 = 5$ j. $7 - 3 = 4$
c. $9 - 6 = 3$ g. $7 - 4 = 3$ k. $8 - 3 = 5$
d. $6 - 2 = 4$ h. $12 - 6 = 6$ l. $7 - 5 = 2$

Part 4
a. $46 < 53$
b. $16 = 16$
c. $57 > 48$
d. $24 = 24$
e. $117 < 130$
f. $43 > 35$

Part 5
a. $5 + 2 = 7$ g. $8 + 10 = 18$
b. $3 + 6 = 9$ h. $4 + 6 = 10$
c. $4 + 4 = 8$ i. $4 + 2 = 6$
d. $5 + 3 = 8$ j. $6 + 6 = 12$
e. $2 + 7 = 9$ k. $9 + 1 = 10$
f. $3 + 4 = 7$ l. $5 + 6 = 11$

Part 6
a.
$$\begin{array}{r} 1 \\ 46 \\ +26 \\ \hline 72 \end{array}$$
b.
$$\begin{array}{r} 1 \\ 58 \\ +32 \\ \hline 90 \end{array}$$

Part 7   INDEPENDENT WORK
a.
$$\begin{array}{r} \$ 5.28 \\ -2.17 \\ \hline \$ 3.11 \end{array}$$
b.
$$\begin{array}{r} \$20.62 \\ +3.33 \\ \hline \$23.95 \end{array}$$

Connecting Math Concepts   Lesson 88 55

Lesson 88 Side 2   Name _____

Side 2 work is independent.

Part 8
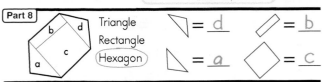
Triangle
Rectangle
Hexagon

$\diagdown = \underline{d}$ $\diagup = \underline{b}$
$\triangle = \underline{a}$ $\diamond = \underline{c}$

Part 9
a. $= 100$
b. $= 37$
c. $= \boxed{}$ one dollar
d. $= \boxed{95}$ one dollar

Part 10
a. $\underline{57} + \underline{6} = 63$ b. $49 + \underline{6} = 55$
c. $\underline{96} + \underline{8} = 104$ d. $9 + \underline{5} = 14$

Part 11
a.
$$\begin{array}{r} 1 \\ 100 \\ 70 \\ +4 \\ \hline 174 \end{array}$$
b.
$$\begin{array}{r} 300 \\ 10 \\ +8 \\ \hline 318 \end{array}$$
c. $60 + 2 = 62$
d. $400 + 0 + 7 = 407$
e. $200 + 50 + 0 = 250$

56   Lesson 88   Connecting Math Concepts

Connecting Math Concepts   Answer Key 357

Lesson 89 Side 1   Name _____

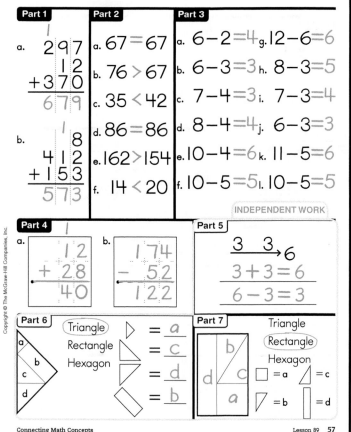

Part 1
a.
```
   1
  297
   12
 +370
  679
```
b.
```
      1
        8
    412
   +153
    573
```

Part 2
a. 67 = 67
b. 76 > 67
c. 35 < 42
d. 86 = 86
e. 162 > 154
f. 14 < 20

Part 3
a. 6−2=4   g. 12−6=6
b. 6−3=3   h. 8−3=5
c. 7−4=3   i. 7−3=4
d. 8−4=4   j. 6−3=3
e. 10−4=6   k. 11−5=6
f. 10−5=5   l. 10−5=5

INDEPENDENT WORK

Part 4
a.
```
  12
 +28
  40
```
b.
```
  174
 − 52
  122
```

Part 5
3 3 →6
3+3=6
6−3=3

Part 6
Triangle ▷ = a
Rectangle = c
Hexagon = d
= b

Part 7
Rectangle (circled)
□ = a   ◺ = c
▽ = b   ▯ = d

Connecting Math Concepts   Lesson 89   57

Lesson 89 Side 2   Name _____

Side 2 work is independent.

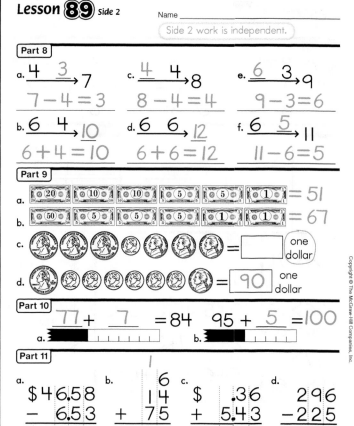

Part 8
a. 4 3 →7   7−4=3
b. 6 4 →10   6+4=10
c. 4 4 →8   8−4=4
d. 6 6 →12   6+6=12
e. 6 3 →9   9−3=6
f. 6 5 →11   11−6=5

Part 9
a. = 51
b. = 67
c. = [ ] one dollar
d. = 90 one dollar

Part 10
a. 77 + 7 = 84
b. 95 + 5 = 100

Part 11
a.
```
 $46.58
 − 6.53
 $40.05
```
b.
```
     6
    14
  + 75
    95
```
c.
```
 $ .36
 + 5.43
 $ 5.79
```
d.
```
  296
 −225
   71
```

58   Lesson 89   Connecting Math Concepts

Lesson 90 Side 1   Name _____

Part 1
a. 7−4=3   d. 3+3=6   g. 8−5=3
b. 6+5=11   e. 9−3=6   h. 3+4=7
c. 10−5=5   f. 4+2=6   i. 8−4=4

Part 2
a. 315 > 296
b. 84 < 92
c. 602 = 602
d. 55 > 48
e. 93 < 101
f. 41 = 41

Part 3
a.
```
  57
 −23
  34
```
b.
```
   1
  16
 +26
  42
```
c.
```
  257
 −204
   53
```

Part 4
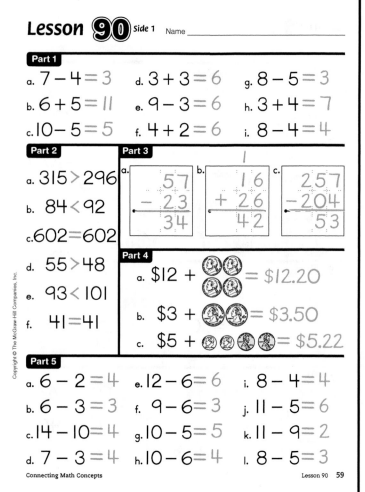
a. $12 + = $12.20
b. $3 + = $3.50
c. $5 + = $5.22

Part 5
a. 6−2=4   e. 12−6=6   i. 8−4=4
b. 6−3=3   f. 9−6=3   j. 11−5=6
c. 14−10=4 g. 10−5=5   k. 11−9=2
d. 7−3=4   h. 10−6=4   l. 8−5=3

Connecting Math Concepts   Lesson 90   59

Lesson 90 Side 2   Name _____

Side 2 work is independent.

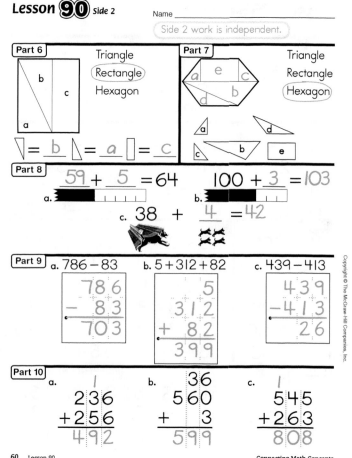

Part 6
Rectangle (circled)
▽ = b   ◺ = a   ▯ = c

Part 7
Hexagon (circled)

Part 8
a. 59 + 5 = 64
b. 100 + 3 = 103
c. 38 + 4 = 42

Part 9
a. 786−83
```
  786
 − 83
  703
```
b. 5+312+82
```
    5
  312
 + 82
  399
```
c. 439−413
```
  439
 −413
   26
```

Part 10
a.
```
   1
  236
 +256
  492
```
b.
```
   36
  560
 +  3
  599
```
c.
```
   1
  545
 +263
  808
```

60   Lesson 90   Connecting Math Concepts

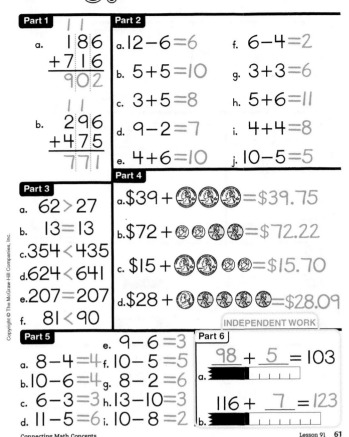

Part 1

a. $\overset{11}{\underset{902}{\begin{array}{r}186 \\ +716\end{array}}}$

b. $\overset{11}{\underset{771}{\begin{array}{r}296 \\ +475\end{array}}}$

Part 2

a. $12-6=6$ f. $6-4=2$
b. $5+5=10$ g. $3+3=6$
c. $3+5=8$ h. $5+6=11$
d. $9-2=7$ i. $4+4=8$
e. $4+6=10$ j. $10-5=5$

Part 3

a. $62 > 27$
b. $13 = 13$
c. $354 < 435$
d. $624 < 641$
e. $207 = 207$
f. $81 < 90$

Part 4

a. $\$39 + $ 🪙🪙🪙 $= \$39.75$
b. $\$72 + $ 🪙🪙🪙 $= \$72.22$
c. $\$15 + $ 🪙🪙🪙 $= \$15.70$
d. $\$28 + $ 🪙🪙🪙🪙🪙 $= \$28.09$

Part 5

e. $9-6=3$
a. $8-4=4$ f. $10-5=5$
b. $10-6=4$ g. $8-2=6$
c. $6-3=3$ h. $13-10=3$
d. $11-5=6$ i. $10-8=2$

Part 6 INDEPENDENT WORK

a. $98 + \underline{5} = 103$
b. $116 + \underline{7} = 123$

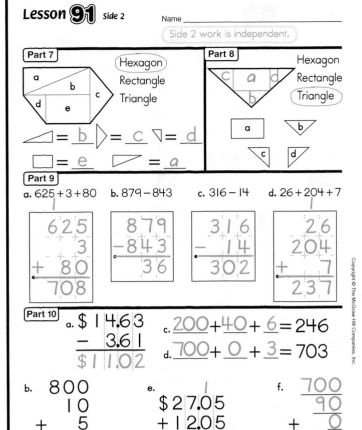

Part 7

Hexagon
Rectangle
Triangle

◺ = $\underline{b}$ ◸ = $\underline{c}$ ◹ = $\underline{d}$
▭ = $\underline{e}$ ◿ = $\underline{a}$

Part 8

Hexagon
Rectangle
(Triangle)

Part 9

a. $625 + 3 + 80$

$\begin{array}{r}625 \\ 3 \\ +80 \\ \hline 708\end{array}$

b. $879 - 843$

$\begin{array}{r}879 \\ -843 \\ \hline 36\end{array}$

c. $316 - 14$

$\begin{array}{r}316 \\ -14 \\ \hline 302\end{array}$

d. $26 + 204 + 7$

$\begin{array}{r}26 \\ 204 \\ +7 \\ \hline 237\end{array}$

Part 10

a. $\begin{array}{r}\$14.63 \\ -3.61 \\ \hline \$11.02\end{array}$

c. $200 + 40 + 6 = 246$
d. $700 + 0 + 3 = 703$

b. $\begin{array}{r}800 \\ 10 \\ +5 \\ \hline 815\end{array}$

e. $\begin{array}{r}\$27.05 \\ +12.05 \\ \hline \$39.10\end{array}$

f. $\begin{array}{r}700 \\ 90 \\ +0 \\ \hline 790\end{array}$

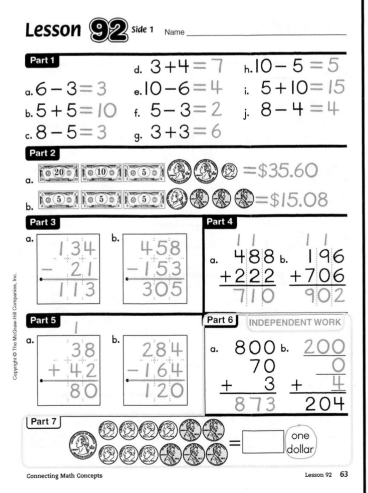

Part 1

a. $6-3=3$ d. $3+4=7$ h. $10-5=5$
b. $5+5=10$ e. $10-6=4$ i. $5+10=15$
c. $8-5=3$ f. $5-3=2$ j. $8-4=4$
             g. $3+3=6$

Part 2

a. 💵💵💵 🪙🪙🪙 $= \$35.60$
b. 💵💵💵 🪙🪙🪙🪙 $= \$15.08$

Part 3

a. $\begin{array}{r}134 \\ -21 \\ \hline 113\end{array}$

b. $\begin{array}{r}458 \\ -153 \\ \hline 305\end{array}$

Part 4

a. $\overset{11}{\underset{710}{\begin{array}{r}488 \\ +222\end{array}}}$

b. $\overset{11}{\underset{902}{\begin{array}{r}196 \\ +706\end{array}}}$

Part 5

a. $\begin{array}{r}38 \\ +42 \\ \hline 80\end{array}$

b. $\begin{array}{r}284 \\ -164 \\ \hline 120\end{array}$

Part 6 INDEPENDENT WORK

a. $\begin{array}{r}800 \\ 70 \\ +3 \\ \hline 873\end{array}$

b. $\begin{array}{r}200 \\ 0 \\ +4 \\ \hline 204\end{array}$

Part 7

🪙🪙🪙🪙🪙🪙🪙🪙🪙🪙🪙 $=$ [one dollar]

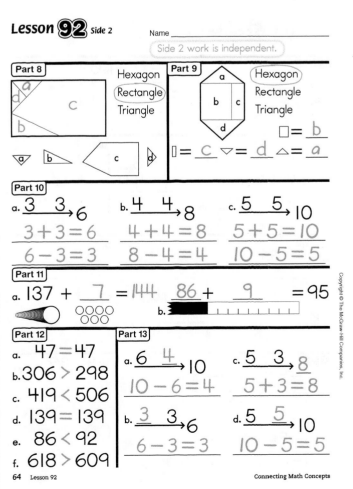

Part 8

Hexagon
(Rectangle)
Triangle

Part 9

Hexagon
Rectangle
Triangle

□ = $\underline{b}$
◫ = $\underline{c}$ ▽ = $\underline{d}$ △ = $\underline{a}$

Part 10

a. $\underset{3+3=6}{\underset{6-3=3}{3 \quad 3 \rightarrow 6}}$

b. $\underset{4+4=8}{\underset{8-4=4}{4 \quad 4 \rightarrow 8}}$

c. $\underset{5+5=10}{\underset{10-5=5}{5 \quad 5 \rightarrow 10}}$

Part 11

a. $137 + \underline{7} = 144$ $\underline{86} + \underline{9} = 95$
b.

Part 12

a. $47 = 47$
b. $306 > 298$
c. $419 < 506$
d. $139 = 139$
e. $86 < 92$
f. $618 > 609$

Part 13

a. $\underset{10-6=4}{6 \quad 4 \rightarrow 10}$

c. $\underset{5+3=8}{5 \quad 3 \rightarrow 8}$

b. $\underset{6-3=3}{3 \quad 3 \rightarrow 6}$

d. $\underset{10-5=5}{5 \quad 5 \rightarrow 10}$

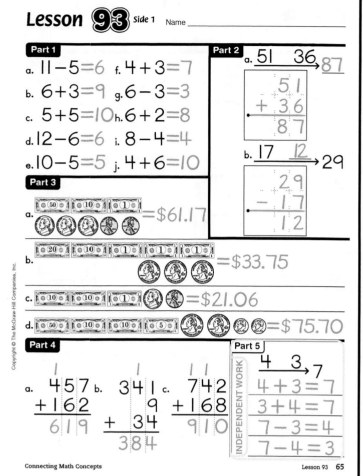

Lesson 93 Side 1 Name _____

Part 1
a. 11−5=6   f. 4+3=7
b. 6+3=9   g. 6−3=3
c. 5+5=10   h. 6+2=8
d. 12−6=6   i. 8−4=4
e. 10−5=5   j. 4+6=10

Part 2
a. 51 36 →87
```
   51
 + 36
   87
```
b. 17 12 →29
```
   29
 − 17
   12
```

Part 3
a. =$61.17
b. =$33.75
c. =$21.06
d. =$75.70

Part 4
a. 457
 +162
  619
b. 341
  +9+34
  384
c. 742
 +168
  910

Part 5 (INDEPENDENT WORK)
4 3 →7
4+3=7
3+4=7
7−3=4
7−4=3

Connecting Math Concepts   Lesson 93 65

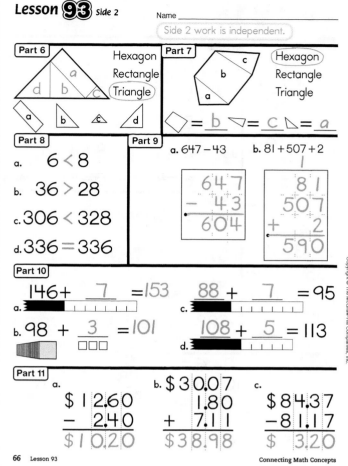

Lesson 93 Side 2 Name _____
Side 2 work is independent.

Part 6 Hexagon Rectangle (Triangle)
Part 7 (Hexagon) Rectangle Triangle
◇ = b   ▷ = c   ◁ = a

Part 8
a. 6 < 8
b. 36 > 28
c. 306 < 328
d. 336 = 336

Part 9
a. 647−43 = 604
b. 81+507+2 = 590

Part 10
a. 146 + 7 = 153
b. 98 + 3 = 101
c. 88 + 7 = 95
d. 108 + 5 = 113

Part 11
a. $12.60 − 2.40 = $10.20
b. $30.07 + 1.80 + 7.11 = $38.98
c. $84.37 − 81.17 = $3.20

66 Lesson 93   Connecting Math Concepts

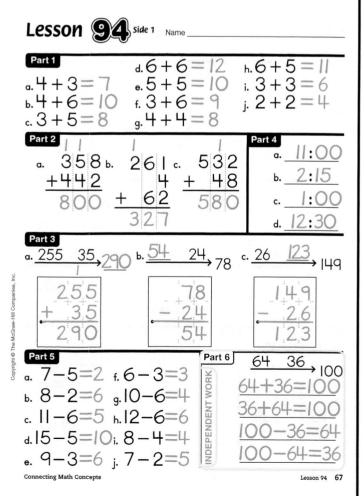

Lesson 94 Side 1 Name _____

Part 1
a. 4+3=7   d. 6+6=12   h. 6+5=11
b. 4+6=10 e. 5+5=10   i. 3+3=6
c. 3+5=8   f. 3+6=9   j. 2+2=4
           g. 4+4=8

Part 2
a. 358
 +442
  800
b. 261
 +4+62
  327
c. 532
  +48
  580

Part 4
a. 11:00
b. 2:15
c. 1:00
d. 12:30

Part 3
a. 255 35 →290   255+35=290
b. 54 24 →78   78−24=54
c. 26 123 →149   149−26=123

Part 5
a. 7−5=2   f. 6−3=3
b. 8−2=6   g. 10−6=4
c. 11−6=5   h. 12−6=6
d. 15−5=10 i. 8−4=4
e. 9−3=6   j. 7−2=5

Part 6 (INDEPENDENT WORK)
64 36 →100
64+36=100
36+64=100
100−36=64
100−64=36

Connecting Math Concepts   Lesson 94 67

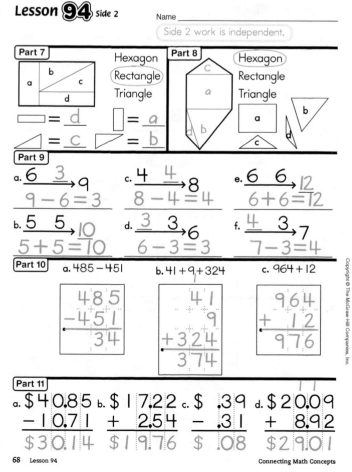

Lesson 94 Side 2 Name _____
Side 2 work is independent.

Part 7 Hexagon (Rectangle) Triangle
▭ = d   ▯ = a   ◁ = c   ▷ = b
Part 8 (Hexagon) Rectangle Triangle

Part 9
a. 6 3 →9   9−6=3
c. 4 4 →8   8−4=4
e. 6 6 →12   6+6=12
b. 5 5 →10   5+5=10
d. 3 3 →6   6−3=3
f. 4 3 →7   7−3=4

Part 10
a. 485−451 = 34
b. 41+9+324 = 374
c. 964+12 = 976

Part 11
a. $40.85 − 10.71 = $30.14
b. $17.22 + 2.54 = $19.76
c. $.39 − .31 = $.08
d. $20.09 + 8.92 = $29.01

68 Lesson 94   Connecting Math Concepts

Lesson 95 Side 1   Name _____

Part 1

a. $10 - 6 = 4$
b. $8 - 4 = 4$
c. $9 - 3 = 6$
d. $11 - 5 = 6$
e. $6 - 3 = 3$
f. $10 - 5 = 5$
g. $7 - 4 = 3$
h. $12 - 6 = 6$
i. $8 - 5 = 3$
j. $10 - 2 = 8$

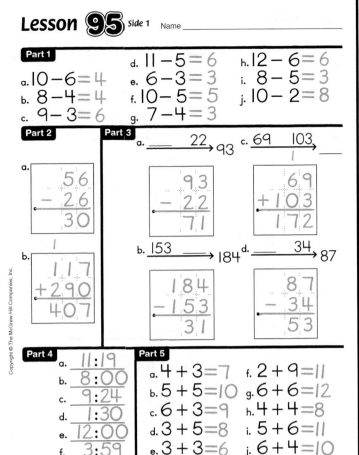

Part 2

a.
$$\begin{array}{r} 56 \\ -\ 26 \\ \hline 30 \end{array}$$

b.
$$\begin{array}{r} \overset{1}{1}17 \\ +290 \\ \hline 407 \end{array}$$

Part 3

a. $\underline{\quad 22\quad} \rightarrow 93$
$$\begin{array}{r} 93 \\ -\ 22 \\ \hline 71 \end{array}$$

c. $\underline{69 \quad 103} \rightarrow$
$$\begin{array}{r} 69 \\ +103 \\ \hline 172 \end{array}$$

b. $\underline{153 \quad\quad} \rightarrow 184$
$$\begin{array}{r} 184 \\ -153 \\ \hline 31 \end{array}$$

d. $\underline{\quad 34\quad} \rightarrow 87$
$$\begin{array}{r} 87 \\ -\ 34 \\ \hline 53 \end{array}$$

Part 4

a. 11:19
b. 8:00
c. 9:24
d. 1:30
e. 12:00
f. 3:59

Part 5

a. $4 + 3 = 7$
b. $5 + 5 = 10$
c. $6 + 3 = 9$
d. $3 + 5 = 8$
e. $3 + 3 = 6$
f. $2 + 9 = 11$
g. $6 + 6 = 12$
h. $4 + 4 = 8$
i. $5 + 6 = 11$
j. $6 + 4 = 10$

Connecting Math Concepts   Lesson 95 **69**

Lesson 95 Side 2   Name _____

Side 2 work is independent.

Part 6

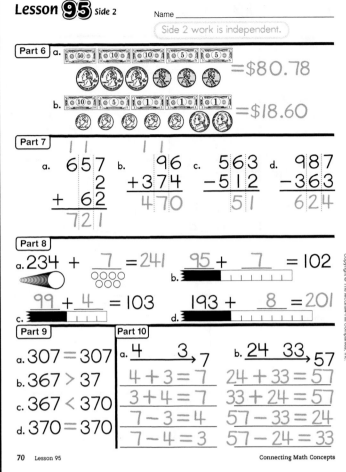

a. = $80.78
b. = $18.60

Part 7

a.
$$\begin{array}{r} \overset{1}{6}5\overset{1}{7} \\ 2 \\ +\ 62 \\ \hline 721 \end{array}$$

b.
$$\begin{array}{r} 96 \\ +374 \\ \hline 470 \end{array}$$

c.
$$\begin{array}{r} 563 \\ -512 \\ \hline 51 \end{array}$$

d.
$$\begin{array}{r} 987 \\ -363 \\ \hline 624 \end{array}$$

Part 8

a. $234 + \underline{7} = 241$
b. $95 + \underline{7} = 102$
c. $99 + \underline{4} = 103$
d. $193 + \underline{8} = 201$

Part 9

a. $307 = 307$
b. $367 > 37$
c. $367 < 370$
d. $370 = 370$

Part 10

a. $\underline{4 \quad 3} \rightarrow 7$
$4 + 3 = 7$
$3 + 4 = 7$
$7 - 3 = 4$
$7 - 4 = 3$

b. $\underline{24 \quad 33} \rightarrow 57$
$24 + 33 = 57$
$33 + 24 = 57$
$57 - 33 = 24$
$57 - 24 = 33$

70 Lesson 95   Connecting Math Concepts

Lesson 96 Side 1   Name _____

Part 1

a. $\underline{32 \quad 12} \rightarrow 44$
$$\begin{array}{r} 32 \\ +12 \\ \hline 44 \end{array}$$

b. $\underline{24 \quad 11} \rightarrow 35$
$$\begin{array}{r} 35 \\ -24 \\ \hline 11 \end{array}$$

c. $\underline{21 \quad 72} \rightarrow 93$
$$\begin{array}{r} 93 \\ -72 \\ \hline 21 \end{array}$$

d. $\underline{50 \quad 27} \rightarrow 77$
$$\begin{array}{r} 50 \\ +27 \\ \hline 77 \end{array}$$

Part 2

a. $8 - 4 = 4$
b. $12 - 10 = 2$
c. $9 - 6 = 3$
d. $10 - 2 = 8$
e. $7 - 3 = 4$
f. $10 - 5 = 5$
g. $8 - 5 = 3$
h. $7 - 6 = 1$
i. $6 - 3 = 3$
j. $8 - 2 = 6$

Part 3

a. 6:00
b. 5:08
c. 12:04
d. 5:00
e. 10:01
f. 11:30

Part 4

a.
$$\begin{array}{r} 48 \\ -27 \\ \hline 21 \end{array}$$

b.
$$\begin{array}{r} 137 \\ -\ 25 \\ \hline 112 \end{array}$$

Part 5

a. $4 + 6 = 10$
b. $5 + 3 = 8$
c. $6 + 5 = 11$
d. $3 + 3 = 6$
e. $6 + 10 = 16$
f. $5 + 5 = 10$
g. $4 + 3 = 7$
h. $6 + 6 = 12$
i. $2 + 5 = 7$
j. $4 + 4 = 8$

Connecting Math Concepts   Lesson 96 **71**

Lesson 96 Side 2   Name _____

Side 2 work is independent.

Part 6

a. $800 + 40 + 7 = 847$
b. $200 + 0 + 9 = 209$

c.
$$\begin{array}{r} 300 \\ 10 \\ +\ 4 \\ \hline 314 \end{array}$$

d.
$$\begin{array}{r} 500 \\ 60 \\ +\ 0 \\ \hline 560 \end{array}$$

Part 7

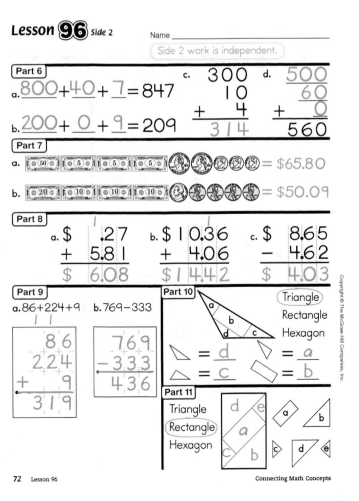

a. = $65.80
b. = $50.09

Part 8

a.
$$\begin{array}{r} \$\ .27 \\ +\ 5.81 \\ \hline \$\ 6.08 \end{array}$$

b.
$$\begin{array}{r} \$10.36 \\ +\ 4.06 \\ \hline \$14.42 \end{array}$$

c.
$$\begin{array}{r} \$\ 8.65 \\ -\ 4.62 \\ \hline \$\ 4.03 \end{array}$$

Part 9

a. $86 + 224 + 9$
$$\begin{array}{r} \overset{1}{8}\overset{1}{6} \\ 224 \\ +\ \ 9 \\ \hline 319 \end{array}$$

b. $769 - 333$
$$\begin{array}{r} 769 \\ -333 \\ \hline 436 \end{array}$$

Part 10

Triangle
Rectangle
Hexagon

⬠ = d   ◁ = a
△ = c   ▱ = b

Part 11

Triangle
(Rectangle)
Hexagon

72 Lesson 96   Connecting Math Concepts

Connecting Math Concepts   Answer Key **361**

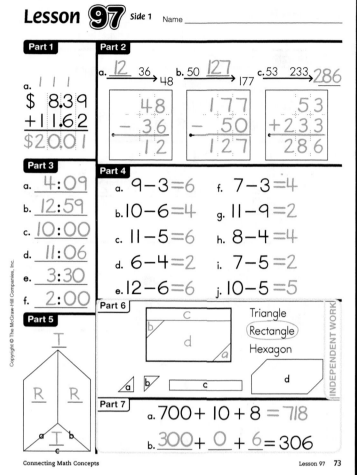

Lesson 97 Side 1   Name _____

Part 1

a. 1 1 1

$$\begin{array}{r} \$\ 8.39 \\ +11.62 \\ \hline \$20.01 \end{array}$$

Part 2

a. 12   36 → 48

$$\begin{array}{r} 48 \\ -36 \\ \hline 12 \end{array}$$

b. 50   127 → 177

$$\begin{array}{r} 177 \\ -50 \\ \hline 127 \end{array}$$

c. 53   233 → 286

$$\begin{array}{r} 53 \\ +233 \\ \hline 286 \end{array}$$

Part 3

a. 4:09
b. 12:59
c. 10:00
d. 11:06
e. 3:30
f. 2:00

Part 4

a. 9 − 3 = 6
b. 10 − 6 = 4
c. 11 − 5 = 6
d. 6 − 4 = 2
e. 12 − 6 = 6
f. 7 − 3 = 4
g. 11 − 9 = 2
h. 8 − 4 = 4
i. 7 − 5 = 2
j. 10 − 5 = 5

Part 5

Part 6

Triangle
(Rectangle)
Hexagon

Part 7

a. 700 + 10 + 8 = 718
b. 300 + 0 + 6 = 306

INDEPENDENT WORK

Copyright © The McGraw-Hill Companies, Inc.

Connecting Math Concepts   Lesson 97   73

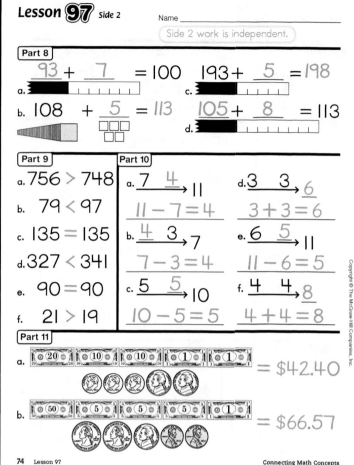

Lesson 97 Side 2   Name _____

Part 8

a. 93 + 7 = 100
b. 108 + 5 = 113
c. 193 + 5 = 198
d. 105 + 8 = 113

Part 9

a. 756 > 748
b. 79 < 97
c. 135 = 135
d. 327 < 341
e. 90 = 90
f. 21 > 19

Part 10

a. 7   4 → 11
   11 − 7 = 4
b. 4   3 → 7
   7 − 3 = 4
c. 5   5 → 10
   10 − 5 = 5
d. 3   3 → 6
   3 + 3 = 6
e. 6   5 → 11
   11 − 6 = 5
f. 4   4 → 8
   4 + 4 = 8

Part 11

a. = $42.40
b. = $66.57

Copyright © The McGraw-Hill Companies, Inc.

74   Lesson 97   Connecting Math Concepts

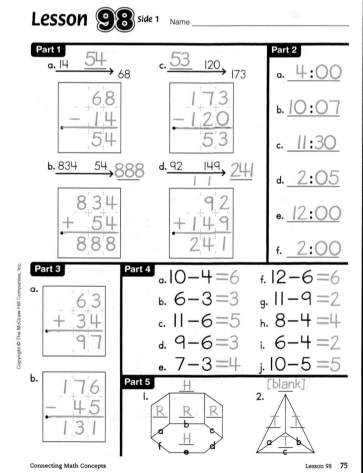

Lesson 98 Side 1   Name _____

Part 1

a. 14   54 → 68

$$\begin{array}{r} 68 \\ -14 \\ \hline 54 \end{array}$$

b. 834   54 → 888

$$\begin{array}{r} 834 \\ +\ 54 \\ \hline 888 \end{array}$$

c. 53   120 → 173

$$\begin{array}{r} 173 \\ -120 \\ \hline 53 \end{array}$$

d. 92   149 → 241

$$\begin{array}{r} 92 \\ +149 \\ \hline 241 \end{array}$$

Part 2

a. 4:00
b. 10:07
c. 11:30
d. 2:05
e. 12:00
f. 2:00

Part 3

a. $$\begin{array}{r} 63 \\ +34 \\ \hline 97 \end{array}$$

b. $$\begin{array}{r} 176 \\ -\ 45 \\ \hline 131 \end{array}$$

Part 4

a. 10 − 4 = 6
b. 6 − 3 = 3
c. 11 − 6 = 5
d. 9 − 6 = 3
e. 7 − 3 = 4
f. 12 − 6 = 6
g. 11 − 9 = 2
h. 8 − 4 = 4
i. 6 − 4 = 2
j. 10 − 5 = 5

Part 5

1.   2. [blank]

Copyright © The McGraw-Hill Companies, Inc.

Connecting Math Concepts   Lesson 98   75

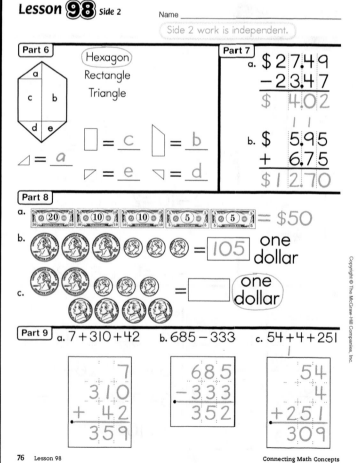

Lesson 98 Side 2   Name _____

Part 6

(Hexagon)
Rectangle
Triangle

□ = c   � = b
◁ = a
▽ = e   ◣ = d

Part 7

a. $$\begin{array}{r} \$\ 27.49 \\ -23.47 \\ \hline \$\ \ 4.02 \end{array}$$

b. $$\begin{array}{r} \$\ 5.95 \\ +\ 6.75 \\ \hline \$12.70 \end{array}$$

Part 8

a. = $50
b. = 105 one dollar
c. = [blank] one dollar

Part 9

a. 7 + 310 + 42

$$\begin{array}{r} 7 \\ 310 \\ +\ 42 \\ \hline 359 \end{array}$$

b. 685 − 333

$$\begin{array}{r} 685 \\ -333 \\ \hline 352 \end{array}$$

c. 54 + 4 + 251

$$\begin{array}{r} 54 \\ 4 \\ +251 \\ \hline 309 \end{array}$$

Copyright © The McGraw-Hill Companies, Inc.

76   Lesson 98   Connecting Math Concepts

Lesson 99 Side 1 Name _____

Part 1
a. $10 - 4 = 6$ e. $8 + 8 = 16$ i. $6 - 3 = 3$
b. $4 + 4 = 8$ f. $10 - 5 = 5$ j. $6 + 3 = 9$
c. $12 - 6 = 6$ g. $7 + 7 = 14$ k. $10 - 6 = 4$
d. $8 - 4 = 4$ h. $14 - 7 = 7$

Part 2 [blank]
1.
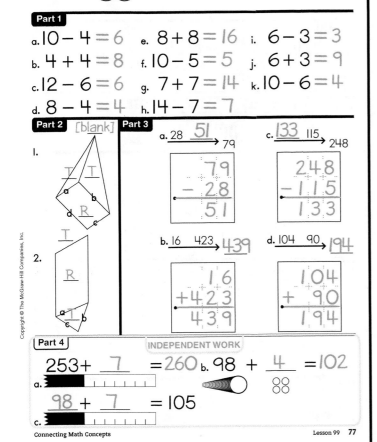
2.

Part 3
a. 28 → 51 → 79
$$\begin{array}{r} 79 \\ -\ 28 \\ \hline 51 \end{array}$$

c. 133 → 115 → 248
$$\begin{array}{r} 248 \\ -115 \\ \hline 133 \end{array}$$

b. 16 → 423 → 439
$$\begin{array}{r} 16 \\ +423 \\ \hline 439 \end{array}$$

d. 104 → 90 → 194
$$\begin{array}{r} 104 \\ +\ 90 \\ \hline 194 \end{array}$$

Part 4   INDEPENDENT WORK
a. $253 + 7 = 260$ b. $98 + 4 = 102$
c. $98 + 7 = 105$

Lesson 99 Side 2   Name _____
Side 2 work is independent.

Part 5
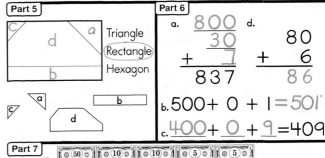
Triangle
(Rectangle)
Hexagon

Part 6
a. $$\begin{array}{r} 800 \\ 30 \\ +\ 7 \\ \hline 837 \end{array}$$
d. $$\begin{array}{r} 80 \\ +\ 6 \\ \hline 86 \end{array}$$
b. $500 + 0 + 1 = 501$
c. $400 + 0 + 9 = 409$

Part 7

a. $= \$81.00$
b. $= \$29.00$

Part 8
a. $85 = 85$
b. $306 < 360$
c. $361 > 316$
d. $503 = 503$
e. $73 > 38$

Part 9
a. $$\begin{array}{r} 1\ 1\ 1 \\ \$42.55 \\ +38.65 \\ \hline \$81.20 \end{array}$$
c. $$\begin{array}{r} 1\ 1 \\ 850 \\ 52 \\ +109 \\ \hline 1011 \end{array}$$
b. $$\begin{array}{r} \$27.43 \\ -24.41 \\ \hline \$\ 3.02 \end{array}$$
d. $$\begin{array}{r} 638 \\ -433 \\ \hline 205 \end{array}$$

Lesson 100 Side 1   Name _____

Part 1
$$300 + 80 = 380$$

Part 2
a. 71 → 404 → 475
$$\begin{array}{r} 475 \\ -\ 71 \\ \hline 404 \end{array}$$

b. 32 → 263 → 295
$$\begin{array}{r} 295 \\ -263 \\ \hline 32 \end{array}$$

c. 54 → 132 → 186
$$\begin{array}{r} 54 \\ +132 \\ \hline 186 \end{array}$$

Part 3 [blank]   [blank]
1.
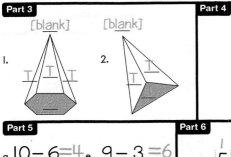
2.

Part 4
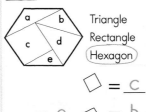

Part 5
a. $10 - 6 = 4$ e. $9 - 3 = 6$
b. $16 - 8 = 8$ f. $12 - 10 = 2$
c. $11 - 5 = 6$ g. $14 - 7 = 7$
d. $6 - 3 = 3$ h. $7 - 4 = 3$

Part 6
a. $$\begin{array}{r} 59 \\ 1 \\ +43 \\ \hline 103 \end{array}$$
b. $$\begin{array}{r} 2 \\ 99 \\ +81 \\ \hline 182 \end{array}$$

Lesson 100 Side 2   Name _____

Part 7
c. $4 - 1 = 3$ f. $7 + 7 = 14$
a. $5 - 3 = 2$ d. $4 + 4 = 8$ g. $11 - 6 = 5$
b. $5 + 5 = 10$ e. $7 - 2 = 5$ h. $3 + 4 = 7$

INDEPENDENT WORK

Part 8
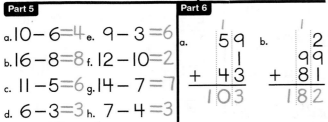
Triangle
Rectangle
(Hexagon)
$\square = c$
$\triangle = e$ $\triangleleft = b$
$\diagup = a$ $\triangledown = d$

Part 9
a. $$\begin{array}{r} 1\ 1\ 1 \\ \$35.96 \\ +\ 5.46 \\ \hline \$41.42 \end{array}$$
c. $$\begin{array}{r} \$43.59 \\ -21.58 \\ \hline \$22.01 \end{array}$$
b. $$\begin{array}{r} 654 \\ -623 \\ \hline 31 \end{array}$$
d. $$\begin{array}{r} 1 \\ 564 \\ +364 \\ \hline 928 \end{array}$$

Part 10
a. 6 → 6 → 12 $12 - 6 = 6$
c. 3 → 3 → 6 $3 + 3 = 6$
e. 4 → 3 → 7 $7 - 4 = 3$
b. 8 → 2 → 10 $10 - 2 = 8$
d. 6 → 3 → 9 $6 + 3 = 9$
f. 5 → 5 → 10 $5 + 5 = 10$

Part 1
a. $\underline{300} + \underline{30} + \underline{1} = \underline{331}$ b. $\underline{500} + \underline{20} + \underline{4} = \underline{524}$

Part 2

a. $7 - 4 = 3$
b. $3 + 3 = 6$
c. $8 - 4 = 4$
d. $9 - 6 = 3$
e. $7 + 7 = 14$
f. $10 - 5 = 5$
g. $6 - 4 = 2$
h. $8 + 8 = 16$

Part 3

a.
$$\begin{array}{r} \$\;\;.\overset{1}{5}1 \\ +\;\;.69 \\ \hline \$\;1.20 \end{array}$$

b.
$$\begin{array}{r} \$\;8.\overset{1}{3}9 \\ +\;8.22 \\ \hline \$16.61 \end{array}$$

c.
$$\begin{array}{r} \overset{2}{8}6 \\ 5 \\ +\;79 \\ \hline 170 \end{array}$$

Part 4
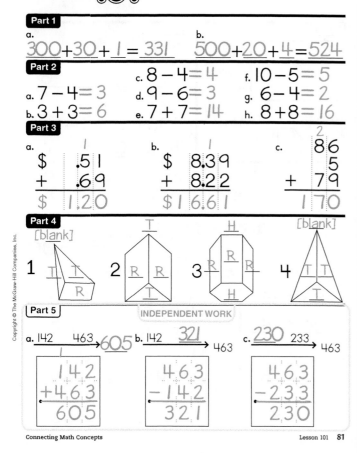

[blank]

1   2   3   4

Part 5   INDEPENDENT WORK

a. $142 \xrightarrow{\;463\;} 605$ b. $142 \xrightarrow{\;321\;} 463$ c. $230 \xrightarrow{\;233\;} 463$

$$\begin{array}{r} \overset{1}{1}42 \\ +463 \\ \hline 605 \end{array} \qquad \begin{array}{r} 463 \\ -142 \\ \hline 321 \end{array} \qquad \begin{array}{r} 463 \\ -233 \\ \hline 230 \end{array}$$

Connecting Math Concepts   Lesson 101   **81**

Side 2 work is independent.

Part 6
a.  $= \$62.57$

b.  $= \$55.46$

Part 7
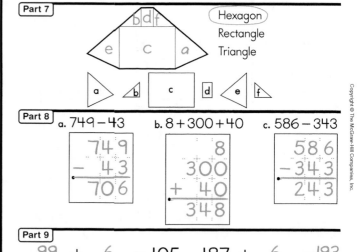

Hexagon
Rectangle
Triangle

Part 8

a. $749 - 43$
$$\begin{array}{r} 749 \\ -\;\;43 \\ \hline 706 \end{array}$$

b. $8 + 300 + 40$
$$\begin{array}{r} 8 \\ 300 \\ +\;\;40 \\ \hline 348 \end{array}$$

c. $586 - 343$
$$\begin{array}{r} 586 \\ -343 \\ \hline 243 \end{array}$$

Part 9

a. $\underline{99} + \underline{6} = 105$ b. $187 + \underline{6} = 193$
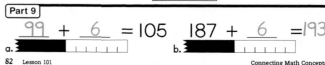

82   Lesson 101   Connecting Math Concepts

Part 1

a.
$$\begin{array}{r} 67 \\ +102 \\ \hline 169 \end{array}$$

b.
$$\begin{array}{r} 137 \\ -\;27 \\ \hline 110 \end{array}$$

c.
$$\begin{array}{r} 25 \\ +35 \\ \hline 60 \end{array}$$

d.
$$\begin{array}{r} 87 \\ -55 \\ \hline 32 \end{array}$$

INDEPENDENT WORK

Part 2

a. $3 + 6 = 9$
b. $6 - 3 = 3$
c. $6 + 5 = 11$
d. $12 - 6 = 6$
e. $8 + 8 = 16$
f. $8 - 3 = 5$
g. $10 - 4 = 6$
h. $14 - 7 = 7$
i. $5 + 5 = 10$
j. $7 - 4 = 3$
k. $7 + 2 = 9$

Part 3
a. $\underline{200} + \underline{40} + \underline{6} = \underline{246}$ b. $\underline{500} + \underline{20} + \underline{5} = \underline{525}$

Part 4

a.
$$\begin{array}{r} \$\;\;.\overset{1}{9}5 \\ \;\;.43 \\ +\;\;.68 \\ \hline \$\;2.06 \end{array}$$

b.
$$\begin{array}{r} \$\;4.\overset{2}{9}\overset{1}{5} \\ \;\;.65 \\ +\;6.42 \\ \hline \$12.02 \end{array}$$

Part 5
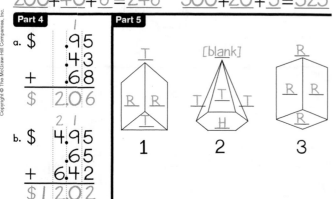

[blank]

1   2   3

Connecting Math Concepts   Lesson 102   **83**

Part 6

a.
$$\begin{array}{r} 214 \\ -\;96 \\ \hline 118 \end{array}$$
$96 \xrightarrow{\;118\;} 214$

c.
$$\begin{array}{r} 385 \\ -214 \\ \hline 171 \end{array}$$
$214 \xrightarrow{\;171\;} 385$

b.
$$\begin{array}{r} 137 \\ +385 \\ \hline 522 \end{array}$$
$137 \xrightarrow{\;385\;} 522$

d.
$$\begin{array}{r} 171 \\ +\;96 \\ \hline 267 \end{array}$$
$171 \xrightarrow{\;96\;} 267$

Part 7

a. $8 - 4 = 4$
b. $10 - 8 = 2$
c. $16 - 8 = 8$
d. $8 - 5 = 3$
e. $6 - 4 = 2$
f. $10 - 6 = 4$
g. $7 - 4 = 3$
h. $10 - 5 = 5$

Part 8

a. $105 < 140$
b. $349 = 349$
c. $68 < 74$
d. $256 > 109$

Part 9   INDEPENDENT WORK

a. $674 - 633$
$$\begin{array}{r} 674 \\ -633 \\ \hline 41 \end{array}$$

b. $4 + 500 + 10$
$$\begin{array}{r} 4 \\ 500 \\ +\;\;10 \\ \hline 514 \end{array}$$

84   Lesson 102   Connecting Math Concepts

---

364   Answer Key   **Connecting Math Concepts**

Lesson 103 Side 1   Name _____

Part 1
c. $6+3=9$ f. $10-5=5$
a. $6+4=10$ d. $11-6=5$ g. $7+7=14$
b. $6-2=4$ e. $3+4=7$ h. $8-4=4$

Part 2
a. (circled) $58 - 46 = 12$
b. $314 + 304 = 618$
c. $75 + 26 = 101$
d. (circled) $539 - 316 = 223$

INDEPENDENT WORK

Part 3
a. $142 + 181 = 323$; $142\ 181 \to 323$
b. $181 - 142 = 39$; $142\ 39 \to 181$
c. $220 - 142 = 78$; $142\ 78 \to 220$
d. $220 + 181 = 401$; $220\ 181 \to 401$

Part 4
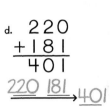 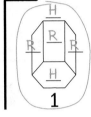 
1   2

Lesson 103 Side 2   Name _____

Part 5
a. $400 + 70 + 3 = 473$
b. $100 + 60 + 5 = 165$

Part 6
c. $10-2=8$ f. $14-7=7$
a. $16-8=8$ d. $7-3=4$ g. $9-7=2$
b. $9-3=6$ e. $10-4=6$ h. $8-3=5$

Part 7
Triangle Rectangle (Pentagon) Hexagon
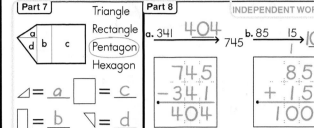
△ = a □ = c
▯ = b ◺ = d

Part 8   INDEPENDENT WORK
a. $341 \quad 404 \to 745$; $745 - 341 = 404$
b. $85 \quad 15 \to 100$; $85 + 15 = 100$

Part 9
a. $96 + 7 = 103$
b. $93 + 7 = 100$

Part 10
a. $8.79 - 5.73 = \$3.06$
b. $\$5.38 + .08 + 4.64 = \$10.10$
c. $\$21.68 - 20.38 = \$1.30$
d. $\$5.58 + 6.51 = \$12.09$

Lesson 104 Side 1   Name _____

Part 1
a. b. c.

Part 2
c. $4+3=7$ f. $5+6=11$
a. $3+5=8$ d. $5+5=10$ g. $4+4=8$
b. $7+7=14$ e. $3+6=9$ h. $4+6=10$

Part 3
a. $300 + 60 + 4 = 364$
b. $400 + 10 + 8 = 418$

Part 4
a. $386 - 73 = 313$; $73\ 313 \to 386$
b. $96 + 156 = 252$; $96\ 156 \to 252$
c. $298 - 96 = 202$; $96\ 202 \to 298$

Part 5
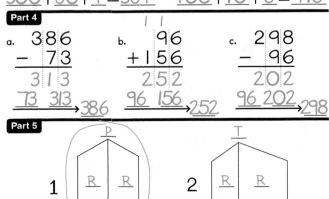
1   2

Lesson 104 Side 2   Name _____

Part 6
c. $16-8=8$ f. $10-5=5$
a. $12-6=6$ d. $9-6=3$ g. $8-2=6$
b. $12-10=2$ e. $6-4=2$ h. $8-4=4$

Part 7   INDEPENDENT WORK
a. 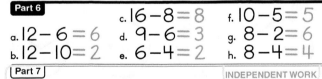 $= \$71.00$
b. 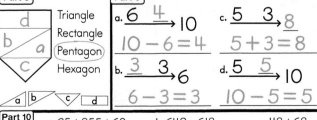 $= \$51.09$

Part 8
Triangle Rectangle (Pentagon) Hexagon
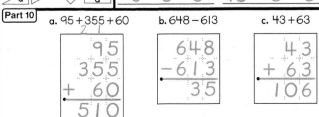

Part 9
a. $6 \quad 4 \to 10$; $10 - 6 = 4$
c. $5 \quad 3 \to 8$; $5 + 3 = 8$
b. $3 \quad 3 \to 6$; $6 - 3 = 3$
d. $5 \quad 5 \to 10$; $10 - 5 = 5$

Part 10
a. $95 + 355 + 60$: $95 + 355 + 60 = 510$
b. $648 - 613 = 35$
c. $43 + 63 = 106$

Part 1

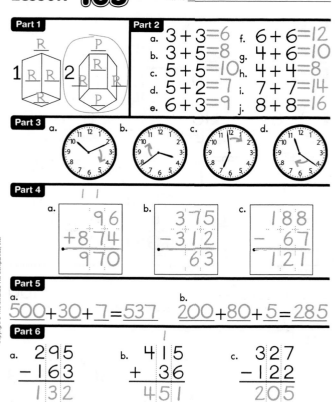

Part 2

a. $3+3=6$	f. $6+6=12$
b. $3+5=8$	g. $4+6=10$
c. $5+5=10$	h. $4+4=8$
d. $5+2=7$	i. $7+7=14$
e. $6+3=9$	j. $8+8=16$

Part 3

a. b. c. d.

Part 4

a.
$$\begin{array}{r} 1\ 1 \\ 96 \\ +874 \\ \hline 970 \end{array}$$

b.
$$\begin{array}{r} 375 \\ -312 \\ \hline 63 \end{array}$$

c.
$$\begin{array}{r} 188 \\ -67 \\ \hline 121 \end{array}$$

Part 5

a. $500+30+7=537$

b. $200+80+5=285$

Part 6

a.
$$\begin{array}{r} 295 \\ -163 \\ \hline 132 \end{array}$$
$\underrightarrow{163\quad 132} 295$

b.
$$\begin{array}{r} 1 \\ 415 \\ +36 \\ \hline 451 \end{array}$$
$\underrightarrow{415\quad 36} 451$

c.
$$\begin{array}{r} 327 \\ -122 \\ \hline 205 \end{array}$$
$\underrightarrow{122\quad 205} 327$

Part 7

	c. $8-5=3$
a. $8-4=4$	d. $14-7=7$
b. $8-2=6$	e. $7-4=3$

f. $10-4=6$
g. $11-6=5$
h. $11-9=2$

INDEPENDENT WORK

Part 8

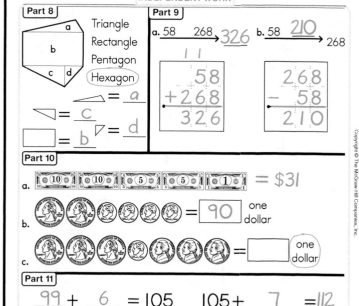

Triangle
Rectangle
Pentagon
(Hexagon)

$\triangleright = \underline{a}$

$\triangleleft = \underline{c}$

$\square = \underline{b}$ $\triangledown = \underline{d}$

Part 9

a. $58 \quad 268 \rightarrow 326$

$$\begin{array}{r} 1\ 1 \\ 58 \\ +268 \\ \hline 326 \end{array}$$

b. $58 \quad 210 \rightarrow 268$

$$\begin{array}{r} 268 \\ -58 \\ \hline 210 \end{array}$$

Part 10

a. 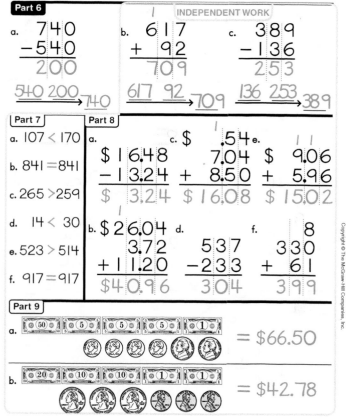 $= \$31$

b. $= \boxed{90}$ one dollar

c. $= \boxed{}$ one dollar

Part 11

a. $99 + \underline{6} = 105$

b. $105 + \underline{7} = 112$

Part 1

a. b. c. d.

Part 2

a. $300+70+2=372$

b. $400+80+4=484$

Part 3

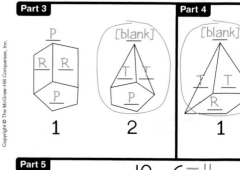

1   2

Part 4

1   2

Part 5

	c. $10-6=4$
a. $12-6=6$	d. $6-3=3$
b. $14-7=7$	e. $11-5=6$

f. $10-2=8$
g. $16-8=8$
h. $10-5=5$

Part 6

INDEPENDENT WORK

a.
$$\begin{array}{r} 740 \\ -540 \\ \hline 200 \end{array}$$
$\underrightarrow{540\quad 200} 740$

b.
$$\begin{array}{r} 617 \\ +92 \\ \hline 709 \end{array}$$
$\underrightarrow{617\quad 92} 709$

c.
$$\begin{array}{r} 389 \\ -136 \\ \hline 253 \end{array}$$
$\underrightarrow{136\quad 253} 389$

Part 7

a. $107 < 170$

b. $841 = 841$

c. $265 > 259$

d. $14 < 30$

e. $523 > 514$

f. $917 = 917$

Part 8

a.
$$\begin{array}{r} \$16.48 \\ -13.24 \\ \hline \$3.24 \end{array}$$

b.
$$\begin{array}{r} 1 \\ \$26.04 \\ 3.72 \\ +11.20 \\ \hline \$40.96 \end{array}$$

c.
$$\begin{array}{r} \$\ .54 \\ 7.04 \\ +8.50 \\ \hline \$16.08 \end{array}$$

d.
$$\begin{array}{r} 537 \\ -233 \\ \hline 304 \end{array}$$

e.
$$\begin{array}{r} 1\ 1 \\ \$9.06 \\ +5.96 \\ \hline \$15.02 \end{array}$$

f.
$$\begin{array}{r} 8 \\ 330 \\ +61 \\ \hline 399 \end{array}$$

Part 9

a. 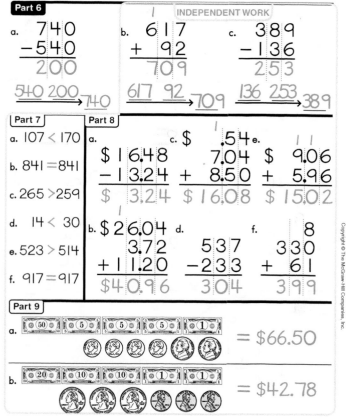 $= \$66.50$

b. $= \$42.78$

Lesson 107 Side 1   Name _____

Part 1

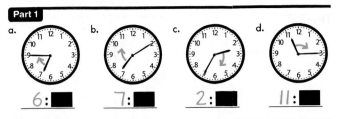

a. 6:██   b. 7:██   c. 2:██   d. 11:██

Part 2

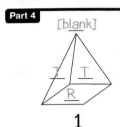 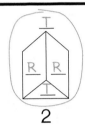 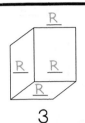

a.
```
  68
- 43
  25
```
b.
```
  1 1
  496
+ 274
  770
```
c.
```
  685
- 613
   72
```

Part 3

a. 300 + 20 + 4 = 324   b. 200 + 50 + 1 = 251

Part 4

[blank]
1   2   3

Connecting Math Concepts   Lesson 107   93

Lesson 107 Side 2   Name _____

Part 5

c. 8 − 5 = 3   f. 8 − 4 = 4
a. 14 − 4 = 10   d. 7 − 3 = 4   g. 16 − 8 = 8
b. 14 − 7 = 7   e. 11 − 5 = 6   h. 10 − 5 = 5

Part 6   INDEPENDENT WORK

a.
```
  1 1
  864
- 534
  330
```
b.
```
   78
+ 288
  366
```
c.
```
  391
- 330
   61
```

534 330 → 864   78 288 → 366   330 61 → 391

Part 7

a. 99 + 5 = 104   b. 98 + 6 = 104

c. 128 + 7 = 135

Part 8

Triangle
Rectangle
Hexagon
Pentagon

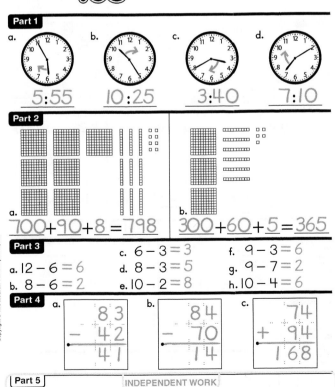

a   b   c   d

Part 9

Triangle
Rectangle
Hexagon
Pentagon

□ = c   ◁ = d   ___ = a
◺ = e   ▭ = b

94   Lesson 107   Connecting Math Concepts

Lesson 108 Side 1   Name _____

Part 1

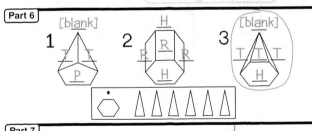

a. 5:55   b. 10:25   c. 3:40   d. 7:10

Part 2

a. 700 + 90 + 8 = 798   b. 300 + 60 + 5 = 365

Part 3

c. 6 − 3 = 3   f. 9 − 3 = 6
a. 12 − 6 = 6   d. 8 − 3 = 5   g. 9 − 7 = 2
b. 8 − 6 = 2   e. 10 − 2 = 8   h. 10 − 4 = 6

Part 4

a.
```
  83
- 42
  41
```
b.
```
  84
- 70
  14
```
c.
```
   74
+ 94
  168
```

Part 5   INDEPENDENT WORK

a. 194 + 8 = 202   b. 99 + 5 = 104

Connecting Math Concepts   Lesson 108   95

Lesson 108 Side 2   Name _____

Side 2 work is independent.

Part 6

[blank]   [blank]
1   2   3

Part 7

a.
```
  86
- 44
  42
```
b.
```
   1
  97
+ 57
  154
```

44 42 → 86   97 57 → 154

Part 8

a. 67 130 → 197
```
  197
-  67
  130
```
b. 84 84 → 168
```
   84
+  84
  168
```

Part 9

a. 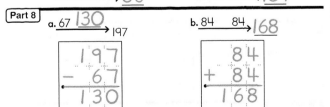 = $66.60

b.  = $42.09

96   Lesson 108   Connecting Math Concepts

Connecting Math Concepts   **Answer Key   367**

Lesson 109 — Side 1   Name _____

Part 1

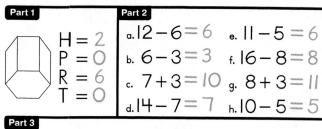

H = 2
P = 0
R = 6
T = 0

Part 2

a. $12 - 6 = 6$ e. $11 - 5 = 6$
b. $6 - 3 = 3$ f. $16 - 8 = 8$
c. $7 + 3 = 10$ g. $8 + 3 = 11$
d. $14 - 7 = 7$ h. $10 - 5 = 5$

Part 3

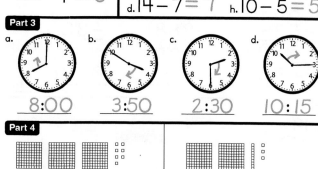

a. 8:00   b. 3:50   c. 2:30   d. 10:15

Part 4

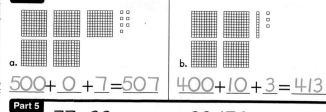

a. $500 + 0 + 7 = 507$ b. $400 + 10 + 3 = 413$

Part 5

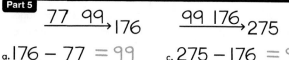

$77 \ 99 \rightarrow 176$ $99 \ 176 \rightarrow 275$

a. $176 - 77 = 99$ c. $275 - 176 = 99$
b. $176 + 99 = 275$ d. $99 + 77 = 176$

Connecting Math Concepts

Lesson 109 — Side 2   Name _____

Side 2 work is independent.

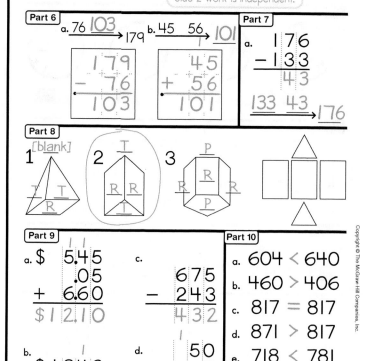

Part 6

a. $76 \ 103 \rightarrow 179$
$\begin{array}{r} 179 \\ -\ 76 \\ \hline 103 \end{array}$

b. $45 \ 56 \rightarrow 101$
$\begin{array}{r} 45 \\ +\ 56 \\ \hline 101 \end{array}$

Part 7

a. $\begin{array}{r} 176 \\ -133 \\ \hline 43 \end{array}$

$133 \ 43 \rightarrow 176$

Part 8

[blank]   1   2   3

Part 9

a. $\begin{array}{r} \$\ 5.45 \\ .05 \\ +\ 6.60 \\ \hline \$12.10 \end{array}$

b. $\begin{array}{r} \$18.49 \\ +10.23 \\ \hline \$28.72 \end{array}$

c. $\begin{array}{r} 675 \\ -243 \\ \hline 432 \end{array}$

d. $\begin{array}{r} 50 \\ 7 \\ +\ 950 \\ \hline 1007 \end{array}$

Part 10

a. $604 < 640$
b. $460 > 406$
c. $817 = 817$
d. $871 > 817$
e. $718 < 781$

Connecting Math Concepts

Lesson 110 — Side 1   Name _____

Part 1

H = 1
P = 0
R = 0
T = 6

Part 2

H = 0
P = 0
R = 3
T = 2

Part 3

$79 \ 268 \rightarrow 347$ $268 \ 347 \rightarrow 615$

a. $615 - 268 = 347$
b. $268 + 79 = 347$
c. $347 - 268 = 79$
d. $347 + 268 = 615$

Part 4

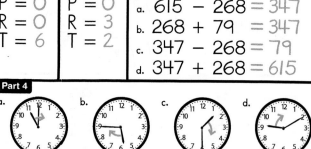

a. 11:00   b. 5:45   c. 1:30   d. 9:10

Part 5

a. $3 + 7 = 10$ c. $7 + 7 = 14$ f. $8 - 5 = 3$
b. $7 - 3 = 4$ d. $11 - 6 = 5$ g. $4 + 4 = 8$
                      e. $11 - 8 = 3$ h. $9 - 3 = 6$

Part 6

a. $\begin{array}{r} 173 \\ -\ 50 \\ \hline 123 \end{array}$

b. $\begin{array}{r} 218 \\ +\ 72 \\ \hline 290 \end{array}$

c. $\begin{array}{r} 278 \\ -165 \\ \hline 113 \end{array}$

Connecting Math Concepts

Lesson 110 — Side 2   Name _____

Part 7

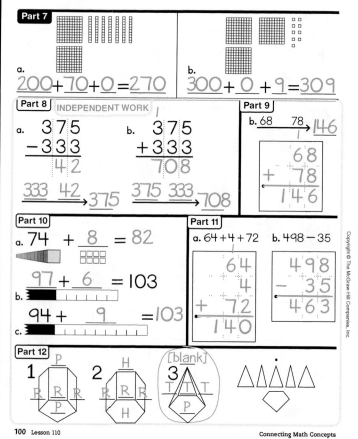

a. $200 + 70 + 0 = 270$ b. $300 + 0 + 9 = 309$

Part 8 INDEPENDENT WORK

a. $\begin{array}{r} 375 \\ -333 \\ \hline 42 \end{array}$

$333 \ 42 \rightarrow 375$

b. $\begin{array}{r} 375 \\ +333 \\ \hline 708 \end{array}$

$375 \ 333 \rightarrow 708$

Part 9

b. $68 \ 78 \rightarrow 146$
$\begin{array}{r} 68 \\ +\ 78 \\ \hline 146 \end{array}$

Part 10

a. $74 + 8 = 82$

$97 + 6 = 103$

$94 + 9 = 103$

Part 11

a. $64 + 4 + 72$
$\begin{array}{r} 64 \\ 4 \\ +\ 72 \\ \hline 140 \end{array}$

b. $498 - 35$
$\begin{array}{r} 498 \\ -\ 35 \\ \hline 463 \end{array}$

Part 12

1   2   3 [blank]

Connecting Math Concepts

Connecting Math Concepts

Lesson 111 — Side 1   Name _____

Part 1

a. 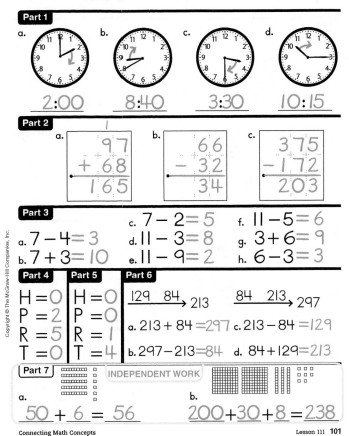 2:00   b. 8:40   c. 3:30   d. 10:15

Part 2

a.
$$\begin{array}{r} 1 \\ 97 \\ +\ 68 \\ \hline 165 \end{array}$$

b.
$$\begin{array}{r} 66 \\ -\ 32 \\ \hline 34 \end{array}$$

c.
$$\begin{array}{r} 375 \\ -172 \\ \hline 203 \end{array}$$

Part 3

a. $7-4=3$
b. $7+3=10$
c. $7-2=5$
d. $11-3=8$
e. $11-9=2$
f. $11-5=6$
g. $3+6=9$
h. $6-3=3$

Part 4

H = 0
P = 2
R = 5
T = 0

Part 5

H = 0
P = 0
R = 1
T = 4

Part 6

$129 \quad 84 \rightarrow 213$
$84 \quad 213 \rightarrow 297$

a. $213+84=297$ c. $213-84=129$
b. $297-213=84$ d. $84+129=213$

Part 7   INDEPENDENT WORK

a. $50+6=56$
b. $200+30+8=238$

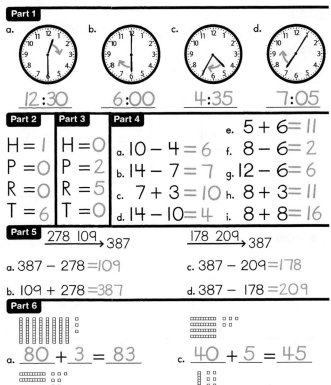

Connecting Math Concepts

Lesson 111 **101**

Lesson 111 — Side 2   Name _____

Side 2 work is independent.

Part 8

Hexagon / (Pentagon) / Rectangle / Triangle

△ = b   ▷ = c
▭ = d   ◺ = a

Part 9

Hexagon / Pentagon / Rectangle / Triangle

a b
c d
e

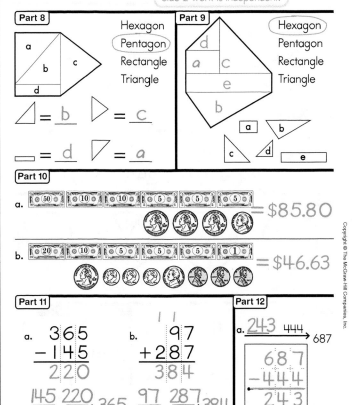

Part 10

a. = $85.80
b. = $46.63

Part 11

a.
$$\begin{array}{r} 365 \\ -145 \\ \hline 220 \end{array}$$
$145 \quad 220 \rightarrow 365$

b.
$$\begin{array}{r} 1\ 1 \\ 97 \\ +287 \\ \hline 384 \end{array}$$
$97 \quad 287 \rightarrow 384$

Part 12

a. $243 \quad 444 \rightarrow 687$
$$\begin{array}{r} 687 \\ -444 \\ \hline 243 \end{array}$$

102 Lesson 111

Connecting Math Concepts

Lesson 112 — Side 1   Name _____

Part 1

a. 12:30   b. 6:00   c. 4:35   d. 7:05

Part 2

H = 1
P = 0
R = 0
T = 6

Part 3

H = 0
P = 2
R = 5
T = 0

Part 4

a. $10-4=6$
b. $14-7=7$
c. $7+3=10$
d. $14-10=4$
e. $5+6=11$
f. $8-6=2$
g. $12-6=6$
h. $8+3=11$
i. $8+8=16$

Part 5

$278 \quad 109 \rightarrow 387$
$178 \quad 209 \rightarrow 387$

a. $387-278=109$
b. $109+278=387$
c. $387-209=178$
d. $387-178=209$

Part 6

a. $80+3=83$
c. $40+5=45$
b. $20+7=27$
d. $10+6=16$

Connecting Math Concepts

Lesson 112 **103**

Lesson 112 — Side 2   Name _____

Side 2 work is independent.

Part 7

a. $84 \quad 26 \rightarrow 110$
b. $84 \quad 103 \rightarrow 187$
c. $233 \quad 36 \rightarrow 269$

$$\begin{array}{r} 1 \\ 84 \\ +\ 26 \\ \hline 110 \end{array}$$

$$\begin{array}{r} 187 \\ -\ 84 \\ \hline 103 \end{array}$$

$$\begin{array}{r} 269 \\ -233 \\ \hline 36 \end{array}$$

Part 8

a. $194+7=201$
b. $188+6=194$
c. $58+4=62$

Part 9

1 [blank]
2 H R R H
3 [blank] T T T H

Part 10

a. $76+85$
$$\begin{array}{r} 76 \\ +85 \\ \hline 161 \end{array}$$

b. $897-423$
$$\begin{array}{r} 897 \\ -423 \\ \hline 474 \end{array}$$

c. $95+5+90$
$$\begin{array}{r} 95 \\ 5 \\ +\ 90 \\ \hline 190 \end{array}$$

104 Lesson 112

Connecting Math Concepts

Connecting Math Concepts

Answer Key **369**

Lesson 113 — Side 1   Name _____

Part 1

a. $60 + 8 = 68$ c. $30 + 9 = 39$

b. $50 + 4 = 54$ d. $80 + 0 = 80$

Part 2

a. $243 - 154 = 89$ c. $243 - 65 = 178$

b. $65 + 89 = 154$ d. $243 - 178 = 65$

Part 3

a. 1:00   b. 4:15   c. 11:05   d. 9:30

Part 4

a. $3 + 8 = 11$ e. $11 - 2 = 9$

b. $7 - 5 = 2$ f. $3 + 2 = 5$

c. $11 - 5 = 6$ g. $8 - 5 = 3$

d. $3 + 4 = 7$ h. $3 + 7 = 10$

Part 5

H = 0
P = 0
R = 6
T = 0

Part 6

H = 0
P = 0
R = 0
T = 4

Connecting Math Concepts

Lesson 113 **105**

Lesson 113 — Side 2   Name _____

Side 2 work is independent.

Part 7

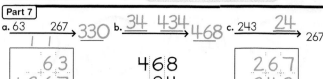

a. 63 267 → 330   b. 34 434 → 468   c. 243 24 → 267

```
  1 1
   6 3
 + 2 6 7
 ───────
   3 3 0
```

```
   4 6 8
  -  3 4
 ───────
   4 3 4
```

```
   2 6 7
 - 2 4 3
 ───────
      2 4
```

Part 8

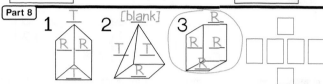

1   2 [blank]   3

Part 9

a. = $50.08

b. = $81.90

Part 10

a.
```
  $ 2 4 . 9 8
  - 1 2 . 9 3
  ───────────
  $ 1 2 . 0 5
```

b.
```
  $    . 0 8
      7 . 6 0
  +   2 . 0 8
  ───────────
  $   9 . 7 6
```

c.
```
    3 7 8
  -   4 4
  ───────
    3 3 4
```

106 Lesson 113

Connecting Math Concepts

Lesson 114 — Side 1   Name _____

Part 1

a. $40 + 7 = 47$ c. $70 + 0 = 70$

b. $10 + 8 = 18$ d. $90 + 6 = 96$

Part 2

a.
```
  $  9 . 2 6
  + 5 . 5 6
  ─────────
  $ 1 4 . 8 2
```

b.
```
  $ 1 8 . 7 9
  - 1 5 . 3 3
  ──────────
  $   3 . 4 6
```

c.
```
  $   7 . 9 6
  -   5 . 9 1
  ──────────
  $   2 . 0 5
```

Part 3

a. 7:20   b. 4:30   c. 12:00   d. 3:45

Part 4

a. $3 + 7 = 10$ c. $3 + 5 = 8$ f. $8 + 3 = 11$

b. $7 + 7 = 14$ d. $3 + 2 = 5$ g. $3 + 3 = 6$

e. $6 + 3 = 9$ h. $2 + 8 = 10$

Connecting Math Concepts

Lesson 114 **107**

Lesson 114 — Side 2   Name _____

Part 5

a. $184 - 108 = 76$ c. $260 - 184 = 76$

b. $184 + 108 = 292$ d. $184 + 108 = 292$

Part 6   INDEPENDENT WORK

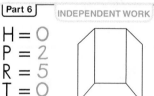

H = 0
P = 2
R = 5
T = 0

Part 7

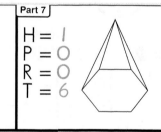

H = 1
P = 0
R = 0
T = 6

Part 8

a. $96 + 5 = 101$

b. $95 + 8 = 103$

c. $49 + 6 = 55$

Part 9

a. 444 420 → 864   b. 203 93 → 296   c. 56 64 → 120

```
    8 6 4
  - 4 4 4
  ───────
    4 2 0
```

```
    2 9 6
  -   9 3
  ───────
    2 0 3
```

```
     5 6
  +  6 4
  ──────
   1 2 0
```

108 Lesson 114

Connecting Math Concepts

Connecting Math Concepts

Lesson 115 Side 1 Name _____

Part 1

a. $10 - 4 = 6$ d. $4 + 4 = 8$ g. $8 + 8 = 16$
b. $14 - 7 = 7$ e. $10 - 7 = 3$ h. $11 - 8 = 3$
c. $7 - 4 = 3$ f. $10 - 5 = 5$ i. $3 + 3 = 6$

Part 2

a. $\underline{93} - 58 = 35$ c. $72 - \underline{58} = 14$
b. $14 + \underline{79} = 93$ d. $58 + 35 = \underline{93}$

Part 3

c. $4 + 6 = 10$ f. $6 + 2 = 8$
a. $5 + 6 = 11$ d. $7 + 2 = 9$ g. $5 + 5 = 10$
b. $7 + 7 = 14$ e. $6 + 6 = 12$ h. $8 + 8 = 16$

Part 4

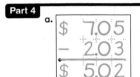

a.
$$\begin{array}{r} \$\ 7.05 \\ -\ 2.03 \\ \hline \$\ 5.02 \end{array}$$

b.
$$\begin{array}{r} \$\ 5.17 \\ +\ 5.42 \\ \hline \$10.59 \end{array}$$

c.
$$\begin{array}{r} 986 \\ -784 \\ \hline 202 \end{array}$$

Part 5 INDEPENDENT WORK

$64 + \underline{7} = 71$ $75 + \underline{8} = 83$

a.
b.

Connecting Math Concepts   Lesson 115 **109**

Lesson 115 Side 2 Name _____

Side 2 work is independent.

Part 6

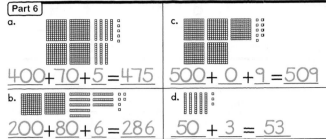

a. $\underline{400} + \underline{70} + 5 = \underline{475}$ c. $500 + \underline{0} + \underline{9} = \underline{509}$

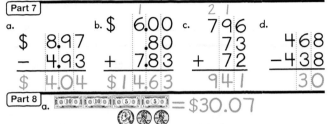

b. $\underline{200} + \underline{80} + \underline{6} = \underline{286}$ d. $50 + 3 = \underline{53}$

Part 7

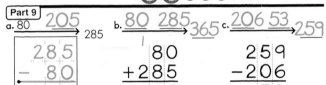

a.
$$\begin{array}{r} \$\ 8.97 \\ -\ 4.93 \\ \hline \$\ 4.04 \end{array}$$

b.
$$\begin{array}{r} \$\ 6.00 \\ +\ 7.83 \\ \hline \$14.63 \end{array}$$

c.
$$\begin{array}{r} 796 \\ +\ 72 \\ \hline 941 \end{array}$$ (with 2 1 carries)

d.
$$\begin{array}{r} 468 \\ -438 \\ \hline 30 \end{array}$$

Part 8

a. = \$30.07

b. = \$67.80

Part 9

a. $80 \xrightarrow{\underline{205}} 285$
$$\begin{array}{r} 285 \\ -\ 80 \\ \hline 205 \end{array}$$

b. $\underline{80}\ 285 \rightarrow 365$
$$\begin{array}{r} 80 \\ +285 \\ \hline 365 \end{array}$$

c. $\underline{206}\ 53 \rightarrow \underline{259}$
$$\begin{array}{r} 259 \\ -206 \\ \hline 53 \end{array}$$

110 Lesson 115   Connecting Math Concepts

Lesson 116 Side 1 Name _____

Part 1

c. $16 - 8 = 8$ f. $10 - 3 = 7$
a. $8 - 4 = 4$ d. $11 - 8 = 3$ g. $11 - 5 = 6$
b. $4 + 6 = 10$ e. $10 + 3 = 13$ h. $7 + 7 = 14$

Part 2

a. $5:30$ b. $12:35$ c. $4:00$ d. $9:20$

Part 3

a. $58 - \underline{31} = 27$ c. $\underline{85} - 58 = 27$
b. $\underline{31} + 58 = 89$ d. $58 - 31 = \underline{27}$

Part 4

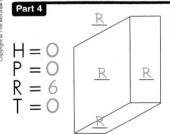

H = O
P = O
R = 6
T = O

Part 5

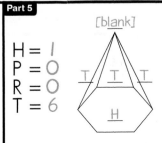

[blank]

H = I
P = O
R = O
T = 6

Connecting Math Concepts   Lesson 116 **111**

Lesson 116 Side 2 Name _____

Part 6

c. $4 + 3 = 7$ f. $6 + 10 = 16$
a. $3 + 5 = 8$ d. $3 + 7 = 10$ g. $6 + 3 = 9$
b. $6 + 4 = 10$ e. $6 + 6 = 12$ h. $2 + 3 = 5$

Part 7 INDEPENDENT WORK

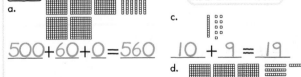

a. $500 + \underline{60} + \underline{0} = 560$ c. $10 + \underline{9} = \underline{19}$

b. $80 + \underline{4} = \underline{84}$ d. $\underline{600} + \underline{70} + 3 = \underline{673}$

Part 8

a. $89 \xrightarrow{12} 101$
$$\begin{array}{r} 89 \\ +\ 12 \\ \hline 101 \end{array}$$

b. $\underline{33}\ 43 \rightarrow \underline{76}$
$$\begin{array}{r} 76 \\ -\ 33 \\ \hline 43 \end{array}$$

c. $34\ \underline{64} \rightarrow 98$
$$\begin{array}{r} 98 \\ -\ 34 \\ \hline 64 \end{array}$$

Part 9

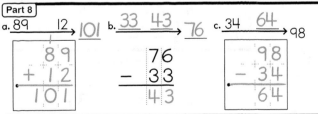

Triangle
Rectangle
Pentagon
Hexagon

Part 10

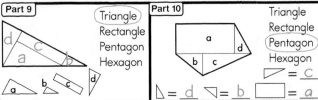

Triangle
Rectangle
(Pentagon)
Hexagon

$\square = \underline{c}$
$\triangle = \underline{d}$ $\square = \underline{b}$ $\square = \underline{a}$

112 Lesson 116   Connecting Math Concepts

Connecting Math Concepts   Answer Key **371**

Part 1
a. 9 − 6 = 3
b. 12 − 6 = 6
c. 8 + 8 = 16
d. 10 − 7 = 3
e. 11 − 6 = 5
f. 4 + 4 = 8
g. 10 − 6 = 4
h. 5 + 5 = 10

Part 2
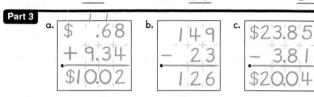
a. 79 − [blank] = 18   c. [blank] − 63 = 18   e. [blank] − 63 = 32
18 [blank] → 79   63 18 → [blank]   63 32 → [blank]

b. [blank] + 31 = 81   d. 58 − [blank] = 17   f. 18 + [blank] = 89
31 [blank] → 81   17 [blank] → 58   18 [blank] → 89

Part 3
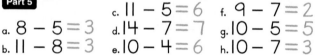
a. $ .68 + 9.34 = $10.02
b. 149 − 2.3 = 12.6
c. $23.85 − 3.81 = $20.04

Part 4
a. 9
b. 7
c. 6

Part 5
a. 8 − 5 = 3
b. 11 − 8 = 3
c. 11 − 5 = 6
d. 14 − 7 = 7
e. 10 − 4 = 6
f. 9 − 7 = 2
g. 10 − 5 = 5
h. 10 − 7 = 3

Connecting Math Concepts   Lesson 117 **113**

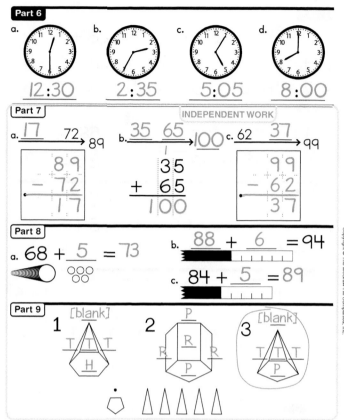

Part 6
a. 12:30
b. 2:35
c. 5:05
d. 8:00

Part 7   INDEPENDENT WORK
a. 17 72 → 89
89 − 72 = 17
b. 35 65 → 100
35 + 65 = 100
c. 62 37 → 99
99 − 62 = 37

Part 8
a. 68 + 5 = 73
b. 88 + 6 = 94
c. 84 + 5 = 89

Part 9
1 [blank]   2   3 [blank]

114 Lesson 117   Connecting Math Concepts

Part 1
a. 6
b. 4
c. 7

Part 2
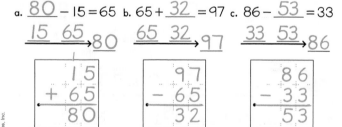
a. 80 − 15 = 65   b. 65 + 32 = 97   c. 86 − 53 = 33
15 65 → 80   65 32 → 97   33 53 → 86
15 + 65 = 80   97 − 65 = 32   86 − 33 = 53

Part 3
a. 11 − 8 = 3
b. 9 − 6 = 3
c. 11 − 6 = 5
d. 14 − 7 = 7
e. 7 − 4 = 3
f. 10 − 7 = 3
g. 10 − 4 = 6
h. 8 − 4 = 4

Part 4
a. 1:40
b. 6:00
c. 8:30
d. 4:15

Connecting Math Concepts   Lesson 118 **115**

Side 2 work is independent.

Part 5 [blank]   H = 0 P = 1 R = 0 T = 5
Part 6   H = 0 P = 0 R = 3 T = 2

Part 7
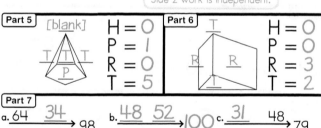
a. 64 34 → 98
98 − 64 = 34
b. 48 52 → 100
48 + 52 = 100
c. 31 48 → 79
79 − 48 = 31

Part 8

a. = $60.60

b. = $67.57

Part 9
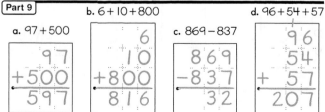
a. 97 + 500 = 597   b. 6 + 10 + 800 = 816   c. 869 − 837 = 32   d. 96 + 54 + 57 = 207

116 Lesson 118   Connecting Math Concepts

Part 1
a. $10 \neq 6 + 5$ c. $9 - 1 = 7 + 1$ e. $5 \neq 8 - 5$
b. $19 \neq 10 - 9$ d. $7 - 2 \neq 5 + 2$ f. $13 + 1 = 15 - 1$

(above a: 11; above c: 8; above d: 5, 7; above e: 8; above f: 3, 14, 14; above b: 1)

Part 2
a. $14 - \underline{} = 9$ b. $\underline{} + 11 = 17$

Part 3
a. $3 + 8 = 11$ c. $4 + 3 = 7$ f. $8 - 4 = 4$
b. $6 - 3 = 3$ d. $10 - 7 = 3$ g. $12 - 6 = 6$
e. $8 + 8 = 16$ h. $9 - 6 = 3$

Part 4
a. $\underline{41} + 27 = 68$ c. $54 + \underline{34} = 88$ e. $88 - \underline{65} = 23$

$\underset{}{27 \;\; 41} \rightarrow 68$ $\underset{}{54 \;\; 34} \rightarrow 88$ $\underset{}{23 \;\; 65} \rightarrow 88$

$\begin{array}{r} 68 \\ -\;27 \\ \hline 41 \end{array}$ $\begin{array}{r} 88 \\ -\;54 \\ \hline 34 \end{array}$ $\begin{array}{r} 88 \\ -\;23 \\ \hline 65 \end{array}$

b. $\underline{90} - 23 = 67$ d. $68 - 23 = \underline{45}$ f. $\underline{50} - 23 = 27$

$\underset{}{23 \;\; 67} \rightarrow 90$ $\underset{}{23 \;\; 45} \rightarrow 68$ $\underset{}{23 \;\; 27} \rightarrow 50$

$\begin{array}{r} 23 \\ +\;67 \\ \hline 90 \end{array}$ $\begin{array}{r} 68 \\ -\;23 \\ \hline 45 \end{array}$ $\begin{array}{r} 23 \\ +\;27 \\ \hline 50 \end{array}$

Connecting Math Concepts   Lesson 119 **117**

Part 5

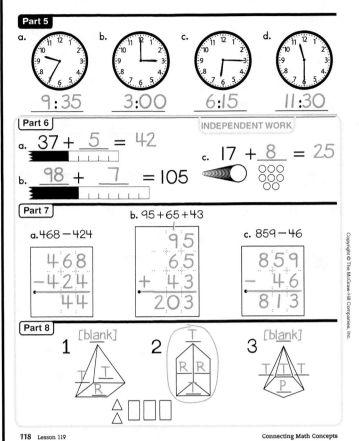

a. 9:35   b. 3:00   c. 6:15   d. 11:30

Part 6   INDEPENDENT WORK

a. $37 + \underline{5} = 42$ c. $17 + \underline{8} = 25$
b. $98 + \underline{7} = 105$

Part 7

a. $468 - 424$ b. $95 + 65 + 43$ c. $859 - 46$

$\begin{array}{r} 468 \\ -424 \\ \hline 44 \end{array}$ $\begin{array}{r} 95 \\ 65 \\ +\;43 \\ \hline 203 \end{array}$ $\begin{array}{r} 859 \\ -\;46 \\ \hline 813 \end{array}$

Part 8

1 [blank]   2   3 [blank]

118 Lesson 119   Connecting Math Concepts

Part 1

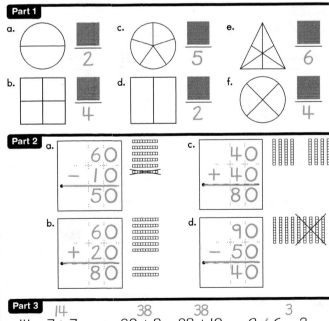

a. 2   c. 5   e. 6
b. 4   d. 2   f. 4

Part 2

a. $\begin{array}{r} 60 \\ -\;10 \\ \hline 50 \end{array}$ c. $\begin{array}{r} 40 \\ +\;40 \\ \hline 80 \end{array}$

b. $\begin{array}{r} 60 \\ +\;20 \\ \hline 80 \end{array}$ d. $\begin{array}{r} 90 \\ -\;50 \\ \hline 40 \end{array}$

Part 3
a. $14 = 7 + 7$ c. $30 + 8 = 28 + 10$ e. $9 \neq 6 - 3$
b. $10 - 1 \neq 9 + 1$ d. $14 - 4 \neq 24 - 10$ f. $15 + 1 \neq 18 - 1$

(above a: 14; above c: 38, 38; above e: 3; above b: 9, 10; above d: 10, 14; above f: 16, 17)

Part 4
a. $9 + \underline{} = 11$ c. $\underline{} + 6 = 10$
b. $\underline{} - 5 = 15$

Connecting Math Concepts   Lesson 120 **119**

Part 5
a. $11 - 3 = 8$ c. $6 - 3 = 3$ f. $7 + 7 = 14$
b. $6 + 3 = 9$ d. $8 - 3 = 5$ g. $10 - 7 = 3$
e. $8 + 3 = 11$ h. $10 - 5 = 5$

Part 6
a. $14 + \underline{52} = 66$ c. $66 - \underline{23} = 43$ e. $79 - \underline{36} = 43$

$\underset{}{14 \;\; 52} \rightarrow 66$ $\underset{}{43 \;\; 23} \rightarrow 66$ $\underset{}{43 \;\; 36} \rightarrow 79$

$\begin{array}{r} 66 \\ -\;14 \\ \hline 52 \end{array}$ $\begin{array}{r} 66 \\ -\;43 \\ \hline 23 \end{array}$ $\begin{array}{r} 79 \\ -\;43 \\ \hline 36 \end{array}$

b. $\underline{80} - 14 = 66$ d. $66 + \underline{13} = 79$ f. $\underline{57} - 14 = 43$

$\underset{}{14 \;\; 66} \rightarrow 80$ $\underset{}{66 \;\; 13} \rightarrow 79$ $\underset{}{14 \;\; 43} \rightarrow 57$

$\begin{array}{r} 14 \\ +\;66 \\ \hline 80 \end{array}$ $\begin{array}{r} 79 \\ -\;66 \\ \hline 13 \end{array}$ $\begin{array}{r} 14 \\ +\;43 \\ \hline 57 \end{array}$

Part 7   INDEPENDENT WORK

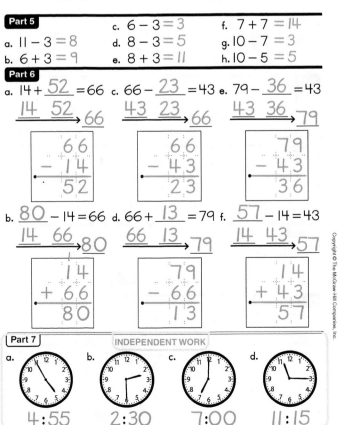

a. 4:55   b. 2:30   c. 7:00   d. 11:15

120 Lesson 120   Connecting Math Concepts

Connecting Math Concepts   Answer Key **373**

Lesson 121 Side 1 Name _____

Part 1

c. $8 - 3 = 5$ f. $7 - 4 = 3$

a. $11 - 3 = 8$ d. $10 - 3 = 7$ g. $7 + 7 = 14$

b. $8 + 8 = 16$ e. $6 + 6 = 12$ h. $10 - 6 = 4$

Part 2

a. $14 + \underline{} = 19$ c. $\underline{} + 8 = 16$

b. $\underline{} - 6 = 11$

Part 3

a. $10 + 6 \overset{16}{\neq} \overset{0}{8 - 8}$ c. $25 + 10 \overset{35}{\neq} \overset{45}{40 + 5}$ e. $9 + 10 \overset{19}{=} \overset{19}{20 - 1}$

b. $9 - 1 \overset{8}{=} 7 + 1$ d. $26 \neq \overset{30}{36 - 6}$ f. $15 + 2 \overset{17}{\neq} \overset{16}{15 + 1}$

Part 4

a.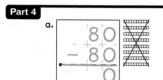
$$\begin{array}{r} 80 \\ -\ 80 \\ \hline 0 \end{array}$$

c.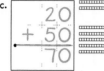
$$\begin{array}{r} 20 \\ +\ 50 \\ \hline 70 \end{array}$$

b.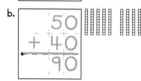
$$\begin{array}{r} 50 \\ +\ 40 \\ \hline 90 \end{array}$$

d.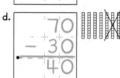
$$\begin{array}{r} 70 \\ -\ 30 \\ \hline 40 \end{array}$$

Connecting Math Concepts

Lesson 121 **121**

Lesson 121 Side 2 Name _____

Part 5

a. $46 - \underline{0} = 46$ c. $46 + \underline{51} = 97$ e. $97 - 24 = \underline{73}$

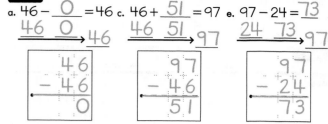

$$\begin{array}{r} 46 \\ -\ 46 \\ \hline 0 \end{array} \qquad \begin{array}{r} 97 \\ -\ 46 \\ \hline 51 \end{array} \qquad \begin{array}{r} 97 \\ -\ 24 \\ \hline 73 \end{array}$$

b. $\underline{92} - 46 = 46$ d. $46 + 24 = \underline{70}$ f. $\underline{121} - 24 = 97$

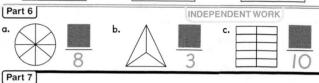

$$\begin{array}{r} 46 \\ +\ 46 \\ \hline 92 \end{array} \qquad \begin{array}{r} 46 \\ +\ 24 \\ \hline 70 \end{array} \qquad \begin{array}{r} 24 \\ +\ 97 \\ \hline 121 \end{array}$$

Part 6 INDEPENDENT WORK

a. 8

b. 3

c. 10

Part 7

a. 6:00

b. 3:20

c. 9:30

d. 11:05

122 Lesson 121

Connecting Math Concepts

Lesson 122 Side 1 Name _____

Part 1

a. $18 + 7 \overset{19}{<} (10 + 9) + 7$ c. $70 + 10 > 70 + \overset{9}{(8 + 1)}$

b. $\overset{10}{(5 + 5)} + 28 = 10 + 28$ d. $15 + \overset{48}{(40 + 8)} < 15 + 49$

Part 2

a. $\dfrac{3}{4}$

b. $\dfrac{2}{2}$

c. $\dfrac{1}{6}$

Part 3

c. $8 + 3 = 11$ f. $14 - 7 = 7$

a. $6 + 3 = 9$ d. $8 - 3 = 5$ g. $12 - 6 = 6$

b. $6 - 3 = 3$ e. $5 + 5 = 10$ h. $10 - 7 = 3$

Part 4

a. $11 - \underline{} = 6$

$\underline{6} \xrightarrow{\text{[blank]}} 11$

b. $\underline{} - 8 = 11$

$\underline{8} \xrightarrow{11} \text{[blank]}$

Part 5

a. $64 + \underline{34} = 98$ b. $\underline{88} - 24 = 64$ c. $64 - 24 = \underline{40}$

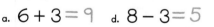

$$\begin{array}{r} 98 \\ -\ 64 \\ \hline 34 \end{array} \qquad \begin{array}{r} 24 \\ +\ 64 \\ \hline 88 \end{array} \qquad \begin{array}{r} 64 \\ -\ 24 \\ \hline 40 \end{array}$$

Connecting Math Concepts

Lesson 122 **123**

Lesson 122 Side 2 Name _____

Part 5

d. $\underline{72} + 26 = 98$ e. $\underline{111} - 23 = 88$ f. $68 + 28 = \underline{96}$

$$\begin{array}{r} 98 \\ -\ 26 \\ \hline 72 \end{array} \qquad \begin{array}{r} 23 \\ +\ 88 \\ \hline 111 \end{array} \qquad \begin{array}{r} 68 \\ +\ 28 \\ \hline 96 \end{array}$$

Part 6 INDEPENDENT WORK

a.
$$\begin{array}{r} 60 \\ +\ 30 \\ \hline 90 \end{array}$$

c.
$$\begin{array}{r} 10 \\ +\ 80 \\ \hline 90 \end{array}$$

b.
$$\begin{array}{r} 80 \\ -\ 10 \\ \hline 70 \end{array}$$

d.
$$\begin{array}{r} 60 \\ -\ 60 \\ \hline 0 \end{array}$$

Part 7

a. 8:25

b. 12:40

c. 7:30

d. 1:00

Part 8

a. $\underline{93} + \underline{5} = 98$ b. $\underline{96} + \underline{8} = 104$

124 Lesson 122

Connecting Math Concepts

---

374 Answer Key

Connecting Math Concepts

Lesson 123 — Side 1 Name _____

Part 1
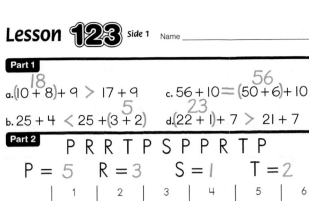

a. $(10+8)+9 > 17+9$ [18]
c. $56+10 = (50+6)+10$ [56]
b. $25+4 < 25+(3+2)$ [5]
d. $(22+1)+7 > 21+7$ [23]

Part 2
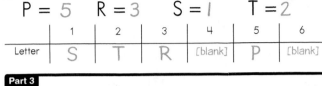

P R R T P S P P R T P

$P = 5$ $R = 3$ $S = 1$ $T = 2$

	1	2	3	4	5	6
Letter	S	T	R	[blank]	P	[blank]

Part 3
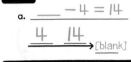 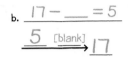

a. $\_\_ - 4 = 14$ [18]
   $\xrightarrow{4 \quad 14}$ [blank]
b. $17 - \_\_ = 5$ [12]
   $\xrightarrow{5 \quad \text{[blank]}} 17$

Part 4

a. triangle $\dfrac{2}{2}$
c. square $\dfrac{0}{2}$
e. octagon $\dfrac{5}{8}$
b. circle $\dfrac{1}{4}$
d. circle $\dfrac{4}{6}$
f. square $\dfrac{4}{4}$

Connecting Math Concepts

Lesson 123 — Side 2 Name _____

Part 5
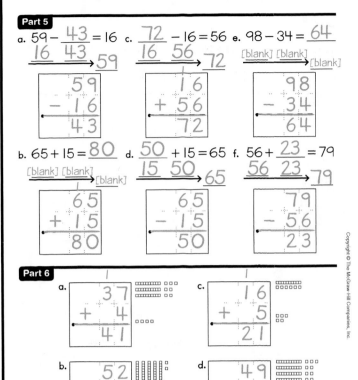

a. $59 - 43 = 16$
$\xrightarrow{16 \quad 43} 59$
box: $59 - 16 = 43$

c. $72 - 16 = 56$
$\xrightarrow{16 \quad 56} 72$
box: $16 + 56 = 72$

e. $98 - 34 = 64$
$\xrightarrow{\text{[blank] [blank]}}$ [blank]
box: $98 - 34 = 64$

b. $65 + 15 = 80$
$\xrightarrow{\text{[blank] [blank]}}$ [blank]
box: $65 + 15 = 80$

d. $50 + 15 = 65$
$\xrightarrow{15 \quad 50}$ [blank]
box: $65 - 15 = 50$

f. $56 + 23 = 79$
$\xrightarrow{56 \quad 23} 79$
box: $79 - 56 = 23$

Part 6
a. $37 + 4 = 41$
c. $16 + 5 = 21$
b. $52 + 20 = 72$
d. $49 + 40 = 89$

Connecting Math Concepts

Lesson 124 — Side 1 Name _____

Part 1
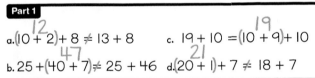

a. $(10+2)+8 \neq 13+8$ [12]
c. $19+10 = (10+9)+10$ [19]
b. $25+(40+7) \neq 25+46$ [47]
d. $(20+1)+7 \neq 18+7$ [21]

Part 2

a. circle $\dfrac{2}{5}$
b. square $\dfrac{1}{2}$
c. square $\dfrac{2}{4}$

Part 3
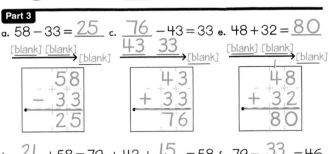 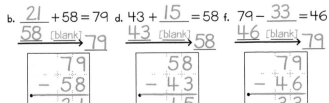

a. $58 - 33 = 25$
$\xrightarrow{\text{[blank] [blank]}}$ [blank]
box: $58 - 33 = 25$

c. $76 - 43 = 33$
$\xrightarrow{43 \quad 33}$ [blank]
box: $43 + 33 = 76$

e. $48 + 32 = 80$
$\xrightarrow{\text{[blank] [blank]}}$ [blank]
box: $48 + 32 = 80$

b. $21 + 58 = 79$
$\xrightarrow{58 \quad \text{[blank]}} 79$
box: $79 - 58 = 21$

d. $43 + 15 = 58$
$\xrightarrow{43 \quad \text{[blank]}} 58$
box: $58 - 43 = 15$

f. $79 - 33 = 46$
$\xrightarrow{46 \quad \text{[blank]}} 79$
box: $79 - 46 = 33$

Connecting Math Concepts

Lesson 124 — Side 2 Name _____

Part 4
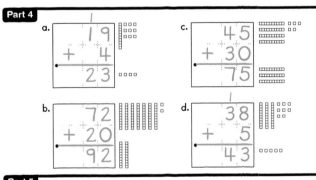

a. $19 + 4 = 23$
c. $45 + 30 = 75$
b. $72 + 20 = 92$
d. $38 + 5 = 43$

Part 5
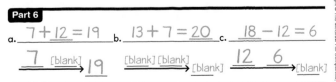

a. $(10+2)+8 < 13+8$ [12]
c. $19+10 = (10+9)+10$ [19]
b. $25+(40+7) > 25+46$ [47]
d. $(20+1)+7 > 18+7$ [21]

Part 6
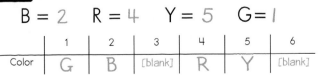

a. $7 + 12 = 19$
$\xrightarrow{7 \quad \text{[blank]}} 19$
b. $13 + 7 = 20$
$\xrightarrow{\text{[blank] [blank]}}$ [blank]
c. $18 - 12 = 6$
$\xrightarrow{12 \quad 6}$ [blank]

Part 7
$B = 2$ $R = 4$ $Y = 5$ $G = 1$

	1	2	3	4	5	6
Color	G	B	[blank]	R	Y	[blank]

Connecting Math Concepts

Lesson 125 Side 1   Name _____

Part 1

a. $80 + 3 = \overset{80}{(70 + 10)} + 3$

$\overset{17}{b. (16 + 1)} + 5 \neq 18 + 5$

c. $8 + \overset{35}{(30 + 5)} \neq 8 + 34$

d. $\overset{16}{(13 + 3)} + 6 = 16 + 6$

Part 2

a. $\dfrac{2}{2}$

b. $\dfrac{5}{6}$

c. $\dfrac{3}{4}$

Part 3

a. $63 + \underline{34} = 97$

$\underline{63}$ [blank] → 97

| 9 7 |
| − 6 3 |
| 3 4 |

b. $38 + 52 = \underline{90}$

[blank] [blank] → [blank]

| 3 8 |
| + 5 2 |
| 9 0 |

c. $\underline{99} - 42 = 57$

$\underline{42}$ $\underline{57}$ → [blank]

| 4 2 |
| + 5 7 |
| 9 9 |

d. $\underline{15} + 42 = 57$

$\underline{42}$ [blank] → 57

| 5 7 |
| − 4 2 |
| 1 5 |

e. $97 - 34 = \underline{63}$

[blank] [blank] → [blank]

| 9 7 |
| − 3 4 |
| 6 3 |

f. $97 - \underline{84} = 13$

$\underline{13}$ [blank] → 97

| 9 7 |
| − 1 3 |
| 8 4 |

Copyright © The McGraw-Hill Companies, Inc.

Connecting Math Concepts

Lesson 125 **129**

Lesson 125 Side 2   Name _____

Part 4

C = 5   T = 2   B = 1   M = 4

Vehicles	1	2	3	4	5	6
	B	T	[blank]	M	C	[blank]

Part 5

a. $80 + 3 = \overset{80}{(70 + 10)} + 3$

$\overset{17}{b. (16 + 1)} + 5 < 18 + 5$

c. $8 + \overset{35}{(30 + 5)} > 8 + 34$

d. $\overset{16}{(13 + 3)} + 6 = 16 + 6$

Part 6

a. $\underline{8 + 8} = 16$

$\underline{8}$ [blank] → 16

b. $\underline{17} - 3 = 14$

$\underline{3}$ $\underline{14}$ → [blank]

c. $\underline{19 - 10} = 9$

[blank] [blank] → [blank]

Part 7 INDEPENDENT WORK

a. 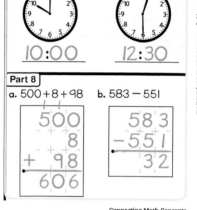 10:00

b. 12:30

Part 8

a. $500 + 8 + 98$

| 5 0 0 |
| 8 |
| + 9 8 |
| 6 0 6 |

b. $583 - 551$

| 5 8 3 |
| − 5 5 1 |
| 3 2 |

Copyright © The McGraw-Hill Companies, Inc.

130 Lesson 125

Connecting Math Concepts

376   **Answer Key**

Connecting Math Concepts

Mastery Test 9

Passing Criteria Table — Mastery Test 9			
Part	Score	Possible Score	Passing Score
1	4 for each problem	12	8
2	2 for each dollar amount	10	8
3	3 for each problem	12	8
4	2 for each problem	16	14
5	4 for each problem	20	16
6	3 for each problem	12	9
7	2 for each equation	18	16
	Total	100	

Mastery Test 9 Name _____

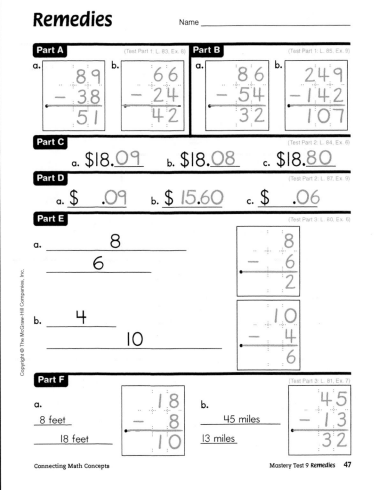

Part 1

a.
$$\begin{array}{r} 89 \\ -38 \\ \hline 51 \end{array}$$
b.
$$\begin{array}{r} 249 \\ -142 \\ \hline 107 \end{array}$$
c.
$$\begin{array}{r} 86 \\ -54 \\ \hline 32 \end{array}$$

Part 2

a. $4.02 b. $15.60 c. $18.08 d. $ .06 e. $ .09

Part 3

a. 4 feet / 10 feet
$$\begin{array}{r} 10 \\ -4 \\ \hline 6 \end{array}$$
c. 27 feet / 16 feet
$$\begin{array}{r} 27 \\ -16 \\ \hline 11 \end{array}$$

b. 18 feet / 8 feet
$$\begin{array}{r} 18 \\ -8 \\ \hline 10 \end{array}$$
d. 13 feet / 45 feet
$$\begin{array}{r} 45 \\ -13 \\ \hline 32 \end{array}$$

Part 4

a. 7 > 5   e. 3 > 2
b. 8 < 9   f. 19 < 20
c. 6 > 3   g. 57 > 48
d. 4 = 4   h. 16 = 16

Connecting Math Concepts   Mastery Test 9 **45**

Mastery Test 9 Name _____

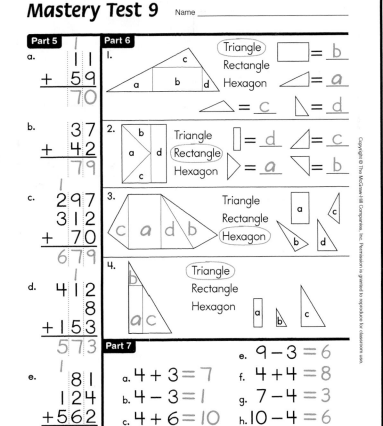

Part 5

a.
$$\begin{array}{r} 11 \\ +59 \\ \hline 70 \end{array}$$

b.
$$\begin{array}{r} 37 \\ +42 \\ \hline 79 \end{array}$$

c.
$$\begin{array}{r} 297 \\ 312 \\ +70 \\ \hline 679 \end{array}$$

d.
$$\begin{array}{r} 412 \\ 8 \\ +153 \\ \hline 573 \end{array}$$

e.
$$\begin{array}{r} 81 \\ 124 \\ +562 \\ \hline 767 \end{array}$$

Part 6

1. Triangle ▭ = b / Rectangle = a / Hexagon = c / = d

2. Triangle / (Rectangle) / Hexagon — = d, = c, = a, = b

3. Triangle / Rectangle / (Hexagon)

4. (Triangle) / Rectangle / Hexagon

Part 7

a. 4 + 3 = 7   e. 9 − 3 = 6
b. 4 − 3 = 1   f. 4 + 4 = 8
c. 4 + 6 = 10   g. 7 − 4 = 3
d. 8 − 4 = 4   h. 10 − 4 = 6
                  i. 7 − 3 = 4

46 Mastery Test 9   Connecting Math Concepts

Remedies Name _____

Part A (Test Part 1: L. 83, Ex. 8)

a.
$$\begin{array}{r} 89 \\ -38 \\ \hline 51 \end{array}$$
b.
$$\begin{array}{r} 66 \\ -24 \\ \hline 42 \end{array}$$

Part B (Test Part 1: L. 85, Ex. 9)

a.
$$\begin{array}{r} 86 \\ -54 \\ \hline 32 \end{array}$$
b.
$$\begin{array}{r} 249 \\ -142 \\ \hline 107 \end{array}$$

Part C (Test Part 2: L. 84, Ex. 6)

a. $18.09   b. $18.08   c. $18.80

Part D (Test Part 2: L. 87, Ex. 9)

a. $ .09   b. $ 15.60   c. $ .06

Part E (Test Part 3: L. 80, Ex. 6)

a. 8 / 6
$$\begin{array}{r} 8 \\ -6 \\ \hline 2 \end{array}$$

b. 4 / 10
$$\begin{array}{r} 10 \\ -4 \\ \hline 6 \end{array}$$

Part F (Test Part 3: L. 81, Ex. 7)

a. 8 feet / 18 feet
$$\begin{array}{r} 18 \\ -8 \\ \hline 10 \end{array}$$

b. 45 miles / 13 miles
$$\begin{array}{r} 45 \\ -13 \\ \hline 32 \end{array}$$

Connecting Math Concepts   Mastery Test 9 *Remedies* **47**

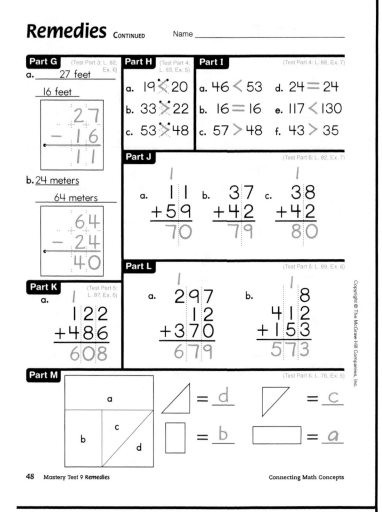

Remedies CONTINUED   Name _____

Part G (Test Part 3: L. 82, Ex. 6)

a.   27 feet

16 feet

$$\begin{array}{r} 2\,7 \\ -\,1\,6 \\ \hline 1\,1 \end{array}$$

b. 24 meters

64 meters

$$\begin{array}{r} 6\,4 \\ -\,2\,4 \\ \hline 4\,0 \end{array}$$

Part H (Test Part 4: L. 83, Ex. 5)

a. 19 < 20

b. 33 > 22

c. 53 > 48

Part I (Test Part 4: L. 88, Ex. 7)

a. 46 < 53   d. 24 = 24

b. 16 = 16   e. 117 < 130

c. 57 > 48   f. 43 > 35

Part J (Test Part 5: L. 82, Ex. 7)

a. $\begin{array}{r} 1 \\ 1\,1 \\ +5\,9 \\ \hline 7\,0 \end{array}$ b. $\begin{array}{r} 3\,7 \\ +4\,2 \\ \hline 7\,9 \end{array}$ c. $\begin{array}{r} 1 \\ 3\,8 \\ +4\,2 \\ \hline 8\,0 \end{array}$

Part K (Test Part 5: L. 87, Ex. 6)

a. $\begin{array}{r} 1 \\ 1\,2\,2 \\ +4\,8\,6 \\ \hline 6\,0\,8 \end{array}$

Part L (Test Part 5: L. 89, Ex. 6)

a. $\begin{array}{r} 1 \\ 2\,9\,7 \\ 1\,2 \\ +3\,7\,0 \\ \hline 6\,7\,9 \end{array}$ b. $\begin{array}{r} 1 \\ 8 \\ 4\,1\,2 \\ +1\,5\,3 \\ \hline 5\,7\,3 \end{array}$

Part M (Test Part 6: L. 76, Ex. 5)

◺ = __d__   ◹ = __c__

▯ = __b__   ▭ = __a__

Copyright © The McGraw-Hill Companies, Inc.

48   Mastery Test 9 *Remedies*   Connecting Math Concepts

Remedies CONTINUED   Name _____

Part N (Test Part 6: L. 78, Ex. 5)

1.   Hexagon
      (Rectangle)
      Triangle

▯ = __d__
▷ = __a__
◺ = __c__
◹ = __b__

2.   Hexagon
      Rectangle
      (Triangle)

▭ = __b__
◸ = __a__
△ = __c__
◺ = __d__

Part O (Test Part 6: L. 87, Ex. 5)

Part P (Test Part 7: L. 88, Ex. 6)

a. 8 − 4 = 4   e. 8 − 6 = 2   i. 11 − 9 = 2

b. 3 − 2 = 1   f. 11 − 6 = 5   j. 7 − 3 = 4

c. 9 − 6 = 3   g. 7 − 4 = 3   k. 8 − 3 = 5

d. 6 − 2 = 4   h. 12 − 6 = 6   l. 7 − 5 = 2

Part Q (Test Part 7: L. 88, Ex. 8)

a. 5 + 2 = 7   e. 2 + 7 = 9   i. 4 + 2 = 6

b. 3 + 6 = 9   f. 3 + 4 = 7   j. 6 + 6 = 12

c. 4 + 4 = 8   g. 8 + 10 = 18   k. 9 + 1 = 10

d. 5 + 3 = 8   h. 4 + 6 = 10   l. 5 + 6 = 11

Copyright © The McGraw-Hill Companies, Inc.

Connecting Math Concepts   Mastery Test 9 *Remedies*   49

Remedy Table — Mastery Test 9				
Part	Test Items	Remedy Lesson	Exercise	Student Material Remedies Worksheet
1	Word Problems (Comparison)	83	8	Part A
		85	9	Part B
		88	2	—
		90	3	—
2	Writing Dollar Amounts	80	2	—
		81	2	—
		84	6	Part C
		87	9	Part D
3	Comparison (Length)	79	2	—
		80	6	Part E
		81	7	Part F
		82	6	Part G
4	Greater Than, Less Than, Equal To	81	4	—
		83	5	Part H
		88	7	Part I
5	Column Addition (Carrying)	82	7	Part J
		86	3	—
		87	6	Part K
		89	6	Part L
6	Shapes (Composition/ Decomposition)	76	5	Part M
		78	5	Part N
		87	5	Part O
7	Facts (Mix)	80	1	—
		82	1	—
		88	1	—
		88	6	Part P
		88	8	Part Q

Mastery Test 10

Passing Criteria Table — Mastery Test 10			
Part	Score	Possible Score	Passing Score
1	2 for each time	16	14
2	4 for each problem	24	16
3	1 for bottom and side faces 2 for each top face	16	14
4	3 for each problem	12	9
5	2 for each problem	20	18
6	3 for each column problem	12	9
	Total	100	

Mastery Test 10 Name _____

Part 1

a. 12:48 b. 2:15 c. 6:00 d. 11:03

e. 1:30 f. 10:00 g. 9:02 h. 7:30

Part 2

a. 24 → 11 → 35

$$\begin{array}{r} 35 \\ -24 \\ \hline 11 \end{array}$$

c. 21 → 72 → 93

$$\begin{array}{r} 93 \\ -72 \\ \hline 21 \end{array}$$

e. 133 → 115 → 248

$$\begin{array}{r} 248 \\ -115 \\ \hline 133 \end{array}$$

b. 16 423 → 439

$$\begin{array}{r} 16 \\ +423 \\ \hline 439 \end{array}$$

d. 28 → 51 → 79

$$\begin{array}{r} 79 \\ -28 \\ \hline 51 \end{array}$$

f. 104 90 → 194

$$\begin{array}{r} 104 \\ +90 \\ \hline 194 \end{array}$$

Part 3

1 [figure] 2 [blank] 3 [figure]

Mastery Test 10 Name _____

Part 4

a. $61 + 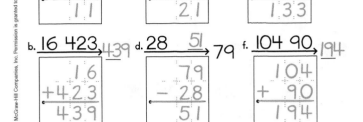 = $61.17

b. $33 + 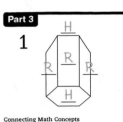 = $33.75

c. = $15.70

d. 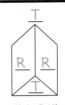 = $21.06

Part 5

a. $6 - 3 = 3$

b. $6 + 3 = 9$

c. $10 + 5 = 15$

d. $10 - 5 = 5$

e. $5 - 5 = 0$

f. $5 + 5 = 10$

g. $10 - 4 = 6$

h. $8 - 4 = 4$

i. $3 + 3 = 6$

j. $8 - 3 = 5$

Part 6

a.
$$\begin{array}{r} 1 \\ 742 \\ +168 \\ \hline 910 \end{array}$$

b.
$$\begin{array}{r} 1 \\ 31 \\ 349 \\ +\ \ 4 \\ \hline 384 \end{array}$$

c.
$$\begin{array}{r} 261 \\ 6 \\ +\ 62 \\ \hline 329 \end{array}$$

d.
$$\begin{array}{r} 1\ 1 \\ 358 \\ +442 \\ \hline 800 \end{array}$$

Remedies Name _____

Part A (Test Part 1: L. 94, Ex. 8)

a. 11:00
b. 2:15
c. 1:00
d. 12:30

Part B (Test Part 1: L. 96, Ex. 7)

a. 6:00
b. 5:08
c. 12:04
d. 5:00
e. 10:01
f. 11:30

Part C (Test Part 1: L. 97, Ex. 8)

a. 4:09
b. 12:59
c. 10:00
d. 11:06
e. 3:30
f. 2:00

Part D (Test Part 2: L. 93, Ex. 6)

a. 51 36 → 87

$$\begin{array}{r} 51 \\ +36 \\ \hline 87 \end{array}$$

b. 17 → 12 → 29

$$\begin{array}{r} 29 \\ -17 \\ \hline 12 \end{array}$$

Part E (Test Part 2: L. 96, Ex. 5)

a. 32 12 → 44

$$\begin{array}{r} 32 \\ +12 \\ \hline 44 \end{array}$$

b. 24 → 11 → 35

$$\begin{array}{r} 35 \\ -24 \\ \hline 11 \end{array}$$

c. 21 → 72 → 93

$$\begin{array}{r} 93 \\ -72 \\ \hline 21 \end{array}$$

d. 50 27 → 77

$$\begin{array}{r} 50 \\ +27 \\ \hline 77 \end{array}$$

Remedies CONTINUED   Name _____

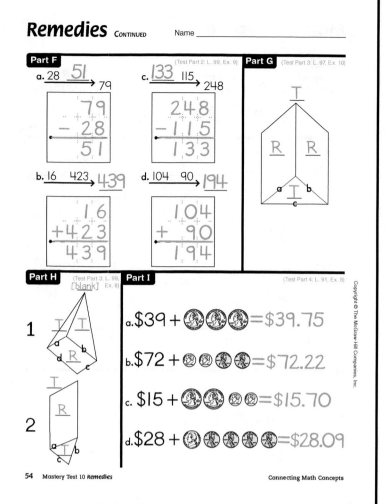

Part F (Test Part 2: L. 99, Ex. 9)

a. 28 → 51 → 79

$$\begin{array}{r} 79 \\ -\ 28 \\ \hline 51 \end{array}$$

c. 133 → 115 → 248

$$\begin{array}{r} 248 \\ -115 \\ \hline 133 \end{array}$$

b. 16 → 423 → 439

$$\begin{array}{r} 16 \\ +423 \\ \hline 439 \end{array}$$

d. 104 → 90 → 194

$$\begin{array}{r} 104 \\ +\ 90 \\ \hline 194 \end{array}$$

Part G (Test Part 3: L. 97, Ex. 10)

Part H (Test Part 3: L. 99, [blank] Ex. 8)

1

2

Part I (Test Part 4: L. 91, Ex. 8)

a. $39 + 🪙🪙🪙 = $39.75

b. $72 + 🪙🪙🪙🪙 = $72.22

c. $15 + 🪙🪙 🪙🪙 = $15.70

d. $28 + 🪙🪙🪙🪙🪙 = $28.09

54 Mastery Test 10 *Remedies*   Connecting Math Concepts

Remedies CONTINUED   Name _____

Part J (Test Part 4: L. 93, Ex. 7)

a. [50][10][1] 🪙🪙🪙🪙🪙 = $61.17

b. [20][10][1][1][1] 🪙🪙🪙 = $33.75

c. [10][10][1] 🪙🪙🪙 = $21.06

d. [50][10][10][5] 🪙🪙🪙🪙 = $75.70

Part K (Test Part 5: L. 92, Ex. 5)

a. 6 − 3 = 3
b. 5 + 5 = 10
c. 8 − 5 = 3
d. 3 + 4 = 7
e. 10 − 6 = 4
f. 5 − 3 = 2
g. 3 + 3 = 6
h. 10 − 5 = 5
i. 5 + 10 = 15
j. 8 − 4 = 4

Part L (Test Part 5: L. 93, Ex. 5)

a. 11 − 5 = 6
b. 6 + 3 = 9
c. 5 + 5 = 10
d. 12 − 6 = 6
e. 10 − 5 = 5
f. 4 + 3 = 7
g. 6 − 3 = 3
h. 6 + 2 = 8
i. 8 − 4 = 4
j. 4 + 6 = 10

Part M (Test Part 6: L. 91, Ex. 5)

a.
$$\begin{array}{r} {\scriptstyle 11} \\ 186 \\ +716 \\ \hline 902 \end{array}$$

b.
$$\begin{array}{r} {\scriptstyle 11} \\ 296 \\ +475 \\ \hline 771 \end{array}$$

Part N (Test Part 6: L. 93, Ex. 8)

a.
$$\begin{array}{r} {\scriptstyle 1} \\ 457 \\ +162 \\ \hline 619 \end{array}$$

b.
$$\begin{array}{r} {\scriptstyle 341} \\ 9 \\ +\ 34 \\ \hline 384 \end{array}$$

c.
$$\begin{array}{r} {\scriptstyle 11} \\ 742 \\ +168 \\ \hline 910 \end{array}$$

Connecting Math Concepts   Mastery Test 10 *Remedies* 55

Remedy Table — Mastery Test 10				
Part	Test Items	Remedy		Student Material Remedies Worksheet
		Lesson	Exercise	
1	Telling Time (Writing Digital Hours)	91	2	—
		94	8	Part A
		96	7	Part B
		97	8	Part C
2	Number Families with 2-Digit and 3-Digit Numbers (Missing Numbers)	92	3	—
		93	6	Part D
		96	5	Part E
		99	9	Part F
3	3D Objects (Faces)	95	2	—
		97	10	Part G
		99	8	Part H
4	Money (Bills and Coins)	88	3	—
		89	4	—
		91	8	Part I
		93	7	Part J
5	Facts (Mix)	89	1	—
		92	5	Part K
		93	5	Part L
6	Column Addition (Carrying)	90	2	—
		91	5	Part M
		93	8	Part N

Mastery Test 11

Passing Criteria Table — Mastery Test 11			
Part	Score	Possible Score	Passing Score
1	1 for bottom and side faces	19	17
	2 for each top face		
	3 for circling the right object		
2	4 for each problem	12	8
3	2 for each problem	16	12
4	3 for each problem	15	12
5	1 for each problem	10	10
6	2 for each problem	8	6
7	2 for each problem	20	18
	Total	100	

Mastery Test 11 Name _____

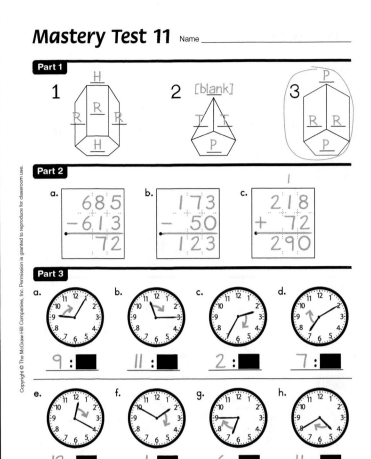

Part 1

Part 2

a.
```
  685
- 613
   72
```

b.
```
  173
-  50
  123
```

c.
```
   218
+   72
   290
```

Part 3

a. 9:■ b. 11:■ c. 2:■ d. 7:■

e. 12:■ f. 1:■ g. 6:■ h. 4:■

Mastery Test 11 Name _____

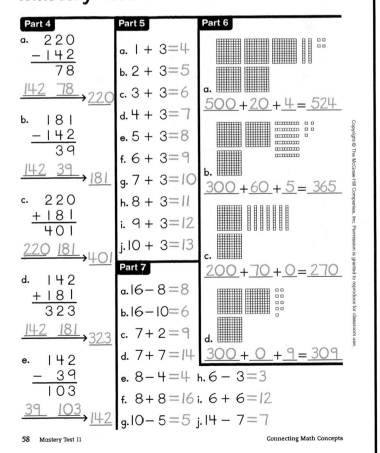

Part 4

a.
```
  220
- 142
   78
```
142 78 → 220

b.
```
  181
- 142
   39
```
142 39 → 181

c.
```
  220
+ 181
  401
```
220 181 → 401

d.
```
  142
+ 181
  323
```
142 181 → 323

e.
```
  142
-  39
  103
```
39 103 → 142

Part 5

a. 1 + 3 = 4
b. 2 + 3 = 5
c. 3 + 3 = 6
d. 4 + 3 = 7
e. 5 + 3 = 8
f. 6 + 3 = 9
g. 7 + 3 = 10
h. 8 + 3 = 11
i. 9 + 3 = 12
j. 10 + 3 = 13

Part 7

a. 16 − 8 = 8
b. 16 − 10 = 6
c. 7 + 2 = 9
d. 7 + 7 = 14
e. 8 − 4 = 4 h. 6 − 3 = 3
f. 8 + 8 = 16 i. 6 + 6 = 12
g. 10 − 5 = 5 j. 14 − 7 = 7

Part 6

a.
500 + 20 + 4 = 524

b.
300 + 60 + 5 = 365

c.
200 + 70 + 0 = 270

d.
300 + 0 + 9 = 309

Remedies Name _____

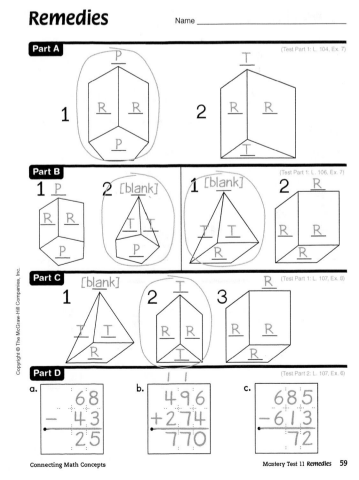

Part A (Test Part 1: L. 104, Ex. 7)

Part B (Test Part 1: L. 106, Ex. 7)

Part C (Test Part 1: L. 107, Ex. 8)

Part D (Test Part 2: L. 107, Ex. 6)

a.
```
  68
- 43
  25
```

b.
```
  496
+ 274
  770
```

c.
```
  685
- 613
   72
```

Remedies CONTINUED   Name _____

Part E
(Test Part 2: L. 108, Ex. 8)

a.
$$\begin{array}{r} 83 \\ -42 \\ \hline 41 \end{array}$$

b.
$$\begin{array}{r} 84 \\ -70 \\ \hline 14 \end{array}$$

c.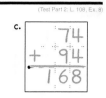
$$\begin{array}{r} 74 \\ +94 \\ \hline 168 \end{array}$$

Part F
(Test Part 2: L. 110, Ex. 8)

a.
$$\begin{array}{r} 173 \\ -50 \\ \hline 123 \end{array}$$

b.
$$\begin{array}{r} 218 \\ +72 \\ \hline 290 \end{array}$$

c.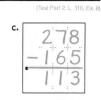
$$\begin{array}{r} 278 \\ -165 \\ \hline 113 \end{array}$$

Part G
(Test Part 3: L. 104, Ex. 3)

a.   b.   c.

Part H
(Test Part 3: L. 106, Ex. 5)

a.   b.   c.   d.

60   Mastery Test 11 *Remedies*   Connecting Math Concepts

Copyright © The McGraw-Hill Companies, Inc.

Remedies CONTINUED   Name _____

Part I
(Test Part 3: L. 107, Ex. 5)

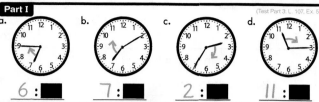

a. 6 : **■**   b. 7 : **■**   c. 2 : **■**   d. 11 : **■**

Part J
(Test Part 4: L. 102, Ex. 9)

a.
$$\begin{array}{r} 214 \\ -96 \\ \hline 118 \end{array}$$
$$\underline{96 \quad 118} \rightarrow \underline{214}$$

b.
$$\begin{array}{r} 137 \\ +385 \\ \hline 522 \end{array}$$
$$\underline{137 \quad 385} \rightarrow \underline{522}$$

c.
$$\begin{array}{r} 385 \\ -214 \\ \hline 171 \end{array}$$
$$\underline{214 \quad 171} \rightarrow \underline{385}$$

d.
$$\begin{array}{r} 171 \\ +96 \\ \hline 267 \end{array}$$
$$\underline{171 \quad 96} \rightarrow \underline{267}$$

Part K
(Test Part 4: L. 103, Ex. 6)

a.
$$\begin{array}{r} 142 \\ +181 \\ \hline 323 \end{array}$$
$$\underline{142 \quad 181} \rightarrow \underline{323}$$

b.
$$\begin{array}{r} 181 \\ -142 \\ \hline 39 \end{array}$$
$$\underline{142 \quad 39} \rightarrow \underline{181}$$

c.
$$\begin{array}{r} 220 \\ -142 \\ \hline 78 \end{array}$$
$$\underline{142 \quad 78} \rightarrow \underline{220}$$

d.
$$\begin{array}{r} 220 \\ +181 \\ \hline 401 \end{array}$$
$$\underline{220 \quad 181} \rightarrow \underline{401}$$

Connecting Math Concepts   Mastery Test 11 *Remedies*   61

Copyright © The McGraw-Hill Companies, Inc.

Remedies CONTINUED   Name _____

Part L
(Test Part 6: L. 101, Ex. 6)

a.
$$300+30+1 = 331$$

b.
$$500+20+4 = 524$$

Part M
(Test Part 6: L. 108, Ex. 6)

a.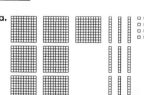
$$700+90+8 = 798$$

b.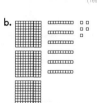
$$300+60+5 = 365$$

Part N
(Test Part 6: L. 110, Ex. 9)

a.
$$200+70+0 = 270$$

b.
$$300+0+9 = 309$$

62   Mastery Test 11 *Remedies*   Connecting Math Concepts

Copyright © The McGraw-Hill Companies, Inc.

Remedies CONTINUED   Name _____

Part O
(Test Part 7: L. 99, Ex. 7)

a. $10-4 = 6$ e. $8+8 = 16$ i. $6-3 = 3$

b. $4+4 = 8$ f. $10-5 = 5$ j. $6+3 = 9$

c. $12-6 = 6$ g. $7+7 = 14$ k. $10-6 = 4$

d. $8-4 = 4$ h. $14-7 = 7$

Part P
(Test Part 7: L. 103, Ex. 4)

c. $6+3 = 9$ f. $10-5 = 5$

a. $6+4 = 10$ d. $11-6 = 5$ g. $7+7 = 14$

b. $6-2 = 4$ e. $3+4 = 7$ h. $8-4 = 4$

Connecting Math Concepts   Mastery Test 11 *Remedies*   63

---

382   **Answer Key**   **Connecting Math Concepts**

	Remedy Table — Mastery Test 11			
Part	Test Items	Remedy		Student Material Remedies Worksheet
		Lesson	Exercise	
1	Decomposition	104	7	Part A
		106	7	Part B
		107	8	Part C
2	Word Problems (Conventional and Comparison)	107	6	Part D
		108	8	Part E
		110	8	Part F
3	Telling Time (Analog Hours)	101	3	—
		104	3	Part G
		106	5	Part H
		107	5	Part I
4	Number Families from Equations	101	5	—
		102	9	Part J
		103	6	Part K
5	Plus 3 Fact Relationships	106	4	—
		109	1	—
6	Place-Value (Pictures)	99	4	—
		101	6	Part L
		108	6	Part M
		110	9	Part N
7	Facts (Mix)	99	2	—
		99	7	Part O
		103	1	—
		103	4	Part P

Mastery Test 12

	Passing Criteria Table — Mastery Test 12		
Part	Score	Possible Score	Passing Score
1	2 for each number	10	8
2	1 for each problem	6	5
3	2 for each correct odd/even number	11	10
4	1 for each problem	6	5
5	1 for each equation	8	7
6	3 for each problem	9	6
7	1 for each equation	20	19
8	2 for each problem	16	12
9	1 for each problem	4	3
10	1 for each equation	10	9
	Total	100	

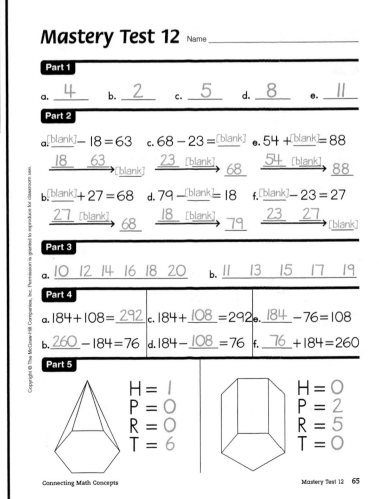

Mastery Test 12 Name_____

Part 1

a. 4   b. 2   c. 5   d. 8   e. 11

Part 2

a. [blank] − 18 = 63   c. 68 − 23 = [blank]   e. 54 + [blank] = 88

$\xrightarrow[\text{[blank]}]{18 \quad 63}$ $\xrightarrow{23 \quad \text{[blank]}} 68$ $\xrightarrow{54 \quad \text{[blank]}} 88$

b. [blank] + 27 = 68   d. 79 − [blank] = 18   f. [blank] − 23 = 27

$\xrightarrow{27 \quad \text{[blank]}} 68$ $\xrightarrow{18 \quad \text{[blank]}} 79$ $\xrightarrow[\text{[blank]}]{23 \quad 27}$

Part 3

a. 10 12 14 16 18 20   b. 11 13 15 17 19

Part 4

a. 184 + 108 = 292   c. 184 + 108 = 292   e. 184 − 76 = 108

b. 260 − 184 = 76   d. 184 − 108 = 76   f. 76 + 184 = 260

Part 5

H = 1
P = 0
R = 0
T = 6

H = 0
P = 2
R = 5
T = 0

Connecting Math Concepts   Mastery Test 12   65

Mastery Test 12 Name _____

Part 6

a. $\underline{88} - 23 = 65$ b. $\underline{21} + 54 = 75$ c. $89 - \underline{26} = 63$

$\underline{23}\quad\underline{65}$ $\underline{54}$ [blank] $\underline{63}$ [blank]
[blank] →   → $\underline{75}$ → $\underline{89}$

```
  23          75          89
+ 65        - 54        - 63
----        ----        ----
  88          21          26
```

Part 7

g. $3 + 7 = 10$ n. $6 - 3 = 3$

a. $3 + 1 = 4$ h. $3 + 8 = 11$ o. $7 - 3 = 4$

b. $3 + 2 = 5$ i. $3 + 9 = 12$ p. $8 - 3 = 5$

c. $3 + 3 = 6$ j. $3 + 10 = 13$ q. $9 - 3 = 6$

d. $3 + 4 = 7$ k. $3 - 3 = 0$ r. $10 - 3 = 7$

e. $3 + 5 = 8$ l. $4 - 3 = 1$ s. $11 - 3 = 8$

f. $3 + 6 = 9$ m. $5 - 3 = 2$ t. $12 - 3 = 9$

Mastery Test 12 Name _____

Part 8

a.    c.   e.   g.

$12:30$ $4:15$ $6:00$ $5:30$

b.   d.   f.   h.

$10:00$ $8:30$ $9:20$ $1:00$

Part 9

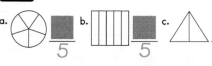

a. $\dfrac{}{5}$ b. $\dfrac{}{5}$ c. $\dfrac{}{2}$ d. $\dfrac{}{4}$

Part 10

d. $8 - 3 = 5$ h. $7 + 7 = 14$

a. $10 - 7 = 3$ e. $11 - 3 = 8$ i. $11 - 8 = 3$

b. $8 + 8 = 16$ f. $7 - 3 = 4$ j. $10 - 3 = 7$

c. $8 + 3 = 11$ g. $7 + 3 = 10$

Remedies Name _____

Part A

(Test Part 1: L. 117, Ex. 7)

a.  $\dfrac{}{9}$ b. $\dfrac{}{7}$ c. $\dfrac{}{6}$

Part B

(Test Part 1: L. 120, Ex. 2)

a.  $\dfrac{}{2}$ c. $\dfrac{}{5}$ e. $\dfrac{}{6}$

b. $\dfrac{}{4}$ d. $\dfrac{}{2}$ f. $\dfrac{}{4}$

Part C

(Test Part 2: L. 117, Ex. 5)

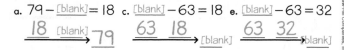

a. $79 - \text{[blank]} = 18$ c. $\text{[blank]} - 63 = 18$ e. $\text{[blank]} - 63 = 32$

$\underline{18}$ [blank] → $\underline{79}$ $\underline{63}\quad\underline{18}$ → [blank] $\underline{63}\quad\underline{32}$ → [blank]

b. $\text{[blank]} + 31 = 81$ d. $58 - \text{[blank]} = 17$ f. $18 + \text{[blank]} = 89$

$\underline{31}$ [blank] → $\underline{81}$ $\underline{17}$ [blank] → $\underline{58}$ $\underline{18}$ [blank] → $\underline{89}$

Remedies Name _____

Part D

(Test Part 2: L. 119, Ex. 7)

a. $\underline{41} + 27 = 68$ c. $54 + \underline{34} = 88$ e. $88 - \underline{65} = 23$

$\underline{27}\quad\underline{41}$ → $\underline{68}$ $\underline{54}\quad\underline{34}$ → $\underline{88}$ $\underline{23}\quad\underline{65}$ → $\underline{88}$

```
  68          88          88
- 27        - 54        - 23
----        ----        ----
  41          34          65
```

b. $\underline{90} - 23 = 67$ d. $68 - 23 = \underline{45}$ f. $\underline{50} - 23 = 27$

$\underline{23}\quad\underline{67}$ → $\underline{90}$ $\underline{23}\quad\underline{45}$ → $\underline{68}$ $\underline{23}\quad\underline{27}$ → $\underline{50}$

```
  23          68          23
+ 67        - 23        + 27
----        ----        ----
  90          45          50
```

Part E

(Test Part 4: L. 113, Ex. 5)

a. $243 - \underline{154} = 89$ c. $\underline{243} - 65 = 178$

b. $\underline{65} + 89 = 154$ d. $243 - 178 = \underline{65}$

Part F

(Test Part 4: L. 114, Ex. 10)

a. $184 - \underline{108} = 76$ c. $\underline{260} - 184 = 76$

b. $184 + \underline{108} = 292$ d. $184 + 108 = \underline{292}$

Part G   (Test Part 5: L. 110, Ex. 4)

H = 1	H = 0
P = 0	P = 0
R = 0	R = 3
T = 6	T = 2

Part H   (Test Part 5: L. 111, Ex. 8)

H = 0	H = 0
P = 2	P = 0
R = 5	R = 1
T = 0	T = 4

Part I   (Test Part 5: L. 112, Ex. 6)

H = 1
P = 0
R = 0
T = 6

H = 0
P = 2
R = 5
T = 0

Part J

a. 79 − [blank] = 18

$\underline{18}$ [blank] → 79

d. 58 − [blank] = 17   (Test Part 6: L. 117, Ex. 5)

$\underline{17}$ [blank] → 58

b. [blank] + 31 = 81

$\underline{31}$ [blank] → 81

e. [blank] − 63 = 32

$\underline{63}$ $\underline{32}$ → [blank]

c. [blank] − 63 = 18

$\underline{63}$ $\underline{18}$ → [blank]

f. 18 + [blank] = 89

$\underline{18}$ [blank] → 89

Part K   (Test Part 6: L. 118, Ex. 6)

a. $\underline{80}$ − 15 = 65

$\underline{15}$ $\underline{65}$ → 80

```
  15
+ 65
────
  80
```

b. 65 + $\underline{32}$ = 97

$\underline{65}$ $\underline{32}$ → 97

```
  97
− 65
────
  32
```

c. 86 − $\underline{53}$ = 33

$\underline{33}$ $\underline{53}$ → 86

```
  86
− 33
────
  53
```

Part L   (Test Part 6: L. 119, Ex. 7)

a. $\underline{41}$ + 27 = 68

$\underline{27}$ $\underline{41}$ → 68

```
  68
− 27
────
  41
```

c. 54 + $\underline{34}$ = 88

$\underline{54}$ $\underline{34}$ → 88

```
  88
− 54
────
  34
```

e. 88 − $\underline{65}$ = 23

$\underline{23}$ $\underline{65}$ → 88

```
  88
− 23
────
  65
```

b. $\underline{90}$ − 23 = 67

$\underline{23}$ $\underline{67}$ → 90

```
  23
+ 67
────
  90
```

d. 68 − 23 = $\underline{45}$

$\underline{23}$ $\underline{45}$ → 68

```
  68
− 23
────
  45
```

f. $\underline{50}$ − 23 = 27

$\underline{23}$ $\underline{27}$ → 50

```
  23
+ 27
────
  50
```

Part M   (Test Part 8: L. 108, Ex. 5)

a. 5:55

b. 10:25

c. 3:40

d. 7:10

Part N   (Test Part 8: L. 109, Ex. 7)

a. 8:00

b. 3:50

c. 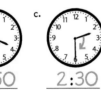 2:30

d. 10:15

Part O   (Test Part 8: L. 113, Ex. 6)

a. 1:00

b. 4:15

c. 11:05

d.  9:30

Part P   (Test Part 8: L. 116, Ex. 6)

a. 5:30

b. 12:35

c.   4:00

d. 9:20

Part Q   (Test Part 9: L. 117, Ex. 7)

a.   9

b.   7

c. 6

Part R   (Test Part 9: L. 118, Ex. 5)

a.   6

b.   4

c. 7

Part S   (Test Part 9: L. 120, Ex. 2)

a.   2

c.   5

e. 6

b.   4

d.   2

f. 4

| Part | Test Items | Remedy | | Student Material Remedies Worksheet |
		Lesson	Exercise	
		117	7	Part A
1	Fractions	119	2	—
		120	2	Part B
2	Writing Number Families from Problems	117	5	Part C
		119	7	Part D
		111	3	—
3	Odd and Even Numbers	114	3	—
		114	5	—
		116	4	—
4	Number Families– Completing Equations with Missing Numbers	113	5	Part E
		114	10	Part F
		110	4	Part G
5	3-D Objects Decompostion	111	8	Part H
		112	6	Part I
		117	5	Part J
6	Number Families from Problems	118	6	Part K
		119	7	Part L
		111	1	—
7	3 Plus, Minus-3 Fact Relationships	113	1	—
		117	1	—
		108	5	Part M
8	Writing Time (Analog Clocks)	109	7	Part N
		113	6	Part O
		116	6	Part P
		117	7	Part Q
9	Fractions (Denominator)	118	5	Part R
		120	2	Part S
		109	2	—
10	Facts (Mix)	110	3	—
		112	4	—

Remedy Table — Mastery Test 12

Remedy Summary—Group Summary of Test Performance

Note: Test remedies are specified in the Answer Key.

	Test 9								Test 10							Test 11								Test 12										
	Check parts not passed							Total %	Check parts not passed						Total %	Check parts not passed							Total %	Check parts not passed										Total %
Name	1	2	3	4	5	6	7		1	2	3	4	5	6		1	2	3	4	5	6	7		1	2	3	4	5	6	7	8	9	10	
1.																																		
2.																																		
3.																																		
4.																																		
5.																																		
6.																																		
7.																																		
8.																																		
9.																																		
10.																																		
11.																																		
12.																																		
13.																																		
14.																																		
15.																																		
16.																																		
17.																																		
18.																																		
19.																																		
20.																																		
21.																																		
22.																																		
23.																																		
24.																																		
25.																																		
26.																																		
27.																																		
28.																																		
29.																																		
30.																																		
Number of students Not Passed = NP																																		
Total number of students = T																																		
Remedy needed if NP/T = 25% or more																																		

Level B Correlation to Grade 1 Common Core State Standards for Mathematics

Operations and Algebraic Thinking (1.OA)

Represent and solve problems involving addition and subtraction.

1. Use addition and subtraction within 20 to solve word problems involving situations of adding to, taking from, putting together, taking apart, and comparing, with unknowns in all positions, e.g., by using objects, drawings, and equations with a symbol for the unknown number to represent the problem.

Lesson	114	115	116	117	119	120	122	123	124	125
Exercise	114.7	115.10	116.6	117.6	119.5	120.5	122.7	123.5	124.7	125.7

Operations and Algebraic Thinking (1.OA)

Represent and solve problems involving addition and subtraction.

2. Solve word problems that call for addition of three whole numbers whose sum is less than or equal to 20, e.g., by using objects, drawings, and equations with a symbol for the unknown number to represent the problem.

This standard is addressed in the following activities of the Student Practice Software:

- **Block 6:** Activity 4
- **Block 6:** Activity 6

Operations and Algebraic Thinking (1.OA)

Understand and apply properties of operations and the relationship between addition and subtraction.

3. Apply properties of operations as strategies to add and subtract. Examples: If 8 + 3 = 11 is known, then 3 + 8 = 11 is also known. (Commutative property of addition.) To add 2 + 6 + 4, the second two numbers can be added to make a ten, so 2 + 6 + 4 = 2 + 10 = 12. (Associative property of addition.)

Lesson	81	82	83	86	91	92	93	94	95	109
Exercise	81.3, IW81.9	82.3	83.2, IW83.10	IW86.10	91.4	92.3	IW93.9	IW94.10	IW95.10	109.2

Lesson	110	111	112	113	121	122	123	124	125
Exercise	110.3	111.4	112.4	113.3	121.2	122.2, 122.4	123.2, 123.3	124.2, 124.6	125.2, 125.6

Operations and Algebraic Thinking (1.OA)

Understand and apply properties of operations and the relationship between addition and subtraction.

4. Understand subtraction as an unknown-addend problem. For example, subtract 10 – 8 by finding the number that makes 10 when added to 8.

Lesson	109	110	111	112	113	114	115	116	117	118
Exercise	109.4	110.2	111.2	112.2	113.5	114.10	115.8	116.7	117.5	118.6

Lesson	119	120	121	122	123	124	125
Exercise	119.5, 119.7	120.5, 120.7	121.8	122.7, 122.8	123.5, 123.7	124.4, 124.7	125.4, 125.7

Operations and Algebraic Thinking (1.OA)

Add and subtract within 20.

5. Relate counting to addition and subtraction (e.g., by counting on 2 to add 2).

Lesson	81	82	83	84	85	86	88	89	90	91
Exercise	IW81.9	IW82.10	IW83.10	IW84.10	IW85.10	IW86.10	IW88.9	IW89.10	IW90.9	IW91.10

Lesson	92	93	95	97	99	101	103	105	107	108
Exercise	IW92.10	IW93.9	IW95.10	IW97.11	IW99.10	IW101.10	IW103.10	IW105.10	IW107.10	IW108.9

Lesson	109	110	111	112	113	114	115	116	117	118
Exercise	109.3	IW110.10	111.3	112.3, IW112.10	113.2	114.3, IW114.11	115.3, IW115.11	116.4	117.3, IW117.10	118.3

Lesson	119	122
Exercise	IW119.9	IW122.9

Operations and Algebraic Thinking (1.OA)

Add and subtract within 20.

6. Add and subtract within 20, demonstrating fluency for addition and subtraction within 10. Use strategies such as counting on; making ten (e.g., 8 + 6 = 8 + 2 + 4 = 10 + 4 = 14); decomposing a number leading to a ten (e.g., 13 – 4 = 13 – 3 – 1 = 10 – 1 = 9); using the relationship between addition and subtraction (e.g., knowing that 8 + 4 = 12, one knows 12 – 8 = 4); and creating equivalent but easier or known sums (e.g., adding 6 + 7 by creating the known equivalent 6 + 6 + 1 = 12 + 1 = 13).

Lesson	81	82	83	84	85	86	87	88	89	90
Exercise	81.1, 81.3, 81.6, IW81.9	82.1, 82.3, IW82.10	83.1, 83.2, 83.4, IW83.10	84.1, 84.3, 84.4, 84.9	85.1, 85.3, 85.5, 85.8	86.1, 86.2, 86.5, 86.7, 86.9, IW86.10	87.1, 87.2, 87.3, 87.4, 87.8, IW87.10	88.1, 88.2, 88.6, 88.8	89.1, 89.2, 89.3, 89.8, IW89.10	90.1, 90.3, 90.4, 90.8

Lesson	91	92	93	94	95	96	97	98	99	100
Exercise	91.1, 91.6, 91.9	92.1, 92.5, IW92.10	93.3, 93.5, IW93.9	94.2, 94.5, 94.9, IW94.10	95.1, 95.5, 95.9, IW95.10	96.2, 96.6, 96.9	97.1, 97.9, IW97.11	98.2, 98.8, IW98.10	99.2, 99.7	100.2, 100.8, 100.10, IW100.11

Lesson	101	102	103	104	105	106	107	108	109	110
Exercise	101.1, 101.7	102.1, 102.5, 102.10	103.1, 103.4, 103.9	104.1, 104.4, 104.8, IW104.9	105.1, 105.4, 105.9	106.1, 106.4, 106.8, IW106.10	107.1, 107.3, 107.9	108.1, 108.3, 108.7	109.1, 109.2, 109.6, IW109.10	110.1, 110.3, 110.7

Lesson	111	112	113	114	115	116	117	118	119	120
Exercise	111.1, 111.4, 111.7	112.1, 112.4, 112.7	113.1, 113.3, 113.7, IW113.9	114.1, 114.4, 114.7, 114.9	115.1, 115.7, 115.9, 115.10, IW115.11	116.1, 116.5, 116.9, IW116.10	117.1, 117.3, 117.4, 117.8	118.1, 118.4, 118.7	119.1, 119.4, 119.5, 119.6	120.1, 120.4, 120.5, 120.6

Lesson	121	122	123	124	125
Exercise	121.1, 121.2, 121.4, 121.6	122.1, 122.2, 122.4, 122.6, 122.7	123.1, 123.2, 123.3, 123.5	124.1, 124.2, 124.6, 124.7	125.1, 125.2, 125.6, 125.7

Operations and Algebraic Thinking (1.OA)

Work with addition and subtraction equations.

7. Understand the meaning of the equal sign, and determine if equations involving addition and subtraction are true or false. For example, which of the following equations are true and which are false? $6 = 6$, $7 = 8 - 1$, $5 + 2 = 2 + 5$, $4 + 1 = 5 + 2$.

Lesson	119	120	121
Exercise	119.4	120.4	121.6

Operations and Algebraic Thinking (1.OA)

Work with addition and subtraction equations.

8. Determine the unknown whole number in an addition or subtraction equation relating to three whole numbers. For example, determine the unknown number that makes the equation true in each of the equations $8 + ? = 11$, $5 = \_\_ - 3$, $6 + 6 = \_\_$.

Lesson	81	82	83	84	85	86	87	88	89	90
Exercise	81.5, 81.6, 81.7, 81.8, IW81.9	82.5, 82.6, 82.7, 82.8, IW82.10	83.6, 83.8, 83.9, IW83.10	84.2, 84.4, 84.5, 84.7, 84.9, IW84.10	85.4, 85.5, 85.6, 85.8, 85.9, IW85.10	86.3, 86.5, 86.7, 86.9, IW86.10	87.3, 87.4, 87.6, 87.8, IW87.10	88.1, 88.4, 88.6, 88.8, IW88.9	89.6, 89.8, 89.9, IW89.10	90.2, 90.4, 90.6, 90.8, IW90.9

Lesson	91	92	93	94	95	96	97	98	99	100
Exercise	91.5, 91.6, 91.9, IW91.10	92.5, 92.7, 92.9, IW92.10	93.5, 93.6, 93.8, IW93.9	94.5, 94.6, 94.7, 94.9, IW94.10	95.5, 95.6, 95.7, 95.9, IW95.10	96.5, 96.6, 96.8, 96.9, IW96.10	97.6, 97.7, 97.9, IW97.11	98.5, 98.7, 98.8, IW98.10	99.7, 99.9, IW99.10	100.5, 100.6, 100.8, 100.9, 100.10, IW100.11

Lesson	101	102	103	104	105	106	107	108	109	110
Exercise	101.6, 101.7, 101.8, IW101.10	102.5, 102.6, 102.7, 102.10, IW102.11	103.4, 103.8, 103.9, IW103.10	104.4, 104.5, 104.6, 104.8, IW104.9	105.4, 105.6, 105.7, 105.8, 105.9, IW105.10	106.2, 106.6, 106.8, 106.9, IW106.10	107.2, 107.6, 107.7, 107.9, IW107.10	108.2, 108.3, 108.6, 108.7, 108.8, 108.9	109.4, 109.6, 109.8, 109.9, IW109.10	110.2, 110.5, 110.7, 110.8, 110.9, IW110.10

Lesson	111	112	113	114	115	116	117	118	119	120
Exercise	111.2, 111.6, 111.7, 111.9, IW111.10	112.2, 112.7, 112.8, 112.9, IW112.10	113.4, 113.5, 113.7, IW113.9	114.6, 114.7, 114.9, 114.10, IW114.11	115.7, 115.8, 115.9, 115.10, IW115.11	116.5, 116.7, 116.9, IW116.10	117.4, 117.5, 117.6, 117.8, IW117.10	118.2, 118.6, 118.7, IW118.9	119.3, 119.6, 119.7, IW119.9	120.3, 120.6, 120.7

Lesson	121	122	123	124	125
Exercise	121.4, 121.7, 121.8	122.6, 122.7, 122.8, IW122.9	123.5, 123.7, 123.8, IW123.9	124.4, 124.5, 124.7	125.4, 125.7, IW125.8

Number and Operations in Base Ten (1.NBT)

Extend the counting sequence.

1. Count to 120, starting at any number less than 120. In this range, read and write numerals and represent a number of objects with a written numeral.

Lesson	81	82	83	84	85	86	87	88	89	90
Exercise	81.1, 81.2, 81.3, 81.4, 81.5, 81.6, 81.7, 81.8, IW81.9	82.1, 82.2, 82.3, 82.4, 82.5, 82.6, 82.7, 82.8, 82.9, IW82.10	83.1, 83.2, 83.4, 83.5, 83.6, 83.7, 83.8, 83.9, IW83.10	84.1, 84.2, 84.3, 84.4, 84.5, 84.6, 84.7, 84.8, 84.9, IW84.10	85.1, 85.2, 85.3, 85.4, 85.5, 85.6, 85.7, 85.8, 85.9, IW85.10	86.1, 86.2, 86.3, 86.5, 86.6, 86.7, 86.8, 86.9, IW86.10	87.1, 87.2, 87.3, 87.4, 87.6, 87.7, 87.8, 87.9, IW87.10	88.1, 88.2, 88.3, 88.4, 88.6, 88.7, 88.8, IW88.9	89.1, 89.2, 89.3, 89.4, 89.6, 89.7, 89.8, 89.9, IW89.10	90.1, 90.2, 90.3, 90.4, 90.5, 90.6, 90.7, 90.8, IW90.9

Lesson	91	92	93	94	95	96	97	98	99	100
Exercise	91.1, 91.2, 91.4, 91.5, 91.6, 91.7, 91.8, 91.9, IW91.10	92.1, 92.3, 92.4, 92.5, 92.6, 92.7, 92.8, 92.9, IW92.10	93.1, 93.2, 93.3, 93.5, 93.6, 93.7, 93.8, IW93.9	94.1, 94.2, 94.5, 94.6, 94.7, 94.8, 94.9, IW94.10	95.1, 95.3, 95.5, 95.6, 95.7, 95.8, 95.9, IW95.10	96.2, 96.5, 96.6, 96.7, 96.8, 96.9, IW96.10	97.1, 97.4, 97.6, 97.7, 97.8, 97.9, IW97.11	98.1, 98.2, 98.3, 98.5, 98.6, 98.7, 98.8, IW98.10	99.2, 99.3, 99.5, 99.6, 99.7, 99.9, IW99.10	100.1, 100.2, 100.3, 100.4, 100.5, 100.6, 100.8, 100.9, 100.10, IW100.11

Lesson	101	102	103	104	105	106	107	108	109	110
Exercise	101.1, 101.2, 101.5, 101.7, 101.8, IW101.10	102.1, 102.4, 102.5, 102.6, 102.7, 102.9, 102.10, IW102.11	103.1, 103.4, 103.5, 103.6, 103.8, 103.9, IW103.10	104.1, 104.4, 104.5, 104.6, 104.8, IW104.9	105.1, 105.4, 105.6, 105.7, 105.8, 105.9, IW105.10	106.1, 106.2, 106.3, 106.4, 106.6, 106.8, 106.9, IW106.10	107.1, 107.2, 107.3, 107.6, 107.7, 107.9, IW107.10	108.1, 108.2, 108.3, 108.5, 108.6, 108.7, 108.8, IW108.9	109.2, 109.4, 109.5, 109.6, 109.7, 109.8, 109.9, IW109.10	110.2, 110.3, 110.5, 110.6, 110.7, 110.8, 110.9, IW110.10

Lesson	111	112	113	114	115	116	117	118	119	120
Exercise	111.2, 111.3, 111.4, 111.5, 111.7, 111.8, 111.9, 111.10	112.2, 112.3, 112.4, 112.5, 112.6, 112.7, 112.8, 112.9, IW112.10	113.3, 113.4, 113.5, 113.6, 113.7, 113.8, IW113.9	114.4, 114.6, 114.7, 114.8, 114.9, 114.10, IW114.11	115.3, 115.4, 115.7, 115.8, 115.9, 115.10, IW115.11	116.5, 116.6, 116.7, 116.9, IW116.10	117.2, 117.4, 117.5, 117.6, 117.7, 117.8, 117.9, IW117.10	118.2, 118.4, 118.5, 118.6, 118.7, 118.8, IW118.9	119.3, 119.4, 119.5, 119.6, 119.7, 119.8, IW119.9	120.2, 120.3, 120.4, 120.5, 120.6, 120.7, IW120.8

Lesson	121	122	123	124	125
Exercise	121.1, 121.2, 121.4, 121.5, 121.6, 121.7, 121.8, IW121.9	122.1, 122.2, 122.3, 122.4, 122.5, 122.6, 122.7, 122.8, IW122.9	123.1, 123.2, 123.3, 123.4, 123.5, 123.6, 123.7, 123.8, IW123.9	124.1, 124.2, 124.3, 124.4, 124.5, 124.6, 124.7, 124.8	125.1, 125.2, 125.3, 125.4, 125.5, 125.6, 125.7, IW125.8

Number and Operations in Base Ten (1.NBT)

Understand place value.

2. Understand that the two digits of a two-digit number represent amounts of tens and ones. Understand the following as special cases: a. 10 can be thought of as a bundle of ten ones — called a "ten." b. The numbers from 11 to 19 are composed of a ten and one, two, three, four, five, six, seven, eight, or nine ones. c. The numbers 10, 20, 30, 40, 50, 60, 70, 80, 90 refer to one, two, three, four, five, six, seven, eight, or nine tens (and 0 ones).

Lesson	81	82	83	84	85	88	89	99	100	101
Exercise	81.5	82.7	83.6	84.5	85.6, IW85.10	IW88.9	IW89.10	IW99.10	100.9	101.8

Lesson	104	108	109	111	112	113	114	115	116	117
Exercise	IW104.9	IW108.9	IW109.10	IW111.10	112.9	113.2, 113.4	114.3, 114.6, IW114.11	115.3, 115.4, IW115.11	116.2, 116.4, IW116.10	117.2, 117.3

Lesson	118	119	120	121	122	123	124
Exercise	118.2, 118.3, IW118.9	119.3	120.3	121.7	IW122.9	123.8, IW123.9	124.5

Number and Operations in Base Ten (1.NBT)

Understand place value.

3. Compare two two-digit numbers based on meanings of the tens and ones digits, recording the results of comparisons with the symbols >, =, and <.

This standard is addressed in the following activities of the Student Practice Software:

- **Block 4:** Activity 4
- **Block 5:** Activity 6

Number and Operations in Base Ten (1.NBT)

Use place value understanding and properties of operations to add and subtract.

4. Add within 100, including adding a two-digit number and a one-digit number, and adding a two-digit number and a multiple of 10, using concrete models or drawings and strategies based on place value, properties of operations, and/or the relationship between addition and subtraction; relate the strategy to a written method and explain the reasoning used. Understand that in adding two-digit numbers, one adds tens and tens, ones and ones; and sometimes it is necessary to compose a ten.

Lesson	81	82	83	84	85	86	87	88	89	90
Exercise	81.1, 81.3, 81.5, IW81.9	82.1, 82.3, 82.7, 82.8, IW82.10	83.2, 83.4, 83.6, IW83.10	84.1, 84.5, 84.9, IW84.10	85.1, 85.6, 85.8, IW85.10	86.1, 86.7, 86.9, IW86.10	87.1, 87.3, 87.8, IW87.10	88.1, 88.8, IW88.9	89.1, 89.3, 89.9, IW89.10	90.1, 90.4, 90.6, IW90.9

Lesson	91	92	93	94	95	96	97	98	99	100
Exercise	91.1, 91.4, 91.6	92.1, 92.3, 92.5, 92.9, IW92.10	93.3, 93.5, 93.6, 93.8, IW93.9	94.2, 94.5, IW94.10	95.1, 95.6, 95.9, IW95.10	96.2, 96.5, 96.9	97.1, IW97.11	98.2, 98.7, IW98.10	99.2, 99.7, IW99.10	100.2, 100.10, IW100.11

Lesson	101	102	103	104	105	106	107	108	109	110
Exercise	101.1, 101.5, 101.7	102.1, 102.5	103.1, 103.4, IW103.10	104.1, 104.4, IW104.9	105.1, 105.4	106.1, 106.4, IW106.10	107.1, 107.3	108.1, 108.3	109.1, 109.2, 109.3, 109.6, IW109.10	110.1, 110.3, 110.7

Lesson	111	112	113	114	115	116	117	118	119	120
Exercise	111.1, 111.3, 111.4, 111.7	112.1, 112.3, 112.4, 112.7, 112.9, IW112.10	113.2, 113.3, 113.4, 113.7, IW113.9	114.4, 114.6, 114.7, 114.9, IW114.11	115.1, 115.3, 115.7, 115.8, 115.9, 115.10, IW115.11	116.1, 116.4, 116.5, 116.7, 116.9, IW116.10	117.1, 117.2, 117.3, 117.4, 117.5, 117.6, IW117.10	118.1, 118.2, 118.3, 118.4, 118.6, IW118.9	119.1, 119.3, 119.4, 119.5, 119.6, 119.7, IW119.9	120.1, 120.3, 120.4, 120.5, 120.6, 120.7

Lesson	121	122	123	124	125
Exercise	121.1, 121.2, 121.4, 121.6, 121.7, 121.8	122.1, 122.2, 122.4, 122.6, 122.7, 122.8, IW122.9	123.1, 123.2, 123.3, 123.5, 123.7, 123.8	124.1, 124.2, 124.4, 124.5, 124.6, 124.7	125.1, 125.2, 125.4, 125.6, 125.7

Number and Operations in Base Ten (1.NBT)

Use place value understanding and properties of operations to add and subtract.

5. Given a two-digit number, mentally find 10 more or 10 less than the number, without having to count; explain the reasoning used.

Lesson	113	114	115	116	117	118	119	120	121	122
Exercise	113.2	114.3	115.3	116.4	117.2, 117.3	118.2, 118.3	119.3	120.3	121.7	IW122.9

Lesson	123
Exercise	IW123.9

Number and Operations in Base Ten (1.NBT)

Use place value understanding and properties of operations to add and subtract.

6. Subtract multiples of 10 in the range 10–90 from multiples of 10 in the range 10–90 (positive or zero differences), using concrete models or drawings and strategies based on place value, properties of operations, and/or the relationship between addition and subtraction; relate the strategy to a written method and explain the reasoning used.

Lesson	109	112	116	117	118	119	120	121	122	123
Exercise	109.3	112.3	116.4	117.2	118.2, 118.3	119.3	120.3	121.7	IW122.9	IW123.9

Measurement and Data (1.MD)

Measure lengths indirectly and by iterating length units.

1. Order three objects by length; compare the lengths of two objects indirectly by using a third object.

Lesson	93	94	95	96	97
Exercise	93.4	94.4	95.4	96.4	97.5

Measurement and Data (1.MD)

Measure lengths indirectly and by iterating length units.

2. Express the length of an object as a whole number of length units, by laying multiple copies of a shorter object (the length unit) end to end; understand that the length measurement of an object is the number of same-size length units that span it with no gaps or overlaps. Limit to contexts where the object being measured is spanned by a whole number of length units with no gaps or overlaps.

This standard is addressed in the following activities of the Student Practice Software:

- **Block 3:** Activity 6
- **Block 6:** Activity 5

Measurement and Data (1.MD)

Measure lengths indirectly and by iterating length units.

3. Tell and write time in hours and half-hours using analog and digital clocks.

Lesson	91	92	93	94	95	96	97	98	100	101
Exercise	91.2	92.4	93.1	94.8	95.8	96.7	97.8	98.6	100.3	101.3

Lesson	102	103	104	105	106	107	108	109	110	111
Exercise	102.3	103.2	104.3	105.5	106.5	107.5	108.5	109.7	110.6	111.5

Lesson	112	113	114	115	116	117	118	119	120	121
Exercise	112.5	113.6	114.8	115.5	116.6	117.9	118.8	119.8	IW120.8	IW121.9

Lesson	122	125
Exercise	IW122.9	IW125.8

Measurement and Data (1.MD)

Measure lengths indirectly and by iterating length units.

4. Organize, represent, and interpret data with up to three categories; ask and answer questions about the total number of data points, how many in each category, and how many more or less are in one category than in another.

Lesson	122	123	124	125
Exercise	122.3	123.4	124.8	125.5

Geometry (1.G)

Reason with shapes and their attributes.

1. Distinguish between defining attributes (e.g., triangles are closed and three-sided) versus non-defining attributes (e.g., color, orientation, overall size); build and draw shapes to possess defining attributes.

Lesson	95	96	97	98	99	100	101	102	103	104
Exercise	95.2	96.3	97.10	IW98.10	99.8	100.7	101.9	102.8	103.7	104.7

Lesson	105	106	107	108	109	110	111	112	113	114
Exercise	105.3	106.7	107.8	108.4, IW108.9	109.5, IW109.10	110.4, IW110.10	111.8, IW111.10	112.6, IW112.10	113.8, IW113.9	114.2, IW114.11

Lesson	115	116	117	118	119
Exercise	115.2	116.3, 116.8, IW116.10	IW117.10	IW118.9	IW119.9

Geometry (1.G)

Reason with shapes and their attributes.

2. Compose two-dimensional shapes (rectangles, squares, trapezoids, triangles, half-circles, and quarter-circles) or three-dimensional shapes (cubes, right rectangular prisms, right circular cones, and right circular cylinders) to create a composite shape, and compose new shapes from the composite shape.

Lesson	81	82	83	84	85	86	87	88	89	90
Exercise	IW81.9	IW82.10	IW83.10	IW84.10	IW85.10	86.4, IW86.10	87.5, IW87.10	88.5, IW88.9	IW89.10	IW90.9

Lesson	91	92	93	94	95	96	97	98	99	100
Exercise	IW91.10	IW92.10	IW93.9	IW94.10	95.2	96.3, IW96.10	97.10, IW97.11	98.9, IW98.10	99.8, IW99.10	100.7, IW100.11

Lesson	101	102	103	104	105	106	107	108	109	110
Exercise	101.9, IW101.10	102.8	103.7, IW103.10	104.7, IW104.9	105.3, 105.10	106.7	107.8, IW107.10	108.4, IW108.9	109.5, IW109.10	110.4, IW110.10

Lesson	111	112	113	114	115	116	117	118	119
Exercise	111.8, IW111.10	112.6, IW112.10	113.8, IW113.9	114.2, IW114.11	115.2	116.3, 116.8, IW116.10	IW117.10	IW118.9	IW119.9

Geometry (1.G)

Reason with shapes and their attributes.

3. Partition circles and rectangles into two and four equal shares, describe the shares using the words halves, fourths, and quarters, and use the phrases half of, fourth of, and quarter of. Describe the whole as two of, or four of the shares. Understand for these examples that decomposing into more equal shares creates smaller shares.

Lesson	117	118	119	120	121	122	123	124	125
Exercise	117.7	118.5	119.2	120.2	121.3, IW121.9	122.5	123.6	124.3	125.3

Standards for Mathematical Practice and Connecting Math Concepts

Connecting Math Concepts: Comprehensive Edition is a six-level series that is fully aligned with the Common Core State Standards for kindergarten through fifth grade. As its name implies, *Connecting Math Concepts* is designed to bring students to an understanding of mathematical concepts by making connections between central and generative concepts in the school mathematics curriculum, as defined by the Common Core State Standards (CCSS) for Mathematical Content. This document illustrates some of those connections as they relate to the eight CCSS for Mathematical Practices. Examples will be provided from each of the six levels of *Connecting Math Concepts (CMC)*, to illustrate how students engage in activities representative of the eight CCSS for Mathematical Practices.

MP1: MAKE SENSE OF PROBLEMS AND PERSEVERE IN SOLVING THEM.

CMC Level C (Grade 2): Comparison Story Problems

Students in the primary grades are just beginning to make sense of situations that can be expressed mathematically. Care must be taken to give them a way of thinking about situations that is not overly simplistic, resulting in a shallow understanding of the circumstance. For example, when considering two values in a statement such as *Sophia is 2 years older than Isabel,* students are likely to interpret the situation as implying addition because of the word *more.* Conversely, they are likely to interpret a statement such as *Isabel is 2 years younger than Sophia* as subtraction. The comparison statements do not imply either operation; they simply identify the larger and smaller of the two values being compared. Given one value, we can find the other by addition or subtraction. If the larger value is given, we subtract 2 (the difference) to find the smaller value. If the smaller value is given, we add the difference to find the larger value.

Students in *CMC* are taught to use this reasoning to make sense of word problems that compare. For example:

> Isabel is 2 years younger than Sophia.
> Isabel is 8 years old. How old is Sophia?

Rather than jumping into a solution attempt, students are taught to analyze and diagram the relationship between the two values, identify and substitute the value given in the problem, and base their solution strategy on the value that is not given. For this example, students first represent the relationship between the two values. Initial letters represent the values. The larger value is written at the end of a number-family arrow, with the smaller value and the difference shown on the arrow:

$$\underrightarrow{2 \qquad I} \; S$$

The value given in the problem replaces the letter in the diagram:

$$\underrightarrow{2 \qquad \overset{8}{I}} \; S$$

The smaller value is given, so we add to find Sophia's age:

$$\underrightarrow{2 \qquad \overset{8}{I}} \; S \qquad \begin{array}{r} 2 \\ + 8 \\ \hline 1\,0 \end{array}$$

Here is the first part of the exercise from Level C Lesson 48, illustrating how the students apply the strategy.

> (Teacher reference:)
>
> a. Heidi has 17 more marbles than Bill has.
>   Heidi has 48 marbles.
>   How many marbles does Bill have?
>
> b. Sarah made 10 more cupcakes than Maria made.
>   Maria made 24 cupcakes.
>   How many cupcakes did Sarah make?
>
> c. Hank's car is 8 years older than Tim's car.
>   Hank's car is 11 years old.
>   How old is Tim's car?
>
> d. Bob has 7 dollars less than Val has.
>   Bob has 31 dollars.
>   How many dollars does Val have?
>
>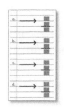
>
> You're going to work problems that tell about people and things. Remember, the first letter of each name is underlined.
>
> b. Touch problem A. ✔
> • Heidi has 17 more marbles than Bill has.
>   Say the sentence. (Signal.) *Heidi has 17 more marbles than Bill has.*
> • What's the letter for Heidi? (Signal.) *H.*
> • What's the letter for Bill? (Signal.) *B.*
>   Heidi has 17 more marbles than Bill has.
> • Make the family with two letters and the number. ✔
>   (Display:)   [48:8A]
>
>   a. $\underset{\longrightarrow}{17 \quad B} H$
>
>   Here's what you should have.
> c. The next sentence in the problem says: Heidi has 48 marbles.
> • Put a number for Heidi in the family. Then figure out how many marbles Bill has. Write that number in the family.
>   (Observe students and give feedback.)
> d. Check your work.
>   (Add to show:)   [48:8B]
>
>   a. $\underset{\longrightarrow}{17 \quad \overset{31}{\cancel{B}}} \overset{48}{\cancel{H}}$ $\begin{array}{r} 48 \\ -17 \\ \hline 31 \end{array}$
>
>   Here's what you should have.
> • The problem asks: How many marbles does Bill have? What's the answer? (Signal.) *31.*

from Lesson 48, Exercise 8

Note that three different types of problems are worked in this problem set:

The smaller-value unknown with "more" language (items a and c), the bigger-value unknown with "more" language (item b), and the bigger-value unknown with "less" language (item d). These types are described in Table 1 in the *Common Core State Standards for Mathematics* document (excerpt below).

Bigger Unknown	Smaller Unknown
(Version with "more"): Julie has three more apples than Lucy. Lucy has two apples. How many apples does Julie Have?	(Version with "more"): Julie has three more apples than Lucy. Julie has five apples. How many apples does Lucy Have?
(Version with "fewer"): Lucy has 3 fewer apples than Julie. Lucy has two apples. How many apples does Julie have? $2 + 3 = ?, 3 + 2 = ?$	(Version with "fewer"): Lucy has 3 fewer apples than Julie. Julie has five apples. How many apples does Lucy have? $5 - 3 = ?, ? + 3 = 5$

Table 1 from the *Common Core State Standards for Mathematics* describes 15 different addition-subtraction situations, all of which students can solve with variations of a number-family strategy by the end of CMC Level C. Foundation work for number families is illustrated in Standard MP7.

MP2: REASON ABSTRACTLY AND QUANTITATIVELY.

CMC Level F (Grade 5): Algebraic Translation

As students progress thorough the levels of *CMC* they make sense of quantities and their relationships through a variety of representations. The earliest representations are pictorial and countable, but by Level C, students frequently use more abstract letter representations for quantities, relationships, and units named in problem situations. By Level E, these problem situations are extended to include the four basic operations, ratio and proportion, and unit conversion.

Students in *CMC Level F* make sense of quantities and their relationships through extensive work with algebraic translation, which they apply to a variety of word-problem contexts, including fractions of a group and probability.

For example:

- 2/3 of the cats are awake. 6 cats are sleeping. How many cats are there?
- There are 8 marbles in a bag; 5 are green. If you take 24 trials at pulling a marble from the bag without looking, how many times would you expect to draw a marble that is not green?

Students are well prepared to *decontextualize* and manipulate symbols independently through initial work with solving letter equations. They also can *contextualize* by attending carefully to the details of the problem to be solved in order to discriminate which specific letter equation a given problem requires.

Early work equips students with the algebraic skills to solve letter equations of the form 2/5 R = M with a substitution for either letter.

Here are the solution steps when R = 10:

$$\frac{2}{5} R = M$$

$$\frac{2}{5} (10) = M$$

$$\frac{20}{5} = M = 4$$

Here are the solution steps when M = 90:

$$\frac{2}{5} R = M$$

$$\frac{2}{5} R = 90$$

$$\left(\frac{5}{2}\right) \frac{2}{5} R = 90 \left(\frac{5}{2}\right)$$

$$R = \frac{450}{2} = 225$$

Students are first taught the definition and application of a reciprocal and the equality principle that states the following: If we change one side of an equation, we must change the other side in the same way.

Here is part of an exercise from Lesson 38.

h. (Display:) [38:3H]

$$\frac{5}{4} P = 10$$

$$1 P =$$

- Read the problem. (Signal.) *5/4 P = 10.* We have to figure out what 1 P equals.
- What do we change 5/4 into? (Signal.) *1.*
- So what do we multiply 5/4 by? (Signal.) *4/5.*
- What do we multiply the other side by? (Signal.) *4/5.*
  (Repeat until firm.)
  (Add to show:) [38:3I]

$$\left(\frac{4}{5}\right) \frac{5}{4} P = \frac{10}{1} \left(\frac{4}{5}\right)$$

$$1 P =$$

i. (Point right.) Say the problem for this side. (Signal.) *10 × 4/5.*
- Raise your hand when you know the fraction answer. ✔
- What's the fraction? (Signal.) *40/5.*
  (Add to show:) [38:3J]

$$\left(\frac{4}{5}\right) \frac{5}{4} P = \frac{10}{1} \left(\frac{4}{5}\right)$$

$$1 P = \frac{40}{5}$$

- Raise your hand when you know the number 1 P equals. ✔
- What does 1 P equal? (Signal.) *8.*
  (Add to show:) [38:3K]

$$\left(\frac{4}{5}\right) \frac{5}{4} P = \frac{10}{1} \left(\frac{4}{5}\right)$$

$$1 P = \frac{40}{5} = \boxed{8}$$

j. Remember, multiply both sides by the reciprocal. Then figure out what the letter equals.

from Lesson 38, Exercise 3

Students have extensive practice with substitution and solving letter equations prior to the introduction of word problem applications. This level of proficiency enables them to manipulate symbols confidently once they have analyzed the problem and represented it with a letter equation.

Students first work with sentences that describe a fraction of a group, and write the letter equation. Here are two sentences and the corresponding equations from Lesson 46:

$\frac{3}{4}$ of the dogs were hungry.

$$\frac{3}{4} d = h$$

$\frac{1}{9}$ of the students wore coats.

$$\frac{1}{9} s = c$$

In subsequent lessons, students work the simplest type of word problem, where a number is given for one of the letters, and students solve for the other letter to answer the question the problem asks.

For example:

- 2/5 of the rabbits were white. There were 25 rabbits. How many white rabbits were there?

$$\frac{2}{5} r = w$$

$$\frac{2}{5} (25) = w = \frac{50}{5} = 10$$

$\boxed{10 \text{ white rabbits}}$

Here's part of an exercise from Level F Lesson 78, where students work a complete problem that asks two questions. They assess their initial solution and refer back to the problem to ascertain which question relates to the letter solution, and which question remains to be answered.

b. Problem A: 2/7 of the dogs were sleeping. 8 dogs were sleeping. How many dogs were awake? How many dogs were there?
- Write the letter equation. Replace one of the letters with a number. Stop when you've done that much. ✔
- Everybody, read the letter equation. (Signal.) *2/7 D = S.*
- Read the equation with a number. (Signal.) *2/7 D = 8.*
  (Display:)   [78:4C]

   a. $\frac{2}{7} d = s$

        $\frac{2}{7} d = 8$

Here's what you should have.
- Work the problem and write the unit name in the answer. Remember to simplify before you multiply. (Observe students and give feedback.)
c. Check your work.
- What does D equal? (Signal.) *28.*
- Which question does that answer? (Signal.) *How many dogs were there?*
  (Add to show:)   [78:4D]

   a. $\frac{2}{7} d = s$

    $\left(\frac{7}{2}\right) \frac{2}{7} d = \overset{4}{\cancel{8}} \left(\frac{7}{\cancel{2}}\right)$

      $d = 28$

    $\boxed{28 \text{ dogs}}$

Here's what you should have.
d. Now figure the answer to the other question. ✔ The other question is: How many dogs were awake?
- Everybody, say the subtraction problem you worked. (Signal.) *28 − 8.*
- What's the answer? (Signal.) *20.*
  (Display:)   [78:4E]

   a. $\frac{2}{7} d = s$

    $\left(\frac{7}{2}\right) \frac{2}{7} d = \overset{4}{\cancel{8}} \left(\frac{7}{\cancel{2}}\right)$

      $d = 28$ $\begin{array}{r} 2\ 8 \\ -\ \ 8 \\ \hline \end{array}$

    $\boxed{28 \text{ dogs}}$ $\boxed{2\ 0 \text{ awake dogs}}$

Here's what you should have.

from Lesson 78, Exercise 4

On later lessons, students work more advanced problems where the problem gives a fraction for one part of the group but gives a number for the other part of the group. For example:

- 2/7 of the dogs were running. 20 dogs were *not* running. How many dogs were running? How many dogs were there in all?

Students must analyze the problem carefully to discriminate it from the earlier problem types that involve only two names (e.g., dogs and running dogs). The number given is for dogs *not running*, so rather than writing the basic equation 2/7 d = r, students write the complementary letter equation 5/7 d = n. (5/7 of the dogs were *not running*.)

Students first practice discriminating between sentence pairs that give two names and those that give three names.

For example:

- 3/8 of the children are boys. There are 15 boys. (2 names: 3/8 c = b)
- 3/8 of the children are boys. There are 15 girls. (3 names: 5/8 c = g)

Here is part of the exercise from Lesson 90, where students solve a complete problem involving three names.

g. (Display:)   [90:7C]

> $\frac{2}{7}$ of the dogs were running. 20 dogs were not running. How many dogs were running? How many dogs were there in all?

New problem: 2/7 of the dogs were running. 20 dogs were not running. How many dogs were running? How many dogs were there in all?
This problem asks two questions. You work it the same way you work the other problems with three names. You write the equation for the name that has a number and solve the equation. That answers one of the questions.

h. Write the equation and solve it. Write the answer with a unit name. Stop when you've done that much.
(Observe students and give feedback.)

i. I'll read the questions the problem asks. You'll tell me which question you can now answer:
How many dogs were running?
How many dogs were there in all?
- Which question can you answer? (Signal.) *How many dogs were there in all?*
- What's the answer? (Signal.) *28 dogs.*

j. (Add to show:)   [90:7D]

> $\frac{2}{7}$ of the dogs were running. 20 dogs were not running. How many dogs were running? How many dogs were there in all?
>
> $$\frac{5}{7}\,d = n$$
> $$\left(\frac{7}{5}\right)\frac{5}{7}\,d = \overset{4}{\cancel{20}}\left(\frac{7}{5}\right)$$
> $$d = 28$$
>
> $$\boxed{28\ \text{dogs}}$$

Here's what you should have.

k. The other question is: How many dogs were running?
Now you can figure out how many dogs were running.
- Raise your hand when you can say the problem. ✔
- Say the problem. (Signal.) *28 – 20.*

l. Figure out the answer and write the unit name—running dogs. ✔
- Everybody, how many dogs were running? (Signal.) *8 running dogs.*

from Lesson 90, Exercise 7

Once students have set up the correct equation, they follow familiar solution steps. They figure out how many dogs there were in all (which is the answer to the second question). They then subtract to figure out the number of dogs that were running.

The strategy builds the habit of a coherent representation of the problem at hand (a familiar equation form), requires students to consider the units for the quantities represented by the letters as they solve the equation, consider both questions the problem asks, and then determine which has been answered and which remains to be answered.

Coherent representation is further strengthened in later applications to new domains such as probability. Students first learn to create a probability fraction based on the composition of the set. This fraction gives the likelihood of a particular object being drawn from the set on any given trial. The following examples from lesson 96 require students to construct probability fractions for given sets of objects and to construct a set of objects given the probability fraction.

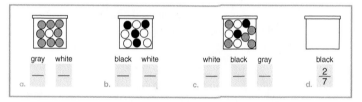

from Workbook Lesson 96, Part 1

For item a, students write the fraction 8/9 for gray and 1/9 for white. For item d, students draw 7 marbles, and shade 2 of them black.

With this background it is a simple transition to work problems that involve trials using the familiar equation form.

Here are two examples and the equations students work from Lesson 98.

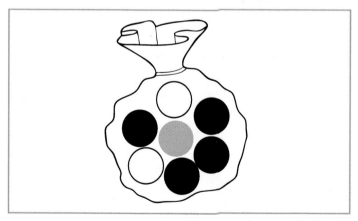

from Textbook Lesson 98, Part 2

- If you took trials until you pulled out four black marbles, about how many trials would you take?

$$\frac{4}{7} t = b$$

$$\frac{4}{7} t = 4$$

- If you take 98 trials, about how many black marbles would you expect to pull out?

$$\frac{4}{7} t = b$$

$$\frac{4}{7} (98) = b$$

Students already have the mathematical tools to solve the equations, and learning a wide range of applications for equation form builds flexibility in the application of those tools.

MP3: CONSTRUCT VIABLE ARGUMENTS AND CRITIQUE THE REASONING OF OTHERS.
CMC Level C (Grade 2): Column Addition/ Subtraction

As students develop mathematically, they build a basis for constructing arguments to justify the mathematical procedures they use, or the steps they take to perform an algorithm. They are able to decide whether variations and deviations from the familiar progression of steps make sense. In a column addition or subtraction problem that involves regrouping, students should display an understanding of the underlying place value that conserves the numbers they are manipulating.

Students in *CMC Level C* work extensively with place value to build the conceptual basis for the steps they take in addition/subtraction regrouping problems, for example:

$$\begin{array}{r} \overset{1}{2}8 \\ +45 \\ \hline 73 \end{array} \quad \text{or} \quad \begin{array}{r} \overset{8}{\cancel{9}}\overset{12}{2} \\ -36 \\ \hline 56 \end{array}$$

Students are familiar with identifying the tens digit and ones digit of two-digit numbers and with the place value of each digit. When applied to column addition, the students simply write a 2-digit answer to the problem for the ones column as the "tens digit" and the "ones digit" in the appropriate order and in the appropriate column. Here's part of an early exercise from Lesson 21.

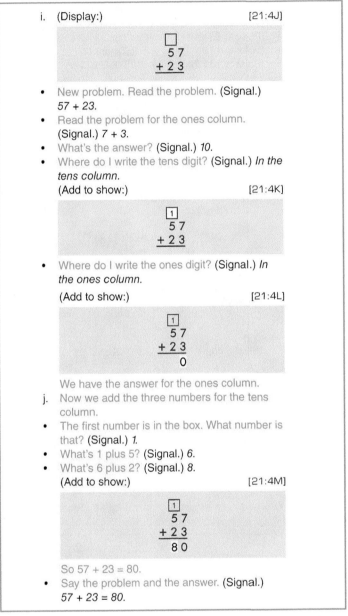

i. (Display:) [21:4J]

$$\begin{array}{r} \square \\ 5\ 7 \\ +\ 2\ 3 \\ \hline \end{array}$$

- New problem. Read the problem. (Signal.) *57 + 23.*
- Read the problem for the ones column. (Signal.) *7 + 3.*
- What's the answer? (Signal.) *10.*
- Where do I write the tens digit? (Signal.) *In the tens column.*
  (Add to show:) [21:4K]

$$\begin{array}{r} \boxed{1} \\ 5\ 7 \\ +\ 2\ 3 \\ \hline \end{array}$$

- Where do I write the ones digit? (Signal.) *In the ones column.*
  (Add to show:) [21:4L]

$$\begin{array}{r} \boxed{1} \\ 5\ 7 \\ +\ 2\ 3 \\ \hline 0 \end{array}$$

  We have the answer for the ones column.
j. Now we add the three numbers for the tens column.
- The first number is in the box. What number is that? (Signal.) *1.*
- What's 1 plus 5? (Signal.) *6.*
- What's 6 plus 2? (Signal.) *8.*
  (Add to show:) [21:4M]

$$\begin{array}{r} \boxed{1} \\ 5\ 7 \\ +\ 2\ 3 \\ \hline 8\ 0 \end{array}$$

So 57 + 23 = 80.
- Say the problem and the answer. (Signal.) *57 + 23 = 80.*

from Lesson 21, Exercise 4

Having practiced this procedure to automaticity for more than 90 lessons, students are ready to discuss the steps they take and critique deviations from the procedure. Here is an exercise from Lesson 114.

a. (Display:) [114:4A]

$$\begin{array}{r} 4\ 6 \\ +\ 3\ 9 \\ \hline \end{array}$$

I'm going to ask you a lot of questions about working this problem.
- Read the problem. (Signal.) *46 + 39.*
- Read the problem for the ones. (Signal.) *6 + 9.*
- What's the answer? (Signal.) *15.*
b. What if I just put 15 in the ones column?
  (Add to show:) [114:4B]

$$\begin{array}{r} 4\ 6 \\ +\ 3\ 9 \\ \hline 1\ 5 \end{array}$$

- Can I do this? (Signal.) *No.*
- What's wrong with this? (Call on a student. Idea: *15 is too many to have in the ones column.*)
- What's the largest number we can have in the ones column? (Call on a student. *9.*)
c. The answer for the ones is 15. Everybody, say the place-value addition for 15. (Signal.) *10 + 5.*
- I can't write **both** digits in the ones column. Where do I write the **ones** digit of 15? (Signal.) *In the ones column.*
- Where do I write the **tens** digit of 15? (Signal.) *In the tens column.*
  (Change to show:) [114:4C]

$$\begin{array}{r} 4\ 6 \\ +\ 3\ 9 \\ \hline 1\ 5 \end{array}$$

- (Point to **1.**) Can I write the tens digit here? (Signal.) *No.*
- Why not? (Call on a student. Idea: *You have to add it to the tens.*)
d. Where do I write it? (Call on a student. Idea: *At the top of the tens column.*)
  (Change to show:) [114:4D]

$$\begin{array}{r} {\scriptstyle 1} \\ 4\ 6 \\ +\ 3\ 9 \\ \hline 5 \end{array}$$

- Everybody, read the problem for the tens. (Signal.) *1 + 4 + 3.*
- Raise your hand when you know the answer. ✔
- What's 1 + 4 + 3? (Signal.) *8.*
  (Add to show:) [114:4E]

$$\begin{array}{r} {\scriptstyle 1} \\ 4\ 6 \\ +\ 3\ 9 \\ \hline 8\ 5 \end{array}$$

- Read the problem we started with and the answer. (Signal.) *46 + 39 = 85.*
- Say the place-value addition for the answer. (Signal.) *80 + 5 = 85.*

from Lesson 114, Exercise 4

A similar development is taught for subtraction. Before students work with column problems that require renaming the top number, student work on the conservation of two-digit values based on place value.

Here's an introductory exercise from Level C Lesson 26.

a. You're going to say the place-value fact for different numbers.
- Say the place-value fact for 53. (Signal.) *50 + 3 = 53.*
- Say the place-value fact for 29. (Signal.) *20 + 9 = 29.*
- Say the place-value fact for 70. (Signal.) *70 + 0 = 70.*
- Say the place-value fact for 71. (Signal.) *70 + 1 = 71.*
- Say the place-value fact for 17. (Signal.) *10 + 7 = 17.*
(Repeat until firm.)

b. (Display:) [26:5A]

$$30 + 6 = 36$$
$$40 + 8 = 48$$
$$70 + 1 = 71$$
$$50 + 3 = 53$$

c. (Point to **30.**) You're going to subtract 10 from this number.
- What's 30 − 10? (Signal.) *20.*
d. (Point to **40.**) You're going to subtract 10 from this number.
- Say the problem. (Signal.) *40 − 10.*
- What's the answer? (Signal.) *30.*

e. (Point to **70.**) You're going to subtract 10 from this number.
- Say the problem. (Signal.) *70 − 10.*
- What's the answer? (Signal.) *60.*
f. (Point to **50.**) You're going to subtract 10 from this number.
- Say the problem. (Signal.) *50 − 10.*
- What's the answer? (Signal.) *40.*
(Repeat until firm.)

g. You've subtracted 10 from the tens number. You're going to add 10 to the ones number.
- (Point to **6.**) What's 10 + 6? (Signal.) *16.*
- (Point to **8.**) What's 10 + 8? (Signal.) *18.*
- (Point to **1.**) What's 10 + 1? (Signal.) *11.*
- (Point to **3.**) What's 10 + 3? (Signal.) *13.*

h. This time you're going to subtract 10 from the tens number and add that 10 to the ones number.

i. (Point to **30.**) What's 30 − 10? (Signal.) *20.*
(Change to show:) [26:5B]

$$30 + 6 = 36$$
$$20$$

- What's 10 + 6? (Signal.) *16.*
(Add to show:) [26:5C]

$$30 + 6 = 36$$
$$20 + 16 = 36$$

- Read the new place-value fact for 36. (Signal.) *20 + 16 = 36.*
j. (Display:) [26:5D]

$$40 + 8 = 48$$

- (Point to **40.**) What's 40 − 10? (Signal.) *30.*
(Add to show:) [26:5E]

$$40 + 8 = 48$$
$$30$$

- What's 10 + 8? (Signal.) *18.*
(Add to show:) [26:5F]

$$40 + 8 = 48$$
$$30 + 18 = 48$$

- Read the new place-value fact for 48. (Signal.) *30 + 18 = 48.*

from Lesson 26, Exercise 5

Students then practice rewriting 2-digit numbers in isolation to show the "new fact." Here's part of the introduction from Lesson 33.

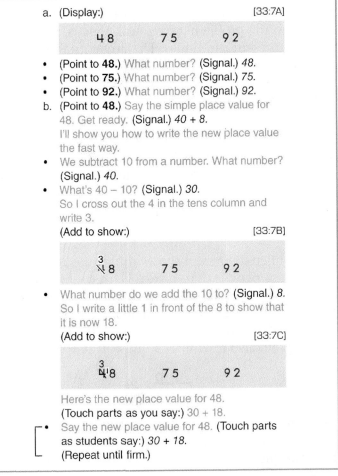

a. (Display:) [33:7A]

48   75   92

- (Point to **48**.) What number? (Signal.) *48.*
- (Point to **75**.) What number? (Signal.) *75.*
- (Point to **92**.) What number? (Signal.) *92.*
b. (Point to **48**.) Say the simple place value for 48. Get ready. (Signal.) *40 + 8.*
I'll show you how to write the new place value the fast way.
- We subtract 10 from a number. What number? (Signal.) *40.*
- What's 40 – 10? (Signal.) *30.*
So I cross out the 4 in the tens column and write 3.
(Add to show:) [33:7B]

³4̶8   75   92

- What number do we add the 10 to? (Signal.) *8.*
So I write a little 1 in front of the 8 to show that it is now 18.
(Add to show:) [33:7C]

³4̶¹8   75   92

Here's the new place value for 48.
(Touch parts as you say:) *30 + 18.*
- Say the new place value for 48. (Touch parts as students say:) *30 + 18.*
(Repeat until firm.)

from Lesson 33, Exercise 7

Given this background and practice that continues for 15 lessons, students have a conceptual framework to make sense of and justify the steps in the subtraction algorithm, which begins on Lesson 43.

After the algorithm is established and practiced for 70 lessons, students also discuss and critique the steps in the algorithm and deviations from the procedure. Here is an exercise from Lesson 115.

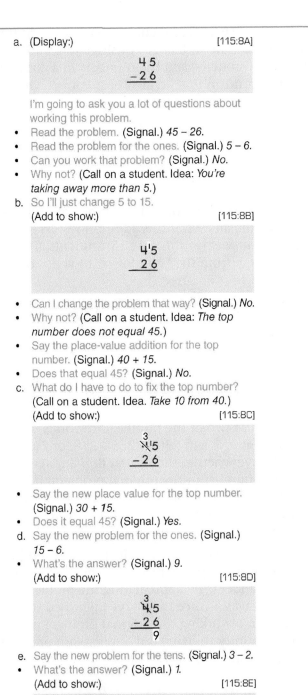

a. (Display:) [115:8A]

$$\begin{array}{r} 45 \\ -26 \\ \hline \end{array}$$

I'm going to ask you a lot of questions about working this problem.
- Read the problem. (Signal.) *45 – 26.*
- Read the problem for the ones. (Signal.) *5 – 6.*
- Can you work that problem? (Signal.) *No.*
- Why not? (Call on a student. Idea: *You're taking away more than 5.*)
b. So I'll just change 5 to 15.
(Add to show:) [115:8B]

$$\begin{array}{r} 4^{\prime}5 \\ 26 \\ \hline \end{array}$$

- Can I change the problem that way? (Signal.) *No.*
- Why not? (Call on a student. Idea: *The top number does not equal 45.*)
- Say the place-value addition for the top number. (Signal.) *40 + 15.*
- Does that equal 45? (Signal.) *No.*
c. What do I have to do to fix the top number? (Call on a student. Idea. *Take 10 from 40.*)
(Add to show:) [115:8C]

$$\begin{array}{r} ^3\!\!\!4^{\prime}5 \\ -26 \\ \hline \end{array}$$

- Say the new place value for the top number. (Signal.) *30 + 15.*
- Does it equal 45? (Signal.) *Yes.*
d. Say the new problem for the ones. (Signal.) *15 – 6.*
- What's the answer? (Signal.) *9.*
(Add to show:) [115:8D]

$$\begin{array}{r} ^3\!\!\!4^{\prime}5 \\ -26 \\ \hline 9 \end{array}$$

e. Say the new problem for the tens. (Signal.) *3 – 2.*
- What's the answer? (Signal.) *1.*
(Add to show:) [115:8E]

$$\begin{array}{r} ^3\!\!\!4^{\prime}5 \\ -26 \\ \hline 19 \end{array}$$

- What's the top number we started with? (Signal.) *45.*
- Say the problem we started with and the answer. (Signal.) *45 – 26 = 19.*

from Lesson 115, Exercise 8

Students in Level C also check answers to addition and subtraction problems. For single-digit addition problems, students add the numbers in a different order to confirm that an answer is correct. Given the problem and (incorrect) answer:

$$\begin{array}{r} 2 \\ 4 \\ + 9 \\ \hline 1\,6 \end{array}$$

Students work the problem from the "bottom up" to check the answer. If this addition results in a different answer (in this case 15) students also work the problem from the "top down" to verify that the answer of 16 is incorrect.

Students also check answers to multi-digit problems using inverse operations. Given the problem and (incorrect) answer:

$$\begin{array}{r} 5\,8 \\ + 2\,1\,8 \\ \hline 2\,6\,6 \end{array}$$

Students use two of the numbers to work a subtraction problem (either 266 − 218, or 266 − 58) and observe whether the subtraction answer is the third number in the original problem. If it is not, they conclude that the answer to the original addition problem is wrong, and they rework the problem to figure out the correct answer.

These strategies enable students to check and justify their answers, or evaluate and critique the work of other students by approaching the problem a "different way." Once they have identified that there is an error, reworking the original problem provides an opportunity to figure out and explain precisely where an error was made.

MP4: MODEL WITH MATHEMATICS.
CMC Level D (Grade 3): Multiplication/ Division Story Problems

Students in *CMC Level D* learn to apply the mathematics they know to a range of word problems that represent situations arising in everyday life. This section will illustrate how students model problems solved by multiplication or division.

Students identify the important quantities in a situation and map their relationship with a number-family diagram showing one name that represents the product, and one name that represents a factor in a multiplicative relationship. This modeling parallels the work with addition/subtraction number families described in MP1 above.

Here's a basic problem:

> **Each box has 6 pencils. There are 24 pencils. How many boxes are there?**

Students represent the relationship between the items named in the problem: boxes and pencils. Initial letters represent the values. The larger quantity is written at the end of the number-family arrow. There are 6 times more pencils than boxes, so *p* is at the end of the number family arrow:

The smaller quantity *(b)* is written on the arrow, and the relationship number (6) is the first number in the diagram.

The completed diagram shows that there are six times more pencils than boxes. With the relationship modeled in this way, when students substitute for the value that is given (24 pencils), they see that they must divide to find the smaller quantity. (Note that the diagram also resembles a division problem).

Conversely, when students substitute for the smaller quantity, they multiply to find the larger quantity. For example:

> **Each box has 6 pencils. There are 50 boxes. How many pencils are there?**

The relationship that students map is the same as above:

Students substitute for b:

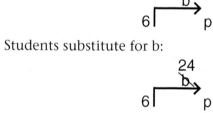

This diagram translates into a multiplication problem to figure out the larger quantity.

$$\begin{array}{r} 2\,4 \\ \times\quad 6 \\ \hline \end{array}$$

By carefully mapping the situation and relating the quantities before deciding which operation to perform, students can reliably solve problems that typically cause confusion for students. In the latter problem, students may read the word *each* and decide that the problem involves multiplication or division, but they are tempted to divide because the numbers given in the problem are from a familiar division fact (24 ÷ 6). By mapping the relationship before they work the problem, students in *CMC* are much less likely to make this mistake.

Students write number families from sentences for several lessons to establish the mapping strategy. They are then ready to work complete problems. Here's part of the exercise from Level D Lesson 72, where students work the following problems:

- Each dime is worth 10 cents. Tom had 8 dimes. How many cents did he have?
- Every chair has 4 legs. There are 12 legs. How many chairs are there?

b. Touch problem A in your textbook. ✔
I'll read it: Each dime is worth 10 cents. Tom had 8 dimes. How many cents did he have?
- Raise your hand when you know which sentence tells about each or every. ✔
- Read that sentence. (Signal.) *Each dime is worth 10 cents.*
- Is the big number dimes or cents? (Signal.) *Cents.*
Yes, there are more cents than dimes.
- Write two letters and a number in the number family.
(Observe students and give feedback.)
(Display:)   [72:6J]

Here's what you should have.
c. The problem gives another number for dimes.
- Look at the problem. Raise your hand when you know the number. ✔
- What's the number? (Signal.) *8.*
- Write the number for dimes in the family. ✔
(Add to show:)   [72:6K]

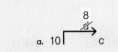

Here's what you should have.
d. Do you multiply or divide to find the missing number? (Signal.) *Multiply.*
- Start with the 2-digit number and say the problem you'll work. (Signal.) *10 × 8.*
- Write the problem next to the number family. Work the problem. Write the unit name in the answer.
(Observe students and give feedback.)
- You worked the problem 10 times 8. What's the whole answer? (Signal.) *80 cents.*
Yes, Tom had 80 cents.
e. Check your work.
(Add to show:)   [72:6L]

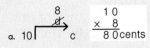

Here's what you should have.

f. Count down 4 lines on your lined paper. Write the letter B and a multiplication number family arrow. ✔
- Touch problem B. ✔
I'll read the problem: Every chair has 4 legs. There are 12 legs. How many chairs are there?
- Raise your hand when you know which sentence tells about each or every. ✔
- Read that sentence. (Signal.) *Every chair has 4 legs.*
- Write two letters and a number in the family.
(Observe children and give feedback.)
(Display:)   [72:6M]

b. 4 ⌐ →c 1

Here's what you should have.
g. The problem gives a number for one of the letters.
Look at the problem. Raise your hand when you know the number. ✔
- What's the number? (Signal.) *12.*
- Is that number for chairs or legs? (Signal.) *Legs.*
- Write that number in the family. ✔
(Add to show:)   [72:6N]

b. 4 ⌐ →c 12
             ⫽

Here's what you should have.
h. Say the problem you'll work to find the number of chairs. (Signal.) *12 divided by 4.*
- Write the problem next to the number family. Work the problem. Write the unit name in the answer.
(Observe students and give feedback.)
- Read the problem and the whole answer. (Signal.) *12 divided by 4 = 3 chairs.*
Yes, there were 3 chairs.
i. Check your work.
(Add to show:)   [72:6O]

b. 4 ⌐ →c 12   3 chairs
             ⫽   4⌐1 2

Here's what you should have.

from Lesson 72, Exercise 6

Note that the same strategy accommodates money problems that convert a coin value into cents and vice versa. (For example: Tom had 80 cents in dimes. How many dimes did he have?)

Later in Level D, students learn to work multiplication and division problems that use the word "times." Here are two items and the student work from Lesson 91.

a. The yellow snake was 9 times as long as the red snake.
The red snake was 10 inches long.
How long was the yellow snake?

b. There were 6 times as many spoons as forks.
There were 66 spoons.
How many forks were there?

Item b results in a division problem. Again, representing the related values in the problem before choosing the operation reduces the likelihood that students will multiply instead of divide because the problem involves the word "times."

Another real-life application for the number-family mapping taught in Level D is unit conversion. Students make a number family for measurement facts. For example:

1 foot equals 12 inches.

Students identify that there are more inches than feet, so the number family relationship is:

Here are the facts that student map in Lesson 110:

- 1 week equals 7 days.
- 1 gallon equals 4 quarts.
- 1 quarter equals 25 cents.
- 1 pound equals 16 ounces.

Here's the first part of the exercise.

> b. You're going to make number families for measurement facts.
> - Touch and read fact A. (Signal.) *1 week = 7 days.*
> - Are there more weeks or days? (Signal.) *Days.*
> - So which is the big number? (Signal.) *Days.*
> - Make the family for the fact about weeks and days.
> (Observe students and give feedback.)
> c. Check your work.
> (Display:)   [110:4A]
>
>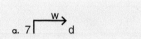
>
> Here's what you should have.
> 7 and weeks are small numbers. Days is the big number.
> d. Read fact B. (Signal.) *1 gallon = 4 quarts.*
> - Are there more gallons or quarts?
> (Signal.) *Quarts.*
> - So which is the name for the big number in the family? (Signal.) *Quarts.*
> - Make the family.
> (Observe students and give feedback.)
> e. Check your work.
> (Display:)   [110:4B]
>
> b. 4⌐—G→ q
>
> Here's what you should have.
> 4 and gallons are small numbers. Quarts is the big number.

from Lesson 110, Exercise 4

On later lessons, students work complete problems. For example:

- 1 nickel equals 5 cents. How many nickels is 150 cents?
- 1 day equals 24 hours. How many hours is 9 days?

Students divide to find the number of nickels in the first item, and multiply to find the number of hours in the second item. Traditionally these problem types cause confusion for elementary students because they are counterintuitive. To solve the problems correctly, students *multiply* when the problem asks about the *smaller* unit, and *divide* when the problem asks about the *larger* unit. The number-family analysis provides students in *CMC* with a consistent and reliable way to tackle these and the full range of other situations that call for multiplication or division.

MP5: USE APPROPRIATE TOOLS STRATEGICALLY.

CMC Level E (Grade 4): Geometry

CMC teaches students self-reliance through the extensive use of conceptual strategies that can be applied using paper and pencil. Students generate solutions to a wide range of problems presented in the Textbook and Workbook. Tasks often move systematically from a workbook (more highly structured) to a textbook setting as students develop the conceptual tools to tackle problems more independently. For example, initial work with area and perimeter appears in the Workbook and only requires students to write the multiplication or addition problem. Here is the Workbook part and student work from Lesson 8.

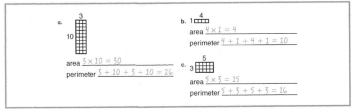

from Workbook Lesson 8 AK, Part 2

In later lessons, students respond to diagrams shown in the Textbook and work much more advanced problems. Below is a set of examples from Lesson 93 Textbook. For some problems, students find the area of a rectangle. For others they find the length of a side of the rectangle. They show their answers with appropriate linear or square units.

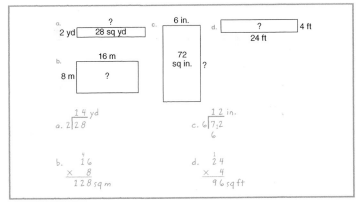

from Textbook Lesson 93 AK, Part 2

The work with area and perimeter culminates with story problems for which students sketch a rectangle to represent the problem. For example:

> a. A farmer wants to put a fence around a rectangular garden. The garden is 132 yards wide and 58 yards long. How many yards of fencing does the farmer need to put a fence around that garden?
>
> b. Jim has enough paint to cover 100 square feet of a wall. The wall he wants to paint is 8 feet tall. If he has just enough paint to cover the wall, how long is the wall?

from Textbook Lesson 126, Part 4

Here is part of the exercise.

> b. Read problem A. (Call on a student.) *A farmer wants to put a fence around a rectangular garden. The garden is 132 yards wide and 58 yards long. How many yards of fencing does the farmer need to put a fence around that garden?*
> • Tell me the length of the longest side of the rectangle. Get ready. (Signal.) *132 yards.*
> • What's the length of the other side? (Signal.) *58 yards.*
> • Make a sketch of the rectangle on your lined paper with the sides labeled. Make sure you label the longest side of your sketch with the biggest length.
> (Observe students and give feedback.)
> (Display:)   [126:6A]

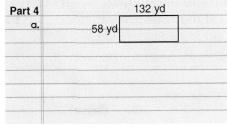

> Here's a sketch of the garden.
> • The question asks: How many yards of fencing does the farmer need to put a fence around that garden? Does that question ask about the area or perimeter? (Signal.) *Perimeter.*
> • Start with 132 and say the problem for finding the perimeter. Get ready. (Signal.) *132 + 58 + 132 + 58.*
> c. Read problem B. (Call on a student.) *Jim has enough paint to cover 100 square feet of a wall. The wall he wants to paint is 8 feet tall. If he has just enough paint to cover the wall, how long is the wall?*
> • The length of one of the sides of the wall is given. What's that length? (Signal.) *8 feet.*
> • Does the problem give another length or the area of the rectangle? (Signal.) *The area.*
> • What's the area of the rectangle? (Signal.) *100 square feet.*
> • Make a sketch of the rectangle on your lined paper with a side labeled and the area labeled.
> (Observe students and give feedback.)
> (Display:)   [126:6B]

> b.   8 ft | 100 sq ft

> Here's a sketch of the wall.
> • The question asks: How long is the wall? Say the problem for finding the length of the wall. Get ready. (Signal.) *100 ÷ 8.*
> d. Write the problem for each word problem below each rectangle and figure out the answer.
> (Observe students and give feedback.)

from Lesson 126, Exercise 6

Students work the addition or division problem and write the answer with a unit name.

Students are also familiar with tools appropriate to their grade, such as a protractor and ruler, which they use to explore angles, rays, line segments, and lines. They use a protractor first to measure given angles and then to construct angles. They use a ruler to complete the angle constructions and to graph straight lines on a coordinate system.

The protractor is introduced in Level E on Lesson 121. Before students physically use a protractor, they work with diagrams that show protractors properly placed. Students are often confused by the two rows of numbers on the protractor, so care is taken to teach students how to discriminate when one row or the other is used to measure an angle. Here is part of the introduction from Lesson 121.

a. In the last lesson, you learned about line segments and when they intersect.
• Think of a triangle. Are the sides of a triangle lines or line segments? (Signal.) *Line segments.*
• Do the line segments **intersect** at the corners or in the middle of a triangle? (Signal.) *At the corners.*
(Repeat until firm.)
b. You're going to learn about a tool that's used to measure angles. The tool is called a protractor. Say **protractor.** (Signal.) *Protractor.*
• What's the name of a tool used to measure angles? (Signal.) *(A) protractor.*
(Display:) [121:5A]

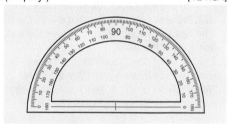

Here's a picture of a protractor. There are two sets of number scales on this protractor. The tens numbers on the inside scale start at zero here (touch 180 over zero) and go to 180 here (touch zero over 180). The tens numbers on the outside scale start at zero here (touch zero over 180) and go to 180 here (touch 180 over zero).
• Both number scales are the same for one number. What number? (Signal.) *90.* (Touch 90.)
• Everybody, what do you use a protractor for? (Signal.) *To measure angles.*
• What's the name of this tool? (Signal.) *(A) protractor.*

c. (Add to show:) [121:5B]

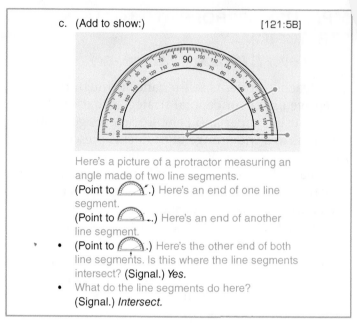

Here's a picture of a protractor measuring an angle made of two line segments.
(Point to ⌐.) Here's an end of one line segment.
(Point to ⌐.) Here's an end of another line segment.
• (Point to ⌐.) Here's the other end of both line segments. Is this where the line segments intersect? (Signal.) *Yes.*
• What do the line segments do here? (Signal.) *Intersect.*

from Lesson 121, Exercise 5

Similarly, students learn the critical features of properly placing a protractor through positive and negative examples. The lines must intersect at the center marker, and one of the lines must go through zero. This instruction ensures that students are able to use the tool accurately. Here are the examples from Lesson 123.

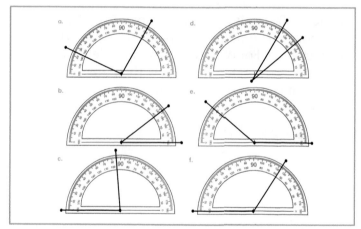

from Textbook Lesson 123, Part 2

Having measured angles for several lessons, students are well prepared to construct angles. Here is part of the exercise from Lesson 128.

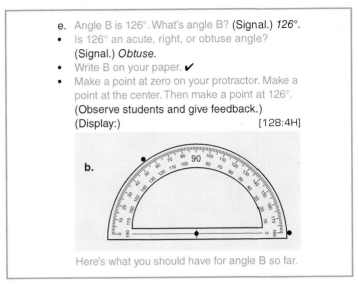

e. Angle B is 126°. What's angle B? (Signal.) *126°.*
• Is 126° an acute, right, or obtuse angle? (Signal.) *Obtuse.*
• Write B on your paper. ✔
• Make a point at zero on your protractor. Make a point at the center. Then make a point at 126°. (Observe students and give feedback.) (Display:) [128:4H]

b.

Here's what you should have for angle B so far.

from Lesson 128, Exercise 4

Students also work with rays, lines, and line segments in the Practice Software. Here are three examples of a task students complete (using a line-drawing tool) to discriminate between a ray, a line, and a segment:

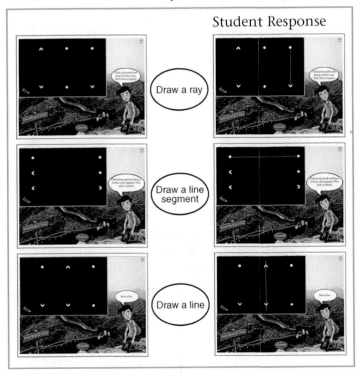

Student Response

Draw a ray

Draw a line segment

Draw a line

Following each example, students then make a line, line segment, or ray on the computer screen that is parallel or perpendicular to the first object they made.

Students also use a ruler to graph lines on the coordinate system. Here's the function table and coordinate system from Lesson 103.

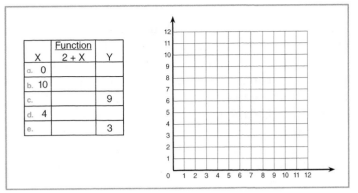

from Workbook Lesson 103, Part 1

Students plot two points, draw the line, and inspect the line to complete the missing X or Y values in the table. Here is the table and coordinate system showing the student work.

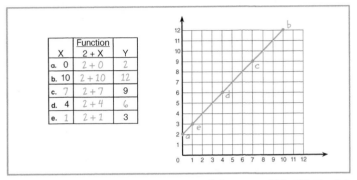

from Workbook Lesson 103 AK, Part 1

Students' development of meaningful pencil-and-paper strategies, their use of grade-level appropriate tools, and their exposure to related concepts in computer-based activities deepens their understanding of the concepts, and prepares them for more advanced applications in later grades.

MP6: ATTEND TO PRECISION.
CMC Level A (Kindergarten): Equality and Equations.

Precise communication in mathematics is necessary at every grade level, beginning in kindergarten. In *CMC Level A*, students are taught to understand the meaning of the mathematical symbols they use, including the equal sign. Students in *CMC Level A* learn a concise and clear definition of equality: "You must have the same number on one side of the equals that you have on the other side." This rule is applied first in the simplest context: 4 = ■ or ■ = 7. Here's part of the introduction:

---

a. (Write on the board:)   [19:6A]

           =

- (Point to =.) Everybody, what's this? (Touch.) *Equals.*
- Is equals a number? (Touch.) *No.*
  Right. It isn't.
- Here's the rule about equals. (Point to the space on the left.) You must have the same number on this side of the equals (point to the right side of the equals) and on this side of the equals.

b. Listen. (Point left.) If we have 4 on this side of the equals, (point right) we must have 4 on this side of the equals.
- (Point left.) Listen: If we have 4 on this side of the equals, (point right) how many must we have on this side of the equals? (Touch.) *4.*
  Yes, 4.
c. (Point right.) If we have 10 on this side of the equals, (point left) how many must we have on this side of the equals? (Touch.) *10.*
- (Point left.) If we have 15 on this side of the equals, (point right) how many must we have on this side of the equals? (Touch.) *15.*
  (Repeat steps b and c until firm.)

d. (Write to show:)   [19:6B]

           4 = ☐

- This says **4 equals box.** What does it say? (Touch.) *4 equals box.*
- There's a number on one side of the equals. What number? (Signal.) *4.*
  Yes, 4.
e. (Point to **4**.) If we have 4 on that side of the equals, (point to box) how many must there be on this side of the equals? (Touch.) *4.*
  Yes, that's the number that goes in the box.
- (Write to show:)   [19:6C]

           4 = 4

(Point to 4 = 4.) Now this says (touch each symbol as you read) 4 equals 4.
- What does it say? (Touch symbols.) *4 equals 4.*
f. (Write on the board:)   [19:6D]

           ☐ = 2

This says **box equals 2.**
- There's a number on one side of the equals. What number? (Signal.) *2.*
- (Point to 2.) If there are 2 on this side of the equals, (point to box) how many must there be on this side of the equals? (Touch.) *2.*
  Yes, that's the number that goes in the box.
- (Write to show:)   [19:6E]

           2 = 2

- (Point to 2 = 2.) What does this say now? (Touch symbols.) *2 equals 2.*

---

from Lesson 19, Exercise 6

By Lesson 25 students have learned to discriminate whether sides are equal by counting lines on one side of the equal sign and comparing that number with a number on the other side of the equal sign. If the sides are not equal, they cross out the equal sign.

Here are 3 examples from the student worksheet and part of the exercise:

$$5 = |||||| \qquad ||||||| = 9 \qquad |||| = 4$$

from Lesson 25, Exercise 8

Through these early exercises, the students learn the precise meaning of the symbol. If the conditions for equality are not met, they cross out the sign.

Next, students learn to complete equality statements, either by drawing lines for a numeral or by writing the numeral for the lines.

$$8 = \qquad ||| =$$

$$= |||||| \qquad = 2$$

This work lays the foundation for the operations that students perform in addition and subtraction. Students learn to modify sides of an equation to make sides equal:

For example:

$$3 = | | | | |$$

Students cross out two lines to make the sides equal:

$$3 = | | | \slashed{|}\slashed{|}$$

Other items show lines crossed out:

$$| | \slashed{|}\slashed{|} = \qquad\qquad | | | | | | \slashed{|}\slashed{|} =$$

Students record the number of lines not crossed out on the other side of the equals.

This work prepares students to incorporate the concept of equality into the subtraction strategy they will learn. For the problem 5 – 1 = ■, students "start with 5 and take away 1."

They draw 5 lines and cross out 1 line:

$$5 - 1 = \blacksquare$$

$$| | | | \slashed{|}$$

They count the lines not crossed out and apply the equality principle: They must have the same number on one side of the equals that they have on the other side. They write 4 in the box to make the sides equal. As each new problem type is introduced for addition or subtraction, the meaning of the equal sign is consolidated as students use the symbol consistently and appropriately

This careful foundation guards against a lack of understanding often displayed by older students who struggle with algebra. These students fail to develop a clear understanding of the equality principle. They may consider the equal sign to be more of a "punctuation mark" in a math statement than a symbol representing a powerful mathematical concept. This causes difficulty when they begin algebra, as the equality principle is the basis for modifying each side of an equation in the same way to conserve equality. The work in early levels of *CMC* prepares students to understand and apply the equality principle in a variety of contexts, in *CMC* and beyond.

MP7: LOOK FOR AND MAKE USE OF STRUCTURE.

CMC Levels B–D (Grades 1–3): Number family Addition/Subtraction Facts.

Students are taught to recognize the structure and organization of the number system in all levels of *Connecting Math Concepts*. They use patterns and relationships to master the basic facts in all four operations. For example, rather than learning the 200 basic addition/subtraction facts as isolated items of information, students learn related sets of facts through the concept of the number family.

The number family $\underset{\longrightarrow}{5 \qquad 2}7$ generates 4 facts:

$$5 + 2 = 7 \qquad 7 - 2 = 5$$
$$2 + 5 = 7 \qquad 7 - 5 = 2$$

Students refer to the largest of the three related numbers as the "big number," and the other two numbers as "small numbers." This standard vocabulary helps students see continuity in the number system as facts are systematically added to their repertoire. Seeing the same structure represented by the number family in many cases (small number + small number = big number; big number – small number = small number) helps them look for and make use of the same structure in new applications. Through verbal and written exercises, students can generate new facts from known facts because they have internalized the structure of the number family.

Here is the first part of an exercise from Level B Lesson 51 that illustrates how students make use of what they know from one family and apply it to a new, related family.

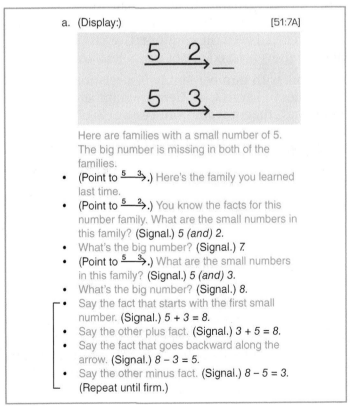

from Lesson 51, Exercise 7

The number family table below incorporates the 200 basic addition and subtraction facts in 55 families. Students in *CMC Levels B–D* are systematically introduced to related sets of facts from columns in the table (as in the example above), rows in the table (e.g., all facts in the top row involve a "small number" of 1) or the diagonal (which shows "doubles").

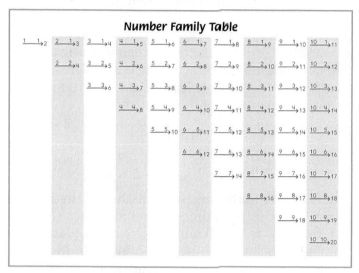

Number Family Table

Here are the number-family items and student work from Level B Lesson 44 Independent Work.

a. 6 2 →8   b. 3 2 →5
6 + 2 = 8   3 + 2 = 5
2 + 6 = 8   2 + 3 = 5
8 − 2 = 6   5 − 2 = 3
8 − 6 = 2   5 − 3 = 2

from Workbook Lesson 44, Part 6

Students are familiar with the number-family structure and generate the four related facts independently. By the end of first grade, students are well practiced with the commutative property, having applied it over the course of many lessons.

The work with number-family fact relationships continues in Levels C–E. Here are number-family items and student work from Lesson 30 of Level C.

Part 1

a. 3 3 →■   b. 4 3 →■   c. 5 3 →■   d. 6 3 →■
3 + 3 = 6   4 + 3 = 7   5 + 3 = 8   6 + 3 = 9
6 − 3 = 3   7 − 3 = 4   8 − 3 = 5   9 − 3 = 6

from Workbook Lesson 30, Part 1

The families are in sequence from the third row of the number-family table. For each item, students recognize that a missing big number requires addition (3 + 3, 4 + 3, 5 + 3, 6 + 3).Once the missing number in the family is identified, the students generate a related subtraction fact. This type of activity not only reduces the memory load for fact memorization, but it also prepares students for later work with inverse operations.

MP8: LOOK FOR AND EXPRESS REGULARITY IN REPEATED REASONING.

CMC Levels C–F: Renaming with Zeros
CMC Level E: Function Tables and the Coordinate System

In *CMC*, students work with repeated reasoning in a variety of situations. They learn to extend and apply what they have learned in one mathematical context to other, related contexts.

Subtraction: Renaming with Zeros

Students in several levels of the program apply the renaming strategy for the tens column to problems with zero in the minuend in an efficient and consistent way.

When renaming from the tens column, students learn a "new place value" for the top number. In the following example, 42 is 4 tens +2. It also equals 3 tens plus 12.

$$\begin{array}{r} \overset{3}{4}\overset{1}{2} \\ - \ 7\,8 \end{array}$$

When a problem requires renaming from two columns, students apply the same reasoning and procedure:

402 is 40 tens plus 2. It also equals 39 tens plus 12.

$$\begin{array}{r} \overset{3\ 9}{4\,0}\overset{1}{2} \\ - \ \ \ 7\,8 \end{array}$$

This same reasoning and procedure can be applied to any number of columns:

$$\begin{array}{r} \overset{3\ 9\ 9}{4\,0\,0}\overset{1}{2} \\ - \ \ \ \ \ 7\,8 \end{array} \qquad \begin{array}{r} \overset{3\ 9\ 9\ 9}{4\,0\,0\,0}\overset{1}{2} \\ - \ \ \ \ \ \ \ 7\,8 \end{array}$$

The repeated reasoning provides students with an efficient algorithm, and a "shortcut" to solve any subtraction problem that requires renaming with zeros. In the last example above, students in *CMC* take one renaming step. In the example below, students not taught the method in *CMC* take four separate renaming steps to work the same problem.

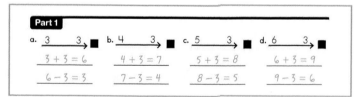

This less efficient method increases the likelihood that students will make an error or become confused.

Function Tables and the Coordinate System

Students in *CMC Level E* apply repeated reasoning to rows of a function table. The function table shows three columns: the X value, the function, and the Y value. For example:

X	Function 3 + X	Y
a. 7		10
b. 0		3
c. 2		
d.		4
e. 5		

from Workbook Lesson 99, Part 3

This table format enables students to see that the same transformation from the X value to the Y value is repeated for each point they will plot on the coordinate system. They also learn that this type of repeated transformation results in points that lie on a straight line. Given two X and Y values, students make two points and plot the line. They then refer to the line to identify where other points lie on the line, based on either the x or y coordinate. Finally, they perform the calculation based on the function rule to verify the accuracy of the points they have identified. Here's part of the exercise from Lesson 99.

(Add to show:)   [99:3D]

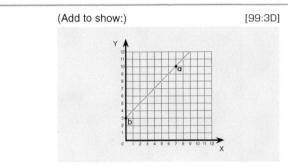

Here's the line. You can see that points A and B are on the line. You're going to figure out what Y equals for point C. Then we'll see if it's on the line.
- Complete row C in the function table. Put your pencil down when you can say the X and Y equation for point C.
  (Observe students and give feedback.)
- Look at row C. Say the equations for X and Y. Get ready. (Signal.) *X = 2 (and) Y = 5.*
  (Add to show:)   [99:3E]

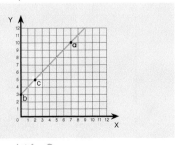

Here's the point for C.
- Is C on the line? (Signal.) *Yes.*

d. We're going to figure out the missing X or Y value for points D and E by using the line.
- Look at row D in the function table and say the equation for Y. Get ready. (Signal.) *Y = 4.*
  To find the point, I go to Y equals 4 on the arrow and go across to the line.
  (Add to show:)   [99:3F]

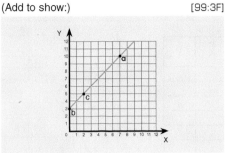

- What does X equal when Y equals 4? (Signal.) *1.*
  (Add to show:)   [99:3G]

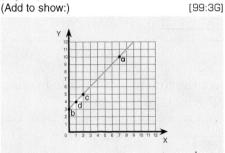

- Write what X equals for row D in the function table. ✔
f. Now we're going to check D to make sure it's right. The function for the table is 3 plus X.
- Raise your hand when you can say the problem for row D. ✔
- Say the problem for row D. (Signal.) *3 + 1.*
- What's the answer? (Signal.) *4.*
- Does Y equal 4 for row D? (Signal.) *Yes.*
  So the values for X and Y are right. Remember, these types of functions tell about all the points on a straight line.

from Lesson 99, Exercise 3

This exercise gives students the opportunity to apply repeated reasoning and also to evaluate the reasonableness of their results. If the point on the line does not correspond to the values in the table, they find the discrepancy and correct it.

On later lessons, students work with multiplication function tables. Here's an example from Lesson 102:

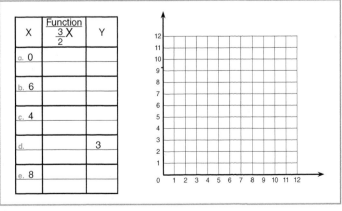

from Workbook Lesson 102, Part 3

Students deepen their understanding of functions and lines by repeating the same reasoning to these examples as they did to the add-subtract functions. They apply the function rule to figure out the y value for points a. and b., plot those points and draw the line.

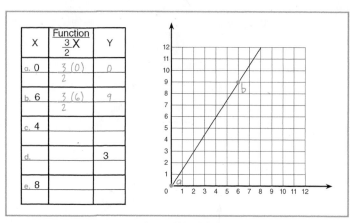

from Workbook Lesson 102, Part 3

They inspect the line to identify the missing x or y values for the remaining points.

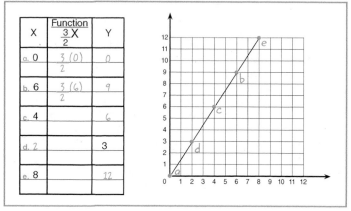

from Workbook Lesson 102, Part 3

Finally, students evaluate their answers for points c through e by multiplying each X value in the table by 3/2.

SUMMARY

For each mathematical concept or strategy introduced in *CMC*, students progress through a careful sequence of examples, move from highly structured activities to independent practice over several lessons, and revisit earlier concepts as they learn new applications or integrate elements of previously taught concepts. The *Comprehensive Edition* of *CMC* includes strategically designed instructional activities that fully meet the Common Core State Standards for Mathematical Content and the Standards for Mathematical Practice in kindergarten through fifth grade.